of the Elements

4B	5B	6B	7B		8B		1B	2B	3A	4A	5A	6A	7A	8A
														1 H
									5 B	6 C	7 N	8 O	9 F	10 Ne
									13 Al	14 Si	15 P	16 S	17 Cl	18 Ar
22 Ti	23 V	24 Cr	25 Mn	26 Fe	27 Co	28 Ni	29 Cu	30 Zn	31 Ga	32 Ge	33 As	34 Se	35 Br	36 Kr
40 Zr	41 Nb	42 Mo	43 Tc	44 Ru	45 Rh	46 Pd	47 Ag	48 Cd	49 In	50 Sn	51 Sb	52 Te	53 I	54 Xe
72 Hf	73 Ta	74 W	75 Re	76 Os	77 Ir	78 Pt	79 Au	80 Hg	81 Tl	82 Pb	83 Bi	84 Po	85 At	86 Rn
104	105	106												

Family Number

2 He

CHEMISTRY TODAY

William F. Kieffer
College of Wooster

Harper & Row, Publishers
New York   Philadelphia   San Francisco   London

Design: Michael Mendelsohn, Pat Brewer
Line Art: Ayxa Art
Cover: Michael Rogondino
Ring Structures and Text Composition: Science Press

Photo Credits: Page 11: Alinari–Art Reference Bureau. Page 64: Doug
Wilson/Black Star. Page 70: Ilka Hartmann/Jeroboam, Inc. Page 82: Con-
solidated Coal Company. Page 102: Flip Schulke/Black Star. Page 121:
United Press International. Page 210: Union Oil Company. Page 239:
NASA. Page 253: PG & E. Page 306: Institute of Energy Conversion, Uni-
versity of Delaware. Page 330: Erich Hartmann/Magnum. Page 350:
Courtesy of the American Museum of Natural History. Page 355: Courtesy
of the American Museum of Natural History. Page 395: National Science
Foundation. Page 396: United Press International. Page 426: Paul Pietzsch/
Black Star. Page 454: Elihu Blotnick/BBM. Page 530: Islander Yacht Com-
pany. Page 533: David Powers/Jeroboam, Inc. Page 561: left: Courtesy
Thomas L. Hayes, Lawrence Radiation Laboratory, University of California, Berkeley; right: Courtesy of Philips Electronic Instruments, Inc.

Library of Congress Cataloging in Publication Data

Kieffer, William Franklin, 1915–
  Chemistry today.

  Includes index.
  1. Chemistry.   I. Title.
QD31.2.K53   540   75-34307
ISBN 0-06-384550-4

Contents

16   An Introduction to the Chemistry of Carbon Compounds 472

17   Giant Molecules 506

18   Giant Molecules in Living Systems 540

Alternative sequences and emphases can be found in the Instructor's Manual.

To the Student

Chemistry is going on all around you, not just in laboratories and factories. Whenever natural or artificial changes in matter occur in our universe, chemistry is involved. You are part of this phenomenon; your body is a living demonstration of the principles of chemistry.

Are you curious about natural events around you and how they demonstrate chemical principles and processes? I hope so, for curiosity will carry you quite far in your study of chemistry. Describing things is not enough: We want to know what things mean.

Can you study and appreciate chemistry without having to approach it from the perspective of someone preparing for a career in science? I think so, and that is why I wrote this book. I believe that you will find enjoyment and satisfaction in discovering how chemistry helps you to tie things together and to understand some of the problems associated with our current technology such as the dangers to our environment.

One concern that you might have as you begin this course is the question of mathematics. Perhaps in the past you have avoided chemistry or other science courses out of fear or distaste for the mathematical terminology that is convenient and efficient for chemical and scientific language. As you read this book, you will find that many fundamental concepts of chemistry need not be expressed in mathematical terms. We must use numbers occasionally, much as you count money or describe the population of a city; and we will use simple ratios, as you do when you use the prices posted in a store to buy several articles. Beyond this, though, I have expressed ideas and information by relying on examples and analogies rather than on mathematics.

Chemistry Today was written in a narrative style. Later chapters draw upon and expand ideas developed in earlier chapters. For example, when you reach Chapter 17, you will learn how chemists can design a synthetic material such as Dacron plastic or Neoprene rubber to have a desirable set of specific properties. You will recognize here an application of the concepts of the structure of matter which we described previously in Chapter 7. Your instructor may prefer to treat some of the book's chapters in an alternative sequence, but still this narrative thread runs through the book. In many places, I refer to earlier chapters. Trace these references back through the book to guide your review. Then you can build new ideas on the foundation of familiar ones—a method typical of the way scientists always work.

I have tried usually to start discussions of a topic with familiar subjects: the weather, rusty tin cans, the food we eat, gasoline in our automobiles, the energy crisis, or environmental problems. Behind each of these commonplace encounters in our daily lives are the basic principles that connect our diverse bits of information and experience by means of scientific laws and theories. Thus, from familiar subjects we can go on to explore how such laws and theories enable us to predict and understand related events. Some of these events occur naturally. Others are consequences of how human beings are constantly changing their world.

Chemistry Today is designed as a self-contained learning tool; you will not need a study guide or other "road map" to grasp the main ideas. The book is designed to help you to readily decide what material is most essential and what is illustrative of the essentials. After all, you make these decisions in every course you take, and with this book, should find it particularly easy.

Do not try to memorize everything. Rather, work to understand ideas and gain a feeling for what they mean. Ask lots of questions—of your instructor, your fellow students, and yourself. Think about how the chemical principles you are studying amplify or clarify something you have read in the newspaper or have watched on television. The glossaries at the end of each chapter will help you review, especially if you re-read the indicated pages where the items were first introduced in context. Test your understanding by doing the end-of-chapter exercises. Some of these reinforce specific ideas in the chapter; others ask you to apply what you have learned to different situations. Suggested answers for the odd-numbered exercises are supplied at the end of the book. Use these to check your progress and then work on the other questions.

Chemistry Today is a metric book—I have used the metric system throughout, providing conversion factors from the English system only in Chapter 2, where we introduce some basic concepts of measurement. The metric system is extremely easy to use, convenient, uniform, and an integral part of scientific language.

You need not feel that your learning the metric system will be a waste of time, limited in use only to your chemistry course. The United States is slowly beginning the process of converting to the metric system. Within ten years, we will join the many nations around the world that have used the metric system exclusively for decades.

Chemistry Today invites you to appreciate as well as understand chemistry. I think you will learn a good deal about this exciting field over the next several weeks. But more than your "book knowledge," I think an understanding and appreciation of chemistry will have some other—more general—benefits. It will challenge you to sharpen your awareness of your surroundings, expand your curiosity, and encourage you to recognize the importance of scientific knowledge as a means for dealing with some of the problems that confront all of us.

I hope your reading *Chemistry Today* will encourage you to be actively involved in what you learn. Tying together experiences and ideas can be just that. I would be especially pleased if you would share your comments and suggestions with me by writing to me in care of my publisher.

William F. Kieffer
Wooster, Ohio
October 1975

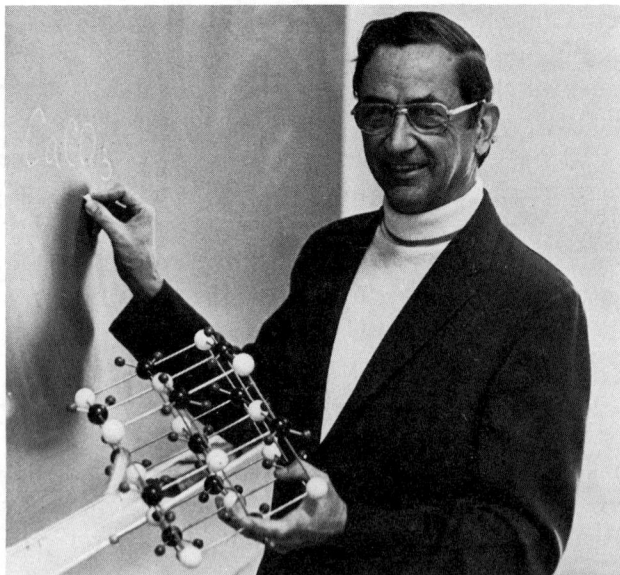

To the Instructor

There are a great many college courses designed to be the non-science major's one formal contact with science. Some courses offered by chemistry departments preserve the conceptual, theoretical glories of chemistry, reluctantly omitting only a few specialized topics. At the other extreme are those courses so devoted to society's problems that the principles of chemistry remain almost totally obscured. *Chemistry Today* was written for those courses that attempt to combine the best features of both of these clearly different approaches. It is a chemistry book, not merely a book about chemistry. It deemphasizes the quantitative aspects of the field in a manner I feel is appropriate to its audience. The minimal mathematical concepts required are presented in brief boxes that reinforce the student's confidence about dealing with numbers in chemistry, while not interrupting the narrative.

I hope that *Chemistry Today* will introduce students to many of the fundamental concepts of chemistry without appearing to be "chemistry for chemistry's sake." Rather, I have stressed their relevance to understanding and interpreting familiar experience. For example, there *is* a connection between the kinetic molecular theory and thunderstorms or scuba diving, between hydrogen bonding and ice skating or heredity. The topics printed in color in the table of contents are those in which applications of chemical concepts are discussed. Applications and theoretical ideas, technology and the science upon which it is based are interwoven into a developing narrative, not segregated into separate chapters. I repeatedly emphasize how the properties of macroscopic matter can be interpreted in relation to the structure and behavior of microscopic entities: electrons, atoms, and molecules. Also emphasized is the consistency in the behavior of these fundamental chemical units, regardless of whether they are in living or inanimate systems.

I have tried also to show the connection between a basic knowledge of fundamental scientific principles and the wisdom needed to approach some of society's problems. For example, policies for the better management of our dwindling resources of potential energy should take into account the fact that fossil fuels are the raw materials for our essential plastics. As another example, I indicate that any attempts to develop and exploit new sources of energy must involve an understanding of how the second law of thermodynamics limits some energy conversions. Also, resource management policies inevitably must compromise conflicting concerns such as cost, safety, convenience, and environmental effects. The interdependence of science and technology is particularly stressed in the chapters "Energy and the Energy Crisis," "Energy from Atomic Nuclei," "Light Energy and Solar Radiation," and "Giant Molecules."

The instructor's manual that accompanies this book provides some additional materials which I hope will aid your use of the text. The *learning objectives, additional exercises,* and *suggested classroom demonstrations* are offered not to replace aspects of your own approach to the course, but rather to suggest pedagogical complements to the nonmathematical, contemporary orientation of the book itself.

Some chemistry professors who are aware of *Chemistry: A Cultural Approach* (Harper & Row, 1971) may ask how that book may relate to the present volume. Both books are directed to an audience of students whose college education focuses on disciplines other than the sciences. Foremost among my reasons for writing both books is a personal conviction that there are many ways to present chemistry to this important audience. *Chemistry Today* is quite distinct from my previous volume, mostly through its emphasis on the current state of chemistry— particularly as it relates to our everyday lives, rather than to its historical development or philosophical implications.

Some interesting circumstances during the years intervening between the publication of my previous book and the present volume influenced my decision to write a book with a different approach. My experience during the summers of 1970, 1971, and 1972 as a member of the writing team for the ChemTeC Project gave me an opportunity for many discussions with chemists teaching in two-year colleges. Our sharing of experiences and opinions led to the preparation of materials for students training to become chemical technicians. I have adopted a similar style of presentation for students who are not headed for science-related careers. As a result of many helpful suggestions on content, coverage, and level brought out in a seminar conducted by R. Wayne Oler of Canfield Press and attended by chemists from colleges in the San Francisco Bay area, I have come up with what I hope you will agree is a fresh approach to the subject.

W.F.K.

Acknowledgments

I am grateful to many people for the help and encouragement that made this book possible: to the trustees, administration, and faculty of The College of Wooster, who granted me a leave for the academic year 1974–1975; to the faculty of the Chemistry Board of the University of California, Santa Cruz, whose hospitality provided facilities for studying and writing; to Elaine Kieffer for her many hours of attention to accuracy and detail in helping to prepare the manuscript; to Edna Ilyin Miller for her assistance in compiling exercises and glossaries; to Donald Marshall of California State University at Sonoma for suggested exercises; to Linda Purrington for her editorial guidance. I am grateful to those whose perceptive comments on early manuscript drafts greatly influenced the present version: Wilbert Hutton, Iowa State University; Mildred Johnson, City College of San Francisco; Ellene T. Contis, Eastern Michigan University; Malcolm Renfrew and Jean'ne Shreve, University of Idaho; O. H. Bezirjian, College of Marin; Guy Buccino, Waterbury State Technical College; Peter Scott, Linn-Benton Community College. In addition, I appreciate the help of those who attended the Canfield seminar and provided the advice, counsel, and encouragement that led to the initial plan and content for the book: Anthony Trujillo, San Joaquin Delta College; Ardas Ozogomonyan, Skyline College; Charles Aldrich, Solano College; Irvin Drew, Laney College; Mildred Johnson, City College of San Francisco; Ned Reed, Merritt College; Julie Bryson, Chabot College; Robert Hubbs, De Anza College.

The imaginative vision, perceptive guidance, and generous encouragement of Wayne Oler deserve special recognition and thanks. They made the preparation of this book not only possible but enjoyable. I add a special note of admiration and gratitude for the great judgment and consummate skill of Malvina Wasserman and Patricia Brewer, who carried this project from initial speculation all the way through to publication in the offices of Canfield Press.

W.F.K.

1

What Is Chemistry All About ?

☐ How does knowing some chemistry enhance our understanding of our natural environment?

☐ Can we appreciate the accomplishments of chemistry without dealing with a lot of numbers?

☐ How is nature's consistency one of the scientist's most valuable tools?

☐ How does chemistry relate to the other sciences?

☐ How is science different from the arts?

☐ Is there a pattern to the method which scientists use to study nature?

☐ How does science express its conclusions?

☐ When does science become technology?

Chemistry is wheat growing, bread baking, food digesting, and human bodies living. Chemistry helps us raise more food and make the drugs that bring us longer, healthier lives. Chemistry is petroleum forming in the earth, gasoline burning in automobile engines, and smog forming in the atmosphere. Chemistry also is cleaning up the environment. Chemistry is turning natural resources into the many articles that enrich our lives and free us from drudgery. Chemistry also helps us conserve our dwindling resources of raw materials, especially resources of fuel. Chemistry is all these things and more. Yet, we are only describing chemistry in terms of examples. Let us turn from examples to generalizations that will help us broaden our definition of chemistry.

A DEFINITION OF CHEMISTRY

Chemistry deals with the composition of materials and the ways in which one substance changes into another. So we can say that *chemistry deals with matter and energy.* Chemistry is a branch of science. The word *science* comes from a Latin word *scio*, which

---

Chemistry, like all science, is a human activity.

---

means "I know." This definition helps us to realize that chemistry, like all science, is a human activity. Chemistry is a part of *human knowledge.* Chemistry is information and ideas about the natural world that people have put together.

Curiosity is the forerunner of knowledge. The mind of a young child is driven by curiosity; that is why children are born scientists. And humans have always been curious about the world around them. When primitive men and women began to remember what their explorations had led them to discover, they began to accumulate knowledge. Language enabled them to share their knowledge with one another. Written symbols allowed them to store and transmit knowledge to others, even to later generations.

But knowledge is more than a mere record of observations. The same curiosity that drove humans to explore the natural world also led them to seek connections between bits of information. Such connections are the ideas, the concepts, that humans use to help *organize* knowledge.

Ideas that connect bits of information are typical of the way scientific knowledge is organized in human minds. For example, the separate observations that grass is green or that the leaves of trees are green is not nearly as important as the realization that some part of most growing plants is green. What is the significance of this realization? On a higher level, this knowledge leads to the question "Why?" By asking why, we are not looking for some purpose behind the presence of green color in plants. Rather, we seek to explain this fact by discovering how it relates to other things we know. Is there some substance responsible for the green color? What role does this substance play in the process of plant growth? Does the fact that growing animals are not green tell us anything?

Questions like these suggest a human activity that goes beyond mere observations of the natural world as we find it. We can do

All are green! Why?

The questions never stop coming.

things to the natural world; we can perform experiments. We can plan a way to ask a particular question by seeing how the natural world responds to some special change we make in it. Each answer nature gives suggests another idea that may link bits of information together. Science has grown by this process. The questions never stop coming. New ideas invented to help interpret observations always lead to further experiments and observations.

TAKING APART AND
PUTTING TOGETHER

Chemistry is one of the *natural* sciences. That means that chemists deal with the natural world. Nature answers chemists' questions by revealing phenomena. A *phenomenon* is a fact or event that can be observed by using our senses. We see, smell, hear, taste, or touch something. We can take something apart and do to the simpler parts the same thing we have done to the whole. But these observations are just the start of chemistry. Chemists also think about phenomena and *invent ideas* that help them interpret what they see, smell, hear, taste, and touch in terms of broad general principles. This process of taking things apart and thinking about how the parts relate to one another is called *analysis.* Analysis is finding out what is in the natural world and understanding how it works.

The knowledge that chemists accumulate by the analysis of nature is useful in another way. Chemists are not just interested in substances and phenomena that occur naturally. By understanding the general principles of the way nature works, chemists can predict and put together new substances or phenomena. The process of putting together naturally occurring materials into new and different ones is called *synthesis.*

Synthesis does not mean "fooling Mother Nature." Synthesis means changing materials into new forms, according to the operation of natural laws. Mother Nature is still in charge. For example, for many years chemists analyzed the structure of such natural fibers as cotton and silk. The knowledge they obtained, along with knowledge from many other investigations, made it possible for them to make such *synthetic* fibers as Dacron, Nylon, and Orlon. Sometimes you may hear these materials called miracle fibers. But there is nothing miraculous about these materials. They are made from coal, petroleum, air, and water, all of which are naturally occurring materials. All synthetic fibers are made by using chemical reactions that Mother Nature uses in one way or another for other purposes.

BUT CHEMISTRY IS SO COMPLICATED! OR IS IT?

Here you are, beginning your study of chemistry. You may feel that you are facing a complicated subject. If you are planning to learn enough chemistry to become a research scientist, your feeling might be justified. But that probably is not your plan. Nor does this book expect it to be. Rather, we hope that this book will help satisfy your curiosity about substances and phenomena you frequently encounter. Many of your everyday experiences illustrate fundamental chemical ideas that are essentially quite simple.

Sometimes simple scientific ideas seem complicated, but only because scientists tend to talk in mathematical language. They do this because the abstract symbols of mathematics are often the most efficient way for people familiar with this kind of expression to communicate. But this kind of language is not really necessary to get the basic ideas across. This book will use words instead of mathematical symbols and manipulations of those symbols.

You also will find many analogies throughout this book. An *analogy* expresses a likeness, a correspondence between a new, unfamiliar idea and familiar ideas. For example, in later chapters we will liken electrons in atoms (probably an unfamiliar idea to you) to waves in water or bullets (familiar ideas). We will count on your recognizing that, although the electron *is* neither a wave

Although the electron is *neither a wave nor a bullet, it does behave in similar ways.*

$$-\frac{\hbar^2}{2m}\left(\frac{\partial^2\psi}{\partial x^2} + \frac{\partial^2\psi}{\partial y^2} + \frac{\partial^2\psi}{\partial z^2}\right) = (E - V)\psi$$

nor a bullet, it does behave in similar ways. This analogy will be more useful to you than the calculus equation a scientist uses to describe an electron most accurately.

Except for using the numbers that allow us to describe things precisely, we will avoid talking in mathematical terms. With that possible hurdle removed, the one important step we ask you to take is to learn how to interpret facts in terms of broad underlying principles.

Here is an example of what we mean by saying that broad general principles can be used to interpret various familiar experiences: You know that a roast cooks faster in a hot oven than in one that is merely warm. Cooking involves chemical reactions. Cooked food looks, feels, and tastes different from uncooked food. Some substances in the food are changed into different ones by the cooking process. This is what we mean by saying that chemical reactions are involved. Faster cooking at a higher temperature means that these chemical reactions go at a more rapid rate when the temperature is higher. The same thing happens when you have a fever. The chemical reactions your body uses to fight disease germs go faster when your temperature rises. You need not know the details of either the chemical reactions of cooking or the way the cells of your body fight invaders. The important point is the way your observations about cooking and fever are connected by the generalization that chemical reactions are speeded up by higher temperatures.

Another way we can look for simplicity rather than complication is to emphasize repeatedly that *the natural world behaves consistently*. The more observations we make, the more questions we ask, the more experiments we do—the more our explorations

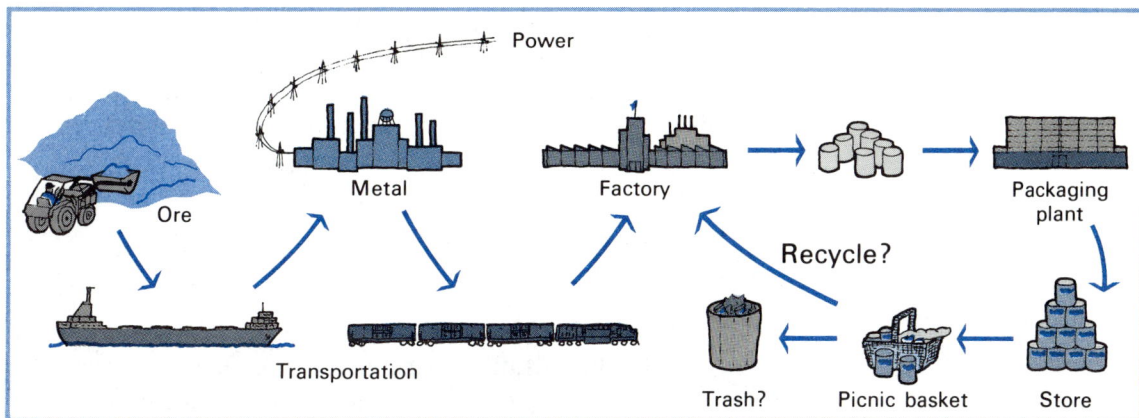

convince us that nature can be counted on, that nature is dependable. The goal of science is to recognize what the consistencies are. Then we can make predictions and plan what to do in a beneficial way. For example, we find that the first principle nature follows is that matter can never be created or destroyed. The significance of this principle for us is that we can never really burn up or throw away the trash and waste of our modern living. We can only convert waste material into something else. Recycling starts matter back on the course of changes that made it useful to us in the first place. And recycling usually puts the material back into its course further along than if we had to start at the raw material stage. Consequently, recycling saves the energy we would have to expend to get fresh material from nature. We may have other reasons for practicing recycling, such as economics or beautifying the landscape. All these reasons for recycling are based on one of nature's laws: Matter cannot be destroyed, it can only be transformed. And transforming matter involves energy.

One of our goals in this book will be to look at substances and the way they act the way a chemist does. We want to discover the

We want to discover the general principles that make interpretations simple rather than complicated.

general principles that make interpretations *simple* rather than *complicated*. Any mystery seems complicated until it is solved. The solution makes it seem simple. This rule also applies to what appear to be mysteries about the natural world. Scientists try to solve such apparent mysteries by discovering the ways nature behaves consistently. Then the answers to the mysteries turn out to be simple. The engine of a Cadillac is more complicated than that of a Volkswagen. Yet a simple speck of dirt in the carburetor of either one causes the same trouble; the engine stops running.

WHAT IS CONSISTENT ABOUT DIVERSITY IN NATURE?

The idea that nature behaves in consistent ways needs to be explored a bit further. Many experiences in life seem to suggest just the opposite, that nature is inconsistent. The wind may or may not blow; the sun may or may not be behind a cloud. Earthquakes, hurricanes, and tornadoes, those special devastating events of nature, are unpredictable. To be sure, scientists are learning more and more about the conditions that precede such events. But so many interacting factors must be considered that predicting with certainty is still a long time into the future. No two living things are ever identical. No two rocks or hills or mountains are alike. Variation and diversity are found everywhere. How can we say there is consistency in the natural world?

Whether or not we recognize the consistency in nature depends on the point of view we take toward objects and events in the natural world. A raindrop or buttercup or grizzly bear are different, but all are made up of matter. And in the chemist's view, matter is made up of atoms. At this point, you may not be sure of what the word *atom* means to a chemist. The discussions in the first chapters of this book will help you find out. But the important idea behind the word is already familiar to you. *Atoms are the tiny building blocks out of which all matter is made.* There are only about a hundred different kinds of atoms. The same way that many of only a few different kinds of bricks or stones can be put together to build a wall, a sidewalk, or a cathedral, so can a raindrop, a buttercup, or a grizzly bear be made of tremendously large numbers of only a few kinds of atoms. Each kind of atom,

Each kind of atom, wherever it is found in the universe, does *behave the same way every time.*

wherever it is found in the universe, *does* behave the same way every time. At the level of atoms, chemists have found that nature never varies. The diversity we find in all the countless objects of the natural world occurs because everything we can see contains so many atoms. The kinds of atoms, their numbers, and the way they are arranged in something are all responsible for the differences we see in objects around us.

CHEMISTRY AND OTHER SCIENCES

Several hundred years ago, all the various branches of science we now know, such as physics, chemistry, and biology, were known as "natural philosophy." Systematic, organized knowledge about the natural world was just beginning to accumulate. Few of the general principles we have referred to were recognized. One human mind could know them all. Leonardo da Vinci, an outstanding example of Renaissance Man, whose curiosity drove him to explore all of nature, would today be called a physicist, a chemist, an engineer, a biologist, and a physician, as well as an artist. Sir Isaac Newton, the intellectual giant of the late seventeenth century whose name we associate with the law of gravity, was a scientist whose interests knew no boundaries. He even invented some of the mathematics he needed to describe how parts of the natural world behave.

Contributions from these and other great minds added to humanity's storehouse of knowledge. Inevitably, the total became too great for any one human mind to handle. Human curiosity explored too many separate frontiers of knowledge. Divisions of knowledge developed, each organizing the information from these various explorations. These divisions evolved into the various branches of science we are familiar with today.

Rather than dwell on the differences among the branches of science, we will emphasize the similarities and interweaving of the various present-day scientific disciplines. Physics deals with knowledge about matter and energy. So does chemistry, which concentrates on trying to understand the transformations matter undergoes. Chemistry and physics constantly merge and overlap. For example, physicists describe an electric current as a flow of tiny charged particles called electrons. Chemists use the same idea of electrons to explain how one kind of matter can change into another.

Biology, too, is the study of matter and energy, as they interact in living things. Although the driving force that makes something alive is still a mystery, many of the chemical reactions involved in the life process are recognized and understood. Here again, nature is found to behave consistently. Atoms, the tiny fundamental particles of matter, make the same combinations and have the same energy exchanges in living cells as they do in test tubes or in factories. Many of the examples in later chapters are taken from biochemistry, the field where biology and chemistry merge.

Medicine, the application of scientific knowledge to problems of human health, has broad interactions with chemistry. Many drugs that physicians prescribe today were unknown even a few years ago. More and more is being learned about the intricate interactions of the complex chemical substances in the human body. Drugs that formerly could be obtained only by extracting them from plants or animals are now being built, atom by atom, in chemical laboratories. Medicine and chemistry are being used to start to solve the puzzles of how the human brain functions. Thus chemistry, by providing the basis for understanding the delicate chemical reactions involved in thoughts and emotions, is beginning to merge with psychology.

So in a sense science is being remolded into the state it was in before it became subdivided. The difference is that problems are now being tackled by teams of specialists rather than only one person. Each specialist brings to the attack the depth of his or her knowledge in a specific area.

THE SCIENCES AND THE ARTS

It is often helpful when starting the study of a new and different branch of knowledge to make comparisons with a more familiar type of learning. You may be more at home with music or literature or art than you are with science. What can you expect to find in chemistry that is similar to these fields, and what is different? What role do imagination and creativity play in science? How can the creativity of a scientist be compared to that of Bob Dylan

Why does science sometimes seem to be so full of cold, hard numbers rather than the warmth of sound or color?

or Kahlil Gibran or Pablo Picasso? Why does science sometimes seem to be so full of cold, hard numbers rather than the warmth of sound or color?

The more you learn about science, the more you become aware of the importance it places on *order*. A scientist looks for order in the natural world. Order is what we mean by the consistencies in nature. When we can see the connections between ideas and bits of information, we feel that we understand them. The scientist, likewise, constantly tries to find the order that can organize knowledge. To be sure, new discoveries are constantly being made. But each new discovery fits into and enlarges the orderly pattern of knowledge about the natural world. The connections between isolated observation mean more than the individual events themselves.

For example, when Sir Alexander Fleming found that some mold, which had accidentally fallen onto a dish, killed a colony of bacteria, he recalled an earlier experience. Years before, he had discovered that tears or mucus also killed bacteria. He had isolated a substance in the mucus that was responsible for the effect. So now he looked for a particular substance coming from the mold. This substance turned out to be different from the substance he isolated previously and many times more powerful. The new substance was penicillin, the antibiotic that has saved countless lives. Researchers since that time have found a whole collection of substances similar to but slightly different from penicillin, which now provide a medical arsenal of weapons to fight disease.

The order so typical of the pattern of scientific knowledge has a parallel in artistic expression. This parallel is the *form* of a work of art. The beauty of a musical composition, a poem, or a painting comes not from haphazard chaos but from its form. The creative genius of the musician, the poet, or the artist is responsible for the special way the individual sounds or words or colors are put together. Harmony, melody, rhythm, and balance of shape and color are not haphazard. Although the song or poem or painting stands unique and alone, its attractiveness to ear or eye depends on the degree to which it expresses the universal underlying principle of form. Thus the scientist, building a theory to bring order to knowledge about the natural world, and the musician, expressing form by arranging sounds in melody and rhythm, are engaged in only slightly different activities.

A more pronounced difference between the sciences and the arts shows up in the kind of language each uses for communication. In science the language is precise. This is why so many numbers are used. Numbers stand for specific quantities. Because nature is consistent and dependable, the descriptions of what is and what happens in the natural world must be expressed in the most consistent and dependable language possible. Words like *big* or *small* are not very useful for a scientific description of a volcano or a germ, a bacterium. The idea of *big* must be expressed in numbers of miles or kilometers to describe the size of

The more you look carefully at a great work of art like da Vinci's Mona Lisa, the more you recognize its exquisite form. A great scientific theory like the atomic theory expresses a comparably exquisite order. Both are the products of human creative imagination.

Beethoven Sonata

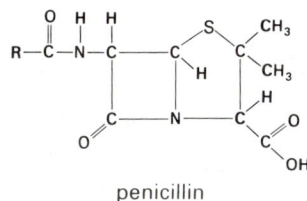

penicillin

A bacterium is small compared to the drop of water in which it lives, but huge compared to the atoms from which it is made.

the volcano. The bacterium is small compared to the drop of water in which it lives but huge compared to the atoms from which it is made. So some fraction of a recognized unit of length, such as a millimeter (a thousandth of a meter, approximately a twenty-fifth of an inch) must be used to describe a bacterium in a useful way. Moreover, scientists all over the world agree on exactly what the terms mean. An electron, the fundamental particle of electricity, is the same in Berkeley or Indianapolis or Singapore. So is the chemical equation that describes how gasoline burns or a storage battery works. There is universality in the language used in science.

A work of art communicates quite differently. Each person who listens to a song reacts or responds in his or her own way, quite possibly in a way different from the way the composer felt when creating the work. A particular musician, singing a particular song, may bring tears to the eye of one listener. The same performance may produce only puzzled detachment in another listener or possibly even anger in a third. This lack of agreement or universality is compensated for by the warmth or personal feeling that the song gives to at least some listeners.

In contrast, a scientific equation may appear to be cold and impersonal. Seldom does anyone get emotional over an equation. $E = mc^2$ does not make people laugh or cry. Yet it stands for a relationship among the quantities energy (E), mass (m), and the velocity of light (c) that every scientist recognizes immediately and accurately. An equation lacks warmth but has universality. It has a definite, special meaning.

Neither art nor science has a better way of communicating; each way is appropriate to its kind of human experience. Realizing this difference and accepting the need for the precision that is appropriate for scientific language is a good way to start your study of science.

SCIENTIFIC KNOWLEDGE ACCUMULATES

One other feature of scientific knowledge makes it somewhat different from literature and the arts. Scientific knowledge is much more *cumulative*. Scientific knowledge piles up or accumulates.

The ideas that emerged to answer yesterday's questions are used today to ask new questions. And answers to those are the basis for tomorrow's queries. New ideas necessarily build on old ideas.

If we imagine Michelangelo, the artistic genius of the Renaissance, in a present-day sculpture class, we would expect him to assume immediately the role of instructor. A comparable imaginary return of Antoine Lavoisier, whose genius started chemistry as a science at the end of the eighteenth century, would place him as a student. He would have a lot of catching up to do before his great mind would be able to use the ideas and information with which modern chemistry students deal. Such is the nature of scientific knowledge.

This emphasis on the cumulative nature of scientific knowledge has a very practical hint for you to follow in studying chemistry. Be sure to understand today's discussion, because tomorrow's will probably build on that understanding. Ask questions. Do the exercises. Be sure that you have ideas clearly in mind before going on. Take small sips day by day, and enjoy the taste rather than risk choking on a big gulp all at once.

THE METHODS OF SCIENCE

Many people who are starting their study of science expect to find some powerful method of thinking that scientists are supposed to use to solve problems. The great success of scientists in making discoveries and adding to the knowledge about the natural world implies some particular "scientific method" that ensures success. It is true that there is a typically scientific approach to problems. Of course, curiosity, imagination, and a skeptical open-mindedness are involved, as they are in the search for knowledge in any field. But the one special feature of the way scientists seek answers to questions is the way they *test* an answer before accepting it. When the imagination of a scientist proposes an answer to a problem, the test that the answer must meet always is in the form of an experiment. Moreover, scientists publish observations and ideas so that they can be tested in laboratories everywhere. This procedure is crucial to the acceptance of any new idea by the scientific community.

True science was not part of the culture of the ancient Greeks. They placed great reliance on rational thinking and logic. But they apparently never felt it necessary to check their conclusions against experience. It was not until the later centuries that the idea of doing an experiment to check the predictions of an assumed answer began to be important. Many of the first true scientists, men living in the Middle Ages, lost their lives in the fires of the Inquisition because they defied the authority of the church as the final test for the validity of knowledge. They turned to the natural world as it exists to see if their ideas were sound rather than to some doctrine of how the natural world should be. Gradually, experience replaced authority as the crucial test for the acceptance of knowledge. When this powerful idea entered human thinking, modern science began.

Those who suggest that there is a special pattern or formula for the way a scientist solves problems propose the following outline:

1. *Recognize* and define the problem.
2. *Observe* the pertinent facts of nature. Often these can be organized into a statement of a *law*, a description of a consistency in the behavior of the natural world.
3. *Make a guess* as to "why" the natural world behaves as it does.
4. *Predict* some additional facts on the basis of the guess.
5. *Test* the predictions by doing experiments. Ask nature some specific questions to see if the predictions are true.
6. If the results of the experiment do not fit the predictions, *start over again* with another guess, prediction, and experiment.
7. When the experiments confirm the predictions, put together the simplest *theory* that connects the ideas of the guess with the outcome of the experiments.

Such a recipe for a step-by-step procedure is a highly idealized and artificial view of how scientists make discoveries. In reality, this particular sequence is seldom followed. The biggest difference in the way discoveries are really made comes in how the problem is recognized. Sometimes the observations, the facts of

Sometimes the facts of nature almost shout out the question "Why?"

nature, almost shout out the question "Why?" For example, we will discuss in Chapters 5 and 6 how the information that is organized by the periodic table of the elements led to the development of ideas about the internal structure of atoms.

In other cases, the process of discovery starts with a purely theoretical bright idea, a guess. Albert Einstein's idea that mass can be converted into energy was found to be true by experiments

performed years after he proposed the idea. Nuclear power reactors now being used in place of power plants that burn up our
resources of coal and petroleum are a very practical consequence
of Einstein's prediction.

Still other discoveries have started from the unexpected results
of an experiment. These are the "impertinent" facts that nature

---

*Then there are the "impertinent" facts that nature brings to
the attention of a scientist.*

---

brings to the attention of a scientist. Teflon, the plastic widely
used to coat cooking utensils, was developed from a totally accidental discovery. A cylinder containing a particular kind of gas
became clogged when the gas turned into a solid.

Figure 1.1 suggests, in outline form, how the various steps of
the formal scientific method are related. The circular path involving observation, guess, prediction, and experiment can be entered
at any stage. Seldom does a first guess work through to an acceptable theory. Usually many laps around the track are required
to complete the race. Furthermore, research scientists seek
knowledge with whatever working habits suit their tempera-
ments. Some plan an experiment carefully and elegantly. Others
are impatient to follow a hunch and immediately start exploring
with the simplest apparatus. All these variations in scientists'
plans of attack lead us to refer to them as the *methods* of science
rather than speak of *the* scientific method.

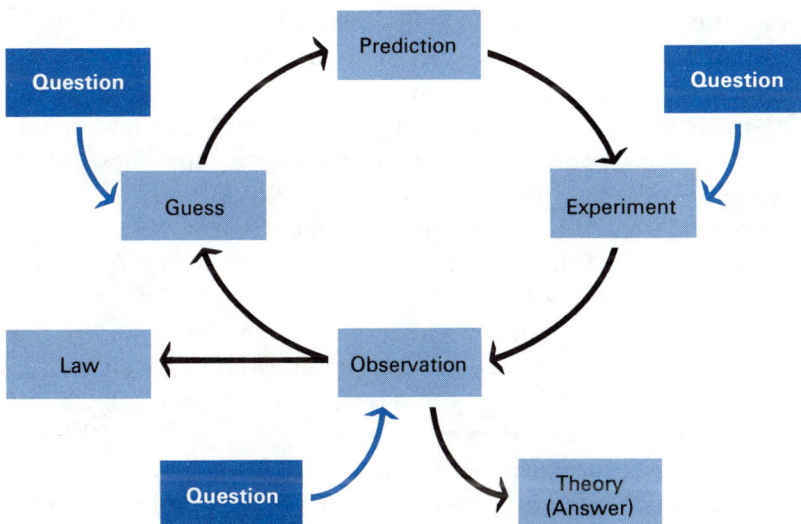

Figure 1.1 An outline of the methods of science.

LAWS AND THEORIES IN SCIENCE

It is appropriate for us to explore, in a bit more detail, what a scientist means by a *law* and a *theory*. In Figure 1.1, which suggests the methods scientists use, *law* appears beside observation; *theory* appears as an answer to the problem. It is important to recognize that laws do not involve the guess as to why nature behaves in a certain way. Both laws and theories are inventions of human minds. Nature "knows" no laws. They are not imposed rules that nature must follow as humans must obey traffic laws for safety and efficiency of travel. *A law is a statement of some consistency in the behavior of the natural world. A theory involves ideas invented to account for the behavior of the natural world.*

Let us use an example of a simple experiment to illustrate the relationship between laws and theories. When you add a few drops of liquid ink to a glass of pure water, you first see streaks of color in the water. If you set the glass aside where it can remain undisturbed, the color gradually diffuses throughout the whole of the liquid. If you return after a day, the whole liquid shows a uniform light shade of the color of the ink you added. Now if you take another glass containing pure water and add a few drops of a solution of water-soluble dye (such as you would use to color cake icing), you observe the same thing happening. After a day, the whole liquid shows a uniform light shade of the dye. If you do this a third time, with a dye of a different color, you observe the same result. You are now in a position to formulate the law: A colored liquid mixes thoroughly and uniformly throughout the clear water if you wait long enough. This phenomenon happens every time. You need not try other colored water-soluble liquids; you can predict with confidence that they will behave the same way under the same circumstances.

Now comes the guess as to why this happens. The law merely states *what* happens; it contains no statement, nor does it even hint as to why. To answer the question "Why?" you have to develop a different kind of idea. Perhaps the ink is made up of tiny submicroscopic fish or microbes that swim away from one another and so spread the color. But certainly this is not a very useful idea. The idea of the spread of color being caused by the movement of living creatures seems preposterous or silly. Why does it appear so? Because you have a store of knowledge based on your previous experiences. For example, you know that no food is required to keep ink and dye alive. You also know that ink and dye do not grow or increase in size as a mold and other

2 minutes   1 day

collections of living cells do. Therefore, you reject this guess by applying the crucial test of experience.

However, you can explore the idea of the ink or dye being made up of tiny inanimate (nonliving) particles. Let us call these particles *molecules*, a word that means "little units." In the next chapter, we will define the term molecule more precisely. For the present, you can assume that these molecules must be moving around randomly in every possible direction. They apparently cannot move far without bumping into the water, because at first the color stayed concentrated in a few locations.

You can readily imagine the clear water as also being made up of molecules. This explains how the molecules of ink and dye can spread. You can find an analogy in the way a group of a dozen people can very easily become separated from one another in a big jostling crowd of spectators leaving a stadium. You can predict that, if a long enough time elapses, the tiny molecules of the ink or dye will thoroughly mix with the water molecules, because their motion is completely random.

This idea also fits with what you would find if you stirred the solution. Your stirring also would move molecules, so that the mixture would become uniform much sooner. The same effect is accomplished if you heat the water. Apparently, the higher the temperature, the faster the molecules move; hence they become thoroughly mixed much more quickly. So your theory states that the ink and dye and water are liquids made up of tiny units or particles we choose to call molecules.

Notice the difference between the law and the theory. The law merely describes what happens. The theory accounts for what happens in terms of molecules. Where did the molecules come from? The *idea* of molecules was invented. You cannot see the individual molecules. In fact, if you put just a very small drop of colored liquid in the glass of clear water, the color of the completely mixed liquid may be too light to see. The molecules and their property of random motion is a theoretical idea. You believe in molecules because, whatever experiment you do, the results can be explained in terms of molecules and their motion.

At this stage, you have to be very careful to resist the temptation to consider your theory "right" and to consider the alternate

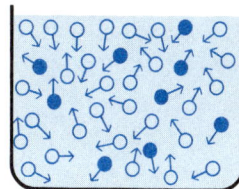

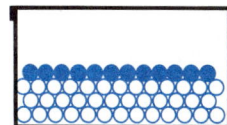

Scientists seldom apply the words right *and* wrong *to theories. Rather, they think of a theory as being more or less useful.*

theory about the swimming microbes "wrong." Scientists seldom apply the words *right* and *wrong* to theories. Rather, they think of a theory as being more or less *useful*. You can tie together and explain a wider range of experience more simply by thinking in terms of molecules and their behavior than in terms of swimming microbes. You will find very soon in our discussions how useful this idea of molecules is. The behavior of many different kinds of matter can be accounted for by using the idea of molecules.

THEORIES CAN BE MODELS

Another way of considering a theory is to think of it as a *model*. In your mind, you have built a working model of a part of the natural world—molecules. This model allows you to test the idea in still another way. You can turn to an analogy. You can build a model of this analogy out of real, decidedly visible things to see how the model performs. Suppose you partially fill a box with white marbles. Then you place a layer of colored marbles on top of them. Close the box with a lid, and shake it. The longer you shake, the more the marbles become mixed up. The harder you shake, the faster they become mixed up. While the marbles are being shaken, they are behaving as you imagine molecules behave. Thus the analogy model helps to make you believe in the theoretical model.

Throughout our study of chemistry, we will often be following this pattern of reasoning. First, we will observe on a *macroscopic* level (*macro* means "big"). We can see the whole, big glass of uniformly colored liquid. Our theoretical interpretation of what happens is on a *microscopic* level (*micro* means "small"). We cannot observe the individual molecules on the microscopic level. Yet we find a great amount of evidence to support the idea that molecules are the tiny units out of which liquids are made.

SCIENCE AND TECHNOLOGY

Up to this point, most of our discussion has dealt with what usually is called "basic" or "pure" science. Scientific research produces new knowledge. As scientists continue to explore, more and more of the way the natural world behaves can be understood and tied together by broad, general principles. But there is another side of science, the *use* of this knowledge to change the natural world, to make things for people to use. *Technology* is the application of scientific knowledge and methods to solving the practical problems of commerce and industry. In the minds of many people, the image of science is limited to this partial aspect. They see "scientific progress" only in terms of the products now available that were unknown a generation ago. An example is the plastics on which we sit, from which we eat, with which we are clothed, and in which much of what we buy is packaged. It is more accurate to consider these and the countless other examples of products from industries based on chemistry as evidence of *technological progress.*

Technology is an influence in modern living that is impossible to ignore. It has been so since early humans discovered the wheel and lever. We have derived so much from nature that some fear we may be altering nature too much. This may be so. But the blame cannot be put on an impersonal technology. Technology, as well as pure science, is a human activity. Technology makes no decisions; humans make decisions. Humans decide to make drugs to cure disease. At other times, they decide to kill many of their neighbors on the planet with devices assembled in their factories. Technology, like all forms of human activity—such as social structures and systems of government—can be used for good or bad ends.

One unfortunate trait that humans often exhibit is short-sightedness in their thinking and planning. "Let the future take care

---

We so enjoy the immediate, obvious benefits of technological progress that we tend not to look very far into the future.

---

of itself!" We so enjoy the immediate, obvious benefits of technological progress that we tend not to look very far into the future. Immediate benefits tend to overshadow possible long-term hazards. The widespread use of chemical substances specifically made to kill insects is an example. These substances, such as DDT, are now found to last a very long time in the environment and to be dangerous to other living things, especially birds. Circumstances such as this serve as warnings for the future. But the warnings should not keep us from further exploration. The more we learn about the chemical properties of the substances available for use, the more information we have for a decision about using them.

Some critics of modern society argue that technological progress should be slowed. Probably a more realistic view is that it should be redirected. Our collective responsibility for easing the contamination of the environment and slowing the use of our natural resources requires us to use technology wisely. Any such decision must be based on knowing what the problems are and what can be done about them.

SCIENCE, TECHNOLOGY, AND PUBLIC POLICY

Anyone who compares our style of living today with that of people living a few generations ago cannot miss recognizing how significant the influence of technological progress has been. We move about and communicate faster; we have virtually wiped out some diseases and become increasingly the victims of others; some of us live more comfortably at the expense of denying some resources to future generations. Large portions of our commerce and industry are based on scientific knowledge quickly turned into products and processes unknown in the recent past. Jet aircraft, plastics, and antibiotics are examples. Our government spends our tax dollars to sponsor scientific and engineering research in such areas as improving agricultural yields and generating nuclear power. Our government also spends our tax dollars to support agencies to regulate our technology. The Federal Food and Drug Administration is an example. Technology and the science behind it touch us at every turn.

In recent years many individuals have begun to ask more and more penetrating questions about the directions our widespread use of technology are taking. Are the tremendous amounts of waste from our factories and our homes causing too great a strain on nature's processes for restoring purity to our air and water? How can we conserve our resources of materials by choosing products that come from abundant rather than limited resources? How can we cut down our energy appetite, which puts

How can we cut down our energy appetite?

such a drain on our supplies of fuel? What alternatives should be developed to replace our present reliance on resources that are in limited supply?

Answers to these questions cannot be simple. A balance among scientific, economic, and social factors must be sought. Turning the answers into action is even more complicated. This is where a public policy comes in. Individuals collectively must decide how their government should act. Whether we approve or not, both research and regulation have increasingly become governmental responsibilities. Governments, although cumbersome in their movements, ultimately do respond to policies advocated by their citizens. Citizens—individual thinking humans—must first decide issues for themselves and then together establish public policies. Such issues are too important either to ignore or to remain uninformed about. Knowing what the scientific principles are and how they are applied in our technology is central to being an informed citizen.

CHEMISTRY TODAY

Essential principles of chemical science and their applications in your daily living are the themes of this book. We cannot cover all of chemistry. Probably many things you are curious about will be omitted. We hope you will search for answers elsewhere. The information and, most of all, the *ideas* you find in these pages should give you a good start toward both understanding what you encounter and prompting your curiosity.

Many articles you read in newspapers and magazines call attention to problems in an age where science and technology have such a great influence on our way of life. Some articles are dire predictions; others are complacently optimistic. Many urge the establishment of long-term policies. We hope that, after reading this book, you can evaluate better what you hear and read in the news and elsewhere. You cannot escape having to make up your mind. We hope what you learn from these pages can be a help!

Glossary

The number in parentheses indicates the text page where you can find the term defined in context.

analysis   the process of breaking substances or ideas down into parts to understand their structure or function (5)

chemistry   the branch of science that studies the forms of matter and energy and the ways in which these forms are interconverted (3)

law   a statement summarizing scientific observations, describing a way in which the natural world is consistent in form or behavior (16) *macroscopic*   big or visible, related to large collections of atoms or molecules (18)

microscopic   small or invisible to the unaided eye, related to individual atoms or molecules (18)

order   the arrangement or pattern that expresses the consistency of the natural world (10)

phenomenon   a fact or event that can be observed by the human senses (4)

science   the collection of human knowledge about the natural world and the theories that explain those observations (3)

synthesis   the process of constructing a substance or concept from its component parts (5)

technology   the application of scientific knowledge to solve practical problems (18)

theory   a statement proposed as a testable explanation of observed facts (16)

Exercises

1.1   Decide whether each of the following is an example of analysis or synthesis.

a.   A chemist finds that a dark-colored liquid is a mixture of three different substances.

b.   You taste a piece of cake and decide that it contains chocolate, nuts, and probably eggs, sugar, and butter.

c.   You are listening to a lecture in a history class. You realize that something your political science professor said yesterday contributes to your understanding of this lecture.

d.   A bricklayer expertly uses several materials in constructing a wall.

e.   On a tour of a research laboratory, you are shown a tall glass column filled with white powder. Liquid is added at the top and drips slowly from the bottom into collecting tubes. "This column gives us good separations from a mixture of materials," says your guide.

f.   A living cell makes a very large molecule by stringing together many small ones.

g.   "The budget breaks down into regular annual expenditures and special expenses for this year only. Annual expenditures can then be broken down into costs of postage, materials, and wages," says the budget director.

h. Each of twenty authors contributes a paper on his or her specialty to a well-rounded report on foreign policy.

1.2 Explain fully why recycling discarded glass or aluminum is an important conservation measure. (Consider both the matter and the energy involved.)

1.3 Explain how you know that the following situations could not occur.

a. When you put two identical ice cube trays full of water in the same freezer, one freezes and the other boils.

b. When you let go of a book, it will either rise into the air or drop to the ground.

1.4 These two general statements are true:

a. Water at sea level always freezes at a certain temperature.

b. When you let go of a book, it always drops.

Are these statements laws or theories?

1.5 Professor Jones argues that his explanation of the cause of a certain phenomenon is better than that of Professor Smith. Is his explanation a law or a theory?

1.6 Which of the following definitions best fits the word *model* as we have used it in this chapter?

a. An exact small replica of a much larger object

b. A mental picture that explains how some phenomenon occurs

c. One of several styles produced by an appliance manufacturer

2 What Is Matter?

☐ What are the different ways in which chemists categorize matter?

☐ Why is the metric system so easy to use?

☐ What is the difference between heat and temperature?

☐ How far can we subdivide a sample of matter without changing its properties?

☐ What kinds of things do chemists do to matter to learn more about it?

☐ Are there laws which describe the ways in which different forms of matter react with one another?

☐ How do chemists relate the way matter *reacts* with the way they think it is *organized*?

☐ How small is an atom?

Matter is one of the central themes in chemistry. So it is appropriate for us to begin our study of chemistry by asking, "What is matter, anyway?" We could go to a dictionary and get an answer: "Matter is the substance of which physical objects are composed." That really does not help us much, because we have merely traded one abstract word, *matter*, for others, such as *substance* or *physical objects*. Perhaps a more pointed question would be "How do chemists describe matter?" That is really what this chapter is all about.

THE IDEA OF CLASSIFYING SUBSTANCES

It is a big task to describe the matter we find everywhere around us. It can be made easier by adopting the idea of classifying the things we are examining. All the time, we simplify our experiences by classifying them. Even the idea of classifying matter is as familiar as the childhood guessing game "animal, vegetable, or mineral?" We put things in categories based on what we know

---

We put things in categories based on what we know about them.

---

about them. What we know about them is based in turn on how we or some other humans describe them. Such descriptions are merely the collected records of how something affects human senses: How does it look or feel or sound or taste or smell?

An example of a classification scheme that chemists find useful in describing matter involves the categories *solid, liquid,* and *gas*. These terms identify what is called the *physical state* of a substance. The physical state of a substance is the form in which we find it. For example, we classify the bread we eat as a solid, the water we drink as a liquid, and the air we breathe as a gas.

We put bread, water, and air in these categories because we use our senses to test them. We look for the characteristic behavior of each substance. We see that a piece of bread has a definite volume (how much space it occupies) and a definite shape. Definite volume and definite shape are properties we associate with solids. Volume and shape are both *physical properties*. Physical properties are those characteristics of something which we can observe without changing it into a different kind of substance. (By contrast, one of the *chemical* properties of bread is its nutritive value. This can be observed only by consuming the bread and digesting it as food. The bread is changed into other substances by this process.)

In a similar manner, we call water an example of a liquid. We observe that a sample of water also has a definite volume but that it does not have a rigid shape. As we tilt a glass to drink the water in it, the shape of the water changes. Definite volume but changeable shape are properties we associate with liquids. If the glass holding the water is tipped on its side, the water flows over the table or the floor. Here, too, is the observation of a physical property, ability to flow, that we associate with liquids.

Our common experiences with gases are more limited than our encounters with various liquids and solids. Few gases can be seen, although the brownish tint of one gas called nitrogen dioxide often colors the air over a smoggy city. Some invisible gases have a characteristic odor. The ammonia gas that escapes from an ammonia-containing solution used for cleaning floors or windows is one example. Only sometimes can we consciously feel gases, as we do when a breeze is blowing. Yet we know that gases are a form of matter. Even the free-falling parachutist knows that, despite his or her joy of experiencing quiet nothingness, when the parachute is opened, it will catch something that supports and slows the fall. A characteristic property of gases is having neither definite volume nor definite shape. A gas uniformly fills any container in which it is placed. Another characteristic of gases is that, if the container is not kept closed, the gas escapes, as the air escapes out of a punctured tire.

This classification scheme of gas, liquid, or solid may be a start, but it is much too broad and general to help us far toward understanding the makeup of matter. Again, our common experience poses questions. The liquid stuff we call water can be changed by heating into steam, which is a gas, or by cooling into ice, which is a solid. We also know that some rocks can be in liquid form, like hot lava in volcanoes. Moreover, we recognize that helium, a gas, must be different from air because balloons filled with helium rise through air. Oil, another liquid, forms a floating slick if it is spilled on the ocean, because it is lighter than water and it mixes very little with water. A solid tree and a solid automobile body certainly behave differently when the car crashes into the tree. Each of these contrasts calls for an explanation. By the time you finish this book, you will have found explanations in terms of the ideas that chemists use to describe matter.

Gases, liquids, and solids differ in the number of particles present in a certain volume. Figure 2.1 suggests this. These particles are what chemists call *molecules*. (Recall that we introduced the term molecule in Chapter 1, when we discussed how two liquids can mix together.) The properties of various liquids and solids depend on how tightly the molecules hold onto one another. In turn, how tightly the molecules hold onto one another depends on the makeup and structure of the molecules. You may naturally wonder, "How have chemists developed this idea of molecules, and what leads chemists to talk in terms of molecular structure? And how do differences in molecular structure produce substances that behave differently?" In later chapters you will discover answers to these questions.

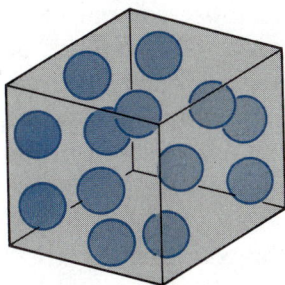

Gas

Molecules moving at random, restricted only by container walls

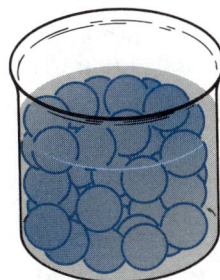

Liquid

Molecules less restricted but confined to liquid. Flow rather than rigidity is typical.

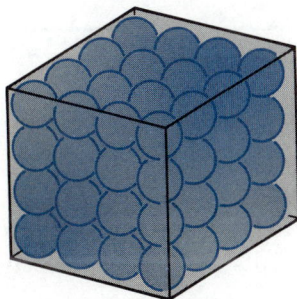

Solid

Molecules restricted to vibrations about fixed points

Figure 2.1   Molecules in solids, liquids, and gases.

PROPERTIES OF SUBSTANCES

The reason the classification scheme of gas, liquid, and solid is so broad and general is that the tests we applied were so loose (can we see it or feel it, does it flow, and so forth?). These are not very specific tests. The next step toward learning to describe matter as chemists do is to recognize what they mean when they speak of the *properties* of a substance. Once chemists know the properties of something, not only can they describe it accurately so that other scientists know what they are talking about, but they can make more precise comparisons between substances. Then, by comparing and contrasting the properties of various substances, they can build theories that correlate (tie together) their information into a more complete understanding of matter.

Let us examine this idea of properties of substances in more detail. We have said that properties are the characteristics of something that we determine by observation. To be useful, the observation must involve some specific test in the form of a measurement. If the results of the measurement can be expressed in numbers, the observation becomes even more useful in making comparisons and drawing conclusions from those comparisons. An analogy is the way that your weight, expressed in pounds or kilograms, is an aid in identifying you among a group of people. (A kilogram, a unit of weight used internationally, is approximately equal to 2.2 pounds.) A physician can use a record of the numerical values of your weight over a period of time as an aid in diagnosing any changes in your health.

Another important idea is involved in the way that scientists identify the properties of substances: They make measurements on samples. Pure water is the same everywhere in the universe. (We say *pure* water, because ocean water is different from water that you can drink. Ocean water has many substances besides water dissolved in it.) But to measure the properties of pure water, they do not have to measure all the water in the universe.

A physician need not take all your blood in order to measure its properties.

A sample is merely a representative part of any whole thing that is a convenient size to work with. A physician need not take all your blood in order to measure its properties. Specific tests to measure specific properties can be made on small quantities, sometimes even on a single drop.

PHYSICAL AND CHEMICAL CHANGES

The contrast between the observations we call physical properties and those we call chemical properties involves what happens to a substance when we make the observation. Observation of a physical property never involves changing the substance into something else. For example, measuring your height or weight does not change you. Similarly, measuring the volume or weight of a sample of solid ice or liquid water or gaseous steam does not change the material, water, into a different substance.

The observation of some physical properties may involve the *physical change* of the material. A physical change only alters the form or physical state of a substance. For example, you may wish to measure the temperature at which some liquid freezes. This is a physical property. You cool the liquid and read on a thermometer the temperature at which the liquid changes to solid. Melting, the reverse physical change, gives you the same liquid substance back again. Observation of physical properties may or may not involve physical changes.

Chemical changes are quite a contrast to physical changes. A *chemical change* converts a substance into something else. The new substance has properties (both physical and chemical) different from the properties of the original substance. For example, milk goes sour, an iron nail becomes covered with rust, and a log of wood burns in a fire. Sour milk contains new substances, which are produced by the bacteria that consumed some of the substances originally in the fresh milk. Oxygen and moisture in the air change the surface of the iron nail. (The chemical name for rust is iron oxide, implying that iron and oxygen have combined.) The properties of rust are different from those of iron: Rust is brown and crumbly, whereas iron is shiny, black, and tough. The wood burns to produce gases, smoke, and ashes. The chemical properties of a substance are described in terms of the chemical changes that it can undergo. Iron can rust; gold cannot. These are contrasting chemical properties.

Iron can rust; gold cannot. These are contrasting chemical properties.

MEASURING THE
QUANTITY OF MATTER

Some physical properties are related to how much there is of an object. Size, or volume, is one of these properties: The more there is of an object, the more space it occupies. The same is true of the weight, or mass, of an object. If it is made up of many individual units, then counting out the number of units gives a measure of its quantity. These different ways of measuring quantity are illustrated by the way you make purchases in a grocery store. You buy a weight of butter, a volume of milk, and a number of eggs. In scientific work, each of these methods also is used to measure amounts. Let us look first at weight and volume. In a later chapter, we will describe how chemists use a very large number as the unit for counting individual tiny items, such as molecules.

First, we should make clear the distinction between mass and weight. The *mass* of an object is a fundamental property representing the quantity of matter in it. The *weight* of an object on the surface of the earth is measured as the force with which gravity pulls the object toward the center of the earth. The force of gravity on an object is proportional to the mass of the object. The force of gravity also depends on the mass of the earth. On another planet or satellite of a different mass, the pull of gravity is different. A 60-pound dog would be much easier to hold as a

---

A 60-pound dog would be much easier to hold as a lap dog on the moon, because it would weigh only 10 pounds there.

---

lap dog on the moon, because it would weigh only 10 pounds there. (The mass of the moon is only one-sixth the mass of the earth.) The force of gravity also decreases with the distance from the center of the earth. Thus an object may weigh less on the top of Mt. Everest or more at the bottom of a deep mine shaft than it weighs at sea level. Far out in space, where the gravitational pull of the earth is nearly zero, it will weigh almost nothing at all. Because the weight of an object or substance changes, but mass does not, scientists prefer to use the term mass to refer to the quantity of a substance.

The standard unit of mass used in scientific work is the *kilogram*. The prefix *kilo* means "one thousand"; a kilogram is 1000 grams. The gram is a more convenient, smaller unit of mass to use when measuring the amounts of mass that are usually encountered in laboratory experiments. From here on, in this book we will usually indicate masses in numbers of grams. For comparison with the familiar English units, you can think of a pound as 454 grams (approximately half a kilogram). To get some idea of the amount of mass in a gram, think of a penny as a mass of 3 grams and a nickel as 5 grams.

The size or *volume* of an object is the amount of space it occupies. Space is a three-dimensional quantity. Consequently, measurements of volume are properly expressed as the cube of a distance measurement. For example, consider the volume of a room that is 4 meters wide, 6 meters long, and 3 meters from floor to ceiling.

penny, 3 g

nickel, 5 g
(2 cm diameter)

$$\text{Volume} = 4 \text{ meters} \times 6 \text{ meters} \times 3 \text{ meters}$$

$$\text{Volume} = (4 \times 6 \times 3) \times (\text{meter} \times \text{meter} \times \text{meter})$$

$$\text{Volume} = \underbrace{72} \qquad \underbrace{(\text{meter})^3}$$

This quantity is read as 72 meters cubed or 72 cubic meters.

The standard unit of length used in scientific work is the meter, so the standard unit of volume is a cubic meter. A meter (39.37 inches) is slightly longer than a yard (36.00 inches). A convenient smaller unit of length is a centimeter ($\frac{1}{100}$ of a meter). A centimeter is slightly shorter than half an inch. A nickel is very close to 2 centimeters in diameter. A cubic meter is a relatively large volume unit. A more convenient unit of volume measurement is the *liter*. It is defined as $\frac{1}{1000}$ of a cubic meter or as 1000 cubic centimeters. For comparison with the familiar English units, you can think of 1 liter as approximately equal to 1 quart (1.000 liter = 1.057 quart).

The system of measurement used in scientific studies is the *metric system.* Except for English-speaking countries, this is also the system in common use. Both England and the United States have passed legislation toward sponsoring a change to the international system. In the near future, you will find more and more occasion to recognize the value of measurements expressed in the metric system. One of the advantages of the metric system is its use of a decimal (multiples of ten) way of designating the relationship among units. A regular system of prefixes has been adopted, which greatly simplifies the designation of numbers. These are listed in the box on the next page. A second box lists the values of some equalities that will help you relate the less familiar metric units to the more familiar English units. Figure 2.2 suggests these comparisons with diagrams.

Decimal Notation

The following table lists the prefixes and abbreviations that are used in all decimal systems of notation.

Multiple of 10	Prefix	Abbreviation Symbol
1,000,000,000	giga	G
1,000,000	mega	M
1000	kilo	k
100	hecto	h
10	deka	da
0.1	deci	d
0.01	centi	c
0.001	milli	m
0.000001	micro	μ
0.000000001	nano	n

Examples of the use of these prefixes with units of mass, length, and volume are

1 kilogram   = 1000 gram

1 kilometer = 1000 meter

1 centimeter = 0.01 meter

1 milliliter = 0.001 liter

Equivalents between English and Metric Systems

The following table lists the numerical values in the English and metric systems for equal lengths, masses, and volumes. Use this table as a reference for estimating conversions from one system to the other.

LENGTH

1.00 inch	=	2.54 centimeter	1.00 centimeter	=	0.394 inch
1.00 yard	=	0.914 meter	1.00 meter	=	1.09 yard
1.00 mile	=	1.61 kilometer	1.00 kilometer	=	0.621 mile

MASS

1.00 ounce	=	28.4 gram	1.00 gram	=	0.0353 ounce
1.00 pound	=	0.454 kilogram	1.00 kilogram	=	2.20 pound

VOLUME

1.00 ounce	=	0.0296 liter	1.00 liter	=	33.8 ounce
1.00 quart	=	0.946 liter	1.00 liter	=	1.06 quart
1.00 gallon (U.S.)	=	3.79 liter	1.00 liter	=	0.264 gallon (U.S.)

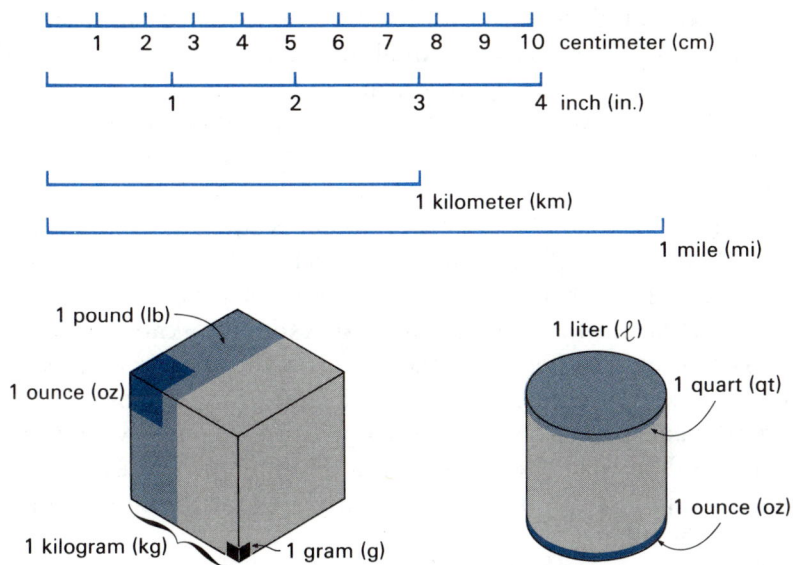

Figure 2.2 Comparisons of metric and English units of measurement.

TEMPERATURE SCALES

Mass and volume are properties of matter that depend on the quantity in the sample. *Temperature* is a different kind of property. The temperature of an object has no connection with how large or small it is. A cup of hot coffee has the same temperature

A cup of coffee has the same temperature as the whole potful.

as the whole potful. Of course, you must add more heat to raise the temperature of a potful of coffee than to raise the temperature of a cupful by the same number of degrees. So heat and temperature cannot be the same thing. Heat is a form of energy; temperature is a property of matter. We will explore the idea of what heat is and how objects gain or lose heat in a later chapter. For example, we will find that heat always flows from a hot body to a cold body. But at this point, we will focus on temperature as a property of matter. How can we measure the *degree* of hotness or coldness?

The familiar device used to measure temperature is a thermometer. Actually, a thermometer indicates how the volume of a fixed amount of liquid changes with temperature. A fixed amount of a liquid, such as mercury or colored alcohol, is held in the bulb of a thermometer. As the liquid warms up, it expands and partially fills a thin tube connected to the bulb. The tube is made with a uniform internal diameter. Consequently, distances along the tube are proportional to the volume of liquid filling it to that point. (The volume of a uniform cylinder is equal to the height of the cylinder multiplied by the cross-sectional area of the cylinder.) To make such a bulb and tube useful as a thermometer, the tube must be marked at points that correspond to known, chosen temperatures. The height of the liquid in the tube is marked at some chosen temperature. Then the height of the enclosed liquid is marked at some other chosen temperature. Equal divisions are marked off between these two points. Then temperatures between these two points can be read by noting how far the enclosed liquid fills the tube.

The two temperatures ordinarily chosen as fixed points to establish a temperature scale are the freezing and normal boiling temperatures of pure water. (The "normal" boiling temperature of water means that the water is boiled under 1 atmosphere pressure. In later chapters, we will discuss what atmospheric pressure is and why a particular pressure must be chosen in order to use boiling water as a standard.) Thus the scale we use to measure temperature is related to the invariable way that something in the physical world behaves.

The temperature scale commonly used in English-speaking countries is the *Fahrenheit scale.* Most household thermometers use this scale. The temperature at which water freezes is given the value of 32 degrees. The temperature at which water boils is given a value 180 degrees higher, 212 degrees. The temperature scale used in scientific work and that used in non-English speaking countries is called the *Celsius scale.* On the Celsius scale, the freezing temperature of water is given the value 0 degree; the boiling temperature of water is given the value 100 degrees. In the near future, as the trend continues toward using the international system in the United States, you will have more and more occasion to use thermometers that are marked with the Celsius scale.

Figure 2.3 is a diagram of a thermometer marked with both Fahrenheit and Celsius temperature scales. Note some of the familiar temperature reference points. The recommended room temperature of 68°F is 20°C. (The symbol ° is used to designate degrees.) Normal body temperature of 98.6°F is 37.0°C.

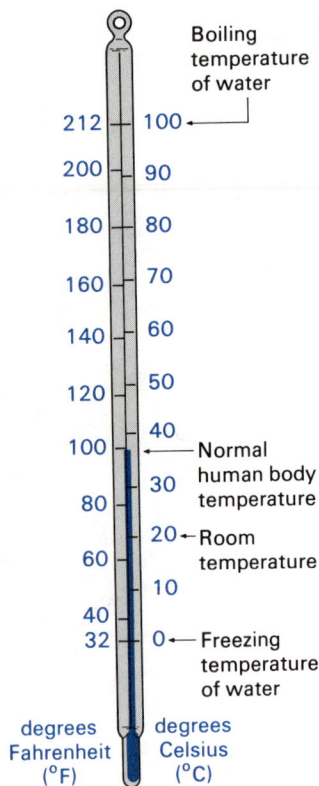

Figure 2.3 Comparison of Fahrenheit and Celsius temperature scales.

A conversion formula can be used to calculate what a reading on one scale corresponds to on the other scale. The formula is

$$°F = \frac{180}{100}°C + 32°$$

Or more simply,

$$°F = 1.8°C + 32°$$

The factor $\frac{180}{100}$ comes from the fact that there are 180 divisions (212−32) on the Fahrenheit scale for the same 100 divisions (100−0) on the Celsius scale between the boiling and freezing temperatures of water. The added number 32 comes from the fact that zero on the Celsius scale corresponds to 32 on the Fahrenheit scale.

A more convenient way to convert a reading on one scale to a reading on the other is to use the graph shown in Figure 2.4. A

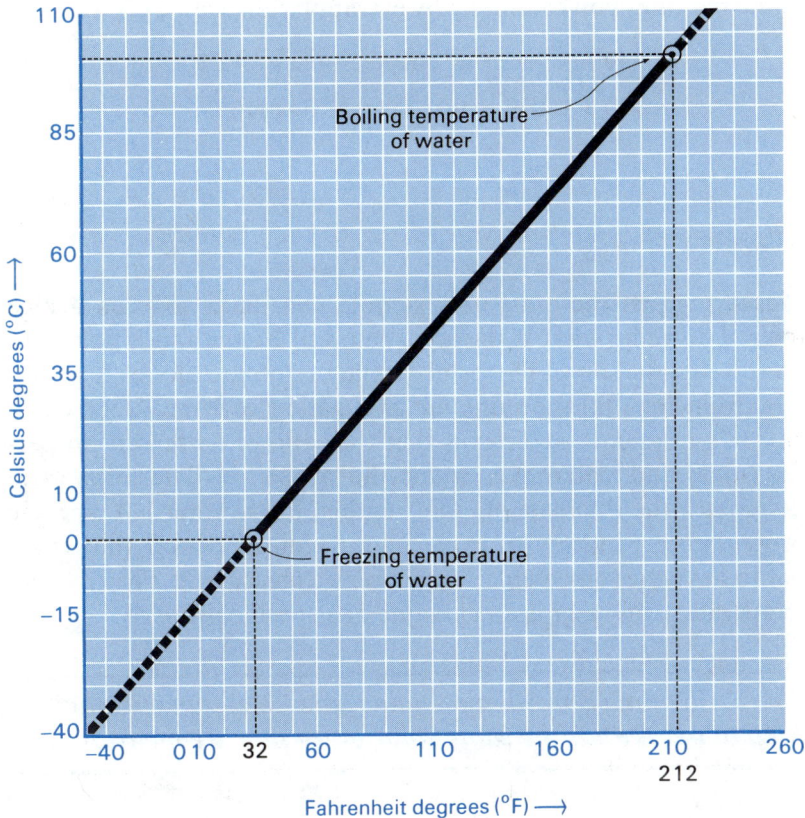

Figure 2.4   Equivalent temperatures in degrees Fahrenheit and degrees Celsius.

graph such as this shows the relationship between the two read-ings. The graph is drawn on a grid of regularly spaced lines. One scale goes across; the other, up and down. Notice that the nu-merical value of each horizontal division is different from the value of each vertical division. A straight line is drawn between the two fixed points (freezing temperature and boiling tempera-ture). Then the corresponding Fahrenheit and Celsius readings can be made by spotting the values for points anywhere along the line. This procedure is called interpolation (*inter* is the Latin word for "between"). Also, if you put a ruler along the line on the graph, you can extend it above or below. This procedure is called extrapolation (*extra* in Latin means "beyond"). Some readings you can make by interpolation are

86°F is the same temperature as 30°C

176°F is the same temperature as 80°C

10°C is the same temperature as 50°F

60°C is the same temperature as 140°F

Some readings you can make by extrapolation are

−40°C is the same temperature as −40°F

230°F is the same temperature as 110°C

DENSITY

Measuring just the mass of a sample can never give us a clue toward deciding what is in it. A pound of butter and a pound of flour weigh the same. This is also true of volume alone. A gallon of gasoline can fill the same jug as a gallon of wine. But one of nature's great consistencies is that the ratio of these two proper-ties (mass and volume) is a specific property. (The term ratio designates the division of one number by another.) A specific property is a property of a particular kind of matter that turns out to have the same value every time we measure it. The freezing temperature of water is an example of a specific property.

The ratio of mass to volume of any sample of a particular kind of matter has the same value every time we measure it.

The ratio of mass to volume is called the *density.*

$$\text{Density} = \frac{\text{mass}}{\text{volume}}$$

For example, at a temperature of exactly 25°C, 1.000 gram (mass) of water occupies 1.003 milliliters (volume). Thus the density of the sample is

$$\text{Density} = \frac{1.000 \text{ g}}{1.003 \text{ ml}} = 0.997 \text{ g/ml}$$

The symbol g/ml is read as "gram per milliliter." Note that, if the sample were 1000 grams mass, its volume would have been 1003 milliliters; the ratio of 1000 g to 1003 ml would still be 0.997 g/ml. This is why we can refer to density as a specific property.

We should also note another part of the above statement: "at a temperature of exactly 25°C." This illustrates an important feature of the scientific way of describing a property. The conditions that alter a property must be specified. A consistency of nature is that the volume of a given mass of something changes when the temperature changes. Usually, volume increases with an increase in temperature. For example, an unlubricated engine bearing, heated by friction, will expand until it sticks tight. We have to say "usually," because liquid water is one of the few substances that contracts (gets smaller) as its temperature increases. The volume of a mass of water decreases as the temperature rises from 0°C to 4°C. Above 4°C up to its boiling point, water expands as the temperature rises. In all cases, however, the volume of a given mass of substance changes as the temperature varies. Thus temperature is called a *variable,* and it must be known if we expect to consider density as a truly specific property.

Pressure is also a variable that affects the volume of a given mass of substance. However, very large changes in pressure are required to produce significant changes in the volume of a sample of liquid water. Small changes in pressure have much more effect on the density of gases. Gases can easily be compressed into a small space, a process that increases their density. For example, the highly compressed gas used to power aerosol spray cans is more dense in the can than it is when it has been released to normal pressures.

Figure 2.5 shows how the density of water changes with temperature over the range from 10°C to 100°C. Here we can see

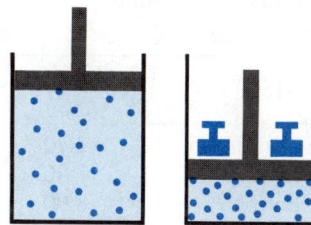

Gas is more dense.

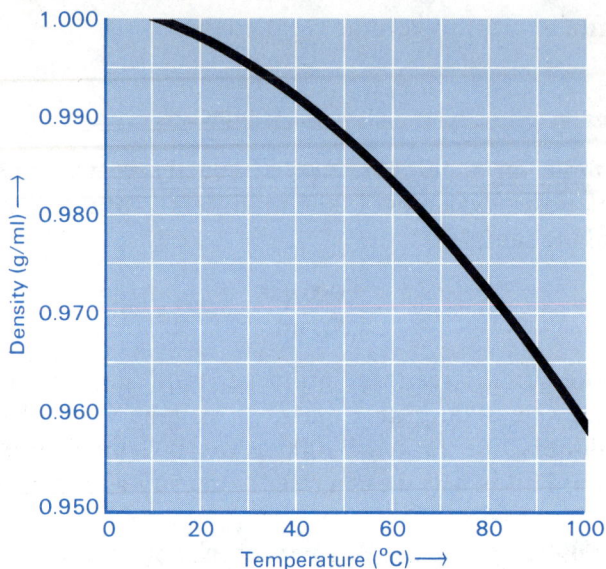

Figure 2.5   Density of water over the temperature range 10° C–100° C.

Table 2.1   Examples of mass, volume, and density of water at various temperatures

Temperature (˚C)	Mass (g)	Volume (ml)	Density (g/ml)
10	1.000	1.000	1.000
50	1.000	1.012	0.988
90	1.000	1.036	0.965

how increasing temperature, which increases the volume, produces a decrease in density. Table 2.1 gives some illustrative figures. As the volume occupied by the same mass (1.000 gram) increases, the density decreases. When the same number is divided by successively larger numbers, the value of the ratio decreases. For example, one-fourth of a dollar is 25 cents; one-tenth of a dollar has less value, 10 cents.

Table 2.2 lists the densities of a few familiar metals. The differences among these numbers illustrate how density is a specific property. If you had a piece of iron and a piece of lead that

Table 2.2

Densities of some metals at 25˚C

Metal	Density (g/ml)
aluminum	2.70
iron	7.90
copper	8.90
lead	11.30
gold	19.30

If you had a piece of iron and a piece of lead that looked alike, you could tell which was which by measuring their densities.

looked alike, you could tell which was which by measuring their densities. Note how high the density of gold is. Counterfeit gold, or gold mixed with other, less-dense metals, can always by identified by measuring the density of the object.

PURE SUBSTANCES AND MIXTURES

We started our discussion of how a scientist describes matter by suggesting that a useful first step is to classify. Any system of classifying is based on the properties of the sample being examined. Items with similar properties go into the same pile or category. Deciding what the categories are to be is an important decision. For example, a library could put all books of the same size

A library could put all books of the same size or color together, but it is more efficient to put books dealing with the same subject together.

or color together. But it is more efficiently organized if books dealing with the same subject are put together. So it is with classifying matter. What is the best way to start sorting out the multitude of samples of matter we encounter?

A classification we can try first is to ask whether a sample is a pure substance or a *mixture*. Most of our everyday experience is with mixtures. Air, rocks, soil, foods, and textiles are examples. Many times it is an advantage to have things come in mixtures. For example, fertilizer for plants or trees is a mixture containing mostly material, such as sand, that has no nutritive value. If the beneficial plant food were applied in a pure form, the plant would be overfed and harm would result. The same is true of many drugs we take. We need such small amounts that the proper dosage has to be mixed with some harmless substance in order to avoid getting too much.

In other cases, the addition of some substance to make a mixture out of what formerly was a pure material, or a less complex mixture, can be harmful. Adulterated foods or beverages are those to which small amounts of harmful substances have somehow been added. The contamination and pollution of the air and water in our environment likewise involves the creation of complex mixtures. The amount of mixing that is constantly going on in the world means that we encounter very few pure substances in our daily living. A few examples of pure substances are table sugar, dry ice, rain water, and aluminum foil.

GUARANTEED FERTILIZER ANALYSIS:

TOTAL NITROGEN (N) 5%
1.6% Ammonical Nitrogen (from Ammonium Sulphate.)
0.15% Water Soluble Nitrogen (from Processed Tankage.)
3.25% Water Insoluble Nitrogen (from Processed Tankage & Bone Meal.)

AVAILABLE PHOSPHORIC ACID (P_2O_5) 10%
(from Superphosphates & Bone Meal.)

SOLUBLE POTASH (K_2O) . . .10%
(from Muriate of Potash)

CALCIUM, expressed as elemental 2%
(from Dolomite, Superphosphate, & Bone Meal.)

MAGNESIUM, expressed as elemental 1%
(from Dolomite.)

SULPHUR, expressed as elemental 2%
(from Ammonium Sulphate, Metallic Sulphates & Sulfides, & elemental Sulphur.)

IRON, expressed as elemental 2%
(from Iron Sulphates, Sulfides & Oxides.)

MANGANESE, expressed as elemental 0.05%
(from Manganese Sulphates, Sulfides & Oxides.)

ZINC, expressed as elemental 0.05%
(from Zinc Sulphates, Sulfides & Oxides.)

Fertilizer is a mixture. The other 67.9% is nonnutrient material.

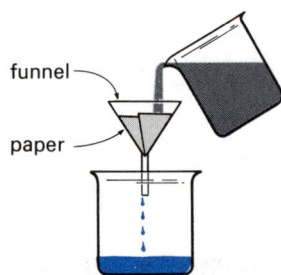

How can we decide whether something is a mixture or a pure substance? Sometimes the decision is easy. We can see that there are several different kinds of things present in some mixtures. For example, the various particles in a sample of soil can be identified by their size, color, hardness or softness, and so on. We can separate them by either picking the pieces out by hand or by shaking the soil in a sieve. We may use a series of sieves with increasingly smaller holes to separate the soil particles according to size. The same principle is used in the process called filtration. The mixture may be a suspension of very fine solid particles in a liquid, as in muddy water. When such a suspension is poured into a funnel holding some filter paper, the water goes through the paper, but the fine particles are retained.

The decision may be more difficult for other mixtures. For example, a mixture of finely ground iron and lead may be made up of metal particles that look very much alike. Yet if you put a magnet into this mixture, the iron will stick to the magnet, leaving the lead behind. *For any mixture, some kind of a procedure can be used to separate the components from one another.* Such separation procedures are *physical methods* based on differences in some physical property of the components of the mixture. For example, filtration works as a separation method when the particles have different sizes. Mixtures are frequently said to be heterogeneous, meaning that they are "made of different parts."

Pure substances, in contrast to mixtures, cannot be separated into different parts by physical methods. No matter how finely you grind up a solid sample, or even if you heat it until it melts or boils, it is a pure substance because every part of it has the same properties. Pure substances are said to be *homogeneous,* meaning that they are "made of only one kind of matter."

THE GRAININESS OF MATTER

Suppose you divide a sample into smaller and smaller parts. Is there some limit to the smallness of the pieces? This question has puzzled philosophers from the time of the ancient Greeks and especially puzzled the scientists who lived in the nineteenth century. The question involves the idea of the "graininess" of matter. If you stand several feet from a stream of flowing sand, what you see is a continuous flow. Yet on closer inspection, you can see the individual grains of sand. Even finely ground sand or flour can be seen under a microscope to be made up of distinct particles.

Modern chemistry has an answer to the question of how far we can go in making a sample into smaller and smaller particles. The answer satisfactorily ties together various bits of information. This idea is broadly referred to as the *molecular theory of matter*. For now, we can define a *molecule* as the smallest unit of any pure substance that can be obtained by physical processes of separation. Physical processes are those that do not change the substance into something else.

The idea of molecules helps us understand why matter can exist as pure compounds or mixtures. *In any sample of a pure substance, all the molecules are alike.* Later we will find that, strictly speaking, not all substances exist as molecules. For example, if

---

If you keep dividing table salt into smaller and smaller pieces, you end up with an equal number of two kinds of particles.

---

you keep dividing table salt into smaller and smaller pieces, you end up with an equal number of two kinds of particles. One kind of particle has a positive electric charge; the other kind has a negative electric charge. However, these particles are about the same size as molecules, so we need not worry about the distinction at this point. We can properly say that *pure substances are homogeneous at the molecular level.* In mixtures, pieces of pure substances are mixed with pieces of other pure substances. In each piece the molecules are alike, but they are different from the molecules in pieces of the other components of the mixture.

Some idea of the size of molecules can be gained from an imaginary "experiment." Suppose you take a glass of water from the ocean and somehow paint the molecules purple, so that you can identify those particular ones. Toss your glassful of purple water molecules back into the ocean, and let them mix uniformly with all the water in all the oceans of the world. Then take another glassful of water from the ocean and look for purple molecules. You would find about 400! There are 400 times as many molecules in a glassful of water as there are glassfuls of water in all the oceans of the world. Here is another analogy: If every molecule of water in a typical snowball were magnified to the size of a pea, there would be enough snow to blanket the whole surface of the earth to a depth that would cover a 90-story building.

There are 400 times as many molecules in a glassful of water as there are glassfuls of water in all the oceans of the world.

IDENTIFYING PURE SUBSTANCES

Separating a sample of a mixture into pure substances is only a first step in the analysis of a sample. The next important step is finding out what the substances are. *Qualitative* analysis merely identifies what is present; *quantitative* analysis determines how much of a substance is present.

The identification step in any analysis can be accomplished in various ways. Some physical property of the pure substance can be measured, and the results can be compared with the values that have previously been measured for known pure substances. Chemists have a tremendous amount of reference data. Handbooks list the physical properties, such as melting or boiling temperatures, of thousands of pure substances. Often the characteristic absorption of light by a substance is used for identification. Foods or drugs are analyzed this way for the presence of small amounts of harmful impurities. The method works on the same principle as the way a piece of colored glass absorbs some of the white light of sunlight and allows the rest of it to go through.

The observation of chemical properties also can be used to identify substances. By chemical properties, we mean the ways that a substance reacts or changes into some other recognizable substance when we do something to it. "Doing something to it" usually involves treating it with some substance that we know will react with it in a certain way. For example, a particular liquid can be mixed with a metal suspected of being either zinc or tin. If gas bubbles form quickly, the sample is reacting as zinc does; tin reacts slowly with that particular liquid.

One kind of chemical property is especially important in further classifying pure substances. This property is whether or not a pure substance can be broken down into simpler substances by a chemical change. If you put some table sugar, a pure substance,

If you put some table sugar in a hot pan on the stove, it will decompose to leave only black carbon.

in a hot pan on the stove, it will *decompose* or break down to leave only black carbon in the pan. Carbon also is a pure substance, but nothing you can do will decompose it further. (You may be able to burn it off the pan by allowing it to react with oxygen in the air to form a different substance, carbon dioxide.)

Sometimes the test for ability to decompose may be more complicated. For example, salt in a hot pan on a stove does not act as sugar does. If the pan is made very hot, the salt may melt, but heating does not make it decompose. But if electric current is passed through the melted salt, it will decompose into two substances, chlorine, a gas, and sodium, a metal. But the process stops there. Neither the chlorine nor the sodium can be decomposed.

ELEMENTS AND COMPOUNDS

The kinds of pure substances that cannot be broken down into simpler substances by chemical reactions are called *elements*. We must be careful to say "by chemical reactions," because later we will discover that even elements can be broken down, by very high-energy processes, into smaller pieces. These are the fundamental particles—such as electrons, protons, and neutrons—of which all types of matter are composed. The number of known elements now stands at 106. Of these, only 88 are found in nature; the other 18 have been manufactured with the aid of high-energy machines.

Almost every element can react with others to form *compounds*. Thus elements are the building blocks, in a chemical sense, of all forms of matter. The list of elements includes many familiar names. Newspapers and magazines tell you about "lead in gasoline," "iron in tired blood," or "calcium in bones." You should realize that the lead in gasoline is not in the same form as the lead in a storage battery. (Incidentally, there is no lead in a "lead" pencil. For many years, pencils have been made instead with graphite, a form of carbon.) The lead in gasoline is chemically combined with other elements to form a lead compound with the chemical name lead tetraethyl. In similar fashion, the iron in blood is in the compound hemoglobin, which is essential to the body's use of oxygen, and not in the form of a nail you can pound into wood.

Table 2.3 lists some of the common, frequently encountered elements and the *chemical symbols* chemists use to designate them. These symbols constitute a kind of chemical shorthand that is used universally. Chemists speaking all languages use the same symbols. Thus, O stands for the English *oxygen*, the German *Sauerstoff*, or the Russian кислород. Later in the chapter we will discuss the full meaning of a symbol. You will find a complete list of elements on the inside back cover of the book.

Table 2.3 Some common elements

Name	Symbol	Role in the Natural World
Aluminum	Al	Light metal; abundant in clays and rocks; many industrial uses
Calcium	Ca	Reactive metal; abundant in clay and rocks; important to life, as in bones and teeth
Carbon	C	Forms millions of compounds; the backbone of living systems; used in petroleum and plastics
Chlorine	Cl	Very reactive nonmetal; abundant in salt; essential to life
Chromium	Cr	Hard, noncorrosive metal; forms many colored compounds.
Copper (L. *cuprum*)	Cu	Important metal; high conductivity; essential trace element for some living systems
Hydrogen	H	Lightest element; very combustible gas; abundant in the compound water; essential to life
Iron (L. *ferrum*)	Fe	Tough metal; used in steel; abundant in rocks and soil; essential to breathing animals
Magnesium	Mg	Light metal; reactive; important to plants in the compound chlorophyll
Manganese	Mn	Tough metal; used in steel; important trace element in living systems
Nitrogen	N	Major constituent of air; essential to all life in the form of proteins
Oxygen	O	Reactive gas; essential for combustion and for breathing in animals; forms many compounds
Phosphorus	P	Reactive nonmetal; abundant in rocks and soil; essential to life
Silicon	Si	Abundant in sand and rocks; forms very stable compounds
Sodium (L. *natrium*)	Na	Very reactive metal; abundant in the compound salt; essential to life
Sulfur	S	Reactive nonmetal; important as source of sulfuric acid; essential to life proteins
Zinc	Zn	Reactive metal; many industrial uses; essential trace element for life

Some of the symbols for elements are direct clues to their names: Cl stands for chlorine; He stands for helium. The names, in turn, sometimes give clues to some property or special attribute of the element. Chlorine exists as a green gas at room temperature, and the name is derived from the Greek word *chloros,* meaning "greenish-yellow." The name helium comes from the Greek word for the sun, *helios*. The element helium was discovered on the sun before it was known to be an earthly element. Other symbols reflect Latin names for elements that were known as specific substances when Latin was the universal language of scholars. For example, Fe, for iron, comes from the Latin *ferrum*.

THE LAWS OF CHEMICAL COMBINATION

We seem to have gone in the decomposition direction about as far as we can go. The idea that elements are pure substances that cannot be further decomposed by chemical changes tells us very little about the vast majority of pure substances, the *compounds,* which are formed by the chemical combination of elements. So let us turn away from the goal of analysis toward the opposite goal, synthesis. In Chapter 1 we introduced the words synthesis and analysis in terms of their broad general meanings. Synthesis means putting together; analysis means pulling apart. Now let us use the term synthesis in a more specific chemical sense. In the chemical sense, synthesis means building a new compound by combining two or more substances. The simplest kind of synthesis is the reaction that results in the chemical combination of two elements to form a compound. Thus the elements carbon and oxygen can combine to form the compound carbon dioxide. Whether this takes place in a furnace (coal is nearly pure carbon) or the pan on your stove in which sugar decomposes or in a chemical laboratory, the reaction is the same. We should properly ask, "How predictable is nature's behavior, not just in this reaction, but in *all* reactions in which elements combine? What are the *laws* of nature that are so powerfully general that they apply to all combinations of elements?"

The first of these laws is the **law of conservation of mass:**

In all chemical reactions, the mass of the products is exactly equal to the mass of the reactants.

Written as an equation for our example of carbon burning,

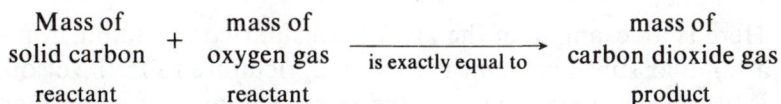

Mass of solid carbon	+	mass of oxygen gas	is exactly equal to →	mass of carbon dioxide gas
reactant		reactant		product

This law is the cornerstone of all chemistry. It seems reasonable and almost as logical as the generalization "you don't get something for nothing." However, it was not until the end of the

You don't get something for nothing.

eighteenth century that the greatest scientists of the age became convinced that it was so. In fact, chemistry became a science only after experimenters believed that this law offered them a checkpoint from which they could interpret their results.

Part of the difficulty for early experimenters was that they did not think mass was an important property. For one thing, they had only very crude balances with which to weigh objects. Moreover, it was hard for them to weigh gases, which are always very much less dense than solids and are sometimes invisible, as is air. So it may not be surprising that the early chemists seemed to miss the significance of confusing results when they did weigh things. When charcoal burned in a dish open to the air, the ashes, the product that was left, weighed less than the charcoal had; when a metal burned in open air, the product weighed more than the metal had. The law of conservation of mass was discovered only after repeated observation of things burning in closed containers and recognition that the mass of all substances in the container is the same before and after the reaction. At this time, chemists also began to regard burning as we do now: When something burns, it undergoes a chemical combination with oxygen from the air.

About this time in history, improvements were made in the devices used for weighing in the laboratory. These improvements made more-accurate and more-precise weighing possible. Most people recognized that a particular compound always contains

When wood burns it loses weight.

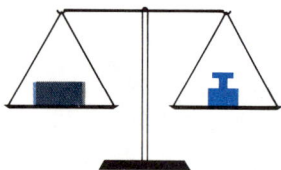

A particular compound always contains the same elements.

the same elements (a qualitative observation). But now enough *quantitative* information began to accumulate so that another of nature's consistencies became apparent. This is the **law of constant composition:**

All samples of a pure compound contain the same elements in the same ratio by mass.

Here is an example of the law of constant composition. Lime is a familiar substance that is used in agriculture to fertilize soil. It is also used to make plaster and cement. Lime is a compound of two elements, calcium and oxygen. In a laboratory, we can easily synthesize lime by heating calcium, a reactive metal, in air.

When metal burns it gains weight.

The calcium readily combines with the oxygen present in the air. The reaction can be written

Calcium + oxygen → lime

If we weigh the calcium we start with and the lime that is formed, we might get these data from two experiments:

burning in air
10 g calcium → 14 g lime

	Experiment I	Experiment II
Mass of lime formed	14.0 g	70.0 g
Mass of calcium reacted	10.0 g	50.0 g

We know that lime resulted from calcium combining with oxygen, so we can calculate how much oxygen reacted. (It is much easier to do this calculation than it would be to measure the mass of oxygen gas by weighing it.) The difference between the mass of solid lime we made and the mass of solid calcium we started with must be the mass of oxygen gas that combined.

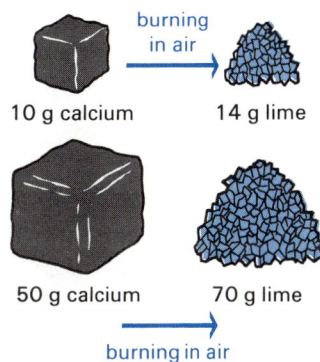
50 g calcium   70 g lime
burning in air

	Experiment I	Experiment II
mass of lime formed	14.0 g	70.0 g
− mass of calcium reacted	− 10.0 g	− 50.0 g
mass of oxygen reacted	4.0 g	20.0 g

What we are doing is applying the law of conservation of mass. According to that law, the mass of the product must equal the sum of the masses of reactants.

We can now go back to the law of constant composition, and set up the ratio we are interested in. Remember, a ratio is one number divided by another. Sometimes it is helpful to express this ratio first in words. Then we can see what the numbers really mean. When we set up the ratios of the masses of calcium and oxygen that react with each other,

	Experiment I	Experiment II
$\dfrac{\text{mass of calcium}}{\text{mass of oxygen}} =$	$\dfrac{10.0\text{ g}}{4.0\text{ g}} = 2.5$	$\dfrac{50.0\text{ g}}{20.0\text{ g}} = 2.5$

Both these ratios have the *same numerical value*, 2.5.

We could do many more experiments, starting each time with a different mass of calcium and measuring the mass of oxygen that combines with it each time. When we calculate the value of the ratio, as for our two examples, we will always get the same value, 2.5. The mass of the calcium is always 2.5 times the mass of the oxygen combined with it in the compound lime.

We have used only one example, the formation of lime, to suggest how a law of nature is discovered. The power of this kind of reasoning in science comes from recognizing how far-reaching and generalized the results are. Once we find enough examples from which to deduce the law, we can then take the next step and have faith that nature always behaves this way. Of course, we have oversimplified; many examples involving many different elements had to be investigated before the law was accepted as being a way nature *always* behaves. Scientists argued about this law for many years during the early part of the nineteenth century. Now, however, it is universally accepted. And as you will see, thinking about a scientific law sometimes leads to a scientific theory.

THE ATOMIC THEORY

A typical scientific attitude is curiosity. Laws are fine and useful, but in the mind of a scientist, they always raise the question "Why is this so?" We have been discussing important consistencies in the way elements undergo chemical reactions. How do these data about the constant composition of compounds fit in with the idea of the "graininess" of matter that we discussed previously? What we are looking for is a mental picture of matter that can tie all this information and imagination together.

Around the beginning of the nineteenth century, John Dalton, an Englishman, put together such a mental picture, which we now call the *atomic theory*. He did not invent the idea of atoms but rather refined an idea that had come down through the ages from the ancient Greeks. (The word *atom* comes from a Greek word that means "cannot be divided.") Dalton suggested a working model for matter at the submicroscopic level. His theory states that

1.   Elements are made up of tiny fundamental particles called *atoms. Fundamental,* as it is used here, means that they cannot be further divided by any chemical changes.
2.   All the atoms of one element are *alike*. Especially important is the idea that they have the same characteristic mass.
3.   Each element has atoms that are *different* from the atoms of other elements. Especially important is the idea that each element has atoms of a characteristic *mass* different from the atoms of all other elements.

4. Atoms *cannot be destroyed* or altered in any way when elements undergo chemical reactions.
5. Atoms of one element *combine* with atoms of other elements to form compounds.
6. In a compound, the *ratio of the number of atoms of one element to the number of atoms of another element is constant.*

THE LAWS OF CHEMICAL COMBINATION IN ATOMIC TERMS

Let us see how these ideas of atoms and the properties that Dalton suggested for them relate to the laws of chemical combination. Here we will be doing what chemists are always doing—making experiments with *macroscopic* amounts of substances. The prefix *macro* means "large," amounts that can be seen, weighed, or otherwise measured with instruments. Yet when it comes to interpreting what happens in a chemical reaction, chemists think in *microscopic* (actually in submicroscopic or "smaller than microscopic") terms. Individual atoms are extremely tiny. A copper penny contains approximately 30,000,000,000,000,000,000,000 atoms. Such a huge number is difficult

A copper penny contains approximately 30,000,000,000,000,-000,000,000 atoms.

to imagine. It is about the number of drops of water in the whole Mediterranean Sea. In spite of their size, we think of these atoms as the kind of particles of which all elements are composed. Most important, elements undergo chemical reactions at the atomic level.

The law of conservation of mass is readily accounted for if matter is made up of atoms that cannot be destroyed. Atoms may be rearranged and combined in various ways in chemical reactions, but no atoms that were in the reactants are ever lost or left out of the products.

The law of constant composition can be explained by recognizing that compounds always contain the same relative numbers of atoms of each element. Thus we can explain our illustration of calcium combining with oxygen if we make two assumptions. First we assume that *one* atom of calcium combines every time

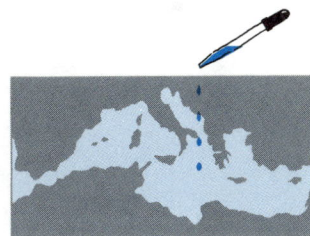

As many drops of water as atoms in a penny.

with *one* atom of oxygen. Then we assume that whatever tiny mass the oxygen atom has, the mass of the calcium atom is 2.5 times as great. The reasoning here is similar to taking a handful of bolts, with a nut on each one, and finding the relative weight of the bolts and the nuts. First, remove the nuts and weigh the pile of bolts. (You get 250 grams.) Then weigh the pile of nuts. (You get 100 grams.) Even though you do not know how many nuts or bolts you have, from the fact that you started with one nut on each bolt, you can conclude that one bolt weighs 2.5 times as much as one nut.

FORMULAS OF COMPOUNDS

Dalton's ideas were a great boon to chemists of the early nineteenth century. The atomic theory made possible some order in the vast amount of data on elements and compounds. One very important advance was the way the composition of compounds could be expressed simply and clearly. Chemists began to have universal agreement that *the symbol of an element stands for one atom of the element.* This agreement meant that the atomic composition of *compounds* could be expressed as a *chemical formula.* Thus we can write the formula for lime, the compound of calcium with oxygen, as CaO: One atom of calcium (Ca) combines with atom of oxygen (O) to produce lime (CaO).

Other facts that puzzled the chemists of Dalton's time could also be accounted for by the atomic theory. In some cases, the same two elements formed distinctly different compounds. For example, copper (Cu) and oxygen (O) were found to form a black compound and a red one. Not only were the colors of the two compounds different, so were the ratios of the masses of the two elements in each. The data are as follows:

	Ratio in words		Ratio of the masses
Black compound:	$\dfrac{\text{mass of copper}}{\text{mass of oxygen}}$	=	$\dfrac{4 \text{ g}}{1 \text{ g}}$
Red compound:	$\dfrac{\text{mass of copper}}{\text{mass of oxygen}}$	=	$\dfrac{8 \text{ g}}{1 \text{ g}}$

The comparison of these two ratios is the important feature that the atomic theory explains. The ratio $^8/_1$ is exactly *twice* the ratio $^4/_1$. This fact is not a violation of the law of constant composition. There are two different compounds, each with its constant ratio for the masses of the two elements in it. Dalton ex-

black

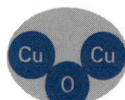

red

plained the fact that one ratio is exactly twice the other as follows: Assume that the mass of one copper atom is four times the mass of an oxygen atom. Then the formula of the black compound is CuO.

$$\text{Black compound:} \quad \frac{\text{mass of copper}}{\text{mass of oxygen}} = \frac{4}{1} = \frac{\text{mass of } \textit{one} \text{ copper atom}}{\text{mass of } \textit{one} \text{ oxygen atom}}$$

formula: CuO

Then the data for the composition of the red compound can be interpreted as follows:

$$\text{Red compound:} \quad \frac{\text{mass of copper}}{\text{mass of oxygen}} = \frac{8}{1} = \frac{2 \times 4}{1} = \frac{\text{mass of } \textit{two} \text{ copper atoms}}{\text{mass of } \textit{one} \text{ oxygen atom}}$$

formula: Cu_2O

According to this line of reasoning, the formula of the red compound is Cu_2O. In writing formulas, the subscript ($_2$), written after the symbol (Cu), means two atoms of the element. The correct formula could just as easily be written OCu_2. The convention chemists follow is to put the symbol of a metal first in the formula. This is just a tradition, like referring to a married couple as "Mr. and Mrs." rather than as "Mrs. and Mr."

At this point, we need not go into all the arguments that had to be settled about the relative masses of atoms in order to write correct formulas for all compounds. Nor can you be expected to know why all formulas are assigned as they are. But at this stage you should recognize that *a symbol stands for one atom of an element and a formula shows how many atoms of what elements are combined in compounds.* The examples in Table 2.4 should make this clear.

Water, H_2O
(lakes, oceans)

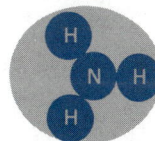

Ammonia, NH_3
(important fertilizer)

Carbon dioxide, CO_2
(gas formed when wood burns; animals breathe out carbon dioxide)

Table 2.4   Some common compounds

Substance	Formula	Makeup of the Compound
water	H_2O	2 atoms of hydrogen 1 atom of oxygen
ammonia	NH_3	1 atom of nitrogen 3 atoms of hydrogen
table sugar	$C_{12}H_{22}O_{11}$	12 atoms of carbon 22 atoms of hydrogen 11 atoms of oxygen
carbon dioxide	CO_2	1 atom of carbon 2 atoms of oxygen

In the many years since Dalton's time, a great deal has been learned about the atoms that he visualized as the chemical units of matter. We now know that atoms have an internal structure. Later in this book we will explore some of the essential ideas about atomic structure. This exploration will give us a theoretical picture of why atoms combine as they do and will further explain why formulas are written the way they are. We will also learn that the techniques now available allow chemists to detect individual atomic events. (Radioactivity, in which some atoms can spontaneously change into others, is one such event.) However, all the increase in scientific knowledge merely refines and elaborates the magnificent concept of atoms that Dalton envisioned. Every bit of modern information makes our belief in atoms more firm.

FORMULAS OF MOLECULES

We have defined a molecule as the smallest piece of a pure substance that we can make by any physical process of subdivision. Although, strictly speaking, not every pure substance is made up of molecules, for the millions that are made up of molecules, we can consider their chemical formulas to be *molecular formulas*. The molecules of compounds all contain definite numbers of different atoms. For example, the chemical shorthand designation for the molecule of water is H_2O; the molecule of table sugar is represented by $C_{12}H_{22}O_{11}$; and so on.

Several elements that are gases at room temperature exist in the form of molecules of two identical atoms. The formula for the molecule of oxygen gas is O_2 (two O atoms in one O_2 molecule). In similar fashion, the formula for the molecule of nitrogen is N_2, of chlorine is Cl_2, and so on. A few gaseous elements and most metals have molecules that are single atoms. The molecular formula for helium is He; for mercury, Hg.

The important point is for you to recognize that formulas express the composition of molecules. Molecular formulas describe composition in terms of what kind of atoms are in a molecule and how many there are of each kind.

Chlorine, Cl_2

Nitrogen, N_2

Helium, He

Mercury, Hg

ATOMIC MASSES

We have seen the importance of Dalton's idea that all atoms of a particular element have a characteristic mass. We also have seen how, even though the early chemists could never hope to weigh

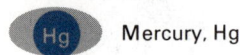

individual atoms, they could accurately learn the *relative* masses of atoms. Our example of calcium combining with oxygen illustrated how this was done. It showed that the mass of a calcium atom is 2.5 times the mass of an oxygen atom.

Scientists habitually share their discoveries; each contributes to the growth of understanding about nature. As more and more data about relative masses piled up, it became obvious that some universally accepted scale of values would be helpful. So in order to know quickly and easily what the other scientists were talking about, they agreed to use some standard value for one element and to express the values for other elements relative to that value. We follow this same system today. For example, one particular kind of carbon atom is arbitrarily given the value 12. (We say "one particular kind of carbon atom" because, as we will find later, Dalton was not quite correct in saying that *all* atoms of an element are *exactly* alike in mass. This apparent complication will be explained when we examine the internal makeup of atoms in a later chapter.) Studies of the mass ratios by which elements combine allow chemists to build a whole table of *relative* values for atomic masses.

This list of values is the *table of atomic mass.* Table 2.5 gives approximate values for the lightest 16 elements. You will notice that there are no *units* of mass, such as grams, pounds, or tons, attached to these numbers. They accurately express only known relative values. (Remember, individual atoms are extremely tiny; the true mass values measured in grams are very inconvenient numbers. Based on modern weighing techniques, the actual masses of atoms are on the order of 0.000,000,000,000,000,-000,000,01 gram.) Hydrogen, the element with the lightest atoms,

Table 2.5   Atomic mass values for the lightest 16 elements

Element	Symbol	Atomic Mass	Element	Symbol	Atomic Mass
Hydrogen	H	1.0	Fluorine	F	19.0
Helium	He	4.0	Neon	Ne	20.2
Lithium	Li	7.0	Sodium	Na	23.0
Beryllium	Be	9.0	Magnesium	Mg	24.3
Boron	B	10.8	Aluminum	Al	27.0
Carbon	C	12.0	Silicon	Si	28.1
Nitrogen	N	14.0	Phosphorus	P	31.0
Oxygen	O	16.0	Sulfur	S	32.0

has an atomic mass value close to 1.0. As you also can see from the values in Table 2.5, many other elements have values close to whole numbers. A more complete list of elements, with more precise atomic mass values, is on the inside cover of this book.

The information in the table of values for atomic masses is very useful to chemists. It allows them to make accurate predictions about the amounts of elements that will combine with one another. By using the information represented by a formula, the atomic mass values can be combined to give comparable relative values for *molecular masses* of compounds. For example, the molecular mass of H_2O is

$$\begin{array}{rcl} 2 \text{ atoms of H} \times 1.0 \text{ (atomic mass of H)} &=& 2.0 \\ +1 \text{ atom of O} \times 16.0 \text{ (atomic mass of O)} &=& +16.0 \\ \hline \text{molecular mass of } H_2O &=& 18.0 \end{array}$$

Chemical reactions can be described in terms of how many molecules of substances are involved. So molecular mass values can be used to predict the amounts of substances either used up in or produced by a chemical reaction.

Summary

We have come a long way toward answering the question of the chapter title, "What Is Matter?" To be sure, we still have a journey ahead; even after finishing this book, we will have only started an answer. But we can see some kind of pattern to the way scientists approach such big questions. We have learned how a classification scheme, to be most useful, must be based on the careful observation of properties. Careful observation means, if at all possible, expressing measurements in quantitative terms. Then certain laws can be stated to describe how at least some kinds of matter behave the same way every time. To interpret these laws, scientists bring into the picture ideas that suggest some kind of a generalized model.

The model in this chapter is the atomic theory. Atoms are the smallest chemical units of elements that combine to form compounds. Accurate values for the relative amount of mass in each kind of atom make it possible for chemists to predict the amounts of substances that will be involved in chemical reactions.

Glossary

The number in parentheses indicates the text page where you can find the term defined in context.

atom   the smallest characteristic unit of an element (48)

atomic mass   the mass of an atom, measured in units defined as a twelfth of the mass of a specific form of carbon (53)

Celsius scale   a temperature scale used in scientific work and in most non–English-speaking countries, in which 0 degree is established as the freezing point of water and 100 degrees is established as the boiling point of water under one atmosphere of pressure (34)

chemical change   the conversion of a substance of specific chemical and physical properties into another substance of different physical and chemical properties (29)

chemical formula   a notation that describes the chemical composition of a compound, specifying the types of atom (elements) present and the number of each (50)

chemical symbol   a one or two letter notation usually derived from the name of an element, which represents one atom of that element (43)

compound   a substance formed by the chemical combination of two or more kinds of elements (43)

density   a characteristic property of a substance, the ratio of the mass to the volume of a sample of the substance (37)

element   one of 106 known pure substances that cannot be broken down into any simpler substances by chemical reactions (43)

Fahrenheit scale   a temperature scale used in English-speaking countries, in which 32 degrees is established as the freezing point of water and 212 degrees is established as the boiling point of water under one atmosphere of pressure (34)

law of conservation of mass   in any chemical reaction, the mass of the products exactly equals the mass of the reactants (45)

law of constant composition   all samples of a pure compound contain the same elements in the same ratio by mass (46)

mass   a measure of the amount of matter in a sample of a substance or a mixture, a constant property of that sample independent of the force of gravity (30)

mixture   a combination of pure substances that can be separated from each other by physical methods (39) *molecule*   the smallest characteristic unit of a substance, formed by the combination of at least two atoms (41)

molecular mass   the sum of the atomic masses of all atoms in a molecule (54)

physical change   a change that alters the form or physical state of a substance without changing its chemical identity (29)

property   a characteristic, identifying fact about a substance (28)

pure substance   a collection of identical molecules that cannot be separated into different substances by physical methods (40)

ratio   the number resulting when one quantity is divided by another, usually the relationship of two variable quantities (36)

variable   a quantity that may have more than one value (37)

volume   the amount of three-dimensional space occupied by an object or sample (31)

weight   a measure of force with which the gravity of the earth or other large body attracts a certain mass. The weight of a sample can vary, but its mass does not (30)

Exercises

2.1   Decide whether the following are physical or chemical changes. Justify your answer.

  a.   Butter melts in a hot pan.

  b.   Butter turns brown in a hot pan.

  c.   Rain changes to sleet.

  d.   A person grows old.

  e.   A silver spoon becomes tarnished.

  f.   A piece of window glass shatters.

  g.   A rubber band stretches.

  h.   An apple rots.

2.2   An astronaut on a journey to the moon spends days in what is described as a "weightless" condition. Could this condition properly be called "massless" also? Justify your answer.

2.3   Scientific language is universal. Suppose that you were to hear a German saying that the *Rauminhalt* of something is 200 liters. Does "Rauminhalt" mean mass, length, or volume?

2.4   The system of currency used in the United States is a decimal system.

   a.   Why do you think we call a penny a "cent"?

   b.   If you call a dollar a "buck," what would you call $1,000,000?

   c.   Rates of taxation are sometimes expressed in "mil" units as a fraction of the dollar value of property. (A one mil tax on assessed value of $1,000 is $1.) Refer to the box on page 32 and decide whether the word mil is derived from the world million or the prefix "milli."

   d.   What prefix would you use with the word dollar to invent a unit to describe the United States' multi-billion dollar national debt?

2.5   In a few years, we may see highway signs giving distances or speeds in kilometers instead of miles. Use the information in the box on page 32 to estimate which of the following are correct.

   a.   The 55 mile-per-hour speed limit will become 34 kilometers per hour, or 88 kilometers per hour?

   b.   A 1985 sign indicates that a town is 160 kilometers away. Does the present-day sign indicate that it is 100 miles or 250 miles away?

2.6   When you buy a half gallon of milk, do you get a little more or a little less than two liters?

2.7   If you weigh 125 pounds, which of the following is your approximate weight in kilograms?

   a.   30 kg   *b.*   57 kg   *c.*   270 kg

2.8   Use Figure 2.4 to complete the following conversions.

   a. $200°F$ = ___°C

   b.   ___°F = $50°C$

   c.   ___°F = $-20°C$

   d. $140°F$ = ___°C

   e.   ___°F = $38°C$

2.9   The Celsius temperature scale is also referred to as the centigrade scale. What is the origin of this name?

2.10   Which properties listed here are always the same no matter what the size of the sample? Which of these properties are specific properties?

a. Mass   b. Volume

c. Melting point   d. Temperature

e. Boiling point   f. Density at a
                          certain temperature

2.11   The density of automobile radiator antifreeze is 1.12 gram/milliliter. How much will one liter weigh?

2.12   One milliliter of gasoline weighs about 0.75 gram. Gasoline and water do not mix, but instead separate into layers when put together. If some water gets into your automobile gasoline tank, will it collect on the bottom of the tank or float on the top of the gasoline? (Hint: Compare the densities of the two liquids.)

2.13   Astronomers believe that some stars, as they get old, do not lose any mass, but rather take up a smaller and smaller volume. Such a star is called a white dwarf. As this shrinking process goes on, does the star's density become greater or smaller?

2.14   Figure 2.5 shows a plot of the density of water from 10°C to 100°C. Use this graph to estimate the value of the density of water at the following temperatures.

a. 30°C   b. 60°C   c. 110°C

2.15   To answer 2.14c, you must make an extrapolation of the line on the graph. Can you be as sure of your answer to 2.14c as you can of your answers to 2.14a and 2.14b? Justify your answer.

2.16   In each of these statements, what factor can you designate as the variable?

a. The annual picnic is always held in April at Crowley Field if the weather permits.

b. The time at which we go to pick up John depends on which train he takes.

c. A college hires faculty members according to its enrollment figures.

d. As long as alumni contributions to the annual drive remain at their present level, a college cannot expand its athletic program.

e. As the price of gasoline rises, people drive less.

2.17 Which of these cases are qualitative analyses and which are quantitative?

a. Sniffing a glass of milk to decide whether it is whole milk or buttermilk

b. Measuring the amount of alcohol that is produced as wine is being made

c. Using various chemical tests to find out whether a coin is genuine or counterfeit

d. Using tests to find out how much copper was used to make a certain sample of sterling silver

e. Finding out that a certain college has both male and female professors

f. Finding out that 10 out of 100 professors at a certain college are female

2.18 Write "chemical process" or "physical process" over each arrow to describe the process that occurs.

a. Mixture ⟶ pure substances

b. Molecule ⟶ atoms

c. Compound ⟶ elements

2.19 Choose one of each pair as the true statement.

a. (1) A sample of a compound contains many molecules.
(2) A sample of a molecule contains many compounds.

b. (1) An element may contain more than one identical atom.
(2) An atom may contain more than one identical element.

c. (1) Every formula has a compound.
(2) Every compound has a formula.

d. (1) A molecule may contain different kinds of elements.
(2) An element may contain different kinds of atoms.

e. (1) A certain kind of atom decomposes to molecules.
(2) A certain kind of molecule decomposes to atoms.

f. (1) Most pure substances are made up of molecules.
(2) Most molecules are made up of pure substances.

2.20   The most common ore of iron is a compound of iron and oxygen called hematite. Ten grams of hematite contains 7 grams of iron.

    a.   What law can you use to decide that 10 grams of hematite contains 3 grams of oxygen?

    b.   What law can you use to decide that one ton (2000 pounds) of hematite yields 1400 pounds of iron?

2.21   Describe in words (how many atoms of what elements?) the makeup of the following molecules.

    a. F_2

    b.   Ne

    c. SiO_2

    d. C_6H_6

    e. P_4O_{10}

    f. $(C_2H_5)_3N$

2.22   Write a formula for each molecule described below.

    a.   Four atoms of carbon and 10 atoms of hydrogen

    b.   One atom of magnesium, one atom of sulfur, and four atoms of oxygen

    c.   Eight atoms of sulfur

    d.   Two atoms of sodium, one atom of carbon, and three atoms of oxygen

    e.   One atom of hydrogen and one atom of chlorine

2.23   At a time just before the French Revolution, a French nobleman, Antoine Lavoisier, proved that diamonds are made of pure carbon, as is charcoal. What do you think he did to prove this? (Hint: He was a very rich man and one of the greatest scientists of his time.)

2.24   You are told that dry ice is pure solid carbon dioxide. Describe in general terms what you would do to test this statement.

2.25   The correct formula for lime, the compound of calcium and oxygen, is CaO. Use the data of the experiments described on page 47 and the assigned value of 16 for the atomic mass of oxygen to calculate the atomic mass of calcium.

2.26   A compound of sulfur and oxygen has the formula SO_2. The mass of sulfur is equal to the mass of oxygen in the compound. Use this information to calculate the atomic mass of sulfur. (The correct answer is listed in Table 2.5.)

3 | What Is a Chemical Reaction?

- ☐ How do chemists clearly represent the literally millions of chemical reactions they study?
- ☐ What chemical reactions occur when the atmosphere is polluted?
- ☐ What does energy have to do with chemical reactions?
- ☐ Do atoms and molecules behave differently in living and nonliving systems?
- ☐ What factors control the speed with which chemical reactions occur?
- ☐ How does nature dispose of the accumulated wastes of living systems?
- ☐ What are the most important chemical reactions on earth?

In Chapter 2, as we developed ideas of atoms, molecules, elements, and compounds, we repeatedly mentioned chemical changes. Chemical change is a broad term that simply means that something changes into something else. The term *chemical reaction* has a more specific, definite meaning. It is the term used to identify a particular chemical happening.

The first thing we have to do to describe a chemical reaction is to be sure we know what substances we are starting with and what substances are produced by the reaction. We also should know that, whenever one substance changes into another, some changes in energy occur. Information about these energy changes should be included to make the description complete. This information often has very practical applications. For example, suppose that you have to buy thousands of tons of fuel to supply energy to run a factory. You need to know how much heat can be released from given amounts of various kinds of fuel. You have to buy fuel in terms of the amount of heat it makes available— not just in terms of the amount of material involved, as you would buy other raw materials for a manufacturing process.

We know from experience that some reactions are very fast, like the explosion of dynamite; other reactions are slow, like the rusting of iron or steel. If we know what changes in conditions can influence the rate of a reaction, we may be able either to speed it up or to slow it down for our advantage. This, too, has very practical importance, as in speeding up a factory's production or stopping a forest fire.

In this chapter, we will first discuss the way that chemists use their language of symbols and formulas to describe chemical reactions. Then we will see what regularities can be used to describe energy changes. Finally, we will tie together some information about the circumstances that influence how fast chemical reactions occur.

WHAT HAPPENS WHEN SOMETHING BURNS?

One way to begin to organize our thinking about chemical reactions in general is to see what we know and can learn about a familiar kind of chemical reaction—burning. Common sense, or what we remember from past experience, is a good place to start. What do we know about burning? First, such things as wood or coal will burn; sand, asbestos, or ashes will not. What is there about the chemical makeup or properties of *combustible* (burnable) things that makes them different from noncombustible things?

Rather than plunging into an attack on this problem, we can do what the scientist often does: make an indirect attack by asking another common-sense question. Why does combustion

Why does combustion stop if we keep air away?

(burning) stop if we keep air away? A fire can be put out by smothering it with sand, flooding it with water, or spraying it with carbon dioxide gas from a "C-O-2" fire extinguisher. All these actions keep air away from a fire. Apparently, something in air is as important as is the fuel in supporting combustion.

Centuries ago, scientists gradually became convinced that oxygen in the air is necessary for combustion. (They became convinced of this idea at about the same time they began to believe in the law of conservation of mass.) We now know that *combustion is the chemical reaction that occurs when oxygen combines rapidly with something.* Heat and light are given off in combustion. The conclusion about the necessary presence of oxygen fits with common-sense experience. If we keep air away, we keep oxygen away. Thus the fire goes out. This idea also explains another pertinent bit of information. A spaceship must carry oxygen in the condensed form of liquid oxygen (called LOX), as well as fuel. Its rocket engines must operate in space where there is no air to provide the necessary oxygen, so it must carry its own oxygen to burn its fuel.

Air is a mixture of gases, the most abundant of which is nitrogen in the form of N_2 molecules. In a typical sample of air at sea level, only 209 out of 1000 gas molecules are oxygen, O_2. At this point, we need only concern ourselves with the oxygen, because the nitrogen and other substances in the air are not involved in burning reactions. At the end of the eighteenth century, experimenters learned this fact by burning candles in the air over water in a closed container. Figure 3.1 suggests this kind of experiment. The flame goes out before the air is all gone. The gases produced by the burning reaction are not part of the air remaining in the container because these product gases have dissolved in the

Burning gasoline + foam
(no air) ⟶ Fire out!

Candle burns in air
over water in a
closed container.

Candle stops burning,
and water rises in
the closed container.

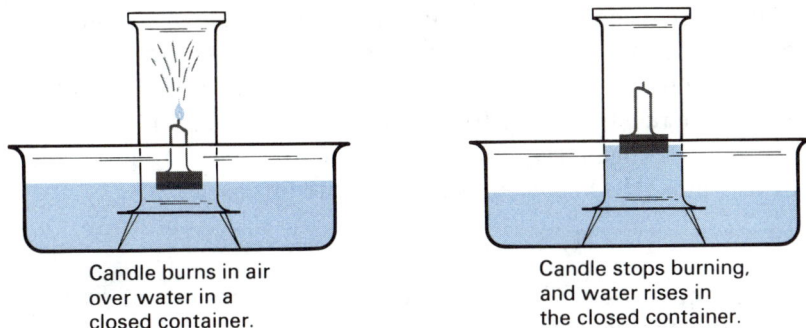

Figure 3.1   Candle burning
in a closed container over
water.

water. If the apparatus is arranged so that water can come into
the container, the water rises only part way. The part of the air
that remains after the burning stops contains gases that do not
support combustion.

These questions and answers about burning lead us toward
questions about the chemical properties of oxygen. Does oxygen
form compounds in other ways? We know that animals must
breathe oxygen to stay alive. Is there a connection between this
fact and the necessary role of oxygen in combustion? The rest of
the chapter will be directed toward answering such questions.
But first we need to learn more about the way that chemical
language expresses ideas. Some of the information about oxygen
and combustion will help to illustrate.

CHEMICAL EQUATIONS

You are already familiar with using formulas to represent pure
substances. The next step is a simple extension of the use of this
kind of chemical shorthand. Formulas are used in a chemical
equation to describe a chemical reaction. We write chemical
equations with an arrow instead of an equal sign.

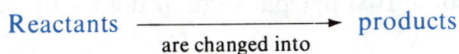

$$\text{Reactants} \xrightarrow{\text{are changed into}} \text{products}$$

In this sense, a chemical equation differs slightly in meaning from
a mathematical equation. We write the equation for the reaction
of carbon with oxygen as

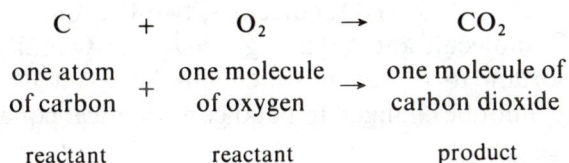

$$C \quad + \quad O_2 \quad \rightarrow \quad CO_2$$

| one atom | | one molecule | | one molecule of |
| of carbon | + | of oxygen | → | carbon dioxide |

| reactant | reactant | product |

Only a few chemical reactions can be represented by chemical equations as simple as this one. Let us illustrate a step-by-step procedure for correctly writing a more typical chemical equation, one for the reaction of hydrogen gas burning in air. The reaction between hydrogen (H_2 molecules) and oxygen (O_2 molecules) forms water (H_2O molecules). If we write this as

$$H_2 \; + \; O_2 \; \rightarrow \; H_2O \quad \textit{Incorrect!}$$

We get into trouble by neglecting that very important first law about the way matter behaves, the law of conservation of mass. We have accounted for the two H atoms on the left (in the molecule H_2) by putting them in a molecule of H_2O on the right. But this formula for water has in it only *one* O atom; there are *two* O atoms in the molecule of O_2 on the left.

There are two tempting ways to handle this atomic bookkeeping, but both are incorrect:

$$H_2 \; + \; O_2 \; \rightarrow \; H_2O_2 \quad \textit{Incorrect!}$$

or

$$H_2 \; + \; O_2 \; \rightarrow \; H_2O \; + \; O \quad \textit{Incorrect!}$$

The trouble with these equations is that they do not tell the truth. In the first, H_2O_2 is the formula for a completely different compound, hydrogen peroxide, that cannot be made by this reaction. In the second, another substance, an atom of oxygen, has been added as a product. This does not happen either!

What we must do is *balance* the equation. We can try to account for the two oxygen atoms in O_2 by writing two product molecules of H_2O on the right.

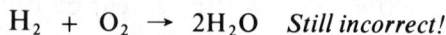

$$H_2 \; + \; O_2 \; \rightarrow \; 2H_2O \quad \textit{Still incorrect!}$$

This equation is incorrect because now our counting is off for H atoms: four on the right, only two on the left. However, because the H atoms come two per package in the H_2 molecules, we can call for two H_2.

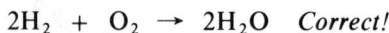

$$2H_2 \; + \; O_2 \; \rightarrow \; 2H_2O \quad \textit{Correct!}$$

We now have correctly accounted for all the atoms on both sides of the arrow. Four H atoms are on the left in two H_2 molecules and on the right in two H_2O molecules; two O atoms are on the left in one O_2 molecule and on the right in two H_2O molecules.

It is important to recognize which numbers can be changed and which cannot be changed to balance a chemical equation. *No*

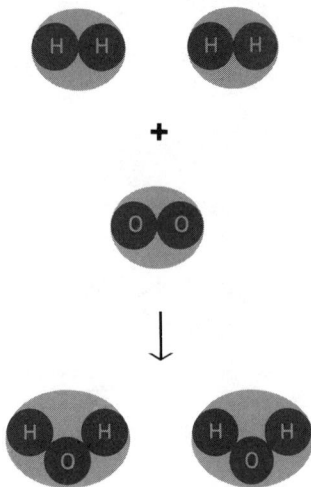

subscript can be changed. Subscript numbers (the small numbers placed below the line) tell how many atoms of each element are in a compound. We must use the correct formula to represent a compound that is the product of a reaction. Changing a subscript changes a formula, and this would mean that the equation is not telling the truth about a reaction. On the other hand, the *coefficients can be changed.* The coefficients are the regular-sized numbers, placed in front of the chemical formulas, that tell how many molecules there are.

We can see the systematic steps for writing a balanced chemical equation with another combustion reaction, that which occurs when natural gas burns. The chief substance in natural gas is the compound *methane.* The formula for the methane molecule is CH_4.

Step 1 Write correct formulas for reactants and products.

$$CH_4 \quad + \quad O_2 \quad \rightarrow \quad CO_2 \quad + \quad H_2O$$
methane   oxygen   carbon   water
                          dioxide

Step 2 Choose *one* reactant molecule, and account for all its atoms among the product molecules by putting in product coefficients.

$$CH_4 + O_2 \rightarrow CO_2 + 2H_2O$$

one C atom ⟶ 2×2

four H atoms ⟶

Step 3 Account for the extra atoms of reactant molecules that are needed for the new number of product molecules.

$$CH_4 + 2O_2 \rightarrow CO_2 + 2H_2O$$

2×2 , 2 O atoms + (2 × 1) O atom

4 O atoms

Another example is the burning of a component of gasoline, C_8H_{18}, to give carbon dioxide, CO_2, and water.

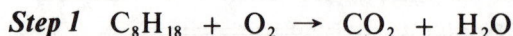

Step 1 $C_8H_{18} + O_2 \rightarrow CO_2 + H_2O$

Step 2 $C_8H_{18} + O_2 \rightarrow 8CO_2 + 9H_2O$

Step 3 $C_8H_{18} + 12.5O_2 \rightarrow 8CO_2 + 9H_2O$

The fractional coefficient 12.5 for O_2 in the last equation does not mean that the reaction requires 12 regular O_2 molecules and half of a split O_2 molecule. Rather, it simply means that the oxygen requirement is 12.5 times as many molecules as the number of C_8H_{18} molecules. Some chemists generally think fractional coefficients are too confusing, so they add a fourth step.

Step 4 Remove fractional coefficients in the balanced equation by *doubling all* coefficients.

$$2C_8H_{18} + 25O_2 \rightarrow 16CO_2 + 18H_2O$$

As you read this book, you will never have to balance complicated chemical equations. However, you will find that equations are the most efficient way to describe chemical reactions. Therefore, you should recognize and read a chemical equation for what it is: a correct counting of how many molecules of what substances are involved in a reaction.

OXYGEN COMBINES WITH METALS

When we see unburned steel beams or metal furniture in the pictures taken after a devastating fire, we may be inclined to think of metals as being able to resist burning. (Objects made of metal may be badly bent and deformed by the heat of the blaze.) However, welders with high-temperature torches actually burn holes in steel. In fact, almost all metals will combine rapidly with oxygen if the temperature is high enough. If a metal is in the form of fine strands or powder, even a flame of moderate temperature is capable of igniting it. A wad of steel wool or a clump of copper trimmings will burn in air if heated in the flame of a gas burner. A ribbon of magnesium metal can be ignited in air by the heat of a gas flame. It will burn so vigorously that the kind of intense light it produces can damage eyes not protected by glasses. These reactions can be represented by the equations

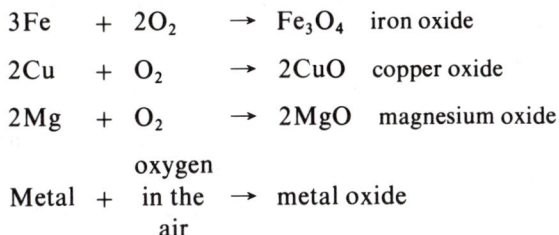

$$3Fe \;+\; 2O_2 \;\rightarrow\; Fe_3O_4 \quad \text{iron oxide}$$

$$2Cu \;+\; O_2 \;\rightarrow\; 2CuO \quad \text{copper oxide}$$

$$2Mg \;+\; O_2 \;\rightarrow\; 2MgO \quad \text{magnesium oxide}$$

$$\text{Metal} \;+\; \begin{array}{c}\text{oxygen}\\\text{in the}\\\text{air}\end{array} \;\rightarrow\; \text{metal oxide}$$

You will note something consistent about the chemical names given to the products of these reactions. They all are called oxides. Whenever chemists identify a compound containing only *two* elements, the name given to the compound ends in the suffix *ide*. If the compound contains oxygen, it is called an *oxide*. Compounds of sulfur with one other element are called *sulfides;* compounds of chlorine with one other element are called *chlorides;* and so on. The other part of the name of the compound identifies the element whose symbol appears first in the formula. Thus the compound with the common name lime, with the formula CaO, is also called calcium oxide. The reaction that forms it can be described by the equation

$$2Ca \;+\; O_2 \;\rightarrow\; 2CaO$$

The reactions of metals with oxygen to form oxides can occur slowly as well as rapidly (as in combustion). The surface of a piece of almost any metal that has been exposed to air is covered

---

Old pennies may look almost black because of the layer of copper oxide on their surfaces.

---

by a thin layer of oxide. A shiny new penny soon darkens, and old pennies may look almost black, because of the layer of copper oxide that forms and increases in thickness.

In some cases, the surface layer of oxide is very tough, and consequently, it protects the metal from further reaction. This protection keeps stainless steel shiny. Stainless steel is an alloy (a mixture of metals) of iron and chromium. An oxide of chromium, Cr_2O_3, forms a layer that is thin enough so that the metal appears shiny through it. Yet it is tight enough to keep more oxygen gas from getting through to cause more reaction. Another metal that behaves this way is aluminum. Aluminum actually is a very reactive metal. (This means that it reacts very readily with a great many substances, including oxygen.) Yet aluminum is an excellent material for cooking utensils, because the oxide film formed by the reaction

$$4Al \ + \ 3O_2 \ \rightarrow \ 2Al_2O_3$$

protects the metal from air, water, acids, or other compounds in foods as they cook.

The oxides on the surface of some metals have the unfortunate property of being porous (they have holes in them) or in the form of flakes or powder. This property means that such metals are not protected by the oxide covering. The holes in the covering oxide film let more oxygen through, which causes more reaction. Iron is an example of such a metal. Rusting of iron can be represented by the chemical equation

$$4Fe \ + \ 3O_2 \ \rightarrow \ 2Fe_2O_3$$

Notice that the product here is a different iron oxide from the Fe_3O_4 formed by burning iron. Fe_3O_4 is black; Fe_2O_3 has the familiar reddish color of rust.

In contrast to what Al_2O_3 does for aluminum, the nonprotective property of Fe_2O_3 is immensely costly to highly developed civilizations that use many articles made of iron. Think of all the rust you see, and remember that rusting, once started, always gets worse. The reaction represented by the above equation costs you

$4Fe \ + \ 3O_2 \longrightarrow 2Fe_2O_3$

The reaction of oxygen and iron costs every person in the United States over $30 per year.

and every other person in the United States over $30 per year. Not only does rusting cost us a lot of money, but it is responsible for our using natural resources of iron ore at a far-too-rapid rate. In a later chapter, we will discuss how scientists' theoretical understanding of rusting can serve as a basis for controlling the rusting process. This is one way that chemistry can make important contributions to conservation.

SOME OXIDES OF NONMETALS ARE ATMOSPHERIC POLLUTANTS

Some elements display none of the properties that we usually associate with metals (solid, shiny, able to be bent or pounded into different shapes, acting as conductors of electricity, and so on). These are the nonmetals. They may be gases like nitrogen (N_2), chlorine (Cl_2), hydrogen (H_2), or oxygen itself (O_2). Some nonmetals are solids, but they are not shiny and tough as metals are. One example is carbon; it is a dull black. Familiar examples of carbon are soot, in the form of powder, and charcoal or coal, either of which can be broken easily into small pieces. Sulfur is another nonmetal. At normal temperatures, sulfur is a yellow solid that can easily be ground into a powder.

Whenever air is used as a source of oxygen for combustion reactions, nitrogen gas (a component of air) is also present in the region of the flame, as occurs inside the cylinders of an automobile or truck engine. Nitrogen and the oxygen not used to burn the gasoline are raised to high temperatures because of the heat released in the burning process. Under these conditions, some of the nitrogen and oxygen combine to form an oxide of nitrogen.

$$N_2 \ + \ O_2 \ \rightarrow \ 2NO \quad \text{nitrogen oxide}$$

The NO molecules are in the gases that escape from the engine exhaust. Once they are mixed with the air of the atmosphere, the NO molecules start a chain of reactions that leads to the formation of smog. (We will discuss these reactions and their products in more detail in a later chapter.) One of the governmental regulations for new automobiles requires a device for decomposing NO molecules back into nitrogen and oxygen, thus reducing the content of smog-producing substances in automobile exhaust.

Another atmospheric pollutant is the oxide of sulfur, SO_2, also a gas. Coal, especially soft coal (bituminous coal), almost always contains some sulfur. So does fuel oil. When the coal or oil is burned, the sulfur in it burns, too. The chemical reaction is

$$S \ + \ O_2 \ \rightarrow \ SO_2 \quad \text{sulfur dioxide}$$

Sulfur dioxide also is a troublemaker in the atmosphere. The SO_2 molecules are the starting materials for a series of reactions that produces sulfuric acid, H_2SO_4. In concentrated form, sulfuric acid is a very dangerous chemical. Even in the very dispersed form of tiny droplets in the atmosphere, sulfuric acid can cause damage to plants, animals, some rocks, and some metals. Consequently, burning any fuel or other substances that contains sulfur can be a source of atmospheric pollution if the product SO_2 is allowed to escape.

Gasoline contains a very small amount of sulfur and sulfur compounds. The devices that have been developed to remove the nitrogen oxide from the exhaust of automobile engines unfortunately do not remove the sulfur dioxide. In fact, they make it even more likely that the very small amount of sulfur in gasoline will be completely converted into SO_2. Thus this smog-prevention device has the effect of trading one pollutant for another.

This smog prevention device has the effect of trading one pollutant for another.

Here is an illustration of the complexity of the problems created by our use of technology. Complex problems require such solutions as removing the last traces of sulfur from all gasoline.

We have already referred to carbon dioxide, another oxide of a nonmetal. Carbon dioxide is formed by the burning of carbon or carbon-containing compounds in the presence of an abundance of air or of pure oxygen gas. Carbon dioxide cannot be considered a pollutant of the atmosphere. Its presence is the result of many natural processes (such as animals breathing). Some natural processes also remove CO_2 from the atmosphere. However, if the supply of oxygen is limited when burning occurs, another gaseous oxide of carbon may be formed. This gas is the compound called carbon monoxide, CO. (*Mono* means "one"; *di* means "two" atoms of oxygen in the formula of the molecule.)

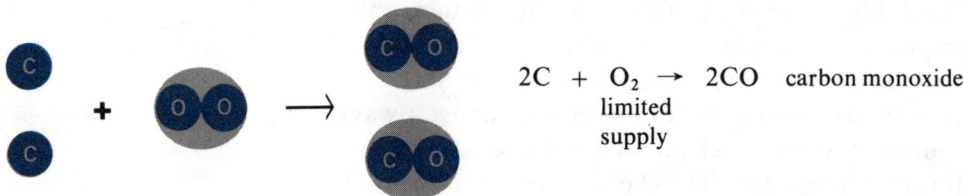

$$2C + O_2 \rightarrow 2CO \quad \text{carbon monoxide}$$

limited
supply

Carbon monoxide is a deadly poison for breathing animals and is consequently an atmospheric pollutant. A pollutant is anything (either matter or heat energy) added to the environment in large enough amounts to affect adversely something that is of value to human beings. Fortunately, most of the carbon monoxide produced by burning gasoline in automobiles or trucks is converted in a relatively short time into the less harmful carbon dioxide by bacteria in the soil and by other reactions in the atmosphere. However, it does constitute a hazard for humans or animals located close to its source, such as in a room where an improperly adjusted gas or oil burner is located or in the vicinity of a busy freeway.

OXYGEN REACTS WITH COMPOUNDS

By this time, you probably have begun to recognize in our discussion an illustration of the cumulative feature of scientific knowledge. We are now using the ideas of atoms and formulas, described in Chapter 2, to express information about chemical reactions. Much of the chemical information in this book is built up in such a cumulative way. You often will find that what appears to you as a new fact is really not so new but it is merely an extension of facts you already know and understand. We can illustrate this general principle by considering how our knowledge of oxygen's reactions with elements relates to the reactions between oxygen and compounds. For example, because we know that copper reacts with oxygen to form copper oxide and that sulfur reacts with oxygen to form sulfur dioxide, we can make a reasonable guess that the compound copper sulfide (CuS) will react with oxygen as follows:

$$2CuS + 3O_2 \rightarrow 2CuO + 2SO_2$$

By applying the same kind of reasoning to what we know about the reactions of carbon and hydrogen with oxygen, we can understand why H_2O and CO_2 appear as products in reactions of O_2 with CH_4 (methane, natural gas) or C_8H_{18} (a typical molecule in gasoline).

$$CH_4 + 2O_2 \rightarrow CO_2 + 2H_2O$$

$$2C_8H_{18} + 25O_2 \rightarrow 16CO_2 + 18H_2O$$

Hundreds of thousands of compounds of carbon and hydrogen exist. You now have enough chemical information at your mental fingertips to write the equations for hundreds of thousands of combustion reactions.

A great many familiar substances contain atoms of carbon, hydrogen, and oxygen. Wood alcohol, CH_3OH, with the chemical name methyl alcohol; grain alcohol, C_2H_5OH, with the chemical name ethyl alcohol; and table sugar, $C_{12}H_{22}O_{11}$, with the chemical name sucrose are examples. Each of these can burn when ignited in air. You can predict that in each case the products are CO_2 and H_2O. The complete equations are

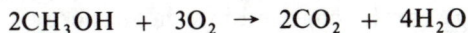

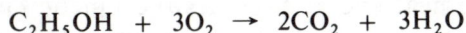

$$2CH_3OH + 3O_2 \rightarrow 2CO_2 + 4H_2O$$

$$C_2H_5OH + 3O_2 \rightarrow 2CO_2 + 3H_2O$$

$$C_{12}H_{22}O_{11} + 12O_2 \rightarrow 12CO_2 + 11H_2O$$

Trees and other plants are made of many complex molecules containing C, H, and O atoms. Wood can be represented by a simplified formula $(C_6H_{10}O_5)_n$. What this means is that the unit $C_6H_{10}O_5$ is hooked together many times over (n is some unknown large number). You can think of n in the way that you think of x in an algebraic equation. In a later chapter, we will discuss why n stands for an unknown *large* number. Even with this complicated formula, you can correctly count up the atoms in the equation for the reaction that occurs when a campfire log burns.

$$(C_6H_{10}O_5)_n \; + \; 6nO_2 \; \rightarrow \; 6nCO_2 \; + \; 5nH_2O$$

ENERGY AND CHEMICAL REACTIONS

So far, we have talked only about the substances, either elements or compounds, that are involved in chemical reactions. No one who has warmed fingers at a campfire, shoveled coal on a fire, or turned on a gas or oil burner to warm a house can miss another important feature of chemical reactions. We make practical use of combustion reactions because they release heat. Heat is a form of energy.

Energy is a word that we read and hear many, many times in the news media, usually in terms of the "energy crisis." A more

A more accurate description of the energy crisis is to call it a crisis in obtaining resources of potential energy.

accurate description of the dilemma is to call it a crisis in obtaining resources of *potential energy*. Potential energy means energy locked into something. The water being held back by a high dam has potential energy because of its position; held high, it has the potential of falling with great force. This potential energy is changed into *kinetic energy* (energy of motion) when the water falls through pipes and turns the blades of a turbine to generate electricity. The energy crisis affects us even more directly as a problem in obtaining supplies of coal, petroleum, or natural gas. Here, too, potential energy is meant. The potential energy in a fuel is "locked into" it by virtue of its chemical composition. This potential energy can be changed into heat energy by combustion reactions. The heat energy becomes the kinetic energy of fast-moving gas molecules at a high temperature. This kinetic energy of motion can be transferred by a jet engine to move an airplane or by the pistons in an automobile engine to turn its wheels or by the blades of a turbine to generate electricity.

If heat is given off when a chemical reaction occurs, the reaction is described by the term *exothermic* (*exo* means "away from" or "out of"; *thermic* refers to heat). *Combustions are exothermic* reactions. When a specific amount of substance is burned, a specific amount of heat is produced. Thus, when 16 grams of natural gas, CH_4, is burned, we can represent the reaction as

$$CH_4 + 2O_2 \rightarrow CO_2 + 2H_2O + 210{,}800 \text{ calories}$$

The unit of measurement for heat is the *calorie*. A calorie is defined as the amount of heat that must be absorbed by 1.0 gram of water to raise its temperature 1.0 degree Celsius. Thus 210,800 calories of heat can raise the temperature of about 2.5 quarts of water from room temperature almost to boiling. (The more familiar dietary calorie used to describe the energy value of foods is 1000 of the calories here defined.)

Heat is absorbed in some other reactions, described as *endothermic* reactions (*endo* means "into"). The formation of nitrogen oxide from nitrogen and oxygen in cylinders of automobile engines is one example of an endothermic reaction. When 28 grams of N_2 combine with oxygen, 43,200 calories of heat are absorbed.

OXYGEN AND LIVING THINGS

Even many centuries ago, the connection between needing air to keep a fire going and needing air to keep living probably was apparent to humans. The first true chemists, those who began to do controlled experiments at about the end of the eighteenth century, also were fascinated by the connection. Even their crudest theories about why a candle flame eventually goes out if it burns in a closed container also tried to account for why a mouse eventually dies under the same conditions. When a true understanding of combustion developed, so too did the realization that breathing animals need oxygen to live. A few deep breaths of oxygen-free air make a person unconscious. A few minutes in such an atmosphere produce permanent brain damage and then death.

Much of our common knowledge contributes to what we know about the necessary role of oxygen in the life process. Perhaps you know how difficult it is to breathe on a high mountain where

running

dead!

the air is "thin" and each lungful provides fewer oxygen molecules. Perhaps you know that the blood in your veins has a darker color than the blood in your arteries has. Almost everyone knows that fish and other underwater creatures also need oxygen. Their gills take the place of other animals' lungs. The oxygen that fish use also is in the form of oxygen molecules. A small amount, but enough, oxygen gas from the air dissolves in the water. (The oxygen atoms in H_2O molecules are too tightly bound to the hydrogen atoms to be useful.) This need for oxygen is why some kinds of lake or stream pollution are so hard on fish. The substances polluting the water react with and hence use up the dissolved oxygen, so that the fish actually suffocate. If you have ever tried to keep the fish population healthy in a tropical fish aquarium, you know how essential it is to have green plants growing in the water to provide oxygen. Oxygen is extremely important to breathing animals. How does all this information tie together?

BREATHING ANIMALS "BURN" FOOD

Suppose that you burn 1 gram of table sugar with pure oxygen in a laboratory experiment. If you do this experiment carefully with the right apparatus, you will be able to measure the heat released by this exothermic reaction. You will find that 1 gram of table sugar releases about 4000 calories when it burns. You also know that burning sugar, $C_{12}H_{22}O_{11}$, will yield CO_2 and H_2O as products of the reaction.

If you eat the sugar instead of burning it, the same products, CO_2 and H_2O, are produced. You get rid of these "waste" products mostly by exhaling them with the air you breathe out of your lungs. The overall chemical reaction is the same, but in the body it takes place in a complicated series of steps. Another important fact should be recognized: In both burning and breathing, the same amount of energy is released (4000 calories per gram of sugar), because that amount of potential energy is locked into every gram of $C_{12}H_{22}O_{11}$ and will be released when the sugar is changed into CO_2 and H_2O. The energy release is the same, regardless of how the reaction occurs or of how many intermediate steps there are. However, there is one important difference: Only a small part of the energy appears as body heat. Most of it is used to drive other chemical reactions involved in living. You may use this energy to wave your hand, to run upstairs, or just to stay alive while you sleep. Food is the fuel that the body's engines burn to make it live and move.

We may well pause at this point to emphasize a general principle: Atoms and molecules have the same chemical properties regardless of where they are. They can be in lifeless rocks or in living things, on the earth or in farthest space. Even though life itself is a mystery and no attempt to generate a living thing out of nonliving things has thus far succeeded, the chemical reactions of life must be consistent with those that occur outside of living things. All science is based on this principle. Biochemistry (chemistry that deals with living things) has gone far in understanding the complicated chemical reactions that occur. Many mysteries remain unsolved, such as what causes cancer and how it can be

We can be sure that the answers will come in the recognizable language of atoms and molecules.

controlled. But whatever clues turn up, and whatever clever detective work is done, we can be sure that the answers will come in the recognizable language of atoms and molecules.

Table sugar, $C_{12}H_{22}O_{11}$, is an example of a type of food called *carbohydrates*. The name comes from the atomic arithmetic in the formulas of all carbohydrates. If we write the formula of sugar as $C_{12}(H_2O)_{11}$, we see that the number of H and O atoms is in the ratio of two to one, as in H_2O. Hence this counting out of atoms suggests that the sugar molecule is a hydrate of carbon. (The word *hydrate* refers to the combination of water with something else. When dehydrated or dried foods are soaked in water, they become hydrated.) However, this assumption is "on-paper chemistry," not really so in the case of the sugar molecule. There is no water in sugar. The carbon atoms are strung together as the backbone of a skeleton, and the H atoms and OH groups of atoms are attached as ribs.

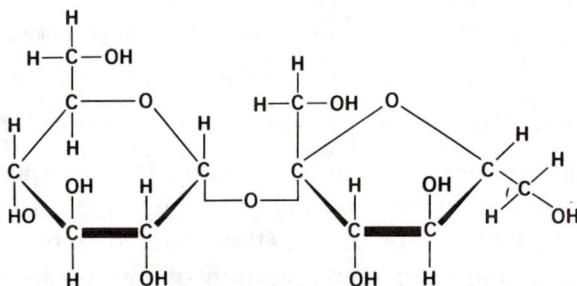

table sugar (sucrose)

In later chapters we will describe in more detail what a structural formula such as this means.

Many carbohydrate molecules are very large. Starch is an example. We can give starch a general kind of formula, $(C_6H_{10}O_5)_n$, as we did earlier for wood. This formula, with the subscript n, means that many $C_6H_{10}O_5$ units are strung together. What the body first accomplishes in a series of steps in digestion is to convert these big carbohydrate molecules into a smaller molecule called glucose, $C_6H_{12}O_6$. Glucose can dissolve in water, whereas starch cannot. (Potatoes, composed chiefly of starch, do not dissolve when you boil them!) Glucose dissolves into the blood through the stomach walls and is carried by the bloodstream to wherever the body needs to have energy released.

Oxygen, the other reactant necessary for the energy-releasing reaction, is also carried by the blood to the scene of the action. Oxygen molecules dissolve into the blood in the capillaries of the lungs. In the blood, the O_2 molecules are attached to molecules of hemoglobin in the red blood cells. Hemoglobin is a large complicated molecule. It contains many atoms of carbon, hydrogen, oxygen, and nitrogen, along with four atoms of iron. The heart of the molecule is the iron atoms, which have an attraction for the oxygen molecules of the air that we breathe into our lungs. However, when the oxygen is needed in a cell to combine with glucose, the hemoglobin gives it up. The bloodstream then carries the hemoglobin back to the lungs to pick up another load of oxygen. Thus, hemoglobin serves as the oxygen carrier. When oxygen is attached to hemoglobin, as in the blood of arteries leading away from the heart and lungs, the blood is bright red. When hemoglobin is not carrying oxygen, as in the blood of veins going back to the heart, the blood has a darker color. Carbon monoxide, CO, is such a deadly poison because it can react with hemoglobin. If the air a person breathes contains carbon monoxide, these CO molecules combine with the person's hemoglobin and thus effectively tie it up. The person's blood is then unable to transport oxygen, and death quickly results.

The equation for the reaction that takes place between oxygen and glucose is

$$C_6H_{12}O_6 + 6O_2 \rightarrow 6CO_2 + 6H_2O + \text{energy}$$

This looks familiar; it is the equation we would expect to write if we were describing the burning of a compound containing carbon and hydrogen. For the reaction that takes place in living cells, it is a simplified representation of one of a complicated series of reactions involved in the overall process called *metabolism*. The term metabolism describes the sum total of chemical processes by which a living organism unlocks the energy in food and synthesizes new tissues.

THE EFFECT OF TEMPERATURE ON REACTION RATE

Many experiences of daily living involve changing the rate of chemical reactions by changing the temperature. A cake bakes faster in a hot oven than it does in an oven that is merely warm. The advantage of a pressure cooker is that, when the tight lid holds in the steam, the temperature can rise above the ordinary boiling temperature of water and consequently cook the food faster. Cooking involves chemical reactions, which take place faster at a higher temperature. Just the opposite effect, slowing down the chemical reactions of decay, is accomplished by putting food in a refrigerator, where the temperature is lower.

All chemical reactions go at a faster rate when the temperature is raised. In many cases, we find that a 10° rise in the Celsius temperature of the reactants doubles the rate of the reaction. The human body reacts to invading bacteria or viruses by raising its

The human body reacts to invading bacteria or viruses by raising its temperature.

temperature. The fever that accompanies sickness speeds up the chemical reactions that the body uses to fight the invaders. Some animals grow faster when kept at a higher temperature. For example, lobsters that require 4 years to mature in the cold ocean off the shore of New England reach the same size in about 18 months in water with a temperature of 70° F (21° C).

All fires burn faster if the heat released by exothermic combustion is unable to escape, so that the temperature rises. This is why a forest fire becomes so devastating and hard to control. In some cases, the phenomenon of *spontaneous combustion* may occur. If a pile of oily rags is left undisturbed, the heat from the slow exothermic reaction of oil or grease reacting with air can be held within the pile. Because the heat cannot escape, the temperature of the rags rises. Eventually the temperature rises to the point where the combination of oxygen and oil is fast enough to make the rags burst into flame.

The opposite effect can be illustrated by blowing out or shaking out the flame of a burning match. The swift current of air moves gas molecules past the flame so that they carry heat away quickly. This process lowers the temperature, and the burning reaction slows to a point where it is no longer a combustion; the flame goes out.

THE EFFECT OF CONCENTRATION ON REACTION RATE

When something burns in air, the molecules of oxygen gas must bump into the substance that is reacting with them. Because only about one-fifth of the molecules in air are oxygen and because *all* the air molecules are flying about, only one-fifth of all the collisions can possibly be effective in making the reaction go. Thus, none of the collisions made by nitrogen gas molecules can possibly result in the burning reaction. However, if a piece of wood or charcoal that is only burning with a dull glow in air is put into pure oxygen, it bursts into flame. The burning reaction speeds up because *every* gas molecule that bumps into the burning piece is an oxygen molecule. You can accomplish the same effect by fanning a smoldering fire. By fanning, you move more air that contains reactive O_2 molecules into the space where reaction can occur.

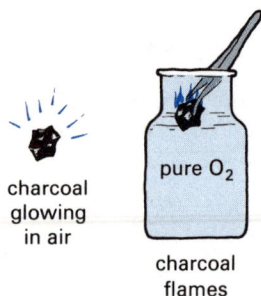

The term *concentration* is used to describe the relative amount of some one thing in a solution or mixture. The concentration of O_2 in air is about one-fifth, or 20%; in pure oxygen gas, the concentration is as high as it can be, 100%. The generalization is: *The rate of a chemical reaction increases when the concentration of a reactant is increased.*

The use of oxygen gas to treat hospital patients with heart or lung disorders also illustrates the effect of concentration on reaction rate. If a person has heart trouble, the effort of normal breathing may be too great. Lung trouble means that a person cannot use the amount of oxygen in ordinary air efficiently enough to live. Such a person is placed under a tent or a mask into which pure oxygen gas is released from a tank. The concentration of O_2 molecules in the air the patient breathes is thus increased. This process makes each breath carry more oxygen into the lungs. So even though patients may not be able to get enough oxygen by breathing ordinary air, they can get enough oxygen to stay alive because of the increased concentration of oxygen in the air that they do breathe.

charcoal
glowing
in air

pure O_2

charcoal
flames

Percentage

The term *percent* is used as a convenient way of expressing fractions of whole things. *Percent* means "parts per hundred." *Cent* means "hundred"; there are 100 cents in a dollar, 100 years in a century, and so on. *Per* means "through," so the word *percent,* or its symbol %, literally means "divide through by 100" or "multiply by $\frac{1}{100}$" to get the true value of the quantity specified. For example, if 40% of a group of people are men, the fraction of the group that is men is 0.40.

$$40\% = 40 \times \boxed{\%}$$

$$\left(\% \text{ is the same as } \frac{1}{100}\right)$$

$$40\% = 40 \times \boxed{\frac{1}{100}}$$

$$\left(40 \times \frac{1}{100} \text{ is the same as } 0.40\right)$$

$$40\% = 0.40$$

Any fraction can be converted into a percentage by multiplying its decimal value by 100%. For example, if approximately one-fifth of the molecules in air are oxygen molecules, the approximate percentage of oxygen in air is 20%.

$$\frac{1}{5} = \text{ the decimal fraction } 0.20$$

If we multiply this number by $100 \times \frac{1}{100}$ we do not change its value.

$$\frac{1}{5} = 0.20 \times 100 \times \boxed{\frac{1}{100}}$$

$$\left(\frac{1}{100} \text{ is the same as } \%\right)$$

$$\frac{1}{5} = 0.20 \times 100 \times \boxed{\%}$$

$$(0.20 \times 100 \text{ is the same as } 20)$$

$$\frac{1}{5} = 20\%$$

It is essential to give the basis for comparison when stating a percentage. Thus in the first example, 40% of the *people* are men; in the second, 20% of the *molecules* are oxygen. But neither of these percentages tells anything, for example, about the fraction of the total mass of the group that is men or the fraction of the mass of an air sample that is oxygen.

You may also recognize why the "No Smoking" rule is tightly enforced in any area where pure oxygen gas is being released. The slow burning of a cigarette or pipe becomes a rapid, vigorous fire with a large flame in pure O_2 gas. Such a flame might ignite anything else that would burn. The presence of a high concentration of oxygen in the surrounding air would produce a flash fire.

Dust explosions are still another example of the concentration effect. Disastrous sudden fires can occur in coal mines. Similarly, wood shavings ignite more easily than a solid log does. If there is more surface for the same amount of material, the gaseous O_2 molecules in the surroundings can collide with more of the substance that will burn. Increasing the surface area of a combustible material has the effect of increasing its concentration, because the reaction with oxygen occurs at the surface. When the burnable material, such as coal or wood, is in the form of dust, the total amount of surface is very large. When the fine particles are dispersed in the air, a great deal of burning can occur extremely rapidly. Thus in a mine where there is coal dust produced by digging the coal out, an accidental spark may start a fire that becomes an explosion. In some modern coal mines, powdered limestone, a rock that will not burn, is sprayed over working

In some coal mines, powdered limestone, a rock that will not burn, is sprayed over working areas.

areas. (Miners emerging from the mine are covered with this white dust rather than black coal dust.) The effect is to dilute the combustible coal dust with noncombustible limestone dust so that the hazard of accidental explosive fire is decreased.

Limestone dust
+
air   No
+   reaction
spark

Coal dust
+
air   EXPLOSION!
+
spark

Which would *you* prefer?

THE EFFECT OF CATALYSTS ON REACTION RATE

The parallel between burning and the body's metabolism of carbohydrates brings up another important generalization about reaction rates. The body is able to convert carbohydrates into CO_2 and H_2O at a temperature far below that necessary for combustion. How is this accomplished? The answer is that the complicated steps of metabolism are helped along by *catalysts*. A catalyst is a substance that can enter into a reaction between

other substances but is set free in its original form after the reaction has taken place. The overall reaction in the presence of a
catalyst is faster than if the catalyst were not there. An analogy is
using a key to open a locked door. You can open and go through
a locked door by the slow process of breaking or picking the lock.
But if you take the correct key from your pocket, you can use it
to turn the lock, go through the door, relock the door, and replace the key in your pocket. The overall reaction, your going
through the door, has been speeded up by the catalyst, your key.

So it is with living things. The living organism makes and uses
enzymes as catalysts for the many chemical reactions of metabolism. Usually a particular enzyme is a catalyst for only one
specific reaction, just as a key is good for only a specific door
lock. These enzymes make it possible for reactions to occur
rapidly, even though the temperature is low. For humans, normal
body temperature is 37° C (98.6° F).

The principle of *catalysis* (the use of catalysts to speed up
chemical reactions) is used very widely in chemical industries.
You can readily recognize the economic advantage of making
products faster. Raising the temperature of reactants to speed
their reactions would require expensive heat energy. And some
reactants may be able to react in several different ways. Undesirable products may be the result of these other "side reactions."
Using a catalyst gives an advantage to a particular desired reaction. Many chemical processes would not be feasible at all if
appropriate catalysts were not used. One very important example
is the manufacture of ammonia, NH_3, from nitrogen, N_2, and
hydrogen, H_2. Over 13 million tons of ammonia are manufactured
in the United States each year. Ammonia is the compound that
provides much of the essential nitrogen to living things by way of
fertilizers and animal feed supplements. Ammonia is also a starting material for the manufacture of explosives. Catalysts are also
important in the manufacture of textiles, rubber, and plastics.

CHANGING THE RATE OF A CHEMICAL REACTION—A SUMMARY

Let us tie together our discussion about reaction rates. If we
change the conditions of a chemical reaction, we change the rate
of the reaction. These changes, referred to as the *factors* that influence reaction rate, are summarized in Table 3.1.

Table 3.1   Factors influencing reaction rate

Factor	Effect
Temperature	Increased temperature increases reaction rate.
Concentration	Increased concentration of reactants increases reaction rate.
Surface	Increased surface of solid reactants increases reaction rate.
Catalysis	The presence of a catalyst increases reaction rate.

These factors also can be changed to slow down reactions. Reaction rate is slower if the temperature is lower, if the reactants are diluted, if the reactants are in large pieces, or if no catalysts are allowed to participate in the reaction.

DECAY—THE ACTION OF MICROORGANISMS

Many humans may not recognize who their best friends are among the millions of living things on this earth. Our best friends are the *microorganisms*, those microscopic animals such as bacteria. If these tiny creatures did not exist, animal waste alone

If these tiny creatures did not exist, animal waste alone would long since have covered the surface of the earth.

would long since have covered the surface of the earth. Dead trees and plants also would leave room for little else! Microorganisms use these waste materials as food. Their metabolism allows them to make use of the potential chemical energy in such substances and to change them into simpler molecules. This is the same principle by which you convert carbohydrates into the simpler molecules CO_2 and H_2O.

Various kinds of microorganisms can use various kinds of substances for food. Each microorganism is able to make its own *enzymes*, which catalyze the many chemical reactions of metabolism. (Enzymes will be described in more detail in a later chapter.) However, there are dominant patterns of preference among the most abundant kinds of microorganisms. (The word preference is not used here in the sense that you may "prefer" steak to spinach but rather in the sense that, if suitable food molecules are not available, the microorganisms die.) Over the eons that microorganisms have been on earth, they have evolved a "preference" for the kinds of molecules that are built up by living things. This is why they have so successfully *degraded* (broken down) so much dead material that formerly was living.

One very abundant kind of microorganism is the *aerobic bacteria*. The name *aerobic* means that they need oxygen for their metabolic reactions. Aerobic bacteria turn carbon atoms of their food into CO_2; they turn the hydrogen and oxygen atoms of their food into H_2O. Much of the *decay* that occurs all around us is accomplished by aerobic bacteria. This transformation shows the essential connection between decay and combustion as similar chemical reactions.

In the past century, human animals have begun to create a different kind of waste. This waste is the trash that consists of many discarded or worn-out articles made of manufactured material. In recent years, the piling up of this waste has become so huge that it is one of our major environmental problems. (In the cities and towns of the United States, the average amount of solid waste is about 6 pounds per person per day.) Not only is the size of the pile the problem, but much of the waste is not *biodegradable*. A biodegradable substance is one that can serve as food for microorganisms; it can be degraded (broken down) by biological processes. A cardboard carton, made from wood pulp, is biodegradable. It decays as readily as a dead log in a forest. A rubber tire or a plastic article, by contrast, is not biodegradable. The same is true of the various kinds of petroleum. Microorganisms can use none of these materials for food. The enzymes used by microorganisms are useless as catalysts for the combination of oxygen with rubber, plastics, or petroleum. What all this means is that a very important factor to be considered by manufacturers of packaging materials that are inevitably discarded is whether the substance is biodegradable or not. No longer can mere convenience or price of raw materials be the determining factors. Otherwise, the nonbiodegradable waste from "civilized" societies may pile up in alarming proportions.

BOD AND EUTROPHICATION

Everyone who reads articles about the environmental problems of lakes or streams encounters the term "BOD." The letters stand for *biological oxygen demand.* When waste materials containing C, H, or O atoms are dissolved or mixed into the water of lakes or streams, the aerobic bacteria that cause them to decay must use oxygen that is dissolved in the water. If there is much such waste, the bacteria will need a lot of oxygen to accomplish the decay. Hence, biological oxygen demand is a measure of how

---

Biological oxygen demand is a measure of how much waste pollutes a sample of water.

---

much waste pollutes a sample of water. This fact is also the reason that installations for the purification of water, such as sewage treatment plants, often spray the water that they are purifying into the air. Some water treatment plants blow air or pure oxygen gas through the contaminated water. This procedure allows more oxygen to dissolve in the water to replenish that used by the bacteria in their work of making the waste materials decay.

Another term related to problems of pollution in lakes or ponds is *eutrophication.* The term eutrophication refers to the situation in which an abundance of added nutrients in a lake or pond causes an abundant growth of plant life, especially simple, single-cell plants called algae. When the algae die, the lake or pond is filled with excess dead algae. The decay of this large amount of dead plant material creates a very large BOD. In such a situation, the demand of the feeding microorganisms for oxygen is greater than the amount of oxygen dissolved in the water. In addition, the process of decay has used up the oxygen that is normally available to fish and other animals that breathe through gills. Consequently, the fish die in the same way that other breathing animals suffocate if deprived of air.

The eutrophication problem is caused by adding something to the water that encourages the very abundant growth of algae. These "somethings" are nutrients or substances that plants need to grow. All plants need some minerals, especially ones containing nitrogen and phosphorus. (That is why compounds of these essential elements are contained in fertilizer.) If a large amount of fertilizer is applied to a field near a pond, some of the fertilizer will be carried from the soil into the pond by rain, and the pond will become overrich in nutrients and ultimately become overrich in plants. Another source of nutrients may be the water that comes from towns or cities. A special problem is caused by the phosphorus compounds used in some laundry detergents. Large amounts of such compounds may enter a lake or pond and thus supply one of the essential nutrients for abundant plant growth. Algae may grow with such profusion that the surface of the lake may appear to be completely covered by such a "bloom." Where eutrophication already has occurred, the trouble can be cut down by controlling what goes into the lake or pond. These controls may be laws that stop people from using such things as phosphate-containing detergents. When the excess plant nutrients are not added to the water, excess plant life does not appear. Consequently, there is less dead plant life in the water, and its BOD stays below the danger point.

PHOTOSYNTHESIS

We have used examples of one reaction, the combination of oxygen with carbohydrates, to learn much about reactions in general. A reaction involves the same energy changes, regardless of

A reaction involves the same energy changes regardless of whether it is fast or slow.

whether it is fast or slow, catalyzed or noncatalyzed, in one step or several. Now we will find out something else about chemical reactions: Often a reaction can be reversed. The products of one reaction can be the reactants of the reverse reaction.

Photosynthesis is the term applied to the process green plants use to take CO_2 and H_2O out of the air and to synthesize carbohydrates, as well as to give off O_2. This is suggested in Figure 3.2.

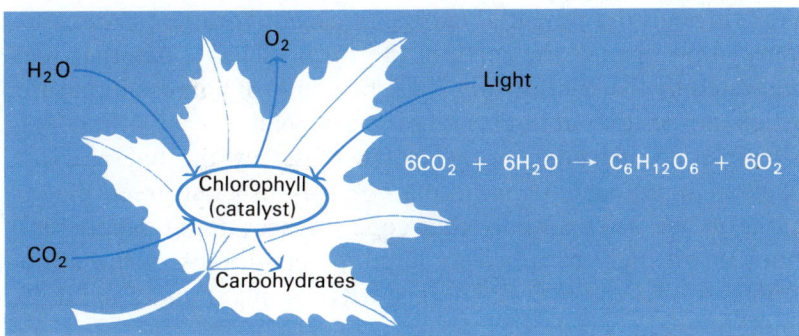

Figure 3.2   Photosynthesis in green plants.

The steps by which plants accomplish this are many and complicated. Not all the steps are understood, but some of the essential features are well known. Two of these features are pertinent to our discussion: Energy, in the form of light, must be absorbed by the plant, and the plant must have *chlorophyll* in its leaves. Chlorophyll is the compound that gives leaves their green color. Chlorophyll must be present to make the reaction go along with all the other enzymes the plant uses. An equation for the reaction is

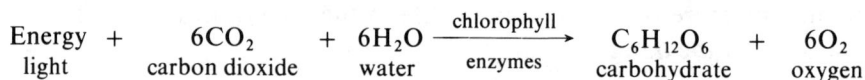

$$\text{Energy} + 6CO_2 + 6H_2O \xrightarrow[\text{enzymes}]{\text{chlorophyll}} C_6H_{12}O_6 + 6O_2$$

Energy   6CO₂   6H₂O   C₆H₁₂O₆   6O₂

light   carbon dioxide   water   carbohydrate   oxygen

This equation should bring to mind the equation we used earlier to represent the overall reaction by which breathing animals accomplish the metabolism of a carbohydrate.

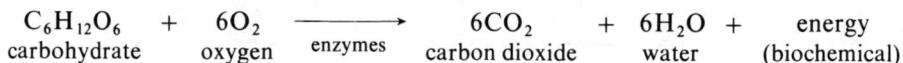

$$C_6H_{12}O_6 + 6O_2 \xrightarrow{\text{enzymes}} 6CO_2 + 6H_2O + \text{energy}$$

carbohydrate   oxygen   carbon dioxide   water   (biochemical)

The numbers of each kind of molecule are the same. So too is the amount of energy involved. But the reactants and products in these two equations are reversed, and the form of the energy is different in the two cases. Biochemical energy is *delivered* by the metabolism of carbohydrates. The animal uses the energy to keep warm and alive. The same amount of energy, in the form of light, must be absorbed by the plant in order to make carbohydrates. One reaction is essentially the reverse of the other.

The relationship between photosynthesis and the metabolism of breathing animals has far-reaching significance. For one thing, the relationship suggests that all animal energy can be traced back through the potential energy in carbohydrates to the sun's energy that reaches the earth in the form of light. Another idea suggested by the relationship is the principle of balance in nature, which implies that animals and plants depend on each other. This principle is illustrated, on a limited scale, by the balance that exists in an aquarium containing not only fish but green plants. Later on, we will examine what consequences might occur if humans unwittingly or unwisely alter the conditions of the atmosphere, so that the overall balance between plant and animal life could be endangered.

Summary

We have explored the chemistry of oxygen combining with various substances to illustrate the full meaning of the term chemical reaction. Along the way, we have seen how chemists use chemical equations as a kind of shorthand for describing chemical reactions.

Chemical reactions may be exothermic (release energy) or endothermic (absorb energy). And the rate or speed of a chemical reaction is important to consider. Adjusting the factors of temperature, concentration, surface area, and presence of catalysts allows us to control the rate of a reaction.

Oxygen combines with some metals, with some nonmetals, and with compounds. Especially important for animals is the combining of oxygen with carbohydrates in foods, because this reaction releases the energy an animal needs to live. The same kind of reaction serves the important function of decay; microorganisms promote decay by eating waste materials.

By changing the conditions, a chemical reaction may be made to go in the reverse direction. The starting materials for photosynthesis in plants (CO_2 and H_2O) are the products of the metabolism of carbohydrates by animals.

Glossary

The number in parentheses indicates the text page where you can find the term defined in context.

aerobic bacteria   bacteria that use oxygen in their metabolic reactions, combining it with hydrogen and carbon from their food to produce water and carbon dioxide (85)

biodegradability   the capacity of a substance to be broken down into smaller substances by natural biological processes, usually involving microorganisms (85)

biological oxygen demand (BOD)   a measure of the amount of biodegradable pollutant matter in water. It is measured as the amount of oxygen that must be used by microorganisms to degrade the matter over a five-day period under controlled conditions. (86)

calorie   the unit of measurement of heat, defined as the amount of heat energy needed to raise the temperature of 1.0 gram of water 1.0 degree Celsius (75)

catalysis   the process of changing the rate of a chemical reaction with a catalyst, a substance that participates in the reaction without itself being permanently changed (83)

chemical equation   a convenient notation in which chemical formulas separated by arrows are used to show the reactant and product substances of a chemical reaction and the proportions of each (65)

combustion   a chemical reaction in which oxygen combines rapidly with some other substance to release heat and light (64)

concentration   the relative amount of one substance in a mixture of substances, which may be expressed as a percentage or a ratio of quantities (80)

endothermic reaction   a chemical reaction that absorbs heat from the surroundings (75)

enzyme   a biological catalyst that controls a specific metabolic reaction in a living organism (83)

eutrophication   the abundant abnormal growth of algae and other plant life, occurring when excess nutrients are present in a body of water (86)

exothermic reaction   a chemical reaction that releases heat to the surroundings (75)

kinetic energy   the energy of a moving body (74)

metabolism   the chemical reactions by which a living organism breaks down molecules from its food to get energy, builds the new molecules it needs for growth, and eliminates waste products (79)

microorganism   a microscopic animal or plant (84)

oxide   a compound that contains oxygen combined with one other element (69)

photosynthesis   the process by which a green plant containing chlorophyll absorbs light energy, takes in carbon dioxide and water, and produces carbohydrates and oxygen (87)

potential energy   the energy stored in a substance or particle because of its position or chemical structure (74)

product   a substance produced by a chemical reaction (65)

reactant   a starting material that undergoes a chemical reaction (65)

spontaneous combustion   burning (or increase of reaction rate) resulting from the rise in temperature in an undisturbed mass of combustible material as an exothermic reaction with oxygen takes place (79)

Exercises

3.1 Why is this chemical equation incorrect?

$$CH_4 + O_2 \rightarrow CO_2 + 2H_2O$$

3.2 Why is this chemical equation incorrect?

$$CH_4 + 2O_2 \rightarrow SO_2 + 2H_2O$$

3.3 In balancing chemical equations, it is important to be able to tell how many of each kind of atom are present. For each case below, account for all the atoms present.
Example: $3CH_4$: 3C and 12H

a.	$4CaCl_2$	b.	$2CO_2$
c.	$10H_2O$	d.	$6SO_3$
e.	$5C_6H_6$	f.	$3CS_2$
g.	$5Al_2O_3$	h.	$2C_4H_8$
i.	$4Fe_3O_4$	j.	$3C_6H_{12}O_6$

3.4 In balancing a chemical equation, why are we allowed to change the coefficients but not the subscripts?

3.5 Why does an airplane or train carry only fuel for its engines, whereas a rocket or spaceship must carry both oxygen and fuel?

3.6 If a landfill in which garbage is dumped is not properly maintained, fires may start inside it. These fires occur more often if rats have dug tunnels from the surface down into the landfill. Give a chemical explanation for these facts.

3.7 What is the difference between exothermic and endothermic reactions?

3.8 Describe the role of hemoglobin as the oxygen carrier in the blood.

3.9 Carbon monoxide molecules form a very stable compound with hemoglobin. (By stable compound, we mean one that does not readily break apart.) Give a chemical explanation for the deadly consequences of breathing air that contains carbon monoxide as well as oxygen.

3.10 The label on a bag of charcoal briquets carries the warning that they should not be burned indoors. Why is this warning necessary?

3.11   Makers of some candies advertise their product as a source of "quick energy" because they put glucose in the candy. Do you think this is truth in advertising? On what chemical evidence do you base your answer?

3.12   There are two ways to cook a roast, says a popular cookbook. One is to cook it for x hours at 325°F and the other for y hours at 400°F. Which is larger, x or y? Why?

3.13   A freezer preserves many foods better than a refrigerator does. Why?

3.14   A person lighting a pipe on a windy day cups hands around the burning match and the pipe bowl. Why?

3.15   Putting out a campfire properly includes spreading out the remains of the fire over a larger area. How does this practice help to insure that the fire will not start again after the campers are gone?

3.16   Explain why a crumpled wad of newspaper can be used as kindling to start a fire while a "log" made of tightly rolled newspaper may burn slowly for as long as an hour.

3.17   A half cup of laundry bleach in a gallon of water makes a solution that will remove a stain faster than a solution containing only one-eighth of a cup of bleach in an equal amount of water. The first solution probably will weaken the fibers of the cloth more than will the second one. Explain these facts in terms of a general principle involving rates of chemical reactions.

3.18   Sugar burns in a flame at a high temperature, but it "burns" in the body at only 37°C. What mechanism does the body have that enables it to accomplish this process at this low temperature?

3.19   A petroleum company advertises that it adds a small amount of a substance called "Platformate" to the gasoline it sells. The claim is that Platformate improves engine performance. What would you guess is the chemical role of Platformate?

3.20   What is meant by the contrasting terms biodegradable and nonbiodegradable? List five or six items found in your trash can. Classify each as biodegradable or nonbiodegradable.

3.21   In what sense is it reasonable to say "microbes are humans' best friends"?

3.22   Animals that are grown for food do not extract all the nutrients from the feed that passes through their systems. Why do some people say that crowded feedlots are a major cause of eutrophication?

3.23   We do not eat all forms of plant life. Why should we worry about widespread pollution damage to green plants that we do not directly depend on for food?

3.24   Why may an orange grower advertise that when you eat his product you are consuming a "package of sunshine"?

4 The Behavior of Gas Molecules

- ☐ What characteristics do all gases share?
- ☐ How do we measure the behavior of gases?
- ☐ How does temperature affect the behavior of gases?
- ☐ What general theory ties together everything we know about the behavior of gases?
- ☐ How can we count the particles in a sample of matter?
- ☐ How does Avogadro's theory relate the macroscopic properties of gases with their microscopic behavior?
- ☐ Why does a helium-filled balloon rise?

The words *gas* and *molecule* are both familiar by now. We have used the word gas to identify one physical form of matter, in contrast to the liquid and solid forms. The word molecule also means something definite. A molecule is the smallest possible individual piece of a pure substance. Even though there are countless numbers of different molecules, each represented by a specific chemical formula, substances in the gaseous form have many properties in common. Our total understanding of matter, then, can be expanded by a detailed examination of these general properties or characteristics of gases.

We will start by pulling together some information from everyday experience, by describing some observations about the properties of gases in general. Then we will examine some of the quantitative laws that describe the behavior of gases. As before, by quantitative, we mean the relationships among the numbers that we get when we measure such properties as the volume, the pressure, or the temperature of a sample of a gas. Finally, we will tie our information together with a theory. This theory is where molecules come into the picture. Remember how we used the idea of atoms that Dalton proposed in his theory to explain the laws of chemical combination? In a similar way, we will use the ideas of molecules and their motions to explain the laws of gas behavior. This theory, called the *kinetic molecular theory,* is one of the truly great creations of the human mind. It is an "elegant" theory, because the ideas it expresses give a logically consistent interpretation for much that we observe.

OBSERVATIONS OF THE PROPERTIES OF GASES

Gases uniformly and completely occupy any container in which they are placed. In a room, the air is the same everywhere, not all collected at the floor by gravity or at the ceiling. (However, samples of air from two greatly different altitudes are distinctly different.) When you blow up a balloon or a bicycle tire, there are

When you blow up a balloon or a bicycle tire, there are no empty corners.

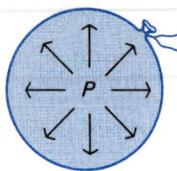

no empty corners. Related to this is the property of *diffusion* of gases. Gases readily mix into one another. The pleasant odor of an orange grove in blossom makes the area for miles around an attractive place to be. The odor of gas from an open but unlit gas jet in one room can be smelled elsewhere in the house. In fact, small amounts of a strong-smelling substance are purposely added to fuel gas so that hazardous leaks can be detected.

Gases also can be *compressed*. The volume of a sample of gas can be made smaller by putting pressure on the container. You can squeeze a filled balloon to make it smaller. When you take your hands off and thus release the pressure, the volume of the balloon returns to its original size. You encounter compressed gases many times in your life. A scuba diver breathes air released from a tank of compressed air strapped on his or her back. Thousands of commercial products, from hair spray to paint to cheese, can be obtained from cans by opening a valve and allowing the compressed gas within the can to push the material out of the container. Cream can be "whipped" into a foam by the bubbles of gas released from such a can. Such cans are often referred to as *aerosol* cans. An *aerosol* is a mist of very fine droplets of liquid dispersed in a gas—for example, droplets of hair spray in air. (In a later chapter, we will describe some of the possible bad effects on the environment that may result from the accumulation in the upper atmosphere of the gases used to propel aerosols out of cans.)

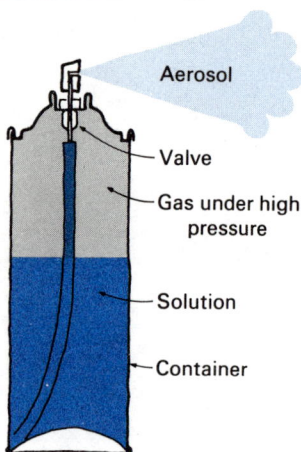

Gases exert pressure on the walls of a container holding them. This pressure remains constant as time goes on. This constant pressure is why the balloon that you squeeze returns to its original size when you stop squeezing it. You need not keep checking the pressure in an automobile tire unless you suspect a leak. The pressure of expanding gases from the combustion of gasoline moves the pistons in the cylinders of an automobile engine. All explosions occur because some extremely rapid chemical reaction has produced gases that exert tremendous pressures.

Adding more gas to a container that is already full increases either the volume or the pressure, or both. You blow up an air mattress either by adding your breath to the air already in it or by connecting it to a pump that pushes more air into it.

Aerosol

Valve

Gas under high pressure

Solution

Container

Another characteristic property of gases is that they appear to be a "dilute" form of matter. Liquids and solids are more "concentrated" forms of matter. We can see and feel the surfaces or edges of liquids and solids; there is more "stuff" in a given space than in a gas. Put in more scientific terms, the density (the mass of a unit of volume) of a gas is low relative to the density of a liquid or a solid.

The low density of gases has a very practical consequence when we try to measure the mass of a gas sample. The weight of 1 liter of room air is only 1.2 grams. This is the amount of air in a cube about 4 inches on each edge, approximately 1 quart in volume. The same volume of water at room temperature weighs 997 grams. So, in order to be at all precise when we weigh gases, we must use a balance or weighing device capable of measuring small masses accurately.

Another problem makes it difficult to measure the mass of a gas sample. We must start with an empty container, weigh it, fill it with a gas, and then weigh it again. The mass of the gas sample is the difference between the weights of the full and empty containers. To empty a container, we must connect it to a vacuum pump, a machine that can remove the gas. We say it creates a *vacuum,* a space in which there is no matter. It is virtually impossible to remove every bit of gas from a container, but a good vacuum pump leaves very little.

All these observations about gases suggest that it might be convenient and useful to discover what relationships exist among the properties of gases so that we can make quantitative predictions. Does some law connect the mass, volume, temperature, and pressure of a gas sample? Indeed, there is such a law. It relates the measured values of these properties. You are already familiar with methods of measuring mass, volume, and temperature. Now let us learn more about pressure and how it is measured.

PRESSURE AND HOW IT IS MEASURED

The word *pressure,* as used in everyday language, conveys the idea of pushing. We say that we feel "under pressure" to get something done. A push means the application of some force. Force, in scientific language, is defined as that which changes the state of rest or motion of a body. You apply a force on something that is standing still to make it move. More force will make it move faster. And the more mass a body has, the more force that is needed to set it in motion. The formula that defines force is

$$\text{Force} = \text{mass} \times \text{acceleration}$$

Acceleration is the change in the speed with which a body is moving. (You step on the accelerator of your automobile to make it go faster.) The most familiar force that we experience is gravitational force, the push toward the center of the earth that acts on all objects. We call this force the *weight* of the object. An object that is free to fall accelerates (changes its speed) toward the center of the earth. If we hold something so that it does not fall, we must exert a force on it that is opposite and equal to the force of gravity.

Pressure is defined as force per unit area. For example, the pressure of the earth's atmosphere at sea level is 14.7 pounds on every square inch of surface.

$$\text{Pressure} = \frac{\text{force}}{\text{area}}$$

The importance of the "per unit area" part of this definition can be illustrated as follows: If you stand on one foot, the pressure that your foot must withstand is twice what it is when you have two feet on the ground. Your weight, the total force of gravity, is pushing on only half the area, the sole of *one* foot instead of two. If you can manage to "stand" on one hand, the pressure is even greater, because the area of your one hand is smaller yet. You can imagine that supporting your total weight on one fingertip (a much smaller area), if it could be done, would probably break your finger, because the pressure would be so great.

$$\frac{\text{force}}{\text{area}} = \text{pressure}$$

$$\frac{\text{same force}}{\text{smaller area}} = \text{larger pressure}$$

The pressure of the atmosphere is measured with a device called a *barometer.* It was invented by an Italian named Evangelista Torricelli in 1643. A mercury barometer is made as follows (see Figure 4.1): A glass tube, at least 76 cm long, is sealed shut at one end and filled completely with mercury. The unsealed end is covered with a finger, and the tube is inverted so that the open end is under the surface of a pool of mercury in an open dish. The column of mercury in the tube falls to a definite level

Hand can hold book plus pencil (small pressure).

Hand cannot hold pencil plus book (great pressure).

Glass tube filled with mercury

Mercury in dish

Vacuum

Atmospheric pressure

Glass tube is inverted with open end under the surface of the mercury in the dish.

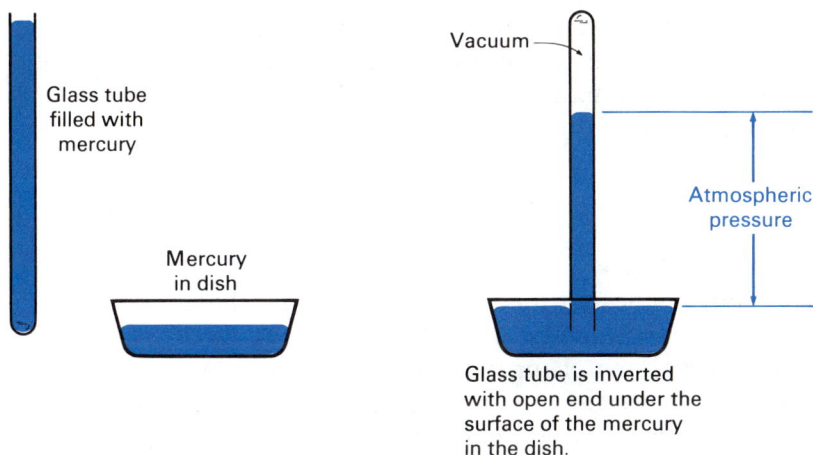

Figure 4.1   How a barometer is made.

above the mercury surface in the dish. A vacuum has been created in the tube above the mercury. Nothing is in this space, because the mercury has dropped out of it. Thus the weight of the mercury alone creates a pressure at the bottom of the tube on an area equal to its cross-sectional area. This pressure is balanced by the weight of a column of atmosphere pushing down on each equal area of the mercury surface in the dish outside the tube.

Everyone who has ever heard or read a weather report knows that the atmospheric pressure changes from time to time. These changes are indicated by the height of the mercury column in the barometer tube. The barometric pressure atop a mountain is less than that at sea level because there is less atmosphere above a mountaintop than there is above the earth at sea level. The average height of the column of mercury at sea level is close to 76.0 cm, or 760 mm of mercury. (Measured in the English system, this is a column 29.9 inches high.) This average is chosen as the standard unit of *1.0 atmosphere*. A unit of barometric pressure, a *torr,* is often used in science laboratories. One torr is equal to 1.0 mm of mercury. Thus 1.0 atmosphere equals 760 torr.

A great variety of gases are used in very large quantity, not only in laboratories, but in factories and shops. Hospitals use large quantities of oxygen and some other gases that are anesthetics. Large quantities of gases usually are sold in heavy tanks capable of withstanding pressures equivalent to hundreds of atmospheres. Such pressures cannot be measured conveniently by such devices as glass tubes full of mercury. Instead, gauges are used in which the gas pushes against a heavy spring or metal plate. Such gauges are usually marked off in units of pounds per

square inch (lb/in²). They always read pressures *above* 1 atmosphere because such high-pressure tanks are considered "empty" when the gas remaining in them is at 1 atmosphere pressure. (No more gas will flow out.)

THE PRESSURE–VOLUME LAW FOR GASES

We start our search for regularities in the behavior of gases by choosing one property, the volume of a gas sample, and seeing how the volume changes when the pressure on the sample is changed. However, pressure is not the only condition that can be varied to affect the gas volume. A change in temperature also causes the volume to change. A blown-up balloon put in a refrigerator will get smaller. So, if we want to do experiments to find a

A blown-up balloon put in a refrigerator will get smaller.

relationship between volume and pressure only, we must do so at some constant (unchanging) temperature. It is also rather obvious that we cannot work with a leaky balloon. The amount of gas in our sample has to be constant.

When we work with a constant amount of gas at a constant temperature, we find that doubling the pressure on the sample makes the volume half what we started with. Tripling the pressure on the sample (increasing it three times) decreases the volume to one-third of its original size. We could go on making many measurements; always we would find that increasing the pressure decreases the volume. Or if we decrease the pressure, the volume increases. The general rule is that volume is *inversely* proportional to the pressure. An inverse proportion is one in which an increase in one thing causes a decrease in the other. "The more time passes, the less you remember" expresses an inverse relationship between time and recall.

A law of nature always can be expressed in quantitative language (numbers). Let us take some laboratory data, the numerical values of our measurements, and arrange them in a table (see Table 4.1). When we do a little arithmetic with these numbers, we find an interesting consistency. The product of the volume times the pressure is always the same, a constant number. This then is the law we are looking for. Remember, a scientific law tells how something in nature behaves every time (constant behavior).

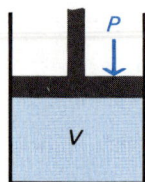

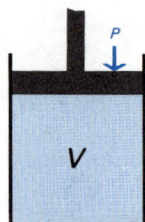

P

V

P

V

Lower
pressure

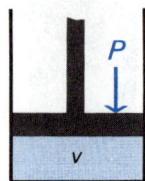

P

v

Higher
pressure

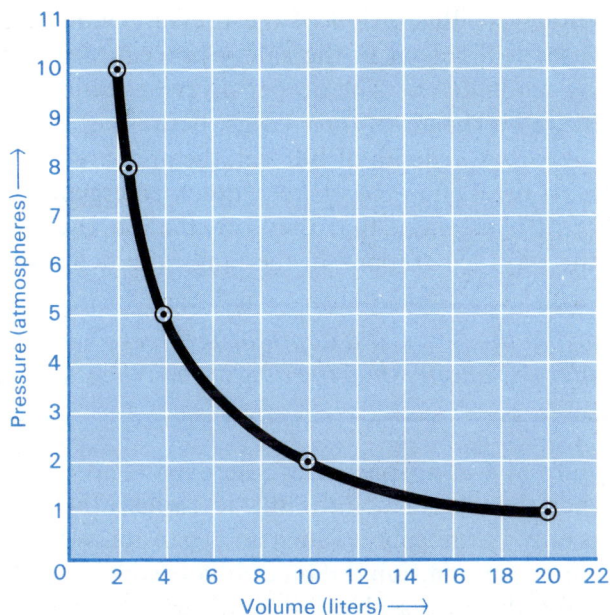

Figure 4.2   Pressure–volume data for a gas sample at constant temperature.

Table 4.1   Experimental data for the pressure and volume of a gas sample at constant temperature

Volume (liter)	Pressure (atmosphere)	$V \times P$ (liter-atmosphere)
20.0	1.0	20.0
10.0	2.0	20.0
5.0	4.0	20.0
4.0	5.0	20.0
2.5	8.0	20.0
2.0	10.0	20.0

Figure 4.2 shows how the data look when plotted in the form of a graph. The pressure of the sample is large when the volume is small. As the value of the pressure becomes smaller, the value of the corresponding volume becomes larger.

This law was first discovered by Robert Boyle in the middle of the seventeenth century. **Boyle's law** is stated as follows:

For a sample of gas at constant temperature, *the volume is inversely proportional to the pressure.*

or

For a sample of gas at constant temperature, *the product of volume and pressure is constant.*

A very practical illustration of the pressure–volume relationship for gases is involved in the safety procedure that must be followed by scuba divers. The deeper divers descend, the greater is the pressure on the air in their lungs, because of the weight of the water above. At a depth of 100 feet, the pressure is about 4 atmospheres. If divers have to make a quick emergency ascent to the surface, it is essential that they breathe out the air in their lungs on the way. Air trapped in the lungs at a depth of 100 feet

Air trapped in the lungs at a depth of 100 feet would expand to four times its volume when the lungs reach the surface.

(at a pressure of 4 atmospheres) would expand to four times its volume when the lungs reach the surface, where the external air pressure is only 1 atmosphere. Such an expansion would result in serious injury. An additional danger involves the fact that some nitrogen from the air dissolves in the diver's blood when the air pressure is increased. Because the pressure suddenly decreases when the diver ascends quickly, the nitrogen gas may bubble out of the blood in the capillary blood vessels. Then the expansion of the bubble with decreased pressure may cause a block in the blood circulation or breakage of the blood vessels.

Boyle's Law: An ascending scuba diver must exhale!

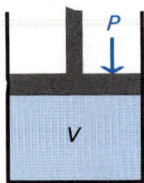

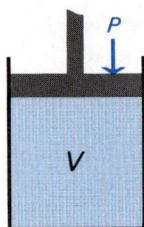

THE TEMPERATURE–VOLUME LAW FOR GASES

Our experiments to discover a law relating changes in the volume of a given amount of gas to changes in temperature must be done at constant pressure. The reasoning behind imposing the condition of constant pressure is the same that we used before. If we want to find how changes in temperature cause changes in volume, then everything else that might affect the volume must not be allowed to change. When we do such an experiment, we find that an increase in temperature causes an increase in volume. Or a decrease in temperature causes a decrease in volume. These results suggest what is called a *direct proportion*. An increase in one thing causes an increase in the other. For example, "the longer you stay awake, the sleepier you become" expresses a direct proportion.

When we arrange the data obtained by an experiment on a sample of gas in the form of a table, no neat numerical relationship is apparent (see Table 4.2). The volume decreases with tem-

P / V

P / V

Higher temperature

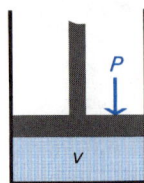

P / v

Lower temperature

perature, but the volume certainly is not zero when the temperature is zero degrees on our thermometer. Let us treat the data in another way to see if some relationship is more easily seen. Let us plot the data on a graph of volume versus temperature. When this is done, as in Figure 4.3, we find that we can draw a straight line through the data points. The straight line can be extended, or extrapolated, to the value zero on the volume axis.

Table 4.2   Experimental data for the volume and Celsius temperature of a gas sample at constant pressure

Volume (liter)	Temperature (°C)
10.0	0
13.6	100
17.3	200
21.0	300
24.6	400

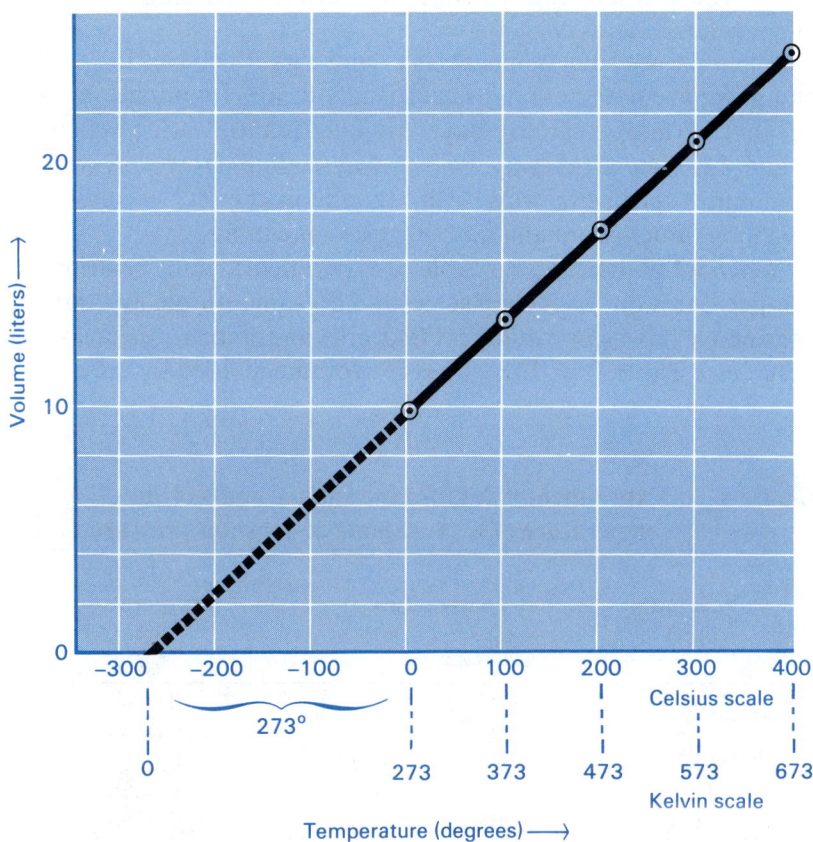

Figure 4.3   Volume–temperature data for a gas sample at constant pressure.

This procedure of extrapolating the line on our graph leads to an interesting conclusion. The volume of our gas sample would have a value of zero if the temperature had a value of $-273°$ on our Celsius thermometer. Of course, this is an imaginary situation. A gas, like any other sample of matter, can never have a zero volume. However, at higher temperatures, the gas behaves in such a way that we can predict this imaginary state of affairs. Consequently, we can say that our real gas is behaving as an *ideal gas* would. If we want to say that the volume of a gas sample is directly proportional to the temperature, we must use a thermometer that gives temperature on a scale that the gas "believes in." The horizontal axis of Figure 4.3 indicates such a scale. If we add the number 273 to every temperature reading on the Celsius scale, we have such a "gas behavior" scale. In this way, we can tie together our imaginary *ideal* gas with the behavior of *real* gases. This scale of temperature that can be used to predict correctly how gases behave is called the Kelvin scale. Lord Kelvin was a famous Scottish physicist who first proposed this scale in 1854. The temperature measured as zero on the Kelvin scale is often referred to as *absolute zero*.

When we use such language as saying a gas "believes in" something, we are not speaking scientifically at all. Inanimate matter (a gas) certainly cannot have human qualities. If a scientist argues his science this way, he is in real trouble. However, we can sometimes speak this way, with "tongue in cheek," to suggest a meaning, much as an analogy suggests a meaning.

When we go back to our table of experimental data, rewrite the temperatures on the Kelvin scale, and do some simple arithmetic, we find we have some numbers that give an interesting consistent result (see Table 4.3). The values for volume divided by the values

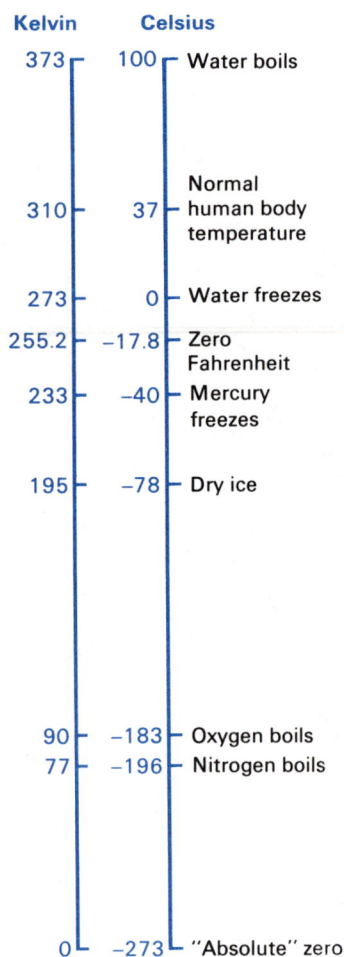

Kelvin	Celsius	
373	100	Water boils
310	37	Normal human body temperature
273	0	Water freezes
255.2	−17.8	Zero Fahrenheit
233	−40	Mercury freezes
195	−78	Dry ice
90	−183	Oxygen boils
77	−196	Nitrogen boils
0	−273	"Absolute" zero

Table 4.3   Experimental data for the volume and Kelvin temperature of a gas sample at constant pressure

Volume (liter)	Temperature		$\dfrac{\text{Volume}}{\text{Temperature}}\left(\dfrac{\text{liter}}{°K}\right)$
	(°C)	(°K)	
10.0	0	273	0.0366
13.6	100	373	0.0366
17.3	200	473	0.0366
21.0	300	573	0.0366
24.6	400	673	0.0366

of the corresponding Kelvin temperatures give a constant. (For example, 10.0 liter/273°K = 0.0336 liter/°K or 24.7 liter/673°K = 0.0336 liter/°K.) This is the law, the statement of constant behavior, that we have been looking for. It is called Charles' law, again named in honor of its discoverer, J. A. C. Charles, a contemporary of John Dalton. **Charles' law** is stated as follows:

> For a sample of gas at a constant pressure, *the volume is directly proportional to the Kelvin temperature.*

or

> For a sample of gas at constant pressure, *the volume divided by the Kelvin temperature is constant.*

A corollary to Charles' law is observed when a sample of gas is held at a constant volume and its temperature is changed. This is the generalization we can make from our experience with samples of gases heated or cooled in closed containers so that

---

The pressure of the air in an automobile tire increases if the tire becomes hot.

---

the volume is kept constant. For example, the pressure of the air in an automobile or truck tire increases if the tire becomes hot from the flexing of the rubber on a long trip. The corollary law is

> For a sample of gas at constant volume, *the pressure is directly proportional to the Kelvin temperature.*

or

> For a sample of gas at constant volume, *the pressure divided by the Kelvin temperature is constant.*

A practical illustration of the application of this law involves the warning on aerosol cans against tossing them into a fire. Suppose that a can is constructed so that it will hold together when the pressure of the gas inside is increased to 4 atmospheres. After the contents of the can are exhausted, the remaining propellant gas in a cool can may exert a pressure of 1.5 atmospheres. Will it explode in a fire where the temperature is over 600°C? We can answer by estimating how much the pressure will increase. Normal room temperature is about 27°C or 300° Kelvin (27 + 273 = 300). We can choose a convenient numerical value for the fire temperature of 900° Kelvin (627 + 273 = 900). So we see that, because the Kelvin temperature is three times greater in the fire, the pressure of the gas in the can will increase three times, to 4.5 atmospheres ($1.5 \times 3 = 4.5$). This pressure exceeds the strength of the can, so we can expect an explosion.

THE KINETIC MOLECULAR THEORY

These laws that describe how gases respond to changes in pressure and temperature, in addition to all the descriptive observations about gases, are the clues we can use to develop a theory about gases. The theory we are looking for has to tie together all the information about gases. We want to explain the behavior of gases by using the theory. What we do when we develop a theory is to build a mental model. By *mental* we mean a model based on ideas rather than on things we can see and handle. Such ideas are called *postulates*

Toward the end of the nineteenth century, scientists gradually developed a consistent theory about gases. This came to be known as the *kinetic molecular theory*. The word *kinetic* means "motion." Kinetic energy is the energy an object has when it is moving from one location to another. The kinetic molecular theory is concerned only with the kinetic energy possessed by gas molecules. (Molecules can have other forms of energy, such as vibrations within the molecules. They also can rotate on their own axis, like the earth.) The postulates of the kinetic molecular theory are as follows:

1. *Gases consist of molecules in ceaseless random motion.* The molecules move in all directions and never stop.
2. *The distances between molecules are very large compared to the size of the molecules themselves.* This statement means that the volume occupied by the molecules is a very small fraction of the volume of the container in which the gas is held. This situation is like having a few grains of sand in a bucket; the bucket still holds almost as much water as if the sand were not there.
3. *The molecules make perfectly elastic collisions with one another and with the container walls.* This statement means that the molecules do not lose any energy of motion as they bounce about. Billiard balls make elastic collisions as they bounce off one another. They exchange kinetic energy in the process. Eventually, they slow down because of friction with

the tabletop. Gas molecules, by contrast, keep on moving and making elastic collisions with one another.

4. If the molecules make perfectly elastic collisions, *they must have no attractive forces on one another.* There are no cohesive forces to make gas molecules stick to one another.

5. *The average kinetic energy of the molecules is proportional to the Kelvin temperature of the gas.* If heat is added to the gas, its temperature rises. The heat energy goes into increasing the kinetic energy of the gas molecules; it makes the molecules move faster.

The mental model we build, based on these postulates, is one for an *ideal gas.* The molecules have mass but no volume. The molecules have no attractive forces on one another, no cohesive forces pulling them together. The only energy the molecules have is kinetic energy that is proportional, on the average, to the Kelvin temperature of the gas. We must start with some misgivings about this model. We certainly know that gas molecules *do* have volume and that gas molecules must have cohesive forces that cause gases to condense into liquid form if the temperature is low enough. Nevertheless, we can imagine conditions under which real gases would behave as an ideal gas would. We have already seen that this is true for the temperature–volume relationship defined by Charles' law. We also can imagine that, if the molecules are far apart, as they are in a gas sample under low pressure (large volume), or are moving very rapidly, as they do at high temperatures, the attractive forces of one molecule for another

This would be like trying to grasp someone who is running fast just beyond your arm's reach.

will not be very effective. (This would be like trying to grasp someone who is running fast just beyond your arm's reach.)

What we have here is an illustration of the way scientists can build very useful models to explain *some* phenomena. In this case, the ideal gas that was invented by the kinetic molecular theory gives us a model with which we can explain the behavior of gases. The essential idea of molecules in motion serves that purpose very satisfactorily. In later chapters, which describe the general features of liquids and solids, we will see how the theory can be extended and modified to provide equally satisfactory explanations for the behavior of these condensed forms of matter.

THE KINETIC MOLECULAR THEORY
EXPLAINS THE PROPERTIES OF GASES

Let us now see how the behavior of gas molecules, as described by the kinetic molecular theory, explains the properties of gases and the gas laws. First, we can visualize how a gas exerts pressure. The molecules are constantly bombarding the walls of the container. An analogy is what happens if you use a piece of light cardboard as a lid for a pan in which you are making popcorn.

What happens if you use a piece of light cardboard as a lid for a pan in which you are making popcorn?

As the kernels pop and strike the lid, the collisions create a pressure (a force on the area of the lid) that cause the cardboard to fly off the pan.

Boyle's law can be explained by recognizing that, if the same number of molecules are crowded into less space, they make more collisions with the container walls. So a decrease in volume is related to an increase in pressure. The idea that molecules of a gas have no attractions for one another also is needed to account for Boyle's law. If the molecules acted like little magnets, so that they stuck to one another when they collided, the molecules would begin to clump together. There would be fewer bouncing around free, so that the walls of the container would be hit less often. The pressure would decrease. Because the pressure does remain constant when the volume is constant, the gas molecules must be assumed to have no appreciable attractive forces toward one another.

The ability of gases to diffuse easily through one another also can be explained. Molecules can easily find the great spaces between other molecules as they move about. An analogy is the way you can move rapidly through a sparsely populated room by running between and around other people. In a similar manner, the molecules of a perfume move between molecules of air, so that the odor of the perfume can be detected at a distance from the wearer.

The relationships between temperature and volume or pressure of a gas also can be understood in terms of the model. Molecules move faster as their temperature rises. Consequently, they make more collisions with one another and with the container walls in a given time. This increase in the number of collisions increases

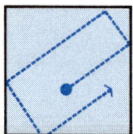

In a smaller volume, each molecule makes more collisions with the walls in a given time.

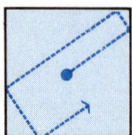

A faster moving molecule makes more collisions with the walls in a given time.

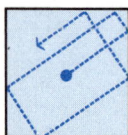

the pressure on the walls, if the volume is held constant. Or the increase in pressure pushes the container walls back against a constant external pressure, so that the volume increases.

AVOGADRO'S THEORY

One additional piece of evidence needs to be considered before we can have a complete model for the gaseous form of matter. This evidence is not apparent in the ordinary experience of daily living, but it jumps at us as soon as we begin to do any laboratory experiments with gases. For example, we can take a volume of oxygen gas at a certain temperature and measure its pressure. Then we can take the same volume of any other gas at the same temperature and measure its pressure. The two pressures will be the same. According to the model proposed by the kinetic molecular theory, the equality of the pressures means that the two

The equality of the pressures means that the two samples must contain the same number of molecules, making the same number of collisions with the walls in a given time.

samples must contain the same number of molecules, making the same number of collisions with the walls in a given time. A nice logical conclusion!

This idea that *equal volumes of gases under the same conditions of temperature and pressure contain equal numbers of molecules* was first proposed by the Italian chemist Amadeo Avogadro, shortly after Dalton proposed his atomic theory at the beginning of the nineteenth century. We can properly say that Avogadro "invented" the idea of molecules, just as Dalton "invented" the idea of atoms. Avogadro was the first clearly to define the molecule as a combination of atoms. He developed his idea to explain the way in which some gases react chemically with each other. Although we will not go into the details of how he came to his conclusion, we should note that his was a brilliant idea. It certainly started the young science of chemistry on the right track. The fact that the later, more elaborate, and complete kinetic molecular theory provides such a logical explanation for Avogadro's guess about molecules in gases is a compliment to his greatness as a scientist. Accordingly, he is honored by the theory being given his name.

MEASURING MOLECULAR MASSES
OF GAS MOLECULES

Combining the ideas of Avogadro's theory with those of the kinetic molecular theory makes it possible to tie together two ways of measuring the amount of gas in a sample. Suppose that we take 1.0 liter of oxygen (O_2) gas in one bottle and 1.0 liter of another gas (the same volume) in another bottle. Both samples are at the same pressure and temperature. Avogadro's theory lets us assume that the same number of gas molecules is in each bottle, because the bottles are the same volume. This measurement expresses the "amount" of gas in terms of numbers of molecules.

We can weigh each of the bottles empty and then full of gas and take the difference in each case. In this way, we can measure the mass of each gas in the bottles. This measure is our familiar one for the "amount" of any kind of matter. The 1.0 liter of O_2 gas weighs 1.30 grams; the 1.0 liter of the unknown gas weighs 1.80 grams. (Let us represent the unknown gas by the imaginary formula Z.) Now let us use the laboratory data and see what conclusions we can draw from the numbers.

We do not know how much one O_2 molecule weighs, but we do know that each one has the same mass. This idea is the logical consequence of Dalton's idea that each atom of an element has the same mass. So identical molecules, composed of the same number and kind of atoms, will have the same mass. We give this mass the symbol M_{O_2}. Likewise, we do not know what number of molecules are in the bottle, so we give that number the symbol n. Then we can say

$$\text{Total mass of } O_2 \text{ molecules } = nM_{O_2} = 1.30 \text{ gram.}$$

In similar manner, we can express the information about the unknown gas Z as follows:

$$\text{Total mass of Z molecules } = nM_Z = 1.80 \text{ gram.}$$

Now divide the total mass of Z by the total mass of O_2.

$$\frac{\text{total mass of Z molecules}}{\text{total mass of } O_2 \text{ molecules}} = \frac{\cancel{n}M_Z}{\cancel{n}M_{O_2}} = \frac{1.80 \text{ \cancel{gram}}}{1.30 \text{ \cancel{gram}}} = 1.38$$

We can cancel the n's because Avogadro's theory says they are the same number. We can cancel the labels for grams because they are the same thing. So now we have the conclusion that the ratio of the mass of one Z molecule (M_Z) to the mass of one O_2

molecule (M_{O_2}) is 1.38.

$$\frac{\text{mass of } \textit{one} \text{ Z molecule}}{\text{mass of } \textit{one} \text{ O}_2 \text{ molecule}} = \frac{M_Z}{M_{O_2}} = 1.38$$

In Chapter 2 we introduced the idea that formulas are the combination of symbols. Each symbol stands for an atom that has a characteristic relative atomic mass. (For example, the atomic mass of oxygen is 16 times as much as the atomic mass of hydrogen.) So now we can combine the idea of formulas with the idea of atomic masses to calculate *molecular masses*. These molecular mass numbers are the *relative* values for the mass of one molecule compared to the mass of another molecule. The atomic mass of the O atom is 16; this gives O_2 a molecular mass of $2 \times 16 = 32$. The numbers in our data can now be used to calculate the molecular mass of Z molecules. Because one Z molecule is 1.38 times as heavy as one O_2 molecule, the molecular mass of one Z molecule is simply $1.38 \times 32 = 44$.

This method is the one that chemists use to find the molecular mass of any gas. Additional information is needed to determine a gas's formula. For example, additional information might tell us that Z gas is composed only of carbon and oxygen. We already know that carbon forms two oxides, carbon monoxide (CO) and carbon dioxide (CO_2). So then, if we add up the formula masses for each of these possibilities,

$$\underset{\substack{\text{one atom} \quad \text{atomic} \\ \text{of C} \quad \text{mass of C}}}{\text{Molecular mass}} = \underset{\substack{\text{one atom} \quad \text{atomic} \\ \text{of C} \quad \text{mass of C}}}{(1 \times 12)} + \underset{\substack{\text{one atom} \quad \text{atomic} \\ \text{of O} \quad \text{mass of O}}}{(1 \times 16)} = 28$$

$$\text{Molecular mass} = (1 \times 12) + (2 \times 16) = 44$$
$$\text{of } CO_2$$

The second number, 44, agrees with the number that was calculated from the data. Consequently, we can conclude that Z gas must be carbon dioxide, CO_2.

This kind of calculation is another example of what we described as chemists' observing *macro* (large) amounts of substances but interpreting on a *micro* (very small) level. An analogy may help you to recognize this method. Suppose we have a large bus in which each of the many seats is occupied by a 200-pound football player. We can weigh the bus empty and loaded and get the weight of all the football players. Then we can put kindergarten children, each of the same weight, in every seat of the bus. Again, the weight of the bus full minus the weight of the bus empty will give us the weight of all the children. The football players' total weight is 5 tons. The children's, 1 ton. We have the same number (*n*) of children and football players.

$$\frac{\text{mass of } n \text{ children}}{\text{mass of } n \text{ football players}} = \frac{1 \text{ ton}}{5 \text{ ton}}$$

$$\frac{\text{mass of } one \text{ child}}{\text{mass of } one \text{ football player}} = \frac{1}{5}$$

So because we know that a child weighs only ⅕ as much as a 200-pound football player, the mass of one child is ⅕ × 200 pounds = 40 pounds.

Nowhere in solving this problem do we need to know how many seats are in the bus. However, we must use the same bus for both weighings and make sure that every seat is filled. We cannot compare the weights of the occupants of two different buses of different seating capacity. Rather, we always use a "standard bus." Then we, or anyone else, can compare weights of the full bus with the value we got the one time we filled it with football players of known weight.

The constant number of bus seats in the analogy represents the number of molecules in each of the two bottles of gas in our laboratory experiment. The number of gas molecules in two bottles of equal volume is the same only if the temperature and pressure are the same. If these conditions change, the number of molecules in a given volume of a gas changes. The choice of some "standard bus" in the analogy is matched by chemists' choice of a volume chosen at *standard conditions* of pressure and temperature. Once these are established, then we can be sure that the masses of equal volumes of various gases will be the masses of equal numbers of molecules of those gases. The ratio of these two quantities also is the ratio of the molecular masses of the two gases.

STANDARD CONDITIONS (STP)

The set of conditions that has been adopted as the *standard temperature and pressure (STP) is a pressure of 1.0 atmosphere (760 torr) and a temperature of 0°C (273°K).* These conditions are chosen because they are conveniently set up in a laboratory. The temperature of a mixture of solid ice and liquid water is 0°C. (The melting point of ice is 0°C.) The average pressure at sea level is 1 atmosphere.

If we choose to work with a gas sample at any other set of pressure and temperature conditions, we can use a combination of Boyle's and Charles' laws to calculate the volume that our sample would have if it were measured at STP.

THE MOLE—THE CHEMIST'S UNIT OF AMOUNT

We have seen that one characteristic of the way scientists work is that they choose arbitrarily assigned standards. This is the same as establishing a vocabulary for a language. If we agree to use the atomic mass scale, then we accept the value 32 for the molecular mass of the O_2 molecule. For example, if we find that a molecule weighs half as much as an oxygen molecule, we can say that its molecular mass is $32 \times \frac{1}{2} = 16$. Likewise, we agree that the set of standard conditions is 0°C and 1.0 atmosphere. Thus when we make measurements, we can describe exactly what we have done to other scientists, because we all "talk the same language."

"Talking the same language" allows us to repeat one another's experience, to compare results, and to draw the same conclusions.

Moreover, this "talking the same language" allows us or anyone else to repeat one another's experience, to compare results, and to draw the same conclusions.

One other important agreement has been made by all chemists. *An amount of substance that contains the same number of molecules has been defined. The unit of this amount is the* mole. The word *mole* comes from a Latin word that means "heap," or "pile." The word *molecule* is derived from this and means "a little heap or pile" (in the same sense that the word *cigarette* implies "a little cigar"). We will explore how the mole idea developed and then something of the way chemists use the idea.

We derive the definition of the mole by applying agreed-upon arbitrary standards to measurements of pure substances as we find them in the natural world. Let us illustrate: Suppose we take a sample of oxygen gas that weighs exactly 32 grams. This choice involves two arbitrary standards, the definition of a gram as the unit of mass and the choice of 32, the assigned value for the molecular mass of oxygen based on the atomic mass scale. Then, we measure the volume of this amount of O_2 gas at STP. The value we get is 22.4 liters. Then, if we take a bottle that has a volume of

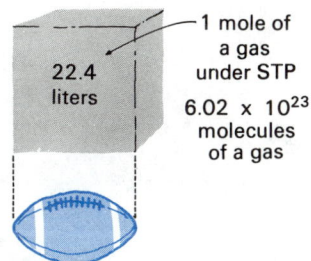

22.4 liters

1 mole of a gas under STP

6.02×10^{23} molecules of a gas

22.4 liters and fill it with any other gas at STP, we know we will have as many molecules of the other gas as we had in the 22.4 liters (STP) of O_2 gas. However, this number of molecules of another gas, for example, natural gas, will not weigh 32 grams. Natural gas is the compound called methane; its formula is CH_4. The molecules of CH_4 do not have the same molecular mass as O_2 molecules do. The molecular mass of CH_4 is 16, so we find that 22.4 liters (STP) of CH_4 weighs 16 grams. We can summarize several examples as follows:

32 grams of O_2			1 mole of O_2
28 grams of N_2			1 mole of N_2
16 grams of CH_4	occupy 22.4 liters (STP)	contain the same number of molecules	1 mole of CH_4
44 grams of CO_2			1 mole of CO_2
168 grams of an unknown gas whose molecular mass is 168			1 mole of an unknown gas whose molecular mass is 168

The definition of a mole is

$$1 \text{ mole } = \begin{cases} \text{a number of grams of a substance equal to its molecular mass} \\ \text{or 22.4 liters (STP) of a gaseous substance} \\ \text{or a certain number of molecules} \end{cases}$$

AVOGADRO'S NUMBER

We have one very important experiment yet to do. The measurements on masses and volumes (STP) of gases have not told us how many molecules are in that 22.4-liter flask. Rather than merely talking about "the same number" of molecules, we want to know what that number is. Experiments that involve the counting of individual molecules are not easy to do. However, the determination of this very important quantity in nature has been accomplished in various ways. The number of molecules in a mole is now known with very high precision. It is 602,204,500,-000,000,000,000,000 or, written more simply, 6.02×10^{23}. This number is referred to as *Avogadro's number*

Exponential Notation

Exponential notation is an efficient and convenient way of writing numbers, especially those that are very large or very small. This system expresses a number as the product of two factors (two parts that are multiplied by each other to give the value of the number). One of these factors we shall call the *numerical factor*; the other is the *exponential factor*. Thus, the number 472 can be written as follows:

$$472 = 4.72 \quad \times \quad \boxed{100}$$
$$472 = 4.72 \quad \times \quad \boxed{10^2}$$

$\}-$ (100 is the same as 10^2)

$\underbrace{}$ numerical factor $\qquad$ $\underbrace{}$ exponential factor

When we write numbers whose value is less than unity, we follow the same procedure:

$$0.0472 = 4.72 \quad \times \quad \boxed{0.0100}$$

$\left(0.0100 \text{ is the same as } \dfrac{1}{100} \right)$

$$0.0472 = 4.72 \quad \times \quad \boxed{\dfrac{1}{100}}$$

$\left(\dfrac{1}{100} \text{ is the same as } \dfrac{1}{10^2} \right)$

$$0.0472 = 4.72 \quad \times \quad \boxed{\dfrac{1}{10^2}}$$

$\left(\dfrac{1}{10^2} \text{ is the same as } 10^{-2} \right)$

$$0.0472 = 4.72 \quad \times \quad \boxed{10^{-2}}$$

$\underbrace{}$ numerical factor $\qquad$ $\underbrace{\phantom{10^{-2}}}$ exponential factor

The exponential factor involves the number 10 raised to the appropriate power. Thus, 10^2 means 10×10 or 100; 10^{-2} means $\dfrac{1}{10} \times \dfrac{1}{10}$ or $\dfrac{1}{100}$.

Exponential Notation

The following list suggests the general relationships between multiples of 10 written as conventional numbers and written in exponential notation.

$$1000 = 1.0 \times 10 \times 10 \times 10 = 1.0 \times 10^3$$

$$100 = 1.0 \times 10 \times 10 = 1.0 \times 10^2$$

$$10 = 1.0 \times 10 = 1.0 \times 10^1$$

$$1.0 = 1.0 \times \frac{10}{10} = 1.0 \times 10^0$$

$$0.10 = \frac{1.0}{10} = \frac{1.0}{10^1} = 1.0 \times 10^{-1}$$

$$0.010 = \frac{1.0}{10 \times 10} = \frac{1.0}{10^2} = 1.0 \times 10^{-2}$$

$$0.0010 = \frac{1.0}{10 \times 10 \times 10} = \frac{1.0}{10^3} = 1.0 \times 10^{-3}$$

The power of 10 indicates where the decimal point is located when writing the number in the conventional fashion. Thus, 4.0×10^5 means 4 followed by five zeros to the left of the decimal point, or 400,000. The negative value of the exponent locates the decimal point for numbers smaller than unity. Thus, the number 4.0×10^{-5} means 4 located in the fifth position to the right of the decimal point, or 0.000040.

In the case of Avogadro's number, you can count up 23 numbers, including zeros, following the first number, 6.

$$6.022045 \times 10^{23} = 602,204,500,000,000,000,000,000$$

A less precise value for this number is 6.02×10^{23}, read as "six-point-zero-two times ten to the twenty-third power."

You should not be surprised to find such a special number with such great importance in your introduction to chemistry. We use many special numbers for counting things in our everyday lives. We count eggs by the dozen (12), years by centuries (100), or pencils by the gross (144). These units for counting are accepted and are convenient. Avogadro's number is often referred to as the "chemist's dozen."

Avogadro's number, the number of items in a mole, 6.02×10^{23}, is a number that is a convenient unit for counting such tiny items as atoms or molecules. You should always remember that

---

Avogadro's number is often referred to as the "chemist's dozen."

---

the mole is the same number of "things." The mass of a mole of these "things" depends on what they are. For example, a mole of O_2 weighs 32 grams; a mole of CO_2 weighs 44 grams. You would find, if you compared the weight of a dozen eggs with the weight of a dozen oranges, that you would likewise have different numbers for the mass of the same number of different things, because one orange weighs more than one egg.

The numbers 6.02×10^{23} or 22.4 liters may appear to be peculiar numbers to have such importance in chemistry. The reason is that, when humans decided on an arbitrary scale that gives the number 32 for the value of the molecular mass of O_2 and defined the gram as the unit of mass, we had to accept what nature gives us when we calculate with these standards. Scientists study the natural world. Consequently, the natural world "has the last word," even in the language of scientists.

THE MOLE CONCEPT FOR GASES—A SUMMARY

A summary of the way in which the concept of the *mole* ties our information together can be outlined as follows:

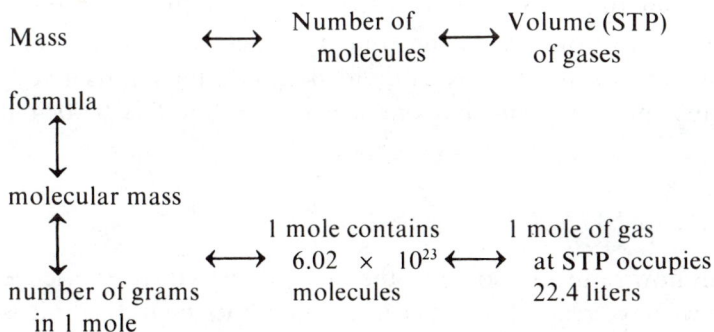

Mass $\longleftrightarrow$ Number of molecules $\longleftrightarrow$ Volume (STP) of gases

formula
$\updownarrow$
molecular mass
$\updownarrow$
number of grams in 1 mole $\longleftrightarrow$ 1 mole contains 6.02×10^{23} molecules $\longleftrightarrow$ 1 mole of gas at STP occupies 22.4 liters

The double-headed arrows in this summary diagram are used to suggest that we can use an item on one side of the arrow to get the item on the other side.

HOW SMALL IS A MOLECULE?

We now have the information to calculate the mass of individual molecules. This is not a very practical problem, but it is interesting to realize what tiny items they are and yet how precisely the values can be determined. For example, we can calculate the mass of one single molecule of ammonia gas as follows: The formula of ammonia is NH_3, so the mass of 1 mole is precisely known as $(1 \times 14.007) + (3 \times 1.008) = 17.031$ grams.

$$\text{Mass of one molecule} = \frac{\text{mass of 1 mole}}{\text{Avogadro's number}}$$

$$= \frac{17.03 \frac{\text{gram}}{\text{mole}}}{6.022 \times 10^{23} \frac{\text{molecule}}{\text{mole}}}$$

$$= 2.827 \times 10^{-23} \frac{\text{gram}}{\text{molecule}}$$

This number can be written 0.000,000,000,000,000,000,000,-02827.

It is important to know such small quantities with high precision. Modern instruments and methods of doing experiments allow us to know indirectly but surely many such vanishingly small values in the micro-world of atoms and molecules.

An analogy may help you to recognize how this kind of calculation is not peculiar or special to science. Suppose you were asked to measure the thickness of one leaf of this book with a ruler graduated in sixteenths of an inch. Impossible? No! You measure the thickness of all the leaves between the covers. Then you count the number of leaves by taking half the number of pages. Then you divide the total thickness by the number of leaves. Thus your crude ruler divided into $\frac{1}{16}$-inch units gives you a value that tells you that one leaf of the book is 0.0034 inch thick.

THE DENSITY OF GASES—BUOYANCY

Let us now return to some of the facts of everyday experience or facts we have read about. Let us see how the theoretical ideas of the kinetic molecular theory and the mole concept allow us to explain or tie these facts together. For example, why does a balloon filled with helium gas float upward while a balloon blown up with your breath falls to the floor? If you put a piece of dry ice (solid CO_2) into a glass of water, you see the vapors roll over the

Multiplying and Dividing Exponential Numbers

The operations of multiplying and dividing exponential numbers follow simple consistent rules.

MULTIPLICATION

Multiply the numerical factors to get the product numerical factor. Add the exponents of 10 in the exponential factors to get the correct power of 10 for the product exponential factor.

EXAMPLES

1. $4.2 \times 10^2 \times 2.0 \times 10^3 = 4.2 \times 2.0 \times 10^{2+3}$
$$= 8.4 \times 10^5$$

2. $3.3 \times 10^2 \times 3.0 \times 10^{-3} = 3.3 \times 3.0 \times 10^{2+(-3)}$
$$= 9.9 \times 10^{-1}$$

3. There are 2000 pounds in 1 ton; 1 pound is 0.454 kilogram. How many kilograms are equal to 1 ton?

$$2.00 \times 10^3 \frac{\text{pound}}{\text{ton}} \times 4.54 \times 10^{-1} \frac{\text{kilogram}}{\text{pound}}$$

$$= 2.00 \times 4.54 \times 10^{3-1} \frac{\text{kilogram}}{\text{ton}}$$

$$= 9.08 \times 10^2 \frac{\text{kilogram}}{\text{ton}}$$

DIVISION

Divide the numerical factors to get the answer numerical factor. Subtract the exponent of 10 in the exponential factor of the denominator from that in the exponential factor of the numerator to get the correct power of 10 for the answer exponential factor.

EXAMPLES

1. $\dfrac{4.2 \times 10^2}{2.0 \times 10^3} = \dfrac{4.2}{2.0} \times 10^{2-3}$
$$= 2.1 \times 10^{-1}$$

2. $\dfrac{3.3 \times 10^2}{3.0 \times 10^{-3}} = \dfrac{3.3}{3.0} \times 10^{2-(-3)}$
$$= 1.1 \times 10^5$$

3. The sun is 93 million miles from the earth. The speed of light is 186,000 miles per second. How long does it take sunlight to reach the earth?

$$\frac{9.3 \times 10^7 \, \text{mile}}{1.86 \times 10^5 \, \frac{\text{mile}}{\text{second}}} = \frac{9.3}{1.86} \times 10^{7-5} \, \text{second}$$

$$= 5.0 \times 10^2 \, \text{second}$$

edge of the glass onto the table rather than float up. Why does this happen?

In Chapter 2 we introduced the idea of the *density* of a sample of matter with the definition

$$\text{Density} = \frac{\text{mass}}{\text{volume}}$$

We now can define the mass of a sample of matter as

$$\text{Mass} = \text{number of molecules} \times \text{mass of one molecule}$$

or

$$\text{Mass} = \text{number of moles} \times \text{mass of one mole}$$

Furthermore, we now can say that, as long as we are dealing with gases, the same number of moles (or molecules) of *any* gas will occupy the same volume. One mole occupies 22.4 liters (STP).

These relationships allow us to calculate the density of any gas at STP if we know the formula of its molecules. For example, helium consists of molecules represented by the formula He. So one mole of He has a mass of 4.0 grams.

$$\text{Density of helium (He)} = \frac{4.0 \frac{\text{gram}}{\text{mole}}}{22.4 \frac{\text{liter}}{\text{mole}}} = 0.179 \frac{\text{gram}}{\text{liter}}$$

Air is a mixture of molecules, so we have to use both the molecular formulas and the composition to calculate an average value for the mass of a mole of air (22.4 liters STP). An approximate value is obtained by considering the composition of air simply as four molecules of N_2 for every one of O_2.

$$
\begin{array}{llll}
4 \text{ mole of } N_2 & = & 4 \times 28.0 \text{ gram} & = & 112.0 \text{ gram} \\
1 \text{ mole of } O_2 & = & 1 \times 32.0 \text{ gram} & = & 32.0 \text{ gram} \\
\hline
5 \text{ mole of air} & & & & 144.0 \text{ gram}
\end{array}
$$

$$1 \text{ mole of air} = \frac{144.0 \text{ gram}}{5 \text{ mole}} = 28.8 \frac{\text{gram}}{\text{mole}} \text{ (average)}$$

$$\text{Density of air (at STP)} = \frac{28.8 \frac{\text{gram}}{\text{mole}}}{22.4 \frac{\text{liter}}{\text{mole}}} = 1.29 \frac{\text{gram}}{\text{liter}}$$

A similar calculation gives us the density of CO_2.

$$\text{Density of } CO_2 \text{ (at STP)} = \frac{44.0 \; \frac{\text{gram}}{\text{mole}}}{22.4 \; \frac{\text{liter}}{\text{mole}}} = 1.96 \; \frac{\text{gram}}{\text{liter}}$$

A comparison of these density values gives us an answer to the question about why the helium-filled balloon rises. It floats in air just as a log or boat floats on water. This example illustrates the buoyancy principle. The boat sinks into the water until the mass of the water that it displaces is equal to the mass of the boat and its load. The helium-filled balloon displaces a volume of air that is much heavier than the balloon plus the helium in it. If we had a 22.4-liter balloon filled with helium, it could support a total mass of

| 28.8 gram | – | 4.0 gram | = | 24.8 gram |
| mass of air that is displaced | | mass of helium | | total mass that could be supported |

If the mass of the balloon itself, plus any other objects hanging to it, were 24.8 grams, the balloon would neither rise nor fall. This is what happens in a helium-filled blimp. The difference in mass between the helium gas and the displaced air gas is the mass of the blimp plus the load it can carry.

We can also understand why the gas from dry ice (CO_2) rolls over the lip of the glass and *down* onto the table. The solid dry ice is changed into CO_2 gas. Considerable water in tiny yet visible droplets is carried along with the bubbles of gas. Consequently, you see a "smoke," but CO_2 itself is invisible, as air is. The density of CO_2 is 1.96 gram/liter (STP), whereas the density of air is 1.29 gram/liter (STP). The heavier gas, CO_2, falls down. The same reasoning explains why the balloon you blow up with your breath cannot rise. You fill it with a mixture of air and a heavier gas, CO_2. (Droplets of liquid water from your breath also add to the mass of the filled balloon.)

Hydrogen gas (H_2), molecular mass 2.0, is the only other common gas, besides helium, with a density small enough that it can be used for its buoyancy effect. In the 1930's, huge lighter-than-air craft, built chiefly in Germany, were inflated with hydrogen. These dirigibles made numerous trans-Atlantic trips. However, hydrogen is extremely combustible, and the danger of its igniting was made tragically evident when the *Hindenburg*, one of the German dirigibles, caught on fire during a landing in New Jersey in 1937.

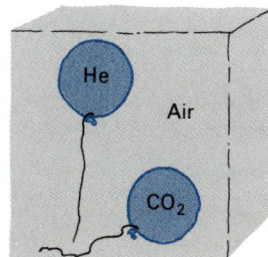

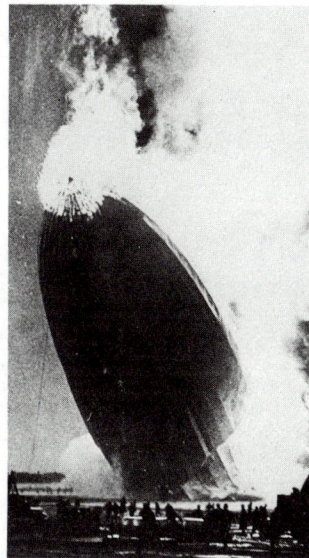

$$2H_2 + O_2 \rightarrow 2H_2O$$

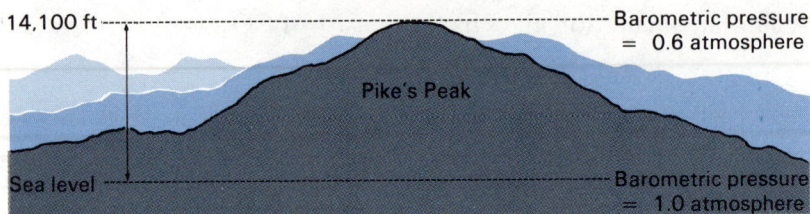

14,100 ft --- Barometric pressure = 0.6 atmosphere

Pike's Peak

Sea level --- Barometric pressure = 1.0 atmosphere

THE ATMOSPHERE, WINDS, AND LIGHTNING

We have seen how the idea that molecules of gases are in ceaseless motion accounts for their filling a container, such as a tank or a room, evenly all over. The same reason is responsible for the gradual thinning of our atmosphere rather than a definite upper limit. Gas molecules diffuse upward against the pull of gravity. But the net effect is a gradual reduction in the number of air molecules in a given volume at greater and greater distances above the surface of the earth. The atmosphere becomes less dense. "Empty" space probably begins almost 1000 miles above the earth.

Near the surface of the earth, the movement of air molecules often appears to be far more organized than the idea of random diffusion would suggest. Winds are caused by the concerted movement of air molecules from regions of high pressure to regions of low pressure. The balmy breeze or the furious hurricane can be likened to the air escaping from a punctured tire. But it can also be said that winds create changes in pressure by moving air molecules around. This chicken-and-egg kind of argument can be avoided if we look for other factors influencing the number of gas molecules in a given space.

When gas molecules absorb heat energy, they move faster, on the average. This effect appears in our measurements as an increase in temperature. The volume occupied by a given number of gas molecules is greater if the temperature is greater. This fact means that the density of a hot gas is less than the density of a cool gas under the same pressure. A large balloon filled with heated air can be buoyant enough to carry passengers. The volume of hot air inside the balloon contains fewer molecules than does an equal volume of cooler air outside.

The same set of conditions arises in regions of the earth's atmosphere. At the equator, the earth and atmosphere receive more heat from the sun's radiation than they do at the poles.

Consequently, the equatorial atmosphere is warmer and has a lower density than the temperate or polar atmospheres do. The consequence of this situation is that the equatorial air tends to rise as it heats up and the temperate or polar air tends to flow over the surface toward the equator. Currents of air such as these are called convection currents. You can see a demonstration of

You can see a demonstration of convection currents in the upward curl of the smoke from a burning cigarette in an ash tray.

convection currents in the upward curl of the smoke from a burning cigarette in an ashtray. The rotation of the earth causes these north–south currents in the northern hemisphere to be twisted toward the east in a complicated but prevailing pattern. And a great complexity of air currents is caused by greater or lesser amounts of heat being given off in different regions. For example, the air is heated more over a desert than over the ocean.

The temperature of the atmosphere generally drops with altitude up to a distance of about 6 miles. This temperature difference near the surface has an important consequence, especially within the first few thousand feet. When the air near the surface is warmed, and becomes less dense, it tends to rise. The cooler, more dense air above then moves downward to replace it.

These vertical convection currents and horizontal winds are very important in carrying away any pollutants that may be added to the atmosphere of a particular region. There are times, however, when cool air moves in close to the surface and stays cool under a layer of warmer air. Such a circumstance is called a *temperature inversion* (cooler air below warmer air instead of the normal arrangement of cooler air above warmer air). Then the difference in densities of the two layers keeps them in place and no convection occurs. The condition may be made worse by the presence of nearby mountains that keep out any horizontal

Temperature inversion

winds. When a temperature inversion occurs, the pollutants in the atmosphere, instead of being blown away or being thinned out by rising, collect in larger and larger amounts. If other conditions, such as abundant sunlight, exist, the consequence is a very bad smog condition. We will examine some of the chemical reactions involving smog-producing pollutants in a later chapter. We can recognize at this point an illustration of the effect of a concentration of reactants on the speed of a chemical reaction. When an inversion layer allows pollutant concentrations to build up, reactions involving them are sure to be speeded up.

Another familiar atmospheric phenomenon that can be explained in terms of the kinetic molecular theory is the loud noise of thunder produced by a lightning flash. The sudden discharge of electric energy heats a narrow column of air to temperatures as high as 10,000° C in about 1 microsecond (one-millionth of a second). This heat so rapidly increases the motion of the molecules that the expansion of the air is an explosion. Other air rushes in to replace the air that has moved out. The thunder sound we hear is the alternate compressing and decompressing of the gases of the atmosphere. The shock is more spread out the farther away we are, so that the nearby crack becomes a long rumble at a distance.

Summary

The gaseous form of matter is characterized by compressing readily, diffusing easily, and having relatively low density. Gases exert uniform pressure on all the walls of any container in which they are held. The quantitative relationships among the properties of pressure, volume, and temperature of a gas sample can be described by laws. Boyle's law states that, at constant temperature, the pressure and volume of a gas sample are inversely proportional. Charles' law states that, at constant pressure, the volume of a gas sample is directly proportional to the Kelvin temperature. The Kelvin temperature is measured as the Celsius temperature plus 273.

The kinetic molecular theory explains the properties and quantitative laws of gases. This theory postulates that gases are made up of molecules in ceaseless random motion. The average kinetic energy of gas molecules is proportional to the Kelvin temperature. The theory describes gas molecules as having no volume and no cohesive attraction forces among them. Thus the model that the theory suggests is one for an ideal gas.

Equal volumes of gases at the same temperature and pressure contain equal numbers of molecules. This idea, originally proposed by Avogadro, leads to the idea of using a definite number of molecules as a unit of measure for amounts of substances. This unit, called the *mole*, is 6.02×10^{23} items. This number, 6.02×10^{23}, is called Avogadro's number and serves as a "chemist's dozen."

Glossary

The number in parentheses indicates the text page where you can find the term defined in context.

absolute zero zero degree on the Kelvin ("gas behavior") temperature scale, the temperature at which the volume of an ideal gas would be theoretically zero, and there is no molecular movement (104)

aerosol a dispersion in a gas of many very small particles (but not individual molecules) of a solid or a liquid (96)

atmosphere (unit) the average barometric pressure at sea level, equal to 760 mm (29.9 inches) of mercury (99)

Avogadro's number the number of particles in a mole, 6.02×10^{23} (114)

Avogadro's theory equal volumes of gases at the same temperature and pressure contain equal numbers of molecules (109)

barometer a device that measures the pressure of the atmosphere. A mercury barometer is made from a sealed glass tube filled with mercury, inverted in a mercury reservoir. (98)

Boyle's law for a sample of gas at constant temperature, the volume is inversely proportional to the pressure; the product of the volume and the pressure is therefore constant (101)

Charles' law for a sample of a gas at constant pressure, the volume is directly proportional to the Kelvin temperature; the ratio of the volume to the Kelvin temperature is therefore constant (105)

compression the process of making a substance or object smaller by applying force (96)

diffusion the mixing or spreading of a sample of one type of molecule through another. It is common in gases, but also occurs in liquids and solids. (96)

direct proportion   the relation of two quantities in which one increases if the other increases, and vice versa (102)

ideal gas   an imaginary, "perfect" gas whose molecules have no volume and no attractive forces for one another (104)

inverse proportion   the relation of two quantities by which one decreases as the other increases, and vice versa (100)

kinetic molecular theory   a theory explaining the consistent behavior of gases: The molecules of gases are far apart, are in ceaseless random motion, and have an average kinetic energy proportional to the Kelvin temperature. (106)

mole   an amount of a substance that contains Avogadro's number (6.02×10^{23}) of chemical units. One mole of a gas occupies a volume of 22.4 liters at STP. (113)

postulate   an idea or "rule of the game" that forms the basis of a mental model or theory (106)

pressure   a measure of the force exerted on a given area, expressed as the ratio of force to area (97)

standard conditions   a set of conditions (0°C or 273° K, 1.0 atmosphere or 760 torr) chosen as a uniform, convenient standard of laboratory measurement (112)

temperature inversion   an atmospheric condition in which a layer of warm air lies over a layer of cool air, creating no convection and trapping any pollutants that are present (123)

torr   a unit of barometric pressure, equivalent to that produced by a column of mercury 1.0 millimeter high (99)

vacuum   a space in which there is no matter. A complete or perfect vacuum is physically impossible. (97)

Exercises

4.1   What property of gases makes it possible for you to smell coffee being prepared in the kitchen while you are still in bed?

4.2   Explain the relationship between force and pressure. Engineers often use units of pounds per square inch (psi). Are these units of force or of pressure?

4.3   A generation ago, fashion suggested that women wear shoes with high "spike" heels. When this happened, some types of floor coverings showed severe wear in the form of "pitting" or many small dents in places where women wearing spike heels frequently walked. Explain why a 120-pound woman wearing spike heels causes more dents in a linoleum floor than does a 250-pound man waring size 12 shoes.

4.4   The atmospheric pressure over the English Channel measures 760 millimeters of mercury on a French barometer. What is the reading of the same 1.0 atmosphere of pressure on an English barometer graduated to read inches of mercury?

4.5   Mercury is 13.6 times as heavy as water. Suppose you tried to make a barometer filled with water instead of mercury. How long would the tube have to be to record 1.0 atmosphere pressure?

4.6   Suppose you are in charge of the decorations for a party. You decide to have helium-filled ballons, each of which can hold 2 liters of helium at 1 atmosphere pressure. How many will you be able to fill from a 20-liter cylinder holding helium at 50 atmospheres pressure?

4.7   In each of the following cases, suppose that all other variables are held constant. Fill in the blanks.

   a. If the pressure increases, the volume _____.

   b. If some gas escapes (the number of molecules decreases), the volume _____.

   c. If the temperature increases, the pressure _____.

4.8   In each of the following cases, is the relationship directly or inversely proportional?

   a. As the temperature rises, you perspire more.

   b. A good sprinter runs 100 meters in 10 seconds. A good distance runner covers 1500 meters in less than 4 minutes. How is the speed of the runner related to the length of the race?

   c. As more nutrients are added to a lake, more algae grow in the water.

   d. The greater the enrollment in a college, the more teachers must be hired.

   e. The more vegetables you grow in your garden, the less you need to spend on food at the market.

4.9   Which of the following is a more nearly elastic collision?

   a. A baseball player hits a home run.

   b. You chop wood with an ax.

4.10   Account for each of the following in terms of the behavior of gas molecules as described by the kinetic molecular theory.

   a.   Gases exert pressure on the walls of a container.

   b.   The pressure of the air in an automobile tire increases if the tire heats up.

   c.   Some weather balloons are sent up to altitudes of several miles. Such balloons are only partially filled with gas on the ground. Why not fill them completely?

   d.   A jet aircraft engine takes in air to supply the oxygen needed to burn its fuel. At an altitude of 30,000 feet, a jet engine has to take in about six times the *volume* of air it does on the ground to provide equal power.

   e.   Gases diffuse through each other very readily.

4.11   We have seen three kinds of zero temperature: $0°K$, $0°C$, and $0°F$. Explain how they are related to one another.

4.12   In the discussion of how a scientist uses the kinetic molecular theory, we suggested that a theory allows you to build a "mental model." In what ways do you think this mental model differs from a "scale model" such as an architect would make for a building he or she was planning?

4.13   Test your understanding of the relationships among the quantities we have discussed in this chapter by filling in the blanks of the following table.

Gas	Mass of One Mole (gram)	Mass of Sample (gram)	Volume STP (liter)	Number of Molecules
A		22.0	11.2	
B		28.0		6.02×10^{23}
C	4.0	8.0		
D	32.0		44.8	
formula: C_2H_6		300		

4.14   A human body is about 80% water, H_2O. How many molecules of water are in a 100-pound person? (Remember: 1.0 pound = 454 gram and 1.0 mole of H_2O has a mass of 18 gram.)

4.15   The breathing of a normal person while standing involves the exchange of about 8.0 liters of air in one minute. At ordinary temperature and pressure, 8.0 liters is about one-third of a mole. Estimate how many air molecules a person moves in and out of the lungs in one minute.

4.16   When grapejuice ferments to wine, CO_2 gas is produced. The fermenting is done in large vats. An unfortunate mouse was found dead on the floor near a vat one morning after the room had been closed up all night. A more fortunate mouse on the rafters near the roof was alive and running. Explain.

4.17   If you had a mole of a gas in which the molecules were as large as grapefruit, which of the following volumes do you think it would fill?

    a.   A railroad freight car

    b.   A billion 100-car freight trains

    c.   A sphere the size of the moon

    d.   A sphere the size of the earth

    e.   Half the known universe

4.18   The actual diameter of a gas molecule is approximately 2×10^{-8} centimeters. If all the molecules in one mole of this gas were placed side-by-side in a row, how many times do you think the line would reach from the earth to the moon and back, a distance of approximately 8×10^5 kilometers?

    a.   Once   *b.*   150 times

    c.   150,000 times   *d.* 1.5×10^{10} times

5 The Periodic Law and Atomic Structure

☐ Can we list all of the elements in a format that reveals interesting patterns of properties?

☐ What can we learn about chemical behavior by locating an element on the periodic table?

☐ What does position on the periodic table tell us about the internal structure of the atoms of an element?

☐ What kinds of particles are atoms made of?

☐ How did chemists learn that the atom is mostly empty space?

☐ Are all samples of an element the same?

W e will start this chapter by looking back to more than a hundred years ago, when chemistry was a youthful science. Just as any young person accumulates a great many bits of information and ideas that need to be sorted out, so chemistry in the mid-nineteenth century was faced with a similar task. Classification of information is always the first step. Then comes a recognition of the generalities and laws that organize the information. Then the questions about why things happen lead to partial answers in the form of theories. Today we have useful theories about the structure of atoms. Let us see where these have come from.

THE CLASSIFICATION OF THE ELEMENTS

During the first half of the nineteenth century, a tremendous amount of chemical information piled up. The chemists of that time found that they could use Dalton's ideas about atoms and Avogadro's ideas about molecules to great advantage.

The great power of these ideas was that they served both as guides for planning investigations and as a way of interpreting results. For example, if a substance was found to have properties different from those of previously known substances, a search was made to see if it contained a new element. Many new techniques of experimenting and methods of purifying substances were invented. The composition of substances was expressed with atomic symbols and molecular formulas. Previously unknown elements were identified and isolated. Each was assigned an atomic mass on the universally adopted standard scale of relative values. During these years, a veritable explosion of chemical information produced a thick encyclopedia indeed.

It was only natural that some similarities among the elements' patterns of chemical behavior would be noticed. As early as 1830, the similarities among the properties of several groups of elements were spotted. These similarities are like the resemblances among members of a human family. Each is an individual in his

Table 5.1   Family resemblances among triads of elements

Element	Symbol	Atomic Mass	Properties in Common
Metals			
Lithium	Li	7	are lustrous, soft; are less dense than water;
Sodium	Na	23	melt below 180° C; react vigorously with water;
Potassium	K	39	form chlorine compounds LiCl, NaCl, KCl
Nonmetals			
Chlorine	Cl	35.5	are colored gases; form diatomic gas molecules Cl_2, Br_2, I_2;
Bromine	Br	80	boil below 185°C; combine with metals;
Iodine	I	127	form sodium compounds NaCl, NaBr, NaI

or her own right, but some properties, such as color of eyes and hair, complexion, and stature, are similar. Thus, groups of three elements, originally called "triads," can be thought of as chemical families because they are chemically similar. Table 5.1 lists the numbers of two such triads. Notice that, even though the atomic masses vary greatly, the members of a triad have characteristics in common.

The number of newly discovered elements increased rapidly during the middle of the nineteenth century. In the 1860's, nearly 70 elements had been identified and their properties studied.

THE PERIODIC LAW

The more scientists study the natural world, the more firmly they are convinced that nature is orderly. One of their primary aims is to be able to find the order in a mass of data and express that order as a law of nature. This goal is important for all of us. Any

Any clearly stated capsule representation of nature helps us understand the natural world.

clearly stated capsule representation of nature helps us understand the natural world. Understanding, in turn, helps us meet the demands of and live in better harmony with nature. Finding the order is often very difficult; the larger the collection of data, the more genius that is required to find the law. In the period of chemical history with which we are dealing, that kind of genius appeared in two scientists, Dmitri Mendeleev, a Russian, and Lothar Meyer, a German. Each worked independently of the other, yet each came to essentially the same conclusion at about the same time, 1869. The law they proposed is called the **periodic law**. The periodic law as stated by Mendeleev and Meyer is:

When the elements are arranged in order of increasing atomic mass, there is a periodic repetition of elements with similar properties.

The meaning of the term "periodic repetition" can be illustrated by an example. Suppose you read a thermometer placed outdoors in the shade in a midwestern city every day at noon. You do this for ten years. You keep a record of your data by plotting the temperature values on the vertical (up and down) axis of the graph and the consecutive months of each year on the horizontal (across) axis. The line connecting the data points has an almost regular rise and fall. Summer readings are high; winter readings are low. This pattern is periodic repetition.

The kind of information that was used as a basis for formulating the periodic law is represented in Table 5.2. Here 16 of the elements that have small values of atomic mass are arranged in order of increasing atomic mass. We start with lithium, atomic mass approximately 7, and go through calcium, atomic mass 40. Hydrogen, the element with the least atomic mass, 1, has been left out because, as we will see later, hydrogen is a special element and need not concern us for the idea that we are describing here. Also omitted are three elements that were not known when the periodic law was proposed. These are the elements discovered since that time: helium (atomic mass 4), neon (atomic mass 20), and argon (atomic mass 40). These elements also fit into the proposed scheme, but they do not need to be considered to show how the periodic law was developed.

The idea of a periodic variation in properties is suggested by Figures 5.1 and 5.2. Figure 5.1 is a graph of the melting temperatures of the elements, arranged in order of increasing atomic mass. Melting temperature is a typical physical property of an element. You can see a periodic rise and fall of melting temperature in the graph. Figure 5.2 suggests a similar periodic repetition of a decidedly chemical property. The number of atoms of chlorine that combine with one atom of an element is plotted against

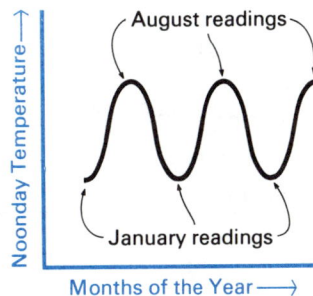

Table 5.2   Some information about the elements of atomic mass 7 through 40

Element	Atomic Mass (approximate)	Melting Temperature (°C)	Formula of Chloride	Chemical Characteristics
Lithium	7	179	LiCl	Highly reactive metal
Beryllium	9	1280	$BeCl_2$	Less reactive metal
Boron	11	2300	BCl_3	Both metallic and nonmetallic properties
Carbon	12	3550	CCl_4	Nonmetal
Nitrogen	14	−210	NCl_3	Less reactive nonmetal
Oxygen	16	−218	OCl_2	Reactive nonmetal
Fluorine	19	−220	FCl	Highly reactive nonmetal
Sodium	23	98	NaCl	Highly reactive metal
Magnesium	24	651	$MgCl_2$	Less reactive metal
Aluminum	27	660	Al_2Cl_6 ($AlCl_3$)	Metal
Silicon	28	1410	$SiCl_4$	Nonmetal
Phosphorus	31	44	PCl_3	Less reactive nonmetal
Sulfur	32	113	SCl_2	Reactive nonmetal
Chlorine	35	−101	Cl_2	Highly reactive nonmetal
Potassium	39	64	KCl	Highly reactive metal
Calcium	40	850	$CaCl_2$	Reactive metal

the atomic mass of the elements. Again, you see a regular rise and fall. (The chloride of aluminum is a special case; the formula for its molecule is Al_2Cl_6. However, the ratio of one Al atom to three Cl atoms, which is represented by Al_2Cl_6, fits into its expected location on the graph.)

The column in Table 5.2 headed "Chemical Characteristics" lists comments that are simplified summaries of the general way the elements behave. We have already noted the distinction between metals and nonmetals. The words "highly reactive," "reactive," and so on are used to suggest the varying tendencies for reaction exhibited by the elements in their pure state. Here again, you will note in the first seven elements, lithium to fluorine, a shift from highly reactive metal to less reactive metal to a gradually increasing tendency to behave as a nonmetal. Then there is an abrupt shift to a repetition of the same cycle in the elements from sodium to chlorine. The cycle starts over again with potassium and calcium but is incomplete as the list ends. So here again we see the demonstration of a periodic repetition in properties like that suggested by Figures 5.1 and 5.2.

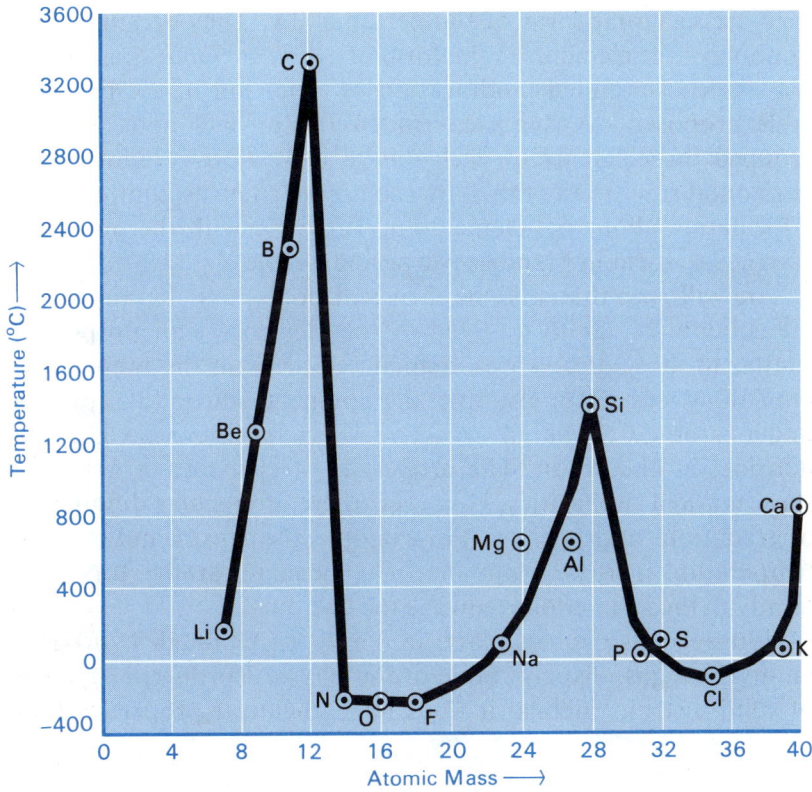

Figure 5.1 Melting temperatures of the elements.

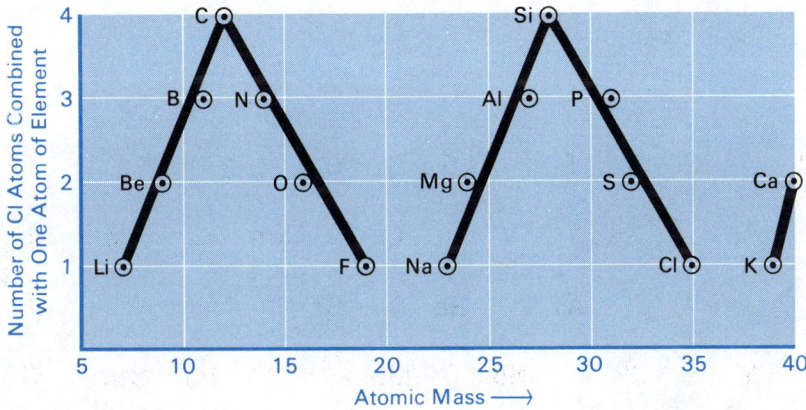

Figure 5.2 Number of chlorine atoms in formulas of various chlorides.

THE PERIODIC TABLE

Both Mendeleev and Meyer went one step further to provide a very useful visual form of the periodic law. They arranged the symbols of the elements in the form of a *periodic table*. Figure 5.3 is a slightly simplified modification of a portion of the periodic table proposed by Mendeleev in 1871. You will note that he grouped the elements of each *family* in *vertical* columns. The horizontal rows in the table are the *periods*. Let us compare the information about the elements in Table 5.2 with the positions assigned to them in Mendeleev's periodic table.

You will note that sodium, Na, and potassium, K, appear in the column for family 1. Each of the elements with properties similar to those of previous elements appears in the family column along with it; for example, the nonmetal silicon, Si, appears under carbon, C, the highly reactive nonmetal chlorine, Cl, under fluorine, F, and so on. The properties of elements in a family show marked similarities. However, there are regular differences or gradations of properties. For example, the density and melting temperature increase from the light elements at the top of a family to the heavier ones toward the bottom.

Hydrogen also has its place: in family 1. Although hydrogen, an invisible gas, displays none of the physical properties typical of solid lustrous metals, it does have chemical properties that

Figure 5.3   A portion of Mendeleev's periodic table, proposed in 1871.

	Family Number →							
Period Number →	1	2	3	4	5	6	7	8
1	H							
2	Li	Be	B	C	N	O	F	
3	Na	Mg	Al	Si	P	S	Cl	
4	K	Ca	*	Ti	V	Cr	Mn	Fe Co Ni
4	Cu	Zn	*	*	As	Se	Br	
5	Rb	Sr	Y	Zr	Nb	Mo	*	Ru Rh Pd
5	Ag	Cd	In	Sn	Sb	Te	I	
6								

*Spaces left open for elements not yet discovered in 1871

make this location in the table appropriate. For example, the formula for the hydrogen chloride molecule is HCl. This is similar in atom-counting to the formulas for the chlorides—LiCl, NaCl, and KCl.

The properties of elements across a period show more striking differences or gradations. For example, across period 3, the first element is sodium, Na, a very reactive metal. Its neighbor, magnesium, Mg, is less reactive as a metal. By the time we move across to phosphorus, P, metallic characteristics have given way to nonmetallic behavior. Sulfur, S, is more nonmetallic, and chlorine, Cl, is a very reactive nonmetal. In Figure 5.2 you also can see a gradation across a period in the number of chlorine atoms that combine with an element.

Another feature of Mendeleev's table is the way he allowed the periods to become longer. Hydrogen is by itself in period 1. Periods 2 and 3 hold seven elements. Periods 4 and 5 are expanded to hold 17 elements. Period 4 starts with potassium, K, a member of family 1. But bromine, Br, the element that belongs with chlorine, Cl, and fluorine, F, in family 7, does not come until the end of an expanded period of 17 elements. Although the reason for this increased number of elements in a period was not clear at the time, the idea of putting similar elements in the same family was strong enough to demand such an arrangement.

The form of the periodic table displayed on the inside cover of this book is a modern modification of the original Mendeleev proposal. It has been expanded still further to include all the 106 elements now known. You will note that the modern table includes one whole family—helium and the other very unreactive gases—that has been discovered since Mendeleev's time. The modern form of the periodic table has the advantage of suggesting not only the similarities among the properties of the elements

The properties of elements in a family show marked similarities.

but also the main features of the structure of their atoms. The structure of the atoms will be the subject of the rest of this chapter and the next. We will see how useful the periodic table is for bringing this information together with predictions of chemical properties.

The similarities and gradations in properties implied by an element's location in the table is very useful to chemists. Mendeleev

A comparison of some of Mendeleev's predictions for the properties of the unknown element below silicon in period 4 with the properties of the element germanium, which was discovered later.

	Mendeleev's Predictions	Observed Property
Atomic mass	72	72.3
Color	dark gray	grayish-white
Density	5.5 g/ml	5.47 g/ml
Formula of chloride	XCl_4	$GeCl_4$
Boiling temperature of chloride	Below 100°C	86°C

himself used these ideas to predict that elements unknown in 1871 would be found to fill the vacancies he left in his table. He predicted the properties of these unknown elements, and his predictions turned out to be very accurate. Since his time, the discovery of many elements has been accomplished by anticipating the properties of an unknown element from the position it should occupy in the table. For example, Marie Curie, the great pioneer in the field of radioactivity research, first identified the element radium (Ra). She knew that it belonged in the family with calcium and barium because the substance she suspected of being an element formed compounds similar to those of calcium and barium.

Chemists constantly use the periodic table to tie together what they know about familiar elements with what they can predict about unfamiliar ones. You, too, can use the ideas of similarity and gradation in properties. Exercises 5.2, 5.3, and 5.4 are illustrative.

ATOMIC STRUCTURE—A THEORY

The periodic law and the periodic table are a great help in classifying and organizing chemical information. But the evidence almost shouts out the question "Why do elements behave in such an orderly pattern?" There must be some regularities in the internal *structure* of atoms that are responsible for the periodic repetition of similar properties when the elements are arranged in order of increasing atomic mass. So we turn from describing a law of nature to discussing the theory that has been produced by inventive human minds.

Looking inside atoms and thus learning about the structure of atoms means exploring nature on a level different from that we have described up to this point. The experiments scientists do in this kind of exploration are much more sophisticated than those we have described thus far. We still cannot "see" the inside of an

We still cannot "see" the inside of an atom as we can see the bones of a hand when we look at the shadows cast by bones in an X ray.

atom as we can see the bones of a hand when we look at the shadows cast by bones in an X ray. Nevertheless, scientists have developed theories of atomic structure that build a very useful mental model of the atom. In this chapter, we will examine this model. We will see what clues scientists have found and how they have followed these clues to solve some of the mysteries about atomic structure. We say "*some* of the mysteries" because the search is never completed. Theories are still only theories; they are human ideas, not actual pictures of nature. Sometimes a theory that ties together some information has to be discarded because new information simply will not fit the interpretation that the theory requires. It is not accurate to call a theory "right" or "wrong." Rather, we have to think in terms of the theory's usefulness. The broader the range of information it ties together, the more acceptable the theory is. The more a theory can predict events or information about the natural world, the more useful it is as a guide to our gaining knowledge. So it is with the theory of atomic structure that we will discuss. We can explain and predict a great deal in terms of the model the theory provides. But rather than say the model is "right," we say it is the best we have. This approach reflects the essential open-mindedness that a scientist must maintain when dealing with the natural world.

One question we can start with to build a model of atomic structure is "Are there fundamental building blocks out of which all atoms are made?" If so, then we may be able to build our model of atoms by assembling these fundamental building blocks in similar or different ways to account for the similarities or differences in the properties of the atoms.

FIRST CLUES TO THE PUZZLE OF ATOMIC STRUCTURE

The first fundamental atomic building block to be suspected was the *electron*. Electrons are the unit particles of an electric current. When an electric current flows through a wire, some electrons are pushed into one end and others are pulled out the opposite end. When Benjamin Franklin flew his famous kite in a lightning storm, electrons came rushing down the wet kite string. When you walk with dry shoes across a dry rug, the charge of static electricity you generate is a pileup, an accumulation, of electrons. When you see the evidence of chemical corrosion on the terminals of a storage battery, you recognize that chemical reactions can be caused by a flow of electrons. This fact was put to good use as early as 1807, when the English scientist Humphrey Davy first isolated the metals sodium and potassium by passing electric current through melted samples of their compounds.

The experiments that proved the electron to be a fundamental particle in all atoms were performed about 1890. These experiments involved the high-voltage discharge of electricity through gases. The apparatus that was used, called a *cathode ray* tube, is diagramed in Figure 5.4. When a gas at very low pressure (about one-millionth of an atmosphere) is in the tube and voltages of about 10,000 volts are used, a stream of negatively charged electrons flies away from the *cathode.* The cathode is the part of the tube connected to the negative terminal of the high-voltage supply. Electric charges of the same sign (− and − or + and +) always repel each other; charges of opposite signs (+ and −) always attract.

The stream of electrons, called a *cathode ray*, is attracted to the *anode*, the part of the tube connected to the positive terminal of the voltage supply. A metal disc with a hole in it is placed in the tube. The disc stops all the electrons except those that go through the hole. In this way, a narrow, well-defined beam of electrons is formed. The course of this beam can be bent or deflected from a straight line by placing a magnet or electric charges on either side of it. In this way, experimenters proved that the beam behaved as electrons were expected to behave.

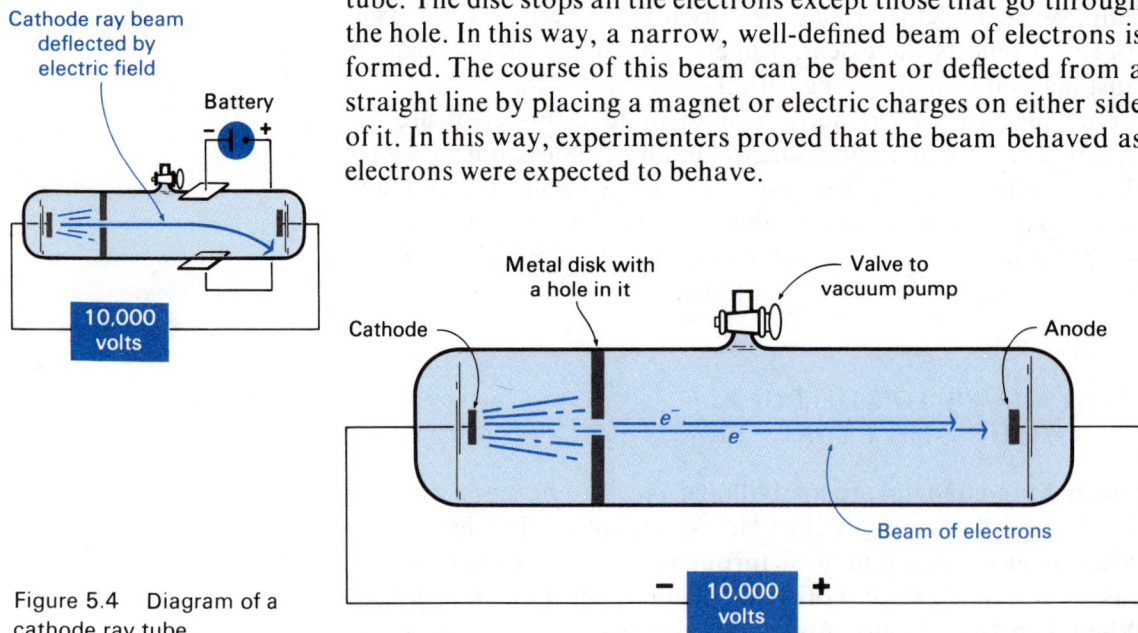

Cathode ray beam deflected by electric field

Battery

10,000 volts

Metal disk with a hole in it

Valve to vacuum pump

Cathode

Anode

Beam of electrons

10,000 volts

Figure 5.4   Diagram of a cathode ray tube.

A modern TV tube is a more elaborate model of a cathode ray tube. The anode is placed in the side of the tube so that the electron beam, moving in a straight line, hits the glass at the end of the tube. The inside surface of the glass is coated with a material that emits light at the spot where the electron beam hits it. The

A modern TV tube is an elaborate model of a cathode ray tube.

TV picture is formed when the electron beam sweeps across the tube surface while the intensity of the beam (the number of electrons in it) is being altered. Thus light and dark areas are produced on the face of the tube.

Another kind of experiment, performed at about the same time, used a modified form of a cathode ray tube to show that a stream of particles also went in the opposite direction. This stream of particles goes toward the negatively charged cathode. Consequently it is called a *positive ray*. Figure 5.5 shows a simple positive ray tube. A narrow beam of positive particles is formed by allowing some of them to go in a straight line through a hole in the cathode. Experimenters can bend this beam off course, too, by placing a magnet or electric charges around it. When they do this, they find an interesting result. The beam is bent in a direction opposite to the way an electron beam would be bent. This result proves that the particles carry a positive electric charge. Moreover, the mass of the positive particles is different when different gases are put in the tube. In contrast, cathode rays always have the same mass, regardless of the gas in the tube. These results suggest that electrons are the same in all kinds of atoms,

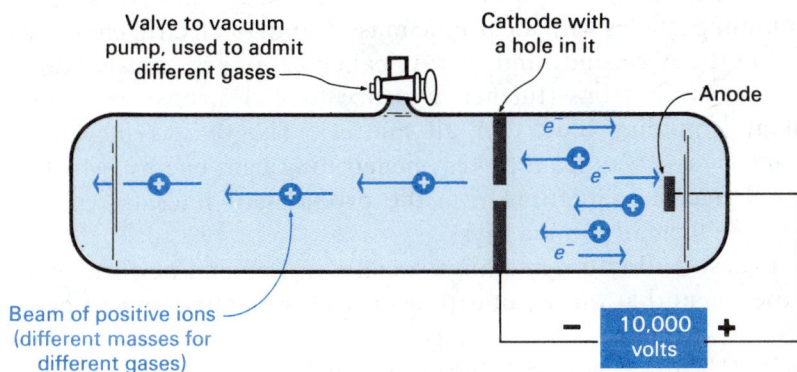

Valve to vacuum pump, used to admit different gases

Cathode with a hole in it

Anode

10,000 volts

Beam of positive ions (different masses for different gases)

Figure 5.5   Diagram of a positive ray tube.

whereas the part of the atom that is left when electrons are removed is different for each kind of atom.

This result was the clue to deciding what happens to the atoms of the gas in the tube. The fast-moving cathode rays collide with atoms and dislodge electrons from the atoms. When a negatively charged electron is removed from a neutral atom, the remaining part of the atom is left with a positive electric charge. Thus the atom, which Dalton saw as an indestructible particle, can be broken into at least two parts. An atom from which one or more electrons has been removed is called a positively charged *ion*. The word *ion* comes from a Greek word that means "to go." An ion, because it carries an electric charge, is attracted toward (goes to) a part of the tube that has an opposite electric charge.

The ions formed in a positive ray tube have very nearly the same mass as the atoms from which they are made. This suggests that the mass of the electron is very small. Indeed, careful measurements show that the mass of the electron is only $1/1837$ of the mass of the hydrogen atom, the lightest kind of atom.

RUTHERFORD'S NUCLEAR MODEL OF THE ATOM

The last decade of the nineteenth century and the first two of the twentieth century were exciting times for scientists. Many phenomena were investigated, and much was learned about how matter interacts with energy in its various forms. One of the discoveries was radioactivity. This phenomenon involves the spontaneous emission of particles of very high energy from some kinds of atoms, such as uranium and radium. Three kinds of rays are given off by radioactive atoms. One of these, called *alpha* rays, is a stream of helium ions. Helium ions are helium atoms (atomic mass 4) that have lost two electrons. Thus alpha rays contain particles with an atomic mass 4 and an electric charge of $+2$ units. A second kind of ray, called *beta* rays, consists of a stream of electrons (further evidence that electrons are fundamental building blocks of all matter). The third type, called *gamma* rays, is a beam of very penetrating light energy. Most of the dangerous radiation from the products of nuclear reactors is in the form of gamma rays.

Ernest Rutherford and his associates in England performed an experiment that turned out to be crucial to the understanding of atomic structure. They arranged a narrow beam of alpha particles (He atoms with $+2$ charge; we can use the symbol He^{++}) so

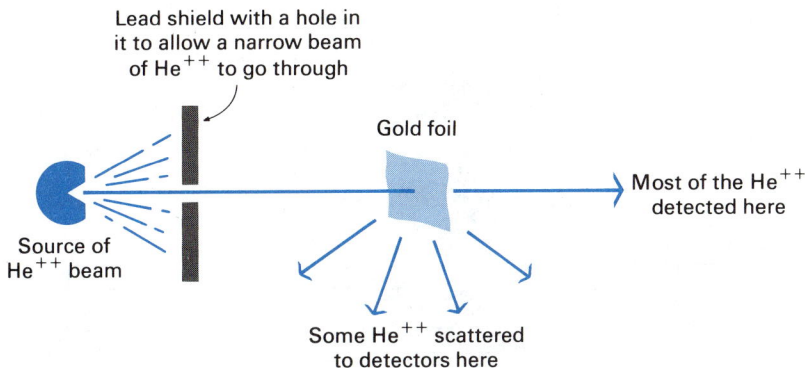

Figure 5.6   Diagram of Rutherford's apparatus in his experiment that led to the discovery of the atomic nucleus.

that it hit a piece of gold foil only 0.0013 cm (0.00051 inch) thick. We can think of this beam as a stream of bullets fired at a target. Figure 5.6 is a diagram of the way the experiment was set up. Detectors were located in various places to find any He^{++} particles that were scattered by the foil. Several remarkable things happened. First, most of the beam went straight through the foil, apparently without hitting anything. This result suggested that, although the atoms of the solid gold must be touching one another, there must be a vast amount of open space for the He^{++} particles to go through. The experiment would be like throwing a handful of sand through a wide-mesh chicken-wire fence made of thin wire thread. Another result was even more surprising. Some of the He^{++} "bullets" were thrown very far off their straight-line course. They were scattered through very large angles. Some even bounced almost straight back toward the "gun." Our sand-through-the-chicken-wire analogy can hardly account for this result. Most sand would go through; some might be deflected. But we would hardly expect any to bounce straight back.

Rutherford's interpretation of these results was to visualize the atom as follows: All the positive electric charge in the atom is concentrated in a tiny *nucleus* that also has almost all the mass of the atom. Figure 5.7 suggests the model Rutherford proposed.

Rutherford's calculations suggested that the diameter of the atom was 10,000 times the diameter of its nucleus.

The model cannot be drawn to scale because Rutherford's calculations suggested that the diameter of the atom was 10,000 times the diameter of its nucleus. Outside the nucleus are the electrons. Because electrons are so small, most of the atom must be

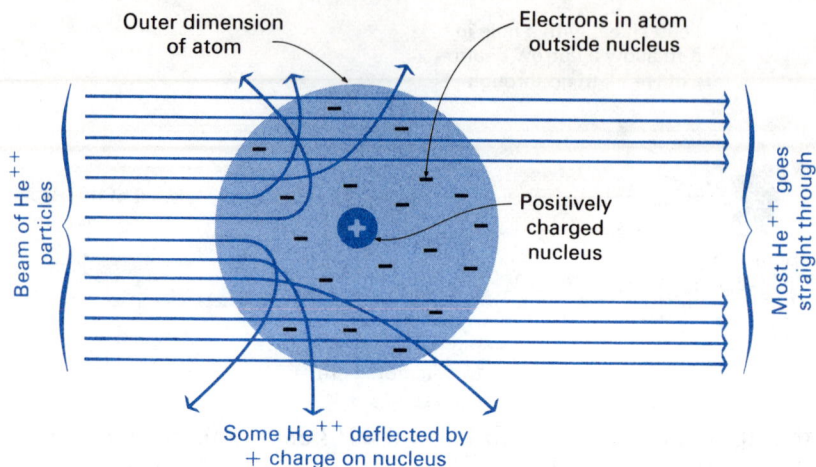

Figure 5.7   Rutherford's nuclear model of the atom.

empty space. This idea accounts for the fact that most of the He^{++} beam goes through. Even if a He^{++} particle did bump an electron, because the He^{++} is more than 7000 times as heavy, the beam would not be deflected any more than a cannonball would be if it hit a ping-pong ball. The scattering of a few He^{++} through large angles and even straight back can be explained by recognizing that, if the nucleus had a large positive electric charge, it would tend to repel the positively charged He^{++} bullets that came close to it.

THE ATOMIC NUMBER OF AN ATOM

Rutherford and his associates went on to do this experiment with many kinds of atoms other than gold in the target. Calculations based on the results always suggested nearly the same *size* for an atom and its nucleus. But the results showed that *each atom has a different characteristic positive charge on its nucleus.* Rutherford called this charge the *atomic number.* Atomic numbers are integers, whole numbers. (For gold, the atomic number is 79, not 69.8 or 79.34.) Because normal atoms are electrically neutral, the number of electrons outside the nucleus must be equal to the atomic number. (The amount of electric charge on an electron is chosen as the unit of electric charge. An atom contains a number of electrons to balance the charge on its nucleus. So the nuclear charge is expressed as a whole or integral number of units of positive charge.)

Other lines of evidence suggest that the atomic number, the positive electric charge on the nucleus, is the feature about atomic structure that establishes the unique identity of every kind of atom. Because the atomic number also establishes the number of electrons in an atom, we can guess that this is a good lead to understanding why atoms have the chemical properties they have. When atoms react with other atoms, only their outer regions interact. Only their outermost electrons can be involved in chemical reactions. It is the number of these outermost electrons and how tightly they are held by an atom that are responsible for the variety of chemical reactivities (the tendency to react) atoms display.

BUILDING BLOCKS IN
THE ATOMIC NUCLEUS

Before we go into more detail about the arrangement of electrons, let us see what more is known about the nucleus of an atom. Because the nucleus of the atom contains almost all the mass and all the positive charge of the atom, the nucleus must be made of fundamental building blocks quite different from electrons. Rutherford and his associates found the first of these building blocks a few years after they had proposed the idea of the atomic nucleus. Their experiments showed that the nucleus of the hydrogen atom is contained in the nuclei of all atoms. This fundamental particle, atomic mass 1.0, electric charge +1, is called a *proton*. The name comes from a Greek word meaning "first rank." This discovery appeals to our logic. It suggests that the atomic number (the number of positive charges on the nucleus) can be accounted for as the number of protons in the nucleus because each proton has one unit of positive charge.

Some years later, in 1932, another nuclear building block was identified. Scientists doing experiments with high-energy machines, the "atom smashers," found that all nuclei, except hydrogen's, contain a particle that has a mass almost identical to the mass of a proton but with no electric charge. This particle is appropriately called a *neutron*

These two kinds of building blocks, protons and neutrons, are all that is needed to account for the mass and charge of an atomic nucleus. However, we should not get the idea that the structure of the nucleus is totally understood merely by counting up these two building blocks. Research on nuclear structure is very far from providing a total answer in full detail. For one thing, the energy balances between electric and gravitational attractions and repulsions among nuclear particles are not the same as those we find for large particles of matter outside atomic nuclei. A vast array of nuclear fragments have been produced by atom-smashing experiments. More are being found every year. But it may be a long time before a fully consistent theoretical model of the atomic nucleus can be put together.

Let us simplify our discussion by introducing a new term, the *mass number*. The mass number is the total number of protons plus neutrons in a nucleus. The mass number also is close to the value of the atomic mass for many elements.

ATOMIC STRUCTURE—A SUMMARY

We can use the idea of counting up numbers of fundamental building blocks to account for the makeup of atoms. Let us tie together the definitions.

	Particle	Mass	Charge
Inside the nucleus	Proton	1	+1
	Neutron	1	0
Outside the nucleus	Electron	$\frac{1}{1837}$	−1

Atomic number $= \begin{cases} \text{Number of protons in the nucleus} \\ \text{or} \\ \text{Number of electrons outside the nucleus} \end{cases}$

Mass number $= \begin{cases} \text{Number of protons plus neutrons in the nucleus} \\ \text{(The mass number is approximately equal to the} \\ \text{atomic mass.)} \end{cases}$

The units of mass are units of the atomic mass scale. The mass of the electron is so small that, for our purposes here, it can be considered as zero. Table 5.3 gives some examples of how these definitions are applied.

ISOTOPES

At about the same time that Rutherford proposed his nuclear model of the atom, a variety of evidence was found to suggest that not every atom of an element has exactly the same mass. For example, careful measurements of the beams of positively

Table 5.3   The number of fundamental particles in atoms

Atom	Atomic Number	Atomic Mass	Mass Number	In the Nucleus		Outside the Nucleus
				Number of Protons	Number of Neutrons	Number of Electrons
H	1	1.0	1	1	0	1
He	2	4.0	4	2	2	2
Be	4	9.0	9	4	5	4
C	6	12.0	12	6	6	6
Si	14	28.1	28	14	14	14
Br	35	79.9	80	35	45	35
U	92	238.0	238	92	146	92

charged atoms in a cathode ray tube filled with neon showed that some neon atoms have a mass of 20, some 22. Even though the masses of these neon atoms are different, all atoms of neon have the same chemical property of being totally inert; neon forms no

Measurements of the beams of positively charged neon atoms showed that some neon atoms have an atomic mass of 20, some 22.

chemical compounds. This existence of atoms with different masses but identical chemical properties holds true for almost all the chemical elements. The name given to these varieties of atoms of one element is *isotopes*. The name comes from Greek words meaning "in the same place." Isotopes of an element belong in the same place in the periodic table.

Table 5.4 adds to our previous summary of definitions for counting atomic building blocks.

Nuclei of atoms

Hydrogen

Helium

Beryllium

$+$ = proton

$\pm$ = neutron

Table 5.4   A comparison of isotopes

Isotopes of One Element Have	
Same	Different
atomic number	mass number
number of protons in nucleus	number of neutrons in nucleus
number of electrons outside nucleus	properties that depend on mass
chemical properties	

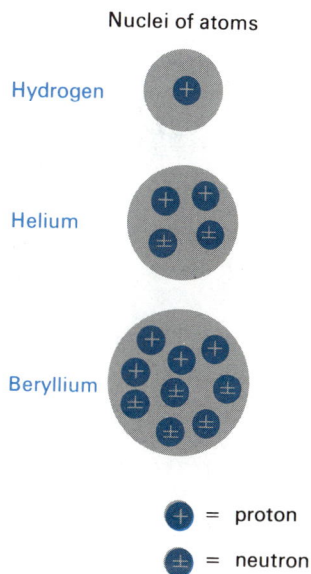

Table 5.5   The number of fundamental particles in two sets of isotopes

Element	Atomic Number	Atomic Mass	Proportion of Isotopes in Natural Abundance	Mass Number	In the Nucleus		Outside the Nucleus
					Number of Protons	Number of Neutrons	Number of Electrons
Cl	17	35.5	75.5%	35	17	18	17
			24.5%	37	17	20	17
Mg	12	24.3	78.7%	24	12	12	12
			10.1%	25	12	13	12
			11.2%	26	12	14	12

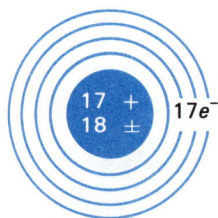

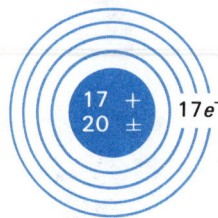

Isotopes of chlorine

Table 5.5 gives some examples of how these definitions can be applied to identify the structure of isotopes.

The idea of isotopes explains a puzzle that may have come to your mind. Many elements have atomic mass values very close to whole numbers, their mass numbers. However, many others decidedly do not. Chlorine, atomic mass 35.5, is a notable example. So is magnesium, atomic mass 24.3. We must recognize that any samples of these elements we find in nature are mixtures of isotopes. The atomic mass of an element is the *average* of the mass numbers of all the isotopes of that element. This situation is just like the course grade that is the average of your grades on several examinations. Suppose you have three grades of 80, one of 100. Your average is $[(3 \times 80) + (1 \times 100)] \div 4 = 85$. So it is with chlorine: Just about three-fourths of its isotopes have mass 35, and one-fourth have mass 37; the average is 35.5.

The proportion of the isotopes in a sample of almost any naturally occurring element is very nearly constant all over the universe. However, it is possible to use various laboratory methods to separate a sample of an element into its various isotopes. It also is possible to use high-energy atom-smashing machines to manufacture one or another isotope of some elements.

ATOMIC NUMBERS AND THE PERIODIC TABLE

Mendeleev stated the periodic law and arranged the periodic table with the elements in the order of their increasing atomic mass. When the idea of atomic number became recognized as defining the distinctive identity of an element, the **periodic law** was modified into the modern form, as follows:

When the elements are arranged in order of increasing atomic number, there is a periodic repetition of elements with similar properties.

Likewise, the place of an element in the periodic table is determined by its atomic number, not by its atomic mass. If you examine the table carefully, along with a list of precise values of atomic mass, you can find a few places where an element has an atomic mass greater than that of its following neighbor. One such example is tellurium, Te, atomic number 52, atomic mass 127.6, which comes ahead of iodine, I, atomic number 53, atomic mass 126.9. Tellurium, Te, certainly displays chemical characteristics that place it with its family members, sulfur, S, and selenium, Se. Likewise, iodine, I, belongs in the family with chlorine, Cl, and bromine, Br. This apparent discrepancy can be explained by recognizing that the atomic mass is an average of the values for all the various isotopes. The mixture of isotopes of tellurium in nature contains a greater proportion of heavier isotopes. Of all the tellurium isotopes, 32% have a mass number of 128 and 34% have a mass number of 130. Any natural sample of iodine contains only one isotope, of mass number 127.

The discrepancy between the mass number 127 and the atomic mass of iodine, 126.9, is a result of the fact that the mass of a proton or a neutron changes slightly from one nucleus to another. Generally speaking, the mass of a proton or neutron in a large nucleus is slightly smaller than that of the corresponding particle in a small nucleus. In very large nuclei, the mass of each particle again increases very slightly.

Thus, we can recognize that the periodic law and periodic table do imply a relationship between the structure of atoms and the properties they exhibit. Mendeleev's law was a statement of a consistency in nature. The law gives no answers to the question "Why?" We look for that answer in a theory. Thus far, our theory suggests that we can phrase the periodic law thus:

When the elements are arranged in order of increasing number of electrons outside the nucleus of the atom, there is a periodic repetition of elements with similar properties.

Our next goal is to see how the theory of atomic structure can account for some periodic, repeating pattern in the arrangement of the electrons. Then we can expect this orderliness of structure to account for the periodic repetition in chemical properties of the elements. This goal will be the focus of our discussion in the next chapter.

Summary

The periodic table, based on the periodic law, is a very useful device for bringing order to information about the properties of the elements. It also makes it possible to predict the properties of unfamiliar elements from information about known elements. The vertical columns of the periodic table contain elements of the

The periodic law and periodic table imply a relationship between the structure of atoms and the properties they exhibit.

same family. The properties of elements in a family show marked similarities. Horizontal rows in the table contain elements of a period. The properties of elements in a period show marked gradations. The original proposals of the periodic law arranged elements in order of increasing atomic mass. The modern form of the periodic law relates more directly to the structure of the atoms; this form of the law arranges the elements in order of increasing atomic number.

Atoms have a central nucleus, whose dimension is only $\frac{1}{10,000}$ the dimension of the atom. The nucleus, made up of protons and neutrons, contains all the positive charge and nearly all the mass of the atom. Outside the nucleus are the electrons, equal in number to the number of protons within the nucleus. All atoms of one element have the same atomic number, equal to a constant number of protons in their nuclei.

Two or more atoms having nuclei with the same number of protons but different numbers of neutrons are known as isotopes. Many naturally occurring elements are mixtures of isotopes. Consequently, the atomic masses of these elements are average values for the mixture of isotopes rather than integral values that would be predicted by counting the expected numbers of protons and neutrons.

Glossary

The number in parentheses indicates the text page where you can find the term defined in context.

alpha ray   a stream of helium nuclei (helium atoms that have lost two electrons), given off by some radioactive nuclei (142)

anode   the positively charged terminal of a voltage supply or battery (140)

atomic number   the number of positive charges (protons) in the nucleus of an atom of an element. It is equal to the number of electrons outside the nucleus of a neutral atom. (144)

beta ray   a stream of rapidly moving electrons, given off by some radioactive nuclei (142)

cathode   the negatively charged terminal of a voltage supply or battery (140)

cathode ray   the stream of electrons moving from the cathode to the anode in a cathode ray tube (140)

electron   the unit particle of negative electric charge, one of the fundamental building blocks of atoms (139)

family (periodic table)   a numbered vertical group (column) of elements that have similar properties because of similarities in their structure (136)

gamma ray   a beam of very high energy light, given off by some radioactive nuclei (142)

ion   an atom that has lost electrons (becoming a positive ion) or gained additional electrons (becoming a negative ion) (142)

isotope   one of the different forms of an element. All forms have the same atomic number and chemical properties but different masses. (147)

mass number   the total number of protons and neutrons in the nucleus, approximately equal to the atomic mass (146)

neutron   a particle with mass almost equal to that of the proton but with no electrical charge; one of the fundamental particles in the atomic nucleus (145)

nucleus   the central body of an atom that contains all of the atom's protons and neutrons (thus all its positive electric charge) and nearly all of its mass (143)

period (periodic table)   a numbered horizontal row of the periodic table, in which the atomic numbers of elements increase from left to right (136)

periodic law (Mendeleev)   when the elements are arranged in order of *increasing atomic mass,* there is a periodic repetition of elements with similar properties (133) *periodic law (modern day)*   when the elements are arranged in order of *increasing number of electrons outside the nucleus of the atom,* there is a periodic repetition of elements with similar properties (149)

periodic table   a table showing the elements in order of increasing atomic number, structured to show the relationships among the elements and summarize information about them (136)

positive ray   a beam of positive ions, moving toward the cathode (negative terminal) in a cathode ray tube (141)

proton   the hydrogen nucleus, the fundamental positively charged particle in atomic nuclei. It has an atomic mass of 1.0 and a charge of +1. (145)

Exercises

5.1   Which of the following are demonstrations of periodic repetition? Justify your answers.

    a.   A clock pendulum moving
    b.   The firing of cylinders in an automobile engine
    c.   A sewing machine sewing
    d.   A balloon leaking gas
    e.   Silver becoming tarnished
    f.   A person running

5.2   Tell whether each of the following sets of three elements is part of a family or part of a period in the periodic table.

    a.   Na, K, Rb   *b.*   C, N, O
    c.   Cu, Ag, Au   *d.*   V, Cr, Mn
    e.   F, Cl, Br   *f.*   Mg, Ca, Sr
    g.   Li, Be, B   *h.*   Na, Mg, Al

5.3   For each of the following sets of two elements, tell which pair probably has the greater difference in properties.

    a.   K and Ca;   K and Rb   *b.*   N and P;   N and O
    c.   N and F;   F and Br   *d.*   C and B;   C and Si

5.4   Marie Curie took advantage of a gradation in properties of the elements in a family to isolate a compound of the new element radium (Ra, atomic number 88). What did she expect the solubility of $RaSO_4$ to be, compared to the solubilities of the similar compounds of the other family members? $CaSO_4$ is quite soluble in water, $SrSO_4$ is less soluble, and $BaSO_4$ is even less soluble.

5.5 Explain the following in terms of the structure of the particles involved.

a. A beam of cathode rays is bent in the opposite direction from that of a beam of positive rays when both go through a magnetic field.

b. All the particles in every beam of cathode rays have the same mass.

c. The particles in different beams of positive rays may have different masses.

5.6 Why is the mass number of an element not always equal to its atomic mass?

5.7 Complete the following table for the various atoms and isotopes in the manner of Tables 5.2 and 5.3.

Atom	Atomic Number	Atomic Mass	Mass Number	In the Nucleus		Outside the Nucleus
				Number of Protons	Number of Neutrons	Number of Electrons
H	1	1.0	1			
N	7	14.0	14			
F	9	19.0	19			
Ca	20	40.1	40			
Sc	21	45.0	45			
Ne	10	20.2	20			
Ne			22			
Ag	47	107.9	107			
Ag			109			

5.8   Consult the periodic table for additional information to enable you to complete the following table in the manner of Tables 5.2 and 5.3.

Atom	Atomic Number	Atomic Mass	Mass Number	In the Nucleus		Outside the Nucleus
				Number of Protons	Number of Neutrons	Number of Electrons
	79		197			
	88	226.0				
K	19	39.1			20	
	19				21	
					4	3
				26	30	
					6	6
					8	6

5.9   The diameter of the earth is about 8000 miles. Imagine the situation in which only the nuclei of all the atoms in the earth were piled together. (In other words, squeeze out all the empty space in atoms.)

    a.   How large a sphere would you have?
    b.   It is impossible to perform this imaginary feat because of the tremendous forces of electrical repulsion that would be produced. Explain.

5.10   Suppose an atom were of such size that its nucleus were the size of a basketball. Suppose two such atoms were touching each other. Which of the following is the best estimate of the distance between the basketball-sized nuclei? Use the ratio Rutherford proposed for the relative size of an atom and its nucleus.

    a.   In the next room
    b.   At opposite ends of a football field
    c.   Nearly two miles away
    d.   Across the United States

5.11   When we trace the development of concepts in science, we often see how the early form of an important basic idea is modified rather than thrown out by later discoveries.

   a.   Illustrate this statement by discussing the topic "The Periodic Law in Chemistry."

   b.   Illustrate this statement by referring to the second postulate of Dalton's atomic theory (all the atoms of one element are *alike*) and present knowledge about the structure of isotopes.

   c.   Do you recognize parallels to this modification of important basic ideas in other kinds of human experience? Consider such topics as "The Role of Women in Society" or "The Political Influence of Labor Unions."

6

The Arrangement of Electrons in Atoms

☐ What happens to electrons when an atom absorbs or gives off energy?

☐ How are electrons distributed in the energy levels of atoms?

☐ How did chemists develop their modern "picture" of the energy levels within atoms?

☐ Is an electron a wave, a particle, or some combination of the two?

☐ How does the number of electrons in the outermost energy level affect the chemical properties of an atom?

☐ What happens when atoms gain or lose electrons?

☐ Can we ever locate an electron exactly within the atom?

W̲hy does a neon sign always glow with a red color? Or a sodium lamp yellow? Everyone knows that a strong electric current and sometimes even a weak one can kill a person. The electric current drastically disrupts the normal body chemistry. But what broad, general principles relate electricity to chemical reactions? Why do some elements, such as helium, form no compounds, whereas other elements, such as chlorine, form hundreds of different compounds? Questions such as these are not very satisfactorily answered just by saying, "That's the way atoms are." We want to know why they

Are there differences in the structures of atoms that can be used to interpret differences in properties?

are that way. We want to go a step further. Are there differences in the makeup, the structure, of atoms that can be used to interpret differences in properties?

In the previous chapter, we found that elements differ in the number of electrons in their atoms. The number of electrons increases in a regular way, one more for each increase in atomic number. But how can this regularity account for the periodic repetition of similar properties among the elements? More than just the *number* of electrons outside the nucleus of an atom must be involved in accounting for the properties of the elements. We properly can expect that some periodic repetition in the arrangement of the atoms' electrons may be responsible for the periodic repetition of similar properties. The feature of the *arrangement* of electrons in atoms is what our theory can supply. That is the subject of this chapter.

ENERGY LEVELS IN ATOMS

Before we can discuss the arrangement of electrons in an atom, we must learn what the basis is for *any* kind of an arrangement or distribution. Usually we think of arrangement or distribution in terms of space, such as the proper arrangement of chess pieces on the spaces of a chessboard to start the game. Likewise, the distribution of population in a country or the distribution of wealth to the population implies knowing *where* the people or the money can be found if we look for them. When we speak of electrons in atoms, the idea of "where" in a geographical sense does not have

When we speak of electrons in atoms, the idea of "where" in a geographical sense does not have much meaning.

much meaning. Rather, we have to shift our thinking to a different frame of reference. This frame of reference is the idea of electrons being in different *energy levels*. Let us see how this idea developed.

Light is a form of energy. We know visible light, the kind of light we can see, is made up of different colors. In more scientific terms, light is thought of as a wave-like phenomenon, and color is more accurately described as the way we see the wavelength of the light. The *wavelength* is the distance between the crests of the wave, or between the troughs. Red light has a longer wavelength than blue light. White light, such as normal daylight or the light from an ordinary electric bulb, is a mixture of light of all visible colors or wavelengths. When white light goes through a glass prism, the mixture is separated and spread out so that we see a *spectrum*. Figure 6.1 summarizes these ideas. The same phenomenon is observed when sunlight goes through a multitude of small raindrops. The spectrum we see is the rainbow. The only additional scientific idea we need for our discussion at this point is: *The energy of light is different for different wavelengths.* The longer wavelength red light has less energy than the shorter wavelength violet light does.

Thus we can explain how our eyes can tell one color from another. The nerve cells of our eyes respond differently to the dif-

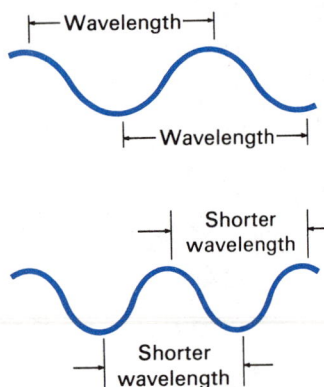

Figure 6.1   A prism separates white light into light of various colors.

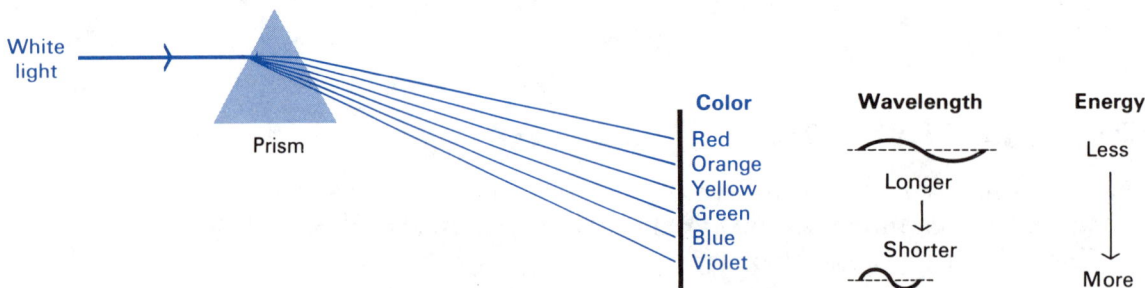

ferent energies of the variously colored light. This idea also is important because it is the basis for using light to investigate the internal structure of atoms. If you sprinkle some table salt in the gas flame of a kitchen stove, you see a brilliant yellow color. You see the same yellow color when a pot of heated water containing some table salt boils over and sprinkles the flame. Table salt is the compound sodium chloride, $NaCl$. You may have seen the same color in the light of some highway illumination lamps or in a special laboratory bulb. These bulbs are filled with sodium in the form of low-pressure gas. In either case, the sodium atoms absorb energy: heat energy in the flame, electric energy in the lamp. We say that the sodium atoms are "excited." The light you see is the energy the atoms emit when they lose the energy that excited them in the first place. We explain this process of an atom's gaining or losing energy by saying that an electron in the atom goes to a higher or lower energy level. This process is represented in Figure 6.2.

AN ELECTRON JUMPS FROM ONE ENERGY LEVEL TO ANOTHER

In case *B* in Figure 6.2, the energy-level jump the electron makes is smaller than in case *A*. Because the amount of energy being emitted is smaller, the light has less energy and a longer wavelength (a different color). We see the same color whenever any sodium atoms lose the energy they have absorbed because *the*

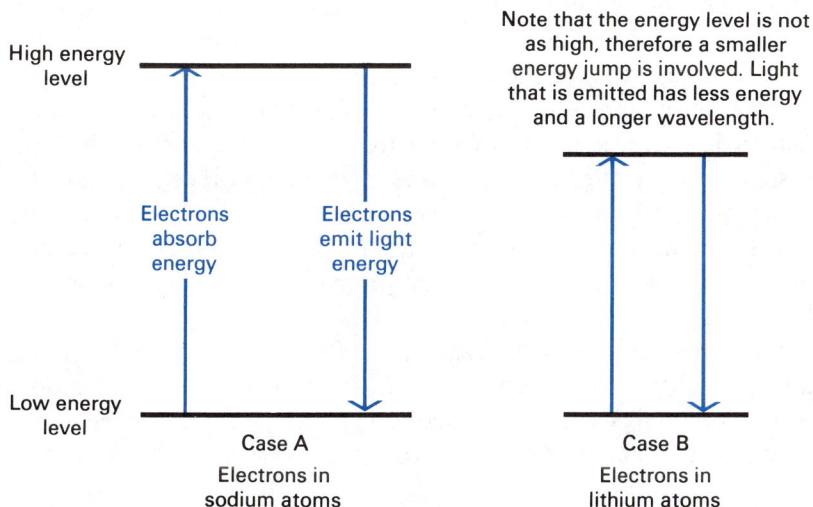

Note that the energy level is not as high, therefore a smaller energy jump is involved. Light that is emitted has less energy and a longer wavelength.

High energy level

Electrons absorb energy

Electrons emit light energy

Low energy level

Case A
Electrons in sodium atoms

Case B
Electrons in lithium atoms

Figure 6.2   Electron jumps between energy levels in atoms.

spacing between energy levels is the same for every sodium atom. If we put some lithium chloride, LiCl, in a flame, we see a brilliant red light. This red light means that the spacing between energy levels in every lithium atom is different from the spacing in sodium atoms. The electron jump involves a different energy, so the emitted light has a different color. This difference corresponds to the contrast between *A* and *B* in the diagram.

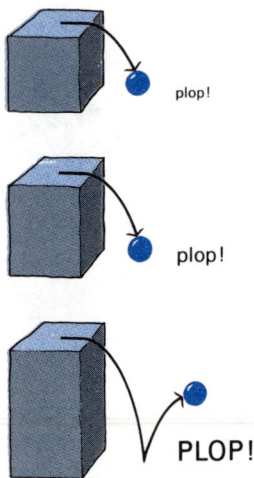

An analogy may help you visualize this model of electrons in energy levels of atoms. If you put a hard ball on the top of a box and allow it to roll off, you hear a "plop" when it hits the floor. When you take the ball from the floor and put it on the top of a box, you add potential energy to the ball (you place it farther from the center of the earth). When it falls back to the floor, some of the potential energy is emitted as the sound you hear. The higher the box off which the ball rolls, the louder the "plop" you hear when it hits the floor. If you hear more sound, you can conclude that the ball fell from a greater height; it lost more potential

plop!

plop!

PLOP!

If you hear more sound, you can conclude that the ball fell from a greater height; it lost more potential energy.

energy. In an excited atom, the electron gains potential energy by going to a higher energy level. When it falls to a lower energy level, the lost potential energy appears as light. The greater the energy of the light, the higher the energy level from which the electron fell.

Now you can answer one of the questions we asked at the beginning of the chapter: "Why does a neon sign always glow with a red color?" The answer is that the electrons in all electrically excited neon atoms make energy jumps of the same size. The light from a neon sign is not just one energy but a mixture of characteristic energies, most of which are in the range of red color. So instead of only one higher and one lower energy level, we have to think of a pattern of energy levels, like steps in an irregular ladder. Every neon atom has the same pattern of steps, so electrons make the same pattern of energy jumps between steps. And the light you see always has the same color.

Our eyes can detect different colors, but they are not able to analyze a beam of mixed light into its components. To accomplish this analysis, a spectroscope is used. A *spectroscope* is an instrument that admits light through a narrow slit; focuses the

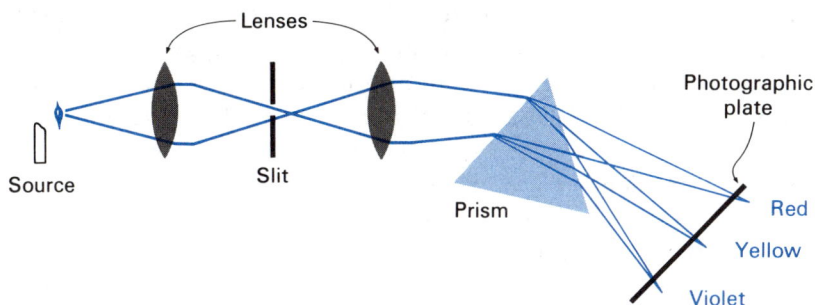

Figure 6.3 Diagram of a spectroscope.

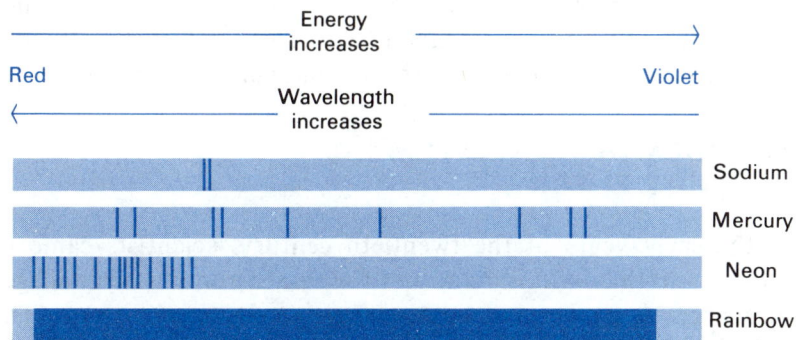

Figure 6.4 The visible spectrum of sodium, mercury, and neon compared to that of a rainbow.

light on a prism, which separates the light into its component colors; and then records the resulting spectrum. Figure 6.3 is a diagram of a spectroscope. The spectrum of white light looks like the rainbow, a continuous gradation of one color into another. In contrast, the spectrum of light from the excited atoms of an element is always a pattern of distinct lines, a pattern of kinds of light with only certain energies, not all energies. If each kind of light means a jump of a particular size, then the presence of various light energies in a spectrum indicates the presence of various distances between an atom's energy levels. Figure 6.4 shows the kind of spectrum produced by excited atoms of sodium, mercury, and neon. Notice that, instead of being a continuous spectrum (light of every wavelength) as a rainbow is, only certain specific wavelengths show up. Sodium exhibits a simple spectrum; the others are more complex. The light emitted by excited mercury atoms looks almost white because it is a mixture of light of various colors.

Each element has a spectrum that is different from that of every other element. What this means is that each element must have its own distinctive pattern of energy-level spacings. Electrons of one element that jump among these energy levels emit a variety of kinds of light. But in another element, electrons jump among energy levels spaced differently, so that a different spectrum results. Consequently, the spectrum of light emitted when atoms of an element are excited can serve as a "fingerprint," a reliable identifying characteristic for that element. Chemists can observe the spectrum of an unknown sample as a method of qualitative analysis for telling what elements are present. Chemists excite the atoms of the sample and then inspect the spectrum of light that is produced to find characteristics of known elements. This method, in principle, is used by analysts in police laboratories or environmental laboratories to detect the presence or absence of some element suspected of being in a sample.

THE IDEA OF A QUANTUM OF LIGHT ENERGY

In the early years of the twentieth century, scientists came to understand that light is made up of separate bundles or packages of energy. A package of light energy is called a *quantum*. (The plural of quantum is *quanta*.) Each quantum has a characteristic energy and a particular wavelength associated with it. We can compare this idea of quanta of light with the idea of atoms of

Light is made up of separate bundles or packages of energy.

matter. Each atom is a separate package of matter with a characteristic mass. If we have a pure sample of an element, all the atoms (packages) are of nearly the same mass. Similarly, if we have light of only one wavelength, all light quanta are of the same energy. The more atoms we have, the heavier is our sample. Similarly, the more quanta we have in a beam of light, the more intense it appears, because it has more energy. If we go to a different element, we find that its atoms have a different characteristic mass. If we go to light of a different wavelength, we find that its quanta have a different energy.

When we apply the quantum idea to a series of things, we recognize that the value of each item in the series increases by definite jumps. The values do not increase gradually or continuously. An analogy is the way our money system is built. Each coin has a definite separate value. The smallest quantum is the penny. Then we have nickels, dimes, quarters, and so on. We do not have coins of in-between value, such as 17.5 cents. You may hear the term "quantum jump" applied to some change that is taking place, such as a shift in public opinion or government policy. This descriptive use of the term means that the change is abrupt and definite rather than gradual.

The idea of light coming in quanta is important to the development of our ideas of atomic structure. It makes possible a connection between the individual atoms in a sample and the characteristics of the light emitted by that sample when it heats up or absorbs electric energy. Each atom absorbs and then emits one quantum of energy when an electron jumps from one of its energy levels to another. So, if all atoms have the same pattern of energy levels, an electron jumping in any one of them will cause the emission of the same quantum of light.

THE BOHR THEORY

The ideas we have been discussing, electrons occupying energy levels and jumping from one level to another, were first suggested as postulates of a theory for atomic structure. The theory was proposed by Niels Bohr, a brilliant Danish scientist who was a pupil of Ernest Rutherford. Bohr went beyond the qualitative picture we have drawn. He tied together the idea of jumps between energy levels with the quantum theory of light. He calculated numerical values for the energy levels of the simplest kind of atom, the hydrogen atom. The hydrogen atom has only one electron outside a nucleus that contains only one proton.

Bohr argued this way. Particles with positive and negative electric charges are attracted toward each other. If they are free to move, they are pulled together. But within the atom, either the rules are different or the electrons are not free to move to the surface of the nucleus. The negatively charged electron is *not* pulled into the positively charged nucleus. Why? Bohr pictured the electron as moving rapidly on a circular orbit around the nucleus, much as the moon moves around the earth. The gravitational pull of the earth and the moon on each other keep the moon from flying off into space; the electric attraction of the positively charged nucleus for the negatively charged electron serves the same purpose. Thus Bohr represented an atom's energy level as an electron moving on a circular path around the nucleus. When the electron jumps to a different energy level, it jumps to another orbit of a different size.

Quanta of money

penny
1 ¢

nickel
5 ¢

dime
10 ¢

quarter
25 ¢

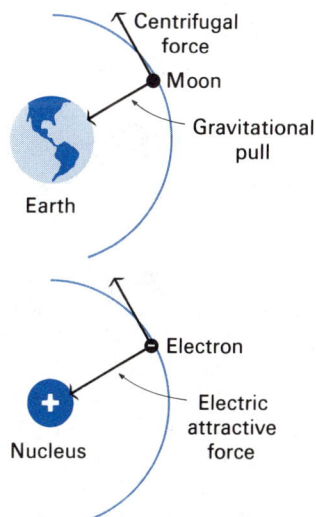

Centrifugal force

Moon

Gravitational pull

Earth

Electron

Electric attractive force

Nucleus

Bohr's next assumption was crucial in explaining why excited atoms emit light of only certain definite quanta. An atom cannot have orbits of any and all sizes for its electrons to use. Only certain sizes are possible; hence, only certain energy levels are possible. Bohr invented a formula to calculate the potential energy of the various possible levels

$$\text{Energy of a level} = \text{“constant”} \times \frac{1}{n^2}$$

In this formula, the symbol n stands for an integer (whole number) called a *quantum number*. The values of n can be 1, 2, 3, 4, ..., ∞. (The symbol ∞ stands for infinity, the largest possible number.) A quantum number is used to identify a particular energy level.

Figure 6.5 is a diagram of the hydrogen atom according to the model Bohr proposed. What made the theory so attractive was that it worked. It explained the spectrum of hydrogen atoms. Bohr calculated the "constant" in his energy-level expression from quantities such as the mass of the electron, its charge, the charge of the proton nucleus, and so on. Then, when he inserted the various values of n, the *differences* between the possible energy levels in the atom corresponded to the various energies of the quanta of light that appeared in the spectrum.

Figure 6.5   Bohr energy-level diagram for hydrogen atoms.

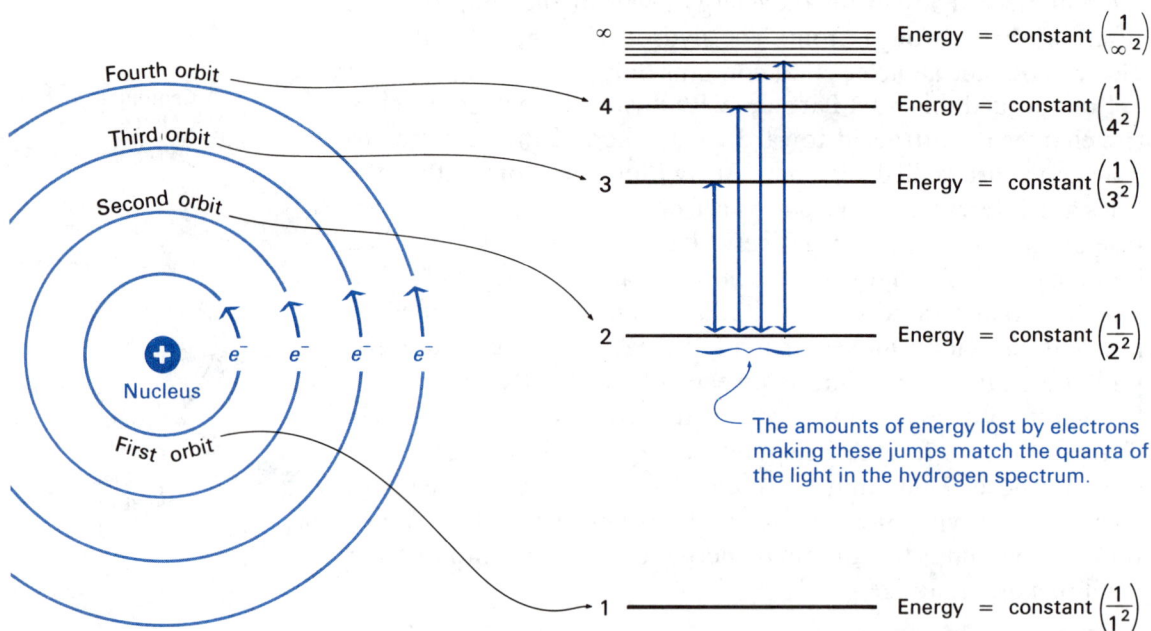

Energy = constant $\left(\frac{1}{\infty^2}\right)$

Energy = constant $\left(\frac{1}{4^2}\right)$

Energy = constant $\left(\frac{1}{3^2}\right)$

Energy = constant $\left(\frac{1}{2^2}\right)$

The amounts of energy lost by electrons making these jumps match the quanta of the light in the hydrogen spectrum.

Energy = constant $\left(\frac{1}{1^2}\right)$

A "BETTER" THEORY OF ATOMIC STRUCTURE

The Bohr model of atomic structure soon proved to be useful only for the simplest atom, hydrogen, with one electron outside a nucleus of one proton. His ideas that electrons occupy energy levels and absorb or emit energy by jumping from one energy level to another were fine. So, too, was his description of these energy levels as orbits for the hydrogen electron. His calculations agreed with the observed spectrum of the hydrogen atom. But when scientists tried to apply these ideas to calculate energy levels in more-complicated atoms containing more than one electron, the theory ran into trouble. For one thing, the spectra are much more complicated. This fact means that the energy-level pattern in such atoms must be much more complicated. Instead of definite single energy levels with a particular quantum number n, each energy level for values of n greater than one apparently is a set of two or more sublevels. Visualizing this set is like visualizing a building with mezzanines or balconies between each of the main floors. The energies of these sublevels also follow patterns, which can be predicted by using more and different quantum numbers in addition to Bohr's n. But Bohr's theory of electron orbits had no postulates to explain this complexity.

We have pointed out before that the test scientists apply to a theory is whether it is useful in explaining a wide variety of information. So after Bohr's theory was proven useful for only a narrow range of information, the search was on for an alternate theory. Scientists wanted to describe energy levels in a way that would include more atoms than the simplest, hydrogen.

At this time, the early decades of the twentieth century, scientists' fundamental ideas about matter and energy were being challenged quite violently. For example, Albert Einstein proposed that matter and energy could be changed from one to the other. Experiments involving the very high energies of atom-smashing machines proved that this was true. Theories about light also had to change. The results of some experiments suggested that quanta of light behave like particles, like a stream of bullets. Other experiments could be explained only if light quanta behave as waves. So a quantum of light energy had to be both a particle and a wave. The crowning blow to any complacency with

A quantum of light energy has to be both a particle and a wave.

older ideas came with a proposal from a young imaginative French scientist, Louis de Broglie. He said that the tiny fundamental particles of matter such as electrons could also behave like waves. His proposal was found to be correct. For example, a beam of rapidly moving electrons can be used in an electron microscope to "see" extremely small objects. The beam of wave-like electrons behaves as does a beam of visible light waves in an ordinary microscope.

So, soon after Bohr proposed his theory, scientists began to construct an alternate theory of atomic structure. This theory starts with the remarkable postulate that *electrons in atoms are waves rather than particles.* Here our troubles begin, because it is very difficult to imagine a "picture" of an atom full of pulsating, vibrating waves instead of neat little electrons on orbits, like satellites going around a planet. A wave is energy. When you see waves in the ocean, you see water rising and falling. The wave is not the water but rather the energy of motion the water has. However, if we go back to the basic idea that all we can really tell about electrons in atoms must be expressed in terms of energy levels (not geographical "locations"), we can see that describing electrons as wave-like energy might prove to be a useful theory.

One interesting property of waves can be illustrated by playing with a length of rope tied to a wall. You can make the rope go up and down in one big loop or wave. Or you can jiggle it faster and make two, three, or even four waves. The important observation is that you can make only *integral numbers* of waves, never 3 $\frac{1}{2}$ or

You can only make integral numbers of waves, never 3 $\frac{1}{2}$ or 4.27 waves!

4.27 waves! Waves such as these, with the two ends of the rope fastened, are called *standing waves.* The number of waves, and hence their lengths within the limits of the rope's ends, are quantized. Quantized means that the series of numbers of waves increases by whole-number jumps, not gradually. This quantization comes from the way waves are; it is a natural phenomenon and not an arbitrary idea of the way you may think the rope *ought* to behave.

The waves you make in the rope are *one*-dimensional waves, along the line between you and the wall. When you strike a drum, the stretched fabric or leather of the drumhead vibrates in standing waves of *two* dimensions, across its surface. These waves, too, must be regulated by some quantum condition. Only certain definite vibrations will fit the dimensions of the drumhead. Waves in *three* dimensions are harder still to imagine. An example is the kind of vibrations you see in a bowl of gelatin when you bump it. The gelatin bounces up and down, back and forth, and side to side. The mathematical formulas that correctly describe three-dimensional waves are complicated, but they have been worked out. When this description is worked out, *three* quantum numbers appear in the formulas, just as the *one* does in your observation with the rope. Here is the remarkable fact that makes scientists believe that the wave theory of electrons in atoms is so useful. The theory is capable of predicting the complicated variety of energy levels that must exist in atoms that contain more than one electron.

We have been careful to say that the theory "is capable of predicting." The mathematical equations of the theory are difficult to solve. Yet approximate solutions agree with observations very well. Moreover, in the past few years, as better and faster electronic computers have become available, more and more accurate calculations have been made using these equations. The theory that describes electrons as waves works much better than the Bohr theory does.

Waves in one dimension; *one* quantum number

Waves in two dimensions; *two* quantum numbers

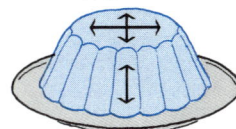

Waves in three dimensions; *three* quantum numbers

PROBABILITY DENSITY INSTEAD OF ORBITS

At this point, you may be saying, "It may be fine for high-powered theory to claim that the best 'picture' of an atom is a complicated mathematical formula, but how am I to visualize an atom?" Fortunately, the mathematics allows us one loophole through which we can slip to come close to imagining what an atom "looks like." This loophole is the idea of probability.

If you set out to steal the honey from a beehive in the middle of a field, you realize that the probability of getting stung by an enraged bee is greater near the hive than at the edge of the field. A similar situation would occur if you could stick your finger inside an atom. You would "feel" more electrons at one place than at another, that is, more as you approach the nucleus. The mathematical equations that describe the electrons' wave behavior in atoms give values for what we can call the probability of finding an electron at a certain place within the atom. This is called the *probability density* of an electron population. This is analogous to what we can do to show the probability density of

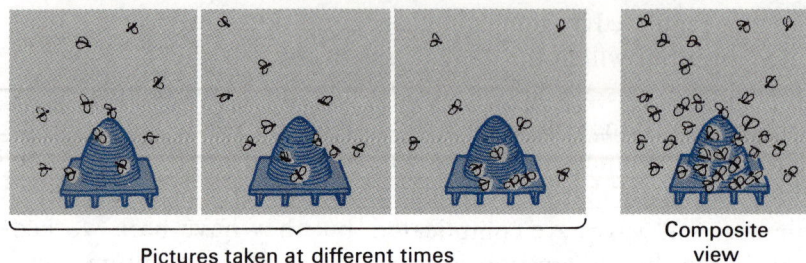

Pictures taken at different times   Composite view

bees around the beehive. If you took a high-speed photograph, you could "stop" the bees in flight. If you took a series of such photographs from the same spot, superimposed the negatives, and then made a print, you would get a photograph that shows, by the density of the images, where bees are likely to be found.

Two types of diagrams are used by chemists to help visualize what an atom "looks like." These are shown in Figure 6.6. In one, the density of the shading suggests that a cross-sectional view of an atom would show a heavier probability density of electrons at some places and a lighter one at others. The second type of diagram suggests a space around the nucleus of the atom in which there is a 90% chance of finding an electron. One type of energy level is visualized as a sphere, the other as a shape like a peanut shell. Other types of energy levels depicted in this manner are even more complex. These different kinds of probability density representations are associated with different kinds of sublevels.

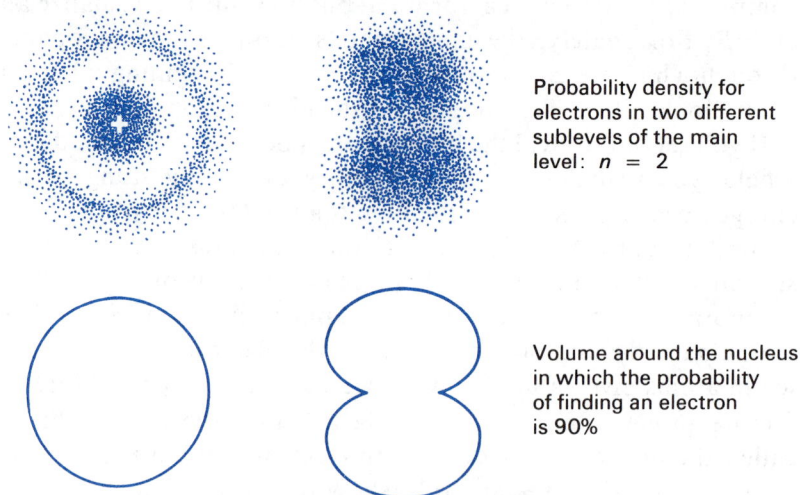

Probability density for electrons in two different sublevels of the main level: $n = 2$

Volume around the nucleus in which the probability of finding an electron is 90%

Figure 6.6   Representations of atoms with electrons in different energy levels.

The situation in which you find yourself may be disappointing. You may want very much to know what an atom looks like. But that is impossible. It is much better to realize this than to believe that some picture suggested by a human theory actually "is" an atom. Our theory expresses in theoretical language a great deal of useful information about atoms. Thus, concepts allow us to realize the connections between otherwise isolated bits of knowledge. This is as far as we can go, indeed as far as we need to go.

ENERGY LEVELS FOR MANY ELECTRONS

Now we can bring two ideas about atomic structure together: locating electrons in energy levels and counting the number of electrons outside the nucleus of an atom (the atomic number). The pattern of energy levels that our theory predicts as possible locations for the one electron in the hydrogen atom is the pattern to follow for the arrangement of more and more electrons in atoms with higher atomic numbers. The hydrogen atom, H (atomic number 1), has only one electron. In the stable, unexcited hydrogen atom, this one electron occupies the lowest possible

In the stable unexcited hydrogen atom, the one electron occupies the lowest possible energy level, just as a stone rolls to the bottom of a gully.

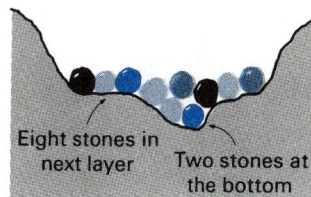

Eight stones in next layer   Two stones at the bottom

energy level, just as a stone rolls to the bottom of a gully. The helium atom, He (atomic number 2), has *two* electrons. The second electron also occupies the lowest possible energy level, just as a second stone rolled into a gully also rolls to the bottom.

A third electron is added in the lithium atom, Li (atomic number 3). This electron also occupies the lowest available energy level. But now imagine that the gully into which you roll stones is narrow at the bottom. If there is room for only two stones, the third one rolls to a position in a second layer, on top of the first layer of two stones. So it is with the third electron in lithium; it begins the occupancy of a second energy level. As you consider each element in order of increasing atomic number (Be, 4; B, 5; C, 6; . . .), one more electron is added to go into the lowest available location in an energy level. When the second level has as many electrons as it can hold, then the next added electron must start the occupancy of a third level. And so on to a fourth, fifth, and as many as are needed to accommodate all the electrons. Rather obvious questions come to mind: "How many electrons can each level hold? Is there any pattern to the arrangement of electrons in each energy level of the stable, unexcited atoms?"

THE PERIODIC TABLE PREDICTS ELECTRON ARRANGEMENT IN ATOMS

To answer these questions, we first must consider the evidence: the periodic repetition of chemical behavior implied by the arrangement of atoms in the periodic table. The manner in which an atom interacts with other atoms must involve only the outermost electrons. An analogy would be fastening two plastic spheres together. You can do this by softening the plastic surface of each with heat or some liquid that dissolves in the plastic and by then pressing the spheres firmly together. Only the surfaces of the two spheres are involved in the joining. In terms of our ideas about atomic structure, we can now say that *the chemical properties of an atom depend on the number of electrons in the outermost energy level of the atom.*

Let us turn to the periodic table to see how the elements with atomic numbers 1 to 20 are arranged. We will look for the numbers that connect the idea of periodic repetition with the idea of electron arrangement. Figure 6.7 is such a table. It is a condensed, pulled-together form of the extended table on the inside cover of the book. In the next section, we will extend and expand to the full table the ideas we develop here. Note also that Figure 6.7 includes the family of helium, He, neon, Ne, and argon, Ar, which were unknown to Dmitri Mendeleev but belong in the periodic table following family 7.

When we note the number of elements in each period, we start to answer the question of how many electrons can occupy each

Number of elements in each period	Family Number: 1	2	3	4	5	6	7	8	Period Number
2	1 H							2 He	1
8	3 Li	4 Be	5 B	6 C	7 N	8 O	9 F	10 Ne	2
8	11 Na	12 Mg	13 Al	14 Si	15 P	16 S	17 Cl	18 Ar	3
	19 K	20 Ca							4

Figure 6.7 A condensed periodic table for the elements of atomic number 1 through 20.

energy level. Let us choose one family of the table as a reference point, family 1. All members of this family have very similar chemical properties, so we can conclude that they must have the same number of electrons in their outermost energy level. This number must be *one* electron because hydrogen is in the family and it has *only one* electron as its total allotment. So we see that this reasoning fits that suggested by the analogy of rolling stones into the gully. The third electron in lithium occupies the first spot in a new energy level. Seven more electrons are added to that level as we proceed across period 2. Then the eleventh electron—which makes sodium (atomic number 11), different from neon (atomic number 10)—goes into the next energy level. The pattern is repeated when we finish period 2 and move on to potassium (atomic number 19), following argon (atomic number 18).

Table 6.1 summarizes the arrangement of the electrons in these first 20 elements of the periodic table. You can compare Table 6.1

Table 6.1   Arrangement of electrons in energy levels of atoms for elements with atomic numbers 1 through 20

Period	Atomic Number	Element	Number of Electrons in Each Energy Level			
			$n = 1$	2	3	4
1	1	H	1			
	2	He	2			
2	3	Li	2	1		
	4	Be	2	2		
	5	B	2	3		
	6	C	2	4		
	7	N	2	5		
	8	O	2	6		
	9	F	2	7		
	10	Ne	2	8		
3	11	Na	2	8	1	
	12	Mg	2	8	2	
	13	Al	2	8	3	
	14	Si	2	8	4	
	15	P	2	8	5	
	16	S	2	8	6	
	17	Cl	2	8	7	
	18	Ar	2	8	8	
4 (start)	19	K	2	8	8	1
	20	Ca	2	8	8	2

with Figure 6.7 to recognize that the family number for each of these elements agrees with the number of electrons in the outermost energy level.

ELECTRON ARRANGEMENT AND THE LONG FORM OF THE PERIODIC TABLE

Let us now turn our attention to the long form of the periodic table, which includes all the elements, not just those of atomic numbers 1 to 20. First, let us count the number of elements in period 4. It now extends to 18 elements rather than eight. We have to go all the way to rubidium, Rb (atomic number 37), before we encounter a metal that behaves as potassium, K, sodium, Na, or lithium, Li, do. Rubidium has the chemical properties that locate it in family 1. Accordingly, we expect that its atoms have one electron in their outermost energy level.

Mendeleev recognized the proper location for rubidium by doubling over the spaces for elements of period 4 (see Figure 5.3). The modern, long form of the table accommodates the extension of period 4 by making a break in the short periods after family 2. Figure 6.8 shows this portion of the modern form of the periodic table. Following calcium, Ca (atomic number 20), ten elements are inserted across the center of the table. Then when we come to gallium, Ga (atomic number 31), we find it located under aluminum, Al (atomic number 13). This procedure then places each following member of the series in a family where it belongs. For example, bromine, Br (atomic number 35), is closely related chemically to chlorine and fluorine, above it. The same is true of krypton, Kr (atomic number 36), which resembles argon and neon.

Figure 6.8   Periods 1, 2, 3, and 4 of the modern form of the periodic table.

Period Number																	
1	1 H																2 He
2	3 Li	4 Be											5 B	6 C	7 N	8 O	9 F	10 Ne
3	11 Na	12 Mg											13 Al	14 Si	15 P	16 S	17 Cl	18 Ar
4	19 K	20 Ca	21 Sc	22 Ti	23 V	24 Cr	25 Mn	26 Fe	27 Co	28 Ni	29 Cu	30 Zn	31 Ga	32 Ge	33 As	34 Se	35 Br	36 Kr

Table 6.2   Arrangement of electrons in the energy levels of atoms for elements in period four of the periodic table

Atomic Number	Element	Number of Electrons in Each Energy Level					Comments
		$n = 1$	2	3	4	5	
18	Ar	2	8	8			end of period 3
19	K	2	8	8	1		A families—electrons added to the
20	Ca	2	8	8	2		outermost energy level
21	Sc	2	8	9	2		B families—electrons added to the
22	Ti	2	8	10	2		next-to-outermost energy level
⋮	⋮			⋮			
30	Zn	2	8	18	2		
31	Ga	2	8	18	3		A families—electrons added to the
⋮	⋮				⋮		outermost energy level
36	Kr	2	8	18	8		
37	Rb	2	8	18	8	1	start of period 5

Table 6.2 indicates the structural feature that is responsible for the extension of period 4. The electron that is added to make scandium, Sc (atomic number 21), different from calcium, Ca (atomic number 20), does not occupy a spot in the outermost energy level but rather is located in the *next-to-the-outermost* energy level. So is the next electron added in the titanium atom, Ti (atomic number 22). This process of adding electrons to the next-to-the-outermost energy level continues until that level holds 18 electrons. Table 6.2 shows that this is the case for zinc, Zn (atomic number 30). Then the regular pattern of having more electrons occupy the outermost energy level resumes. The period ends with the element krypton, Kr (atomic number 36). A new period begins with the next element, rubidium, Rb (atomic number 37), which has one electron in a new, fifth energy level. You can compare the information about electron arrangement from Table 6.2 with the location of the elements in Figure 6.8.

The situation for periods 5, 6, and the incomplete period 7 is similar to that for period 4. We need not go into the full details of the electron arrangement for each atom of these elements. It is

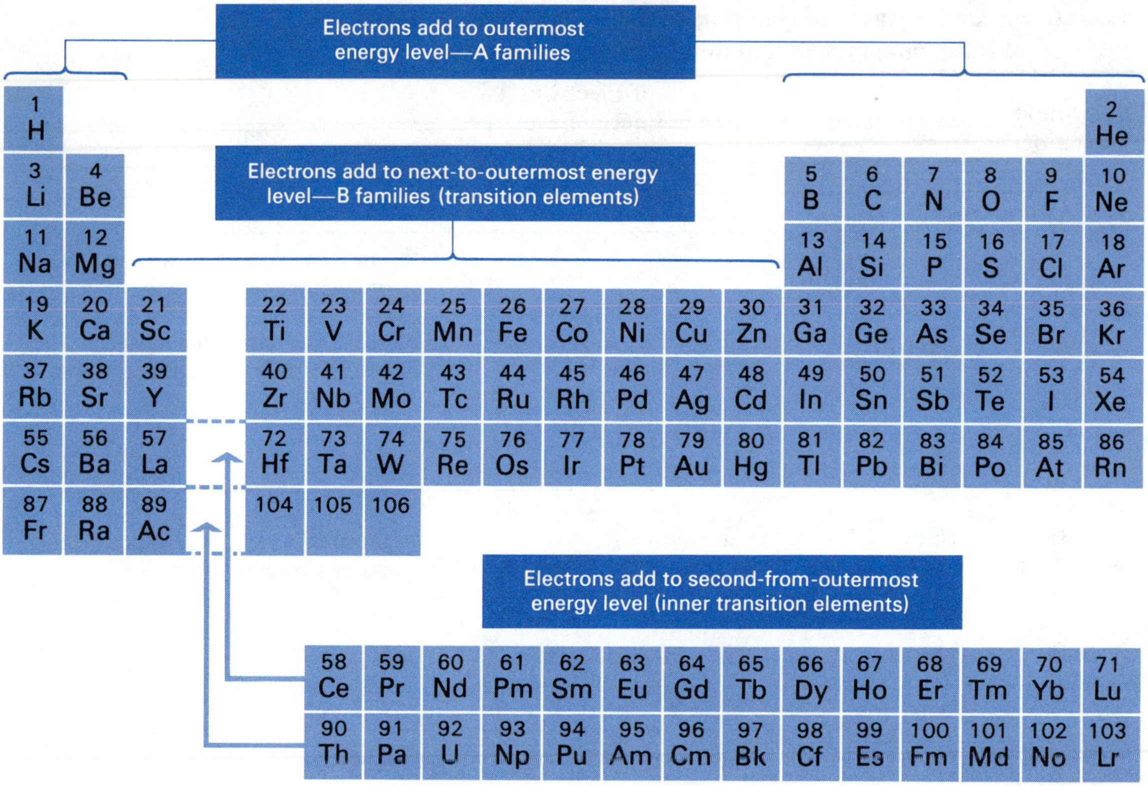

Figure 6.9   Electron arrangement in the atoms of the periodic table.

enough to recognize that one of the great advantages of the periodic table is that it makes this kind of prediction reasonable. Figure 6.9 summarizes the general outline of electron arrangement in all the atoms.

You may note that periods 6 and 7 are extended further. In period 6, 14 elements belong between lanthanum, La (atomic number 57), and hafnium, Hf (atomic number 72). To save space, these elements are set off below the rest of the table. Each atom in these extensions—for example, cerium, Ce (atomic number 58), through lutetium, Lu (atomic number 71)—is different from its preceding neighbor because it has one more electron in the *second-from-the-outermost* energy level.

THE A AND B FAMILIES IN THE PERIODIC TABLE

You probably have noticed that the modern, extended form of the periodic table uses A and B after the numbers designating a family. The reason for this representation is now apparent. The

A families contain not only elements in periods 1, 2, and 3, but also elements in periods 4, 5, 6, and 7. The members of A families in these longer periods are those in which the last electron being added to make one atom different from its preceding neighbor occupies a spot in the outermost energy level. Thus in period 4, gallium, Ga (atomic number 31), adds a third electron in the outermost level, as is the situation for aluminum, Al, and boron, B, above it. In similar fashion, bromine, Br (atomic number 35), has seven electrons in its outermost level, as do the other members of family 7A.

The atoms of the elements in the B families, which appear only in periods 4, 5, 6, and 7, are different. Almost all these atoms hold two electrons in their outermost energy level but have differing numbers of electrons in their next-to-the-outermost energy level. The electrons added to the atoms of the B families go into a different *sublevel* of the next-to-the-outermost energy level. This complexity is better understood in terms of the wave theory than in terms of Bohr's orbits. These elements of the B families are classed as the *transition* elements. The gradations among their properties are very small as we move from one to the next. (In contrast, the gradation among properties of the elements of the A families in any one period are quite large.) The presence of these elements in the long periods provides a transition between the members of families 1A, 2A, and the members of the other A families in the long periods.

THE TRANSITION ELEMENTS

The transition elements are very important in the world we live in. They are all metals. You can identify iron, Fe (atomic number 26), in the middle of the series. The neighboring metals—vanadium, V, chromium, Cr, manganese, Mn, cobalt, Co, and nickel, Ni—all can be mixed with iron to form the various alloys (mixtures of metals) commonly referred to as steel. Tungsten, W (atomic number 74), is used to make the filaments in light bulbs. Copper, Cu (atomic number 29), is widely used in the form of wire and strips to conduct electricity. The "noble" metals—silver, Ag, gold, Au, and platinum, Pt,—are among the transition elements.

Many of the transition elements form compounds that are colored—in contrast to such compounds as table salt, sodium chloride, NaCl, which is a white solid that dissolves in water to form a colorless solution. When the element chromium, Cr (atomic number 24), was discovered, it was named from the Greek word *chroma,* meaning "color," because many of its compounds have a brilliant hue.

Several of the period 4 transition elements are essential to living animals and plants. Their presence is essential in very small amounts. We will describe the importance of these *trace elements* in a later chapter. We have previously noted the need for iron in the hemoglobin of red blood cells. Cobalt atoms form the heart of the important vitamin B_{12}. An enzyme necessary for the function of the liver includes atoms of zinc. The essential role of these transition elements in the life process is related to the fact that their outermost and next-to-the-outermost electrons are in energy levels that are incompletely filled.

The groups of 14 elements in periods 6 and 7, which extend the periods even further, are referred to as the *inner transition elements.* Differences in their electron arrangement are deep-seated. These differences occur in the second-from-the-outermost energy level. Because the differences in electron arrangement are buried so far below the outermost level of electrons, the chemical properties of the inner transition elements are all very much alike.

You will not be expected to remember the details of electron arrangement in all the atoms. However, you can recognize that one of the useful features of the periodic table is that it does suggest to chemists what this electron arrangement is. You can also appreciate that chemists interpret many phenomena in terms of

Our discussion of the connection between the periodic table and electron arrangement is truly a look at what modern chemistry is all about.

the arrangement of electrons in atoms. Thus our discussion of the connection between the periodic table and electron arrangement is truly a look at what modern chemistry is all about.

ONE KIND OF CHEMICAL REACTION: ATOMS GAIN OR LOSE ELECTRONS

We can now take one more step toward connecting the ideas of atomic structure with the observed properties of the elements. Let us start with a consideration of the elements in family 8A. These elements are called the *noble gases.* The majority of the family form no known chemical compounds. They are almost chemically inert except for xenon, Xe (atomic number 54), which can combine rather readily with a few other elements. The term *noble* comes from an old use of the word as it was employed to describe metals. The metals gold and silver were called "noble" because they do not rust or corrode. In contrast, "base" metals, such as copper, lead, and iron, do become discolored with the compounds that readily form on their surface from reactions with air or moisture. Correspondingly, the noble gases tend to be chemically inert. Their inability to react as other atoms do must be the consequence of a common feature of their structure. This feature is their filled outermost energy level.

$$
\begin{array}{llll}
\text{He} & 2 & & \\
\text{Ne} & 2 & 8 & \\
\text{Ar} & 2 & 8 & 8
\end{array}
$$

By contrast, the members of family 7A are very reactive elements. In fact, measurements made on these atoms show that they readily accept electrons from any source to form negatively charged particles. These negatively charged atoms also are called ions, as are atoms that have attained a positive charge by losing electrons. (Recall the use of the term ion in the discussion of positive rays in Chapter 5.) When an atom of an element in family 7A gains an electron, energy is released. For example,

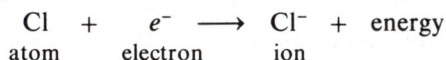

$$
\underset{\text{atom}}{\text{Cl}} \; + \; \underset{\text{electron}}{e^-} \; \longrightarrow \; \underset{\text{ion}}{\text{Cl}^-} \; + \; \text{energy}
$$

This release of energy indicates that the Cl^- ion must be more stable than the Cl atom is. (The Cl^- ion has lower potential energy than the Cl atom does.) We can represent the structures of these species as follows:

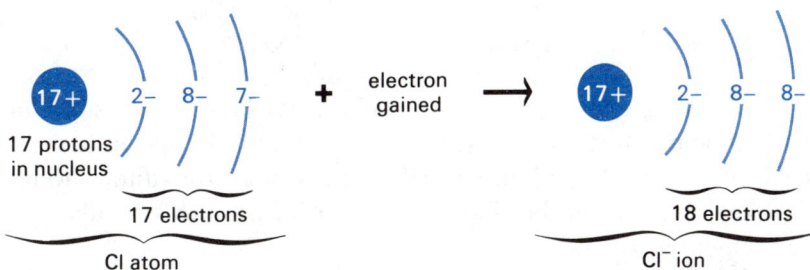

You should note that the Cl^- ion, with a total of 18 electrons, has the same number and arrangement of electrons as does the very stable argon atom (8 electrons in the outermost energy level).

At the other side of the periodic table, we find the elements of family 1A behaving quite differently. Their atoms require the absorption of only a small amount of energy to *lose* an electron. For example,

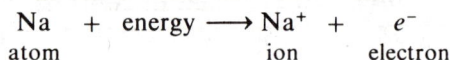

$$Na + energy \longrightarrow Na^+ + e^-$$
$$atom \qquad\qquad\qquad ion \quad\; electron$$

The structural changes are

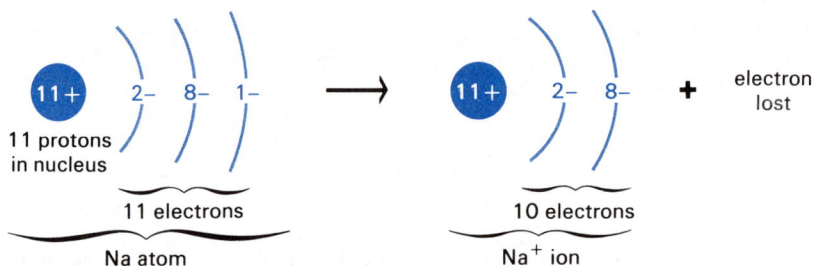

You should note that the Na^+ ion, with a total of 10 electrons, has the same number and arrangement of electrons as does the very stable neon atom (8 electrons in the outermost energy level).

We can predict what to expect if sodium atoms and chlorine atoms react chemically with each other. The sodium atom gives up the one electron it has in its highest energy level and becomes a Na^+ ion with eight electrons in its remaining highest filled energy level. The chlorine atom takes on the electron to completely fill its highest energy level.

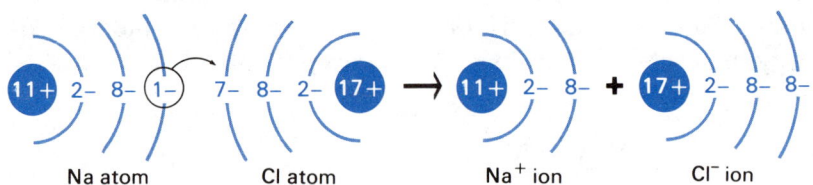

The resulting compound, sodium chloride, NaCl, is a solid in which the positive and negative ions are regularly placed next to each other like the red and black squares of a three-dimensional checkerboard would be. Figure 6.10 shows what this regular arrangement looks like.

This mechanism, the loss and gain of electrons, can be extended to explain why different elements form compounds with different formulas. For example, the element magnesium, Mg

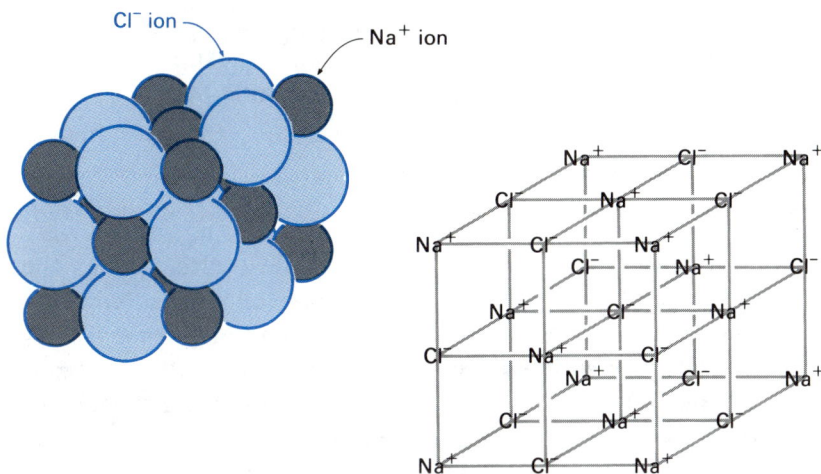

Figure 6.10   The arrangement of the sodium and chloride ions in solid sodium chloride, NaCl.

(atomic number 12), comes right after sodium, Na (atomic number 11). If you refer to Table 6.1, you will find that Mg has *two* electrons in its outermost energy level. This means that *one* Mg atom can supply the needs of *two* Cl atoms.

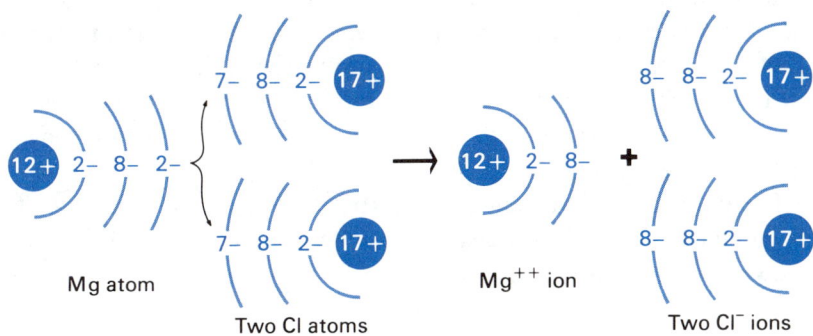

Let us extend these ideas to one more example. What formula do you predict for the compound formed when magnesium atoms react with sulfur atoms? Table 6.1 shows the structures. The prediction is

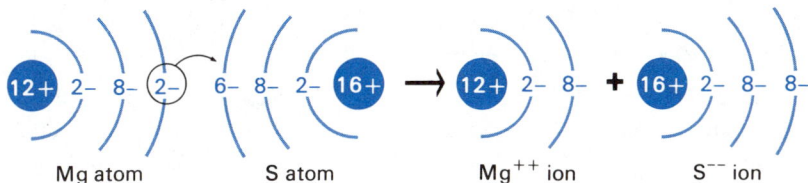

The sulfur atom will accept *two* electrons to gain a filled outermost energy level. One magnesium atom can supply both of these, so the resulting compound has the formula MgS.

PREDICTING THE FORMULAS OF IONIC COMPOUNDS

Chemists designate this ability to gain or lose electrons as the *electrovalence* of the element. The term *valence* means "combining capacity"; it carries no implication of combining *tendency*. By analogy, a person with two free hands has a valence of two; a person with one hand tied behind his or her back has a valence of one. The capacities (two hands or one) tells us nothing about the tendencies to grasp something. Other factors determine how

A person with two free hands has a valence of two; a person with one hand tied behind his or her back has a valence of one.

tightly either person will hold onto something. The term *electro* means that the atom forms an ion with an electric charge. A positive electrovalence means that the atom tends to lose one or more electrons; a negative electrovalence means that the atom tends to gain one or more electrons. The compounds formed by elements that transfer one or more electrons from one atom to another are called *ionic compounds*.

The connection between atomic structure and the position of an element in the periodic table can be used to make the summary shown in Table 6.3.

Table 6.3   The connection between the position of an element in the periodic table and the number of electrons its atoms tend to gain or lose

Family in periodic table	1A	2A	3A	...	6A	7A	8A
Number of electrons in highest energy level	1	2	3	...	6	7	8
Number of electrons an atom tends to *lose*	1	2	3	...			0
Number of electrons an atom tends to *gain*				...	2	1	0
Electrovalence of the element	+1	+2	+3	...	−2	−1	0

You may ask, "What about families 4A and 5A?" The answer is that the atoms of elements in these families do not tend to gain or lose electrons. In fact, most of the members of 3A do not do either of these. The atoms of elements in these middle families undergo many chemical reactions but by a different mechanism. They do not combine into ionic compounds. We will discuss this other mechanism in the next chapter.

One of the characteristics of science as an organized body of knowledge is the great usefulness of the generalizations its theories provide. You can illustrate this by using the above summary to predict the correct formulas of many compounds. If you use the ideas of structure correctly, you can be sure your conclusion about the formula is correct, even though you have never done the experiments. Try Exercise 6.11. In each case, locate the elements in a family of the periodic table and then use the structural information that the location implies to write the correct formula for the ionic compound formed by the combination of the two elements.

Summary

The theory that accounts for the properties of atoms provides a model in which the electrons in an atom are arranged in energy levels. An atom absorbs energy by having its electrons jump to higher energy levels. When an electron falls to a lower energy level, light energy is emitted. Light energy comes in bundles called quanta. Each quantum has a particular energy and also has wave-like properties with a particular wavelength.

Niels Bohr proposed a theory that describes an atom's energy levels as orbits on which electrons revolve around the nucleus. His theory accounts well for the light emitted by excited hydrogen atoms as their electrons jump from level to level. However, a more satisfactory theory describes electrons as waves. The mathematical model provided by this wave theory better accounts for the structure of more-complex atoms, those with more than one electron.

As more and more electrons are added to atoms, with increasing atomic numbers, the arrangement of electrons in energy levels is found to follow a pattern of periodic repetitions. This pattern can be predicted by considering the way in which the periodic table arranges elements in families and periods. The atoms of the A families of the periodic table have electrons added to their outermost energy levels. The atoms of the B families (the transition elements) have electrons added to their next-to-the-outermost energy levels. In two groups of inner transition elements, electrons are added to the second-from-the-outermost energy levels.

Atoms of families 1A and 2A tend to lose electrons when they react with atoms of families 6A and 7A, which tend to accept electrons. The resulting compounds are made up of positively and negatively charged ions and are called ionic compounds. The correct formula of an ionic compound can be predicted from a knowledge of where its constituent elements are located in the periodic table.

Glossary

The number in parentheses indicates the text page where you can find the term defined in context.

atomic energy level   one of several allowed electron states in an atom, in which the electron has a certain potential energy (158)

electrovalence   the tendency of an element to become an ion by losing electrons (positive electrovalence) or gaining them (negative electrovalence) (180)

inner transition elements   the elements in periods 6 and 7 in which the second-from-outermost energy sublevels are being filled, atomic numbers 58 through 71 and 90 through 103 (176)

integral number   a whole number, without fractional parts (166)

ionic compound   a compound made up of ions of opposite charge that attract each other (180)

noble gas   any of the elements in family 8A of the periodic table, which form very few chemical compounds (176)

probability density   a summary of the likelihood that an electron will be found at any given point in the region around an atom's nucleus, usually shown in the form of a diagram (167)

quantum   an indivisible unit or package of energy (162)

quantum number   an integral number designating an energy level or sublevel (164)

spectroscope   a device that analyzes incoming light and records its spectrum (160) *spectrum*   the characteristic pattern of light energies emitted by a sample of matter (158)

standing wave   an unchanging wave motion in a medium, representing a certain quantity of energy (166)

sublevel   a subdivision of an atomic energy level. Sublevels within one energy level have slightly different energies. (175)

trace elements   several elements that living organisms require in very small amounts (176)

transition elements   the metal elements of the B families in periods 4, 5, 6, and 7 of the periodic table, in which the next-to-outermost energy level is not filled to its capacity of electrons (175)

wavelength   the length of one complete wave cycle, from peak to peak or trough to trough; it is inversely proportional to the energy of the wave (158)

Exercises

6.1   When you enter some buildings, you can either walk up a ramp or climb a series of steps. Which of these situations is the better analogy to the way the electrons in atoms exchange energy?

6.2   Use the quantum idea to interpret the following statement: "The 2.1-child family is real only to the Census Bureau."

6.3   To achieve a special effect, a photographer may place a piece of transparent colored glass or plastic over the lens of the camera. What does such a "color filter" do?

6.4   It is possible to give the flames in a fireplace various brilliant colors by sprinkling powdered chemicals on the burning logs. Explain in general terms why this is so and why the chemicals do not make the flames look white.

6.5*a.*   What main ideas of the original Bohr theory of atomic structure are maintained by the modern electron-wave theory of atomic structure?

   b.   What ideas of the Bohr theory are modified by the modern electron-wave theory?

6.6   Imagine an outline diagram of a basketball floor. What do you think the probability density diagram for the location of players would look like for a high-scoring, fast-action game?

6.7   The first energy level for electrons in an atom has no sublevels. As we go to higher energy levels (second, third, fourth, and so on), we find more sublevels in each successive energy level. How is this expansion reflected in the form of the modern periodic table?

6.8   Consider the arrangement of the electrons in each of the atoms in the following pairs of elements. What features are the same in both? What features are different in each?
  a.   Oxygen (8) and sulfur (16)
  b.   Sodium (11) and magnesium (12)
  c.   Iron (26) and cobalt (27)
  d.   Lithium (3) and cesium (55)

6.9   An ion whose symbol is K^+ has 18 electrons. An ion whose symbol is Ca^{++} also has 18 electrons. Since these ions have the same numbers of electrons, why do we not use the same symbol for both?

6.10   Complete the following table. Use the periodic table to supply additional information. The first example is complete as an illustration.

Symbol for Ion	Atomic Number	Number of Protons in Nucleus	Number of Electrons Outside Nucleus	Symbol of the Noble Gas Atom Having the Same Number of Electrons
Na^+	11	11	10	Ne
Ca^{++}	20			
Cs^+	55			
F^-	9			
S^{--}	16			
	35			
		17		
Ba^{++}			54	

6.11 Complete the following table. Refer to the periodic table for necessary information. The first example is complete as an illustration.

	Metal				Nonmetal			Compound
Atom	At. No.	Family	Electro-Valence	Atom	At. No.	Family	Electro-Valence	Formula of Compound
Li	3	1A	+1	Cl	17	7A	−1	LiCl
Na	11			Cl	17			
Na	11			F	9			
Be	4			Cl	17			
Mg	12			F	9			
Mg	12			O	8			
Na	11			S	16			
Ca	20			F	9			
Ca	20			O	8			
Al	13			O	8			Al₂O₃*

*Suggest a reasonable explanation of why this formula is correct.

7 An Introduction to the Structure of Molecules

☐ What can the structure of molecules tell us about the properties of a compound?

☐ Why do some atoms form only one bond, others two, and others even more?

☐ What role do electrons play in the chemical combination of atoms?

☐ Why is carbon a special element?

☐ How do chemists organize the great variety of chainlike and ringlike molecules formed by carbon atoms?

☐ How has chemistry contributed to the conservation of our energy resources?

The logical next step after developing a model for the structure of *atoms* is to see how this model leads to one for the structure of *compounds* formed by atoms. It is important to realize how useful the model of matter at the molecular level is. There is a connection between the observed properties of substances and the structure of the individual molecules. A knowledge of structure enables chemists to build molecules with a desired set of properties. Everyone benefits from drugs that cure disease, fertilizers that enrich barren soil, and synthetic plastics that do not shatter. Even the mistakes we make by not realizing how some things we use end up polluting our air and water may be changed if we learn more about the molecular structure of pollutants. When we know what this structure is, we can predict ways of changing or reacting the molecules to decrease the hazard of their use. Then we have good reason to change our habits and use technology to improve rather than spoil our environment.

Our theme in Chapter 6 was to recognize in broad outline how the number and arrangement of electrons in the outermost energy level of an atom influences its chemical properties. We illustrated this at the end of Chapter 6 with the idea of electrovalence, the capacity some atoms have for gaining or losing one or two electrons. Atoms of family 6A of the periodic table tend to gain two electrons; those of family 7A tend to gain one electron. In both cases, the outermost energy level of their ions becomes filled to capacity by eight electrons. Atoms of family 1A tend to lose one electron; those of family 2A tend to lose two electrons. In both cases, the electrons "peel off," so that what remains in the ion is a completely filled outermost energy level. Consequently, when these two contrasting types of atoms react with each other, electrons are transferred, ions of opposite charge are formed, and the resulting compounds are three-dimensional networks of these ions.

The thought that this cannot possibly be the whole story for the combination of atoms must have crossed your mind. Throughout the first several chapters of this book, we talked about molecules. Certainly all molecules are not formed from positive and negative ions! In ionic compounds, many ions are arranged alternately in big three-dimensional networks, as is the case for NaCl, not grouped into small individual combinations like atoms are in molecules. And some molecules contain identical atoms. Some examples are the molecules H_2, Cl_2, O_2, and N_2. If every chlorine atom tends to gain one electron, we cannot logically expect one chlorine atom to persuade another to give up one electron to form the combination Cl^+Cl^-. We must look for some other mechanism to explain how a Cl_2 molecule is formed. You also may recall that we left out periodic table families 4A and 5A (and most of 3A) in our discussion of electrovalence. These atoms form molecules. We have met some of them in our previous discussions. Methane, CH_4, carbon dioxide, CO_2, ammonia, NH_3, and table sugar, $C_{12}H_{22}O_{11}$, are examples. Are the types of bonds in these molecules different from those we saw in Chapter 6? Obviously, some kind of bond between the atoms hold these molecules together. What kind of mechanism accounts for the structure of these molecules? What role do electrons play in forming these bonds?

ATOMS THAT SHARE PAIRS OF ELECTRONS

A closer look at the number and arrangement of electrons in atoms reveals that some kind of stability appears to be associated with even numbers of electrons. The first energy level in all atoms (except hydrogen atoms) contains two electrons. Noble gases, which are very stable, always have eight electrons (four pairs) in the outermost energy level. Helium, He, has only one pair of electrons. An atom of helium combines with no other atoms. Much additional evidence suggests that a pair of electrons together has more stability or a lower potential energy than two separate electrons. An analogy is the way two compatible roommates can live together more happily and at less cost in money and energy than they can live separately.

If two atoms come together in such a way that they *share a pair of electrons between them,* we can call the resulting stability a *chemical bond.* The atoms are cooperating by each supplying one electron of the pair. In some cases, one atom contributes the pair of electrons to share with another atom. In either case, the resulting bond is called a *covalent bond. The number of pairs of electrons an atom shares is called the covalence of the atom.* (Remember that the term valence refers to the *combining capacity* of an atom.)

The modern theory that considers the electron to be a wave in an atom can be expanded to cover electrons in molecules. The electron wave spreads itself over two atoms instead of just one.

The theory of electrons in atoms can be expanded to cover electrons in molecules.

The two electron waves can be considered to inhabit an energy level for the molecule as a whole. Electrons in these molecular energy levels have lower potential energy and hence are in a more stable arrangement than when they are in energy levels of the separate atoms. A diagram of the probability density for the two electrons in the H_2 molecule looks like Figure 7.1. The figure suggests that the probability of finding an electron is greater between the two nuclei than it is elsewhere around the nuclei. One reason is that an electron is expected to locate where it "feels" the greatest electric attraction of positive charges. An electron located between the nuclei can "feel" the attraction of both, because there it is simultaneously closest to both nuclei. In this sense, the wave theory visualizes a covalent bond as a pair of electrons located between two nuclei.

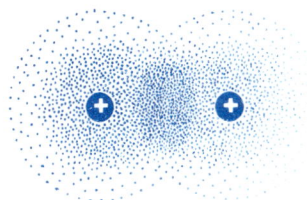

Figure 7.1 The probability density of the two electrons in the H_2 molecule.

ELECTRON-DOT AND BOND-LINE REPRESENTATION OF MOLECULES

We can illustrate the idea of covalence best if we adopt a system of "electron-dot bookkeeping" first suggested by an American chemist, Gilbert N. Lewis, in 1915. The number of electrons in the outermost energy level of an atom is represented by dots surrounding the symbol of the atom, as in Table 7.1. (Remember, you can tell what this number of electrons is for the members of the A families of the periodic table by identifying the family number.)

Table 7.1   Electron-dot representation of atoms

Atom	H	C	N	O	Cl
Family number	1A	4A	5A	6A	7A
Electrons in outermost energy level	H·	·C·	·N·	:O·	:Cl·

Atoms held together in a molecule by a covalent bond are written with the shared pair of dots between the two symbols. Thus for H_2 and Cl_2,

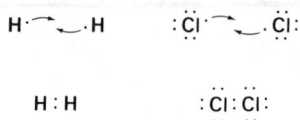

$$H \cdot \smile \cdot H \qquad : \overset{..}{Cl} \cdot \smile \cdot \overset{..}{Cl} :$$

$$H : H \qquad : \overset{..}{\underset{..}{Cl}} : \overset{..}{\underset{..}{Cl}} :$$

In the case of H_2O and NH_3,

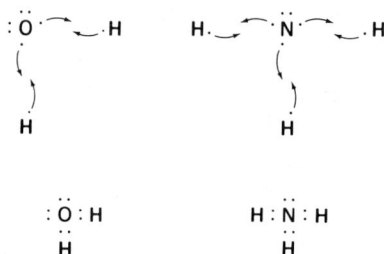

$$: \overset{..}{O} \cdot \smile \cdot H \qquad H \cdot \smile \overset{.}{N} \smile \cdot H$$

$$\overset{\mid}{H} \qquad\qquad \overset{\mid}{H}$$

$$: \overset{..}{O} : H \qquad\qquad H : \overset{..}{N} : H$$
$$\overset{}{H} \qquad\qquad\qquad \overset{}{H}$$

We are deliberately not indicating the electrons of different atoms by different kinds of dots, because all electrons are alike. The electrons in the pair that makes a bond are just like all other electrons except for their location. You also should recognize that trying to represent three-dimensional molecules on a two-dimensional paper surface means that the following all represent the same structure:

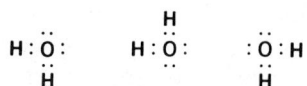

$$\begin{array}{ccc}
 & H & \\
 & \overset{..}{O} & \\
H : \overset{..}{O} : & H : \overset{..}{O} : & : \overset{..}{O} : H \\
\overset{}{H} & & \overset{}{H}
\end{array}$$

A simpler way of showing the covalent bonds in a molecule is to use a line to represent the bond. Such a line always stands for a shared pair of electrons. The unshared pairs of electrons that also may be in the atom need not be shown in this form of representation. Thus H_2, Cl_2, H_2O, and NH_3 also can be shown as

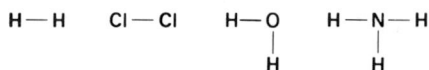

$$\begin{array}{ccccc}
H-H & Cl-Cl & H-O & H-N-H & \\
 & & \overset{\mid}{H} & \overset{\mid}{H} &
\end{array}$$

Table 7.2 compares these two types of representations.

Another important principle appears in this system of representing electrons in atoms and molecules. Each of the two atoms

Table 7.2   Examples of Electron-Dot and Bond-Line Representations of covalent bonds

Substance	Formula	Electron Dots	Bond Lines
hydrogen	H_2	H:H	H—H
methane	CH_4	H:C̈:H (with H above and H below)	H—C—H (with H above and H below)
ammonia	NH_3	H:N̈:H (with H below)	H—N—H (with H below)
water	H_2O	:Ö:H (with H below)	O—H (with H below)
hydrogen chloride	HCl	H:C̈l:	H—Cl
chlorine	Cl_2	:C̈l:C̈l:	Cl—Cl

joined by a covalent bond appears to gain at least a cooperative influence over additional electrons, so that the total number of electrons in the outermost energy level of a bonded atom is equal to that for the atom of the noble gas nearest to it in the periodic table. Each H atom in a H_2 molecule covalent bond is sharing *two* electrons, as are present in the He atom. The C, N, and O atoms in covalent bonds all have *eight* electrons, as does the Ne atom. Each Cl atom in a covalent bond has *eight*, as does the Ar atom. Thus we find again that an atom gains stability by having its outermost energy level filled with eight electrons. The noble gases, with their outermost energy levels already filled, have this stability, so they tend not to react with other atoms.

You may now see why hydrogen is located in two places in the extended form of the periodic table, on the inside cover of the book. Hydrogen's one electron makes its location in family 1A appropriate. But the hydrogen atom also has one less electron than the nearest noble gas, helium. Just as the atoms of all the members of family 7A tend to form one covalent bond by sharing a pair of electrons, so too does the hydrogen atom. Thus hydrogen, H, relates to helium, He, as fluorine, F, relates to neon, Ne, or as chlorine, Cl, relates to argon, Ar.

MULTIPLE BONDS

The elements nitrogen, oxygen, and fluorine all exist as *diatomic molecules* (molecules containing two atoms). Their molecular formulas are similar: N_2, O_2, and F_2. Each of these molecules can be broken apart into atoms by absorbing energy, in the form of either heat or electricity, to break the bond. However, the amounts of energy required vary greatly.

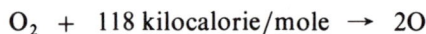

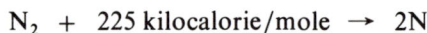

$$F_2 + 36 \text{ kilocalorie/mole} \rightarrow 2F$$

$$O_2 + 118 \text{ kilocalorie/mole} \rightarrow 2O$$

$$N_2 + 225 \text{ kilocalorie/mole} \rightarrow 2N$$

The term kilocalorie/mole refers to the number of kilocalories of energy required to break the bonds in a mole of the molecules. Apparently, the two N atoms are held together most strongly, and the two F atoms, the least strongly. The Lewis electron-dot-bookkeeping system suggests a reason for this difference. The atoms in O_2 and N_2 share more than one pair of electrons. This is often the case with atoms of period 2. Atoms in the other periods only occasionally form *multiple bonds*. When atoms share *two* pairs (four electrons) a *double* bond is created. *Three* shared pairs (six electrons) is represented as a *triple* bond. Table 7.3 shows how the counting is done. In many cases, unlike atoms bond to each other by multiple bonds. Table 7.3 includes a few examples: carbon dioxide, CO_2, familiar to you as dry ice; acetylene, C_2H_2, a gaseous fuel for welders' torches; hydrogen

Table 7.3   Examples of Electron-dot and Bond-line Representations of multiple covalent bonds.

Substance	Formula	Electron Dots	Bond Lines
oxygen	O_2	:O::O:	O=O
nitrogen	N_2	:N::N:	N≡N
carbon dioxide	CO_2	:O::C::O:	O=C=O
acetylene	C_2H_2	H:C::C:H	H—C≡C—H
hydrogen cyanide	HCN	H:C::N:	H—C≡N

cyanide, HCN, a gas that is a deadly poison to breathing animals. Once again, you can see that each atom in a covalent bond gains an influence over enough additional electrons to fill completely its outermost energy level.

THE GEOMETRY OF MOLECULES

The idea that an atom shares additional electrons by forming covalent bonds provides us with more than a game of counting dots. It also allows us to predict what the shape of molecules will be. By shape we mean the form of the space a molecule occupies in three dimensions, not just the way a representation looks on paper, a two-dimensional surface. This is what we mean by the expression "the geometry of molecules." Let us see how our model leads us to this information.

Let us start with the molecule CH_4. If we imagine the carbon in the center of the molecule (the most balanced, symmetrical form the molecule can take), the four hydrogen atoms are on the corners of a tetrahedron (a *four*-sided solid). Figure 7.2 suggests how a model of CH_4 would look if we made one using balls for the atoms and sticks for the bonds. If we recall that a bond represents a pair of electrons, we can see an additional reason for expecting the structure of CH_4 to be a regular (equal-sided) tetrahedron. Each pair of electrons, which have negative charges, tends to repel the other pairs of electrons. So the pairs of electrons are arranged uniformly in space as far from one another as possible. The four bonds made by the central carbon atom of the tetrahedron are at equal angles from one another.

What shape do we expect for the NH_3 molecule? The N atom has five electrons; it makes electron-pair bonds with only three H atoms. The structure we expect is similar to that of CH_4, except that a pair of unused electrons instead of a fourth H atom occupies one of the corners of the tetrahedron. Because our model uses balls for atoms and sticks for bonds, the NH_3 molecule looks like a pyramid (see Figure 7.3).

We move now to the molecule H_2O. The O atom has six electrons; it makes electron-pair bonds with only *two* H atoms. The resulting structure is suggested by Figure 7.4. H_2O is a bent molecule. Two pairs of electrons occupy the other two corners of the imaginary tetrahedron. You can now recognize why we have not represented H_2O in the form H—O—H. Water is not a straight-line molecule.

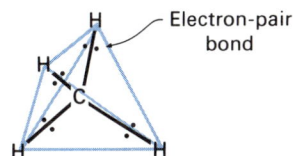

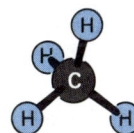

Figure 7.2   The tetrahedral form of the CH_4 molecule.

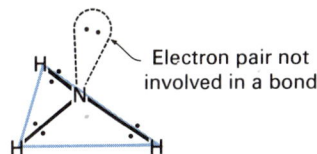

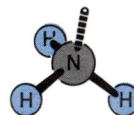

Figure 7.3   The pyramidal form of the NH_3 molecule.

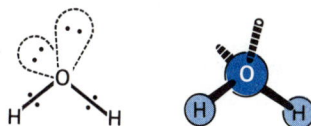

Figure 7.4   The triangular form of the H_2O molecule.

THE SHAPE OF A MOLECULE
DETERMINES ITS PROPERTIES

With these few simple molecular structures in mind, we can illustrate the way chemists interpret some of the properties of substances on the basis of their molecular structures. For example, methane, CH_4, is usually encountered as a gas. It boils at the very low temperature $-164°C$. This means molecules with very little kinetic energy can break away from one another, so we can conclude that the molecules have little attraction for one another. H_2O is usually encountered as a liquid. It boils at $100°C$. This means that its molecules require a large amount of kinetic energy to break away from one another. The molecules must have considerable attraction for one another. How does the difference in the structure of these two molecules account for this contrast in their boiling temperatures?

The bent structure of the H_2O molecule is responsible for many of the interesting properties of water. Let us represent the molecule simply as a slightly bent "potato" (see Figure 7.5). Even though the overall molecule is electrically neutral, the internal "geography" arranges the positive charges of the hydrogen nuclei on one side of the central oxygen atom and some electrons with negative charges on the other side. This unbalanced arrangement of charge within the molecule means that H_2O molecules act as *dipoles* (two sides with different charges). Whenever H_2O molecules are next to one another, the negative side of one dipole experiences an electric attraction for the positive side of a neighboring dipole. A consequence of this dipole attraction is that, in liquid H_2O, where many molecules are close together, all the molecules are held tightly. A large amount of energy must be added to make them break away from one another. You have to heat water to $100°C$ to boil it into gaseous steam.

In contrast to H_2O, the molecule CH_4 has no uneven distribution of the positive and negative charges. You can see that Figure 7.2 would imply a perfectly spherical "potato" with no sides and no uneven charge distribution. Consequently, no dipole attractions exist among CH_4 molecules; they require little energy to break away from one another. Liquid CH_4 boils to a gas at $-164°C$.

Figure 7.5   Unbalanced "geography" of electric charges in the H_2O molecule.

– charge due to electron pairs on oxygen

+ charge due to H nuclei

A LOOK AT THE CHEMISTRY OF CARBON

Carbon is a very special element. Carbon forms more compounds and a greater variety of compounds than any other element. More than two million of these are known. Research reports of new compounds synthesized in laboratories or identified in naturally occurring substances add to the number every day. Virtually all the plastics, synthetic fibers, petroleum products,

Virtually all the plastics, synthetic fibers, petroleum products, medicines, and dyes that you encounter in modern living are carbon-containing compounds.

medicines, and dyes that you encounter in modern living are carbon-containing compounds.

Carbon is the element of central importance to life on the earth. Chains and rings of carbon atoms, along with attached atoms of oxygen, hydrogen, and nitrogen, account for about 99% of the atoms in all living things. In fact, the chemistry of carbon compounds is often referred to as *organic* chemistry. Organic means "relating to or derived from a living organism." In the early part of the nineteenth century, chemists thought that all carbon compounds, except the minerals, such as limestone, $CaCO_3$, could by synthesized only by life processes. Even though as early as 1830 the organic substance urea, known as an animal waste product, had been made from inorganic materials in a laboratory, the name organic remains. An analogy is the way we use the terms "post," "postage," or "postal" in referring to mail. Originally mail was carried by riders or stagecoaches going from one milepost to another along a route. Although this system was abandoned many years ago, the use of the terms remains. *Organic chemistry* is a convenient classification for that branch of chemistry that deals with compounds of carbon. Because so many compounds of carbon exist, organic chemistry is a large field indeed.

The idea that molecules have structure has long helped in bringing organization to the vast information of organic chemistry. The specific idea that carbon is capable of making four bonds and that these bonds are directed toward the corners of a tetrahedron is far older than knowledge about electrons in atoms. The agreement between the older structural ideas and the newer concepts of counting and locating electrons in molecules is typical of the way useful scientific ideas come together to strengthen one

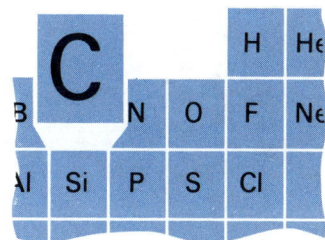

another. Modern theory goes on to calculate the energies associated with electrons in molecules. This calculation provides even more insights into how molecules react and, consequently, into how desired compounds can be made.

Therefore, it is appropriate that our interest in molecular structure takes us into a discussion of how carbon atoms combine. The reason that so many different carbon compounds exist is the carbon atom's unique ability to form very stable bonds with other carbon atoms. One H atom combines with one other to form H_2. In similar manner, two Cl atoms make the stable molecule Cl_2. But that is as far as they go. H_3 or H_4 does not exist. If you look at the electron-dot diagrams, you see why. There is no possibility of forming H_3 (even on paper) because there are not enough electrons to make a bond with a third H atom.

Let us look at the contrast in the way the atoms in a sample of pure carbon bond to one another. A diamond is pure carbon. One of the important properties of diamond is its hardness. You can use a diamond to make a scratch on any other solid you can find. The total value of the diamonds used in grinding tools is far greater than the value of all diamonds used for jewelry. Even your dentist uses a burr that has chips of diamond in it so that he can drill your teeth quickly and efficiently. In diamond, every carbon atom is bound to four other carbon atoms. You can predict from what you know about the carbon atom's structure that each C atom has four other C atoms arranged around it in a tetrahedron. Each C atom uses its four electrons to form covalent bonds with four other C atoms. The result is that diamond is a three-dimensional network of tightly bound C atoms. Any piece of diamond is actually one big molecule. Figure 7.6 suggests how the atoms and the bonds are arranged. This tight bonding of every atom to four others is responsible for the great hardness of diamond. Diamond also is unreactive chemically except at very high temperatures. This property, too, can be explained by realizing that a great deal of energy would have to be added to break the carbon atoms apart so that they could react with some other atoms.

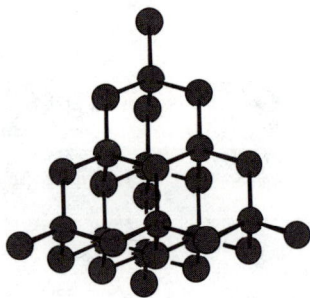

Figure 7.6 The arrangement of carbon atoms and the bonds between them in diamond.

CHAINS OF CARBON ATOMS

Carbon atoms are able to form one or more bonds with other carbon atoms and still use at least one bond to combine with atoms of additional elements. We can see how this ability works if we examine the consistencies in the structures of the various *hydrocarbons.* Hydrocarbons are compounds that contain only atoms of hydrogen and carbon.

The simplest hydrocarbon is a molecule we have discussed, CH_4. The chemical name of this compound is methane. It is the chief constituent of natural gas, the convenient fuel burned in kitchen stoves or house furnaces. Large quantities of natural gas are trapped, under great pressure, in the rocks of some sections of the earth's crust. These supplies are tapped by drilling holes into the rock and controlling the flow of escaping gas through valves and pipes. CH_4 also is formed when some types of bacteria cause dead plant material to decay, as in a swamp. A reasonable speculation is that a similar process eons ago was the source of the natural gas now found trapped in rocks. One of the problems created by our large-scale demand for methane as a fuel is that

Naturally occurring geological events may never again occur to replenish the supply of this natural resource.

naturally occurring geological events may never again occur to replenish the supply of this natural resource. We will be forced to either manufacture methane by other processes or turn to other fuels.

A close chemical relative of CH_4 also is found in some supplies of natural gas. This is the molecule C_2H_6, called ethane. We can see how the structures of these two molecules are related, as follows:

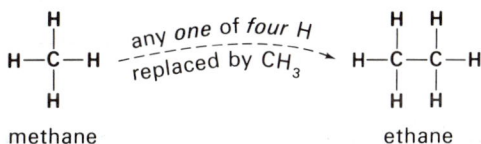

methane   ethane

We use a dashed arrow in this representation because we are using "on-paper" chemistry to represent the structural "picture." The process is an imaginary one, not a true chemical reaction. You should notice several features of this representation. One feature is that we write on the two-dimensional surface of the paper what actually must be a three-dimensional structure. Ball-and-stick models help you to see this better. And any one of the four H atoms in CH_4 is like any other. We end up in C_2H_6 with a stable arrangement in which each C atom has four bonds to it and every H atom has one.

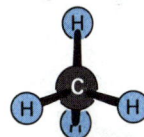

methane

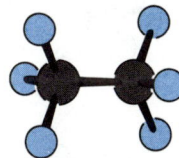

ethane

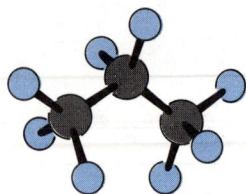

propane

Let us take the next comparable step.

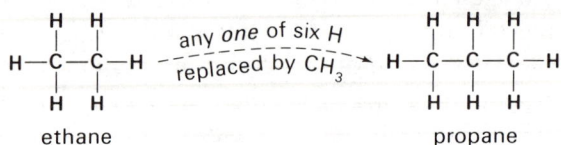

ethane   propane

The molecule C_3H_8 is called propane. Propane occurs in some natural gas deposits. It also occurs dissolved in liquid petroleum or crude oil found in some rocks. Liquid crude oil can be pumped from wells drilled into the rocks and treated in a refinery to make gasoline. A first step in the refining process is to separate the propane. It can then be compressed in tanks and sold in the form of "bottled gas" to be used as a fuel in properly adapted stoves and heaters.

The ball-and-stick models of the molecules show one relationship not clear from the structures on the two-dimensional, paper surface. There is only *one* possible form for the propane molecule. You might think that the following represent different molecules:

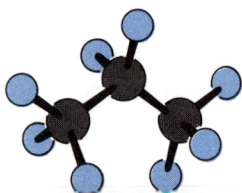

propane

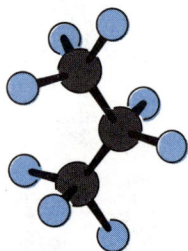

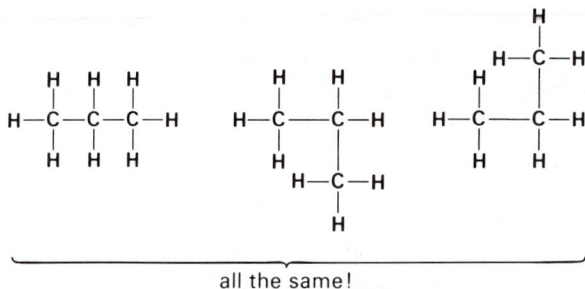

all the same!

Not so! These all represent the same molecule—one in which a string of three C atoms are bonded to one another by single covalent bonds.

ISOMERS

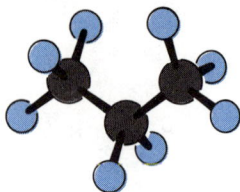

propane

Let us continue our on-paper chemistry game to extend C_3H_8 to the next larger hydrocarbon, C_4H_{10}, called butane. Butane, like propane, also occurs in the mixture of molecules that is crude oil or petroleum. Like propane, it is removed from the mixture in the refining process. When the two gases, propane and butane, are compressed by pressure, they become liquids. Such a mixture is sold in tanks as LPG, *li*quefied *p*etroleum *g*as, for use as a fuel. One popular type of cigarette lighter burns butane. However, unlike propane, butane can take more than one form.

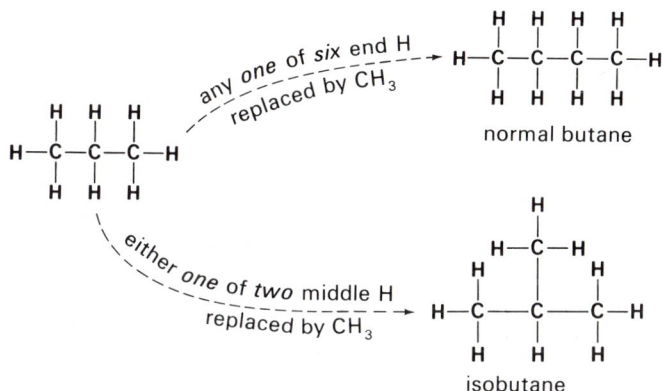

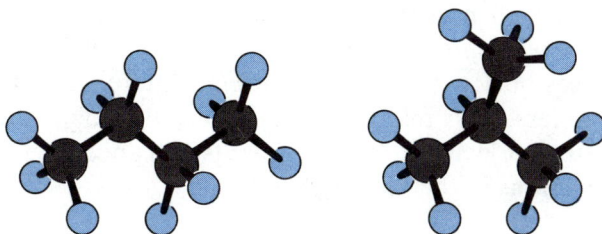

any one of six end H
replaced by CH$_3$

$$H-\overset{\displaystyle H}{\underset{\displaystyle H}{C}}-\overset{\displaystyle H}{\underset{\displaystyle H}{C}}-\overset{\displaystyle H}{\underset{\displaystyle H}{C}}-\overset{\displaystyle H}{\underset{\displaystyle H}{C}}-H$$

normal butane

$$H-\overset{\displaystyle H}{\underset{\displaystyle H}{C}}-\overset{\displaystyle H}{\underset{\displaystyle H}{C}}-\overset{\displaystyle H}{\underset{\displaystyle H}{C}}-H$$

either one of two middle H
replaced by CH$_3$

isobutane

	Normal butane	Isobutane
Melting temperature (°C)	−138	−160
Boiling temperature (°C)	0	−12
Density (g/ml) at −20°C	0.622	0.604

This time we find that *two different* molecules can be constructed. There are two ways to arrange the C atoms to make two possible structures for molecules with the formula C_4H_{10}. One of these is a continuous chain, referred to as a "straight" chain, even though the ball-and-stick models show that it is a zig-zag line. ⌒ The other structure is a branched chain. ⋀ The names given these molecules reflect the difference in structure. *Butane* is C_4H_{10}; of the two possible structures with this formula, *normal butane* is a continuous chain, and *isobutane* is a branched chain. Now you can see how important *structural formulas* are. These representations indicate at a glance what compound is being referred to.

Two or more compounds with the same molecular formula but different structural formulas are called *isomers*. The name means "equal parts" (*iso* plus *mers*). The arrangement of the atoms in each isomer is distinctly different. Consequently, the properties of isomers are different. For example, the boiling temperature of continuous-chain normal butane is 0°C. The boiling temperature of branched-chain isobutane is −12°C.

You can properly expect that this structural game of replacing H with CH_3 can continue. Exercises 7.13 and 7.14 illustrate the complexity that begins to develop with the next larger hydrocarbon, C_5H_{12}. The larger the number of strung-together carbons, the greater the variety of chain-branching that can occur. There are five different isomers of C_6H_{14} and nine different isomers of C_7H_{16}. The predicted number of iosmers of $C_{15}H_{32}$ is over 4000! Only a small fraction of the possible molecules with big formulas has been isolated and identified.

The phenomenon of isomerism (the existence of isomers) is very widespread in organic chemistry. Often very slight differences in structure make a crucial difference in the properties of isomers. This relationship is especially true of the large complicated molecules that living plants and animals use in the chemical reactions of their life processes.

HOMOLOGOUS SERIES

When you line up the formulas of the compounds that you predict by replacing an H atom by a CH_3 group, a regularity becomes apparent. Each formula differs from the previous one by one C and two H.

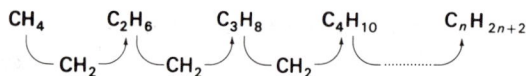

$$CH_4 \quad C_2H_6 \quad C_3H_8 \quad C_4H_{10} \quad C_nH_{2n+2}$$
$$CH_2 \quad CH_2 \quad CH_2$$

The generalized formula for this whole series of compounds is C_nH_{2n+2}. A series like this containing molecules with similar structures is called a *homologous series* (*homo* comes from a Greek word meaning "same"). Many members of the C_nH_{2n+2} homologous series are found in petroleum. Most crude oil consists of medium-sized members of the series, such as C_7H_{16} through $C_{18}H_{38}$, that are liquids at ordinary temperatures. The smaller members, such as CH_4, C_2H_6, and C_3H_8, normally are gases and are dissolved in the liquid mixture. Also present in petroleum are some larger molecules containing as many as 25 or

30 carbon atoms. These hydrocarbons are waxes that have been dissolved in the liquid mixture. The kind of wax used to seal jars of homemade jelly is such a compound. The series of compounds with formulas represented by C_nH_{2n+2} is called the *paraffin series*.

Once again we can see how useful it is to adopt a system of classification. The members of a homologous series, related by similar structures, have similar chemical properties. If one member reacts in a particular way, others in the series can be expected to react that way also.

SUBSTITUTION REACTIONS

If methane, CH_4, is mixed with chlorine, Cl_2, at room temperature, no reaction occurs. However, if energy is added in the form of light or heat, reaction does occur. Let us see what to expect. The only really possible reaction is one in which a Cl atom takes the place of a H atom. The displaced H atom combines with the other Cl atom of the Cl_2 molecule.

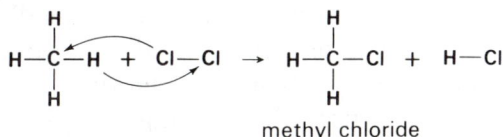

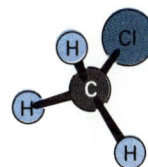

methyl chloride

This type of reaction is called a *substitution reaction*. You can also expect that, if one H in CH_4 can be replaced by Cl, so can the others.

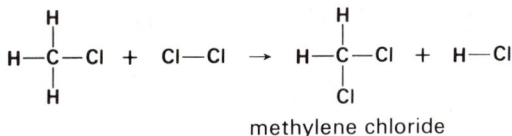

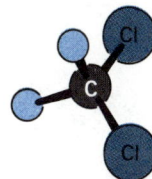

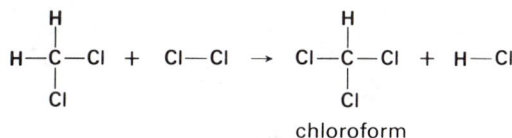

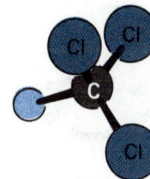

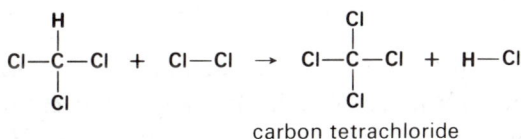

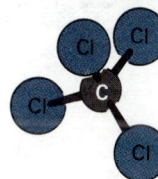

methylene chloride

chloroform

carbon tetrachloride

When CH_4 and Cl_2 react, a mixture of products is formed.

Two of the resulting compounds are ones you probably have encountered. One of these is chloroform, $CHCl_3$. Years ago, a mixture of chloroform gas in air was used as an anesthetic. However, too much chloroform can kill a person or an animal; consequently, its use has been dropped in favor of less hazardous substances. Experiments with animals have indicated that chloroform also may cause cancer. Extremely minute amounts of chloroform have been discovered in some municipal water supplies. It is probably formed by a reaction similar to those we are describing here. Chlorine gas is put in the water to kill dangerous bacteria. Some of it may react with hydrocarbons also in the water to make chloroform. The amounts produced are so small that they probably represent no danger. But the warning is clear, and steps should be taken to make sure no large amounts can occur in water consumed by people.

The other compound you may be familiar with is CCl_4, carbon tetrachloride. It formerly was used in commercial dry-cleaning establishments, and it still can be purchased in small amounts as a spot remover. Carbon tetrachloride also is poisonous. The structural similarity of carbon tetrachloride to chloroform should convince you to heed the warning always to use carbon tetrachloride in a well-ventilated room. If the air you breathe contains even small concentrations of vaporized CCl_4, you are in danger.

MULTIPLE BONDS IN HYDROCARBONS

Two carbon atoms can share more than one pair of electrons between them; they can share two or three pairs. The simplest double-bonded hydrocarbon is C_2H_4, named ethylene or ethene; the simplest triple-bonded hydrocarbon is C_2H_2, named acetylene.

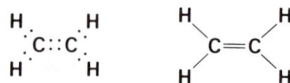

ethene or ethylene

acetylene

Both of these substances are gases at normal temperatures and pressures. Each of these molecules has a geometrical form different from single-bonded ethane, C_2H_6. Ethylene, C_2H_4, is a flat, or *planar,* molecule. In the plane, it is as symmetrical (evenly spaced) as possible. The angles between the two C—H bonds and the C—C bond are equal, one-third of a circle, or 120°. Acetylene, C_2H_2, is a straight-line, or *linear,* molecule.

Just as we found that C_2H_6 has larger relatives, such as C_3H_8, C_4H_{10}, and so on, so also a homologous series of hydrocarbons exist in which two of the carbons are joined by double bonds.

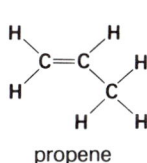

Propene

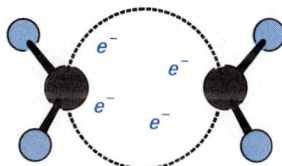

Butene

propene

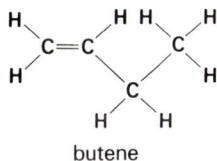

butene

The generalized formula for members of this series is C_nH_{2n}. The names chemists use for them are similar to the ones used for the paraffin series. The final endings *ane* are changed to *ene.* Thus C_3H_8 is propane, C_3H_6 is propene. Some members of this series also are found in petroleum. Large amounts of the smaller molecules, C_2H_4, C_3H_6, and C_4H_8, are produced in the process of refining petroleum.

ADDITION REACTIONS AT DOUBLE BONDS

In contrast to the limited reactivity of the members of the paraffin series, hydrocarbons containing the C=C bond are quite reactive. The members of the paraffin series are referred to as *saturated hydrocarbons;* all the carbon bonds are "saturated" with other atoms. In contrast, the C_nH_{2n} molecules are referred to as *unsaturated hydrocarbons.* Other atoms can add to these molecules to "saturate" the carbon bonds. This difference in reactivity you can connect with the difference in structure. In C_2H_6, all the electrons bonding the molecule together are inside the molecule, but this is not so in the molecule C_2H_4. We can visualize the four electrons involved in the C=C linkage to be so piled up between the two C atoms that they extend in space above and below the plane of the molecule.

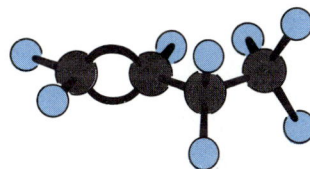

If some other molecule bumps into a C_2H_4 molecule, it may be attracted to this exposed pile of electrons and react by combining and then readjusting the structure to a more stable arrangement. This apparently is what happens. For example, C_2H_4 and Cl_2 react very readily. This reaction is quite different from the reaction between CH_4 and Cl_2.

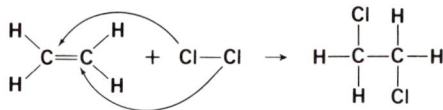

Only one product molecule is formed, $C_2H_4Cl_2$. This type of reaction properly is called an *addition reaction*.

Many other molecules undergo addition reactions with molecules containing C=C bonds. Hydrogen gas, H_2, under pressure in the presence of a catalyst, will so react. Semisolid vegetable fats, such as Crisco, are made from liquid vegetable oils by this process. The vegetable oils contain large molecules in which there are several C=C linkages. Changing the "unsaturated" molecules to "saturated" ones, by adding two H atoms at every C=C, changes the properties of the liquid oil to those of the solid fat.

RINGS OF CARBON ATOMS

In addition to forming continuous or branched chains, carbon atoms can join to form rings. If the end carbon atoms on a continuous chain join together, a *cycloparaffin* is formed. The ring with the greatest stability at ordinary temperatures has six carbon atoms. This molecule C_6H_{12}, is called cyclohexane (*hex* means "six").

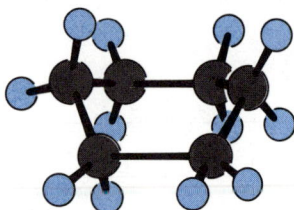

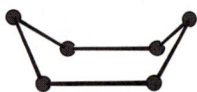

Boat

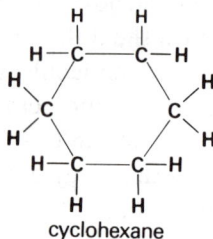

cyclohexane

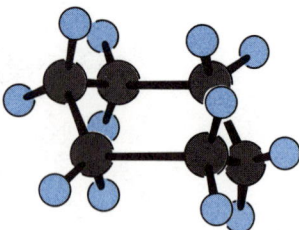

Chair

Cyclohexane

The smallest possible ring has three carbon atoms and is appropriately named cyclopropane, C_3H_6. Cyclobutane, C_4H_8, and cyclopentane, C_5H_{10}, also exist. All these molecules and others with slightly larger rings or rings in which one or more H atoms have been replaced by —CH_3 are found in petroleum.

Another kind of carbon ring is much more frequently encountered. The simplest molecule of this type is benzene, C_6H_6, just as CH_4 is the simplest paraffin hydrocarbon. Benzene is a clear liquid that freezes to a solid at about 5°C. It has a slightly pungent odor and does not mix with water. Early in the nineteenth century, benzene and related compounds were identified in the tars and gases obtained by heating coal without air. Probably because of the variety of tar-like odors these substances have, they were originally classed as *aromatic hydrocarbons,* a name

---

Because of the tar-like odors these substances have, they were originally classed as aromatic hydrocarbons.

---

that is still used. These aromatic ring molecules are the raw materials for many chemical industries. Most dyes and medicines contain molecules with ring structures like that of benzene.

The structure of benzene differs considerably from the structures of all the other hydrocarbons we have discussed. A ring of six C atoms, each with one H atom attached, looks reasonable.

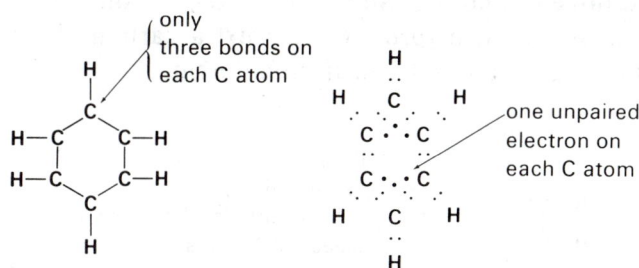

Neither of these representations is correct.

But this ring structure defies our counting up electrons and assigning each to some kind of bond. In the bond-line structure, each carbon atom is shown with only three bonds. In the electron-dot structure, each carbon atom holds one electron that is not assigned to some bond. We could try to get around this problem by giving the carbons of the ring alternate single and double bonds.

However, some chemical evidence argues against this representation. When a reaction substitutes another atom, X, for each of the H atoms on *adjacent* C atoms, the above structure predicts *two* possible isomers, each with different properties.

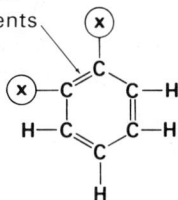

possible structure A   possible structure B

(Note that in A, the two X's are on carbons joined by a *double* bond; in B the two X's are on carbons joined by a *single* bond.) Actually, only *one* product molecule is formed, never two different ones. Only *one* structure is possible for benzene with two substituents for H atoms on adjacent C atoms. *The bonds around the ring must all be equivalent.* There cannot be three bonds of one type and three of another. So we are forced to consider benzene to have a unique structure with the extra carbon electrons in some kind of generalized bond around the ring.

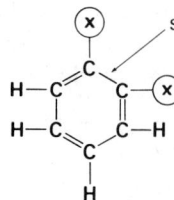

Only *one* structure is possible for benzene with two substituents for H atoms on adjacent C atoms.

A generally agreed-upon convention represents this structure

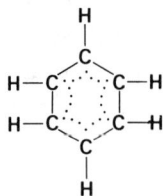

benzene

in the simple form

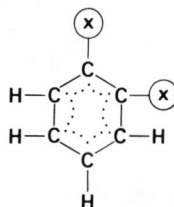

benzene

Note that neither the C atoms nor the H atoms attached to each C need be shown.

Chemists know about a tremendous variety of compounds related to benzene, in which one or more of the H atoms has been replaced by other atoms or groups. One example is toluene, $C_6H_5CH_3$, a substance widely used as a solvent for oil and grease or as a thinner for paints and varnishes.

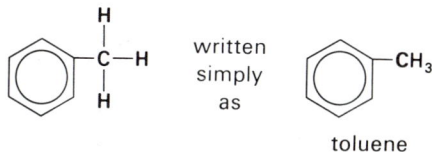

toluene

Because all the H atoms on the benzene ring are equivalent, there is only one possible form for the toluene molecule.

Molecules also can be found in which two or even more rings are fastened together. Naphthalene, $C_{10}H_8$, a substance widely used to keep moths out of stored woolen clothing, is an example. Its structural formula is

naphthalene

THE MANUFACTURE OF GASOLINE

No one living in modern society can fail to appreciate the important role played by petroleum, both in our individual lives and in international trade. Nor can anyone be ignorant of the threat to our future standards and styles of living implied by growing demand in the face of dwindling resources. Everyone must translate that awareness into whatever action he or she can take to lessen the demand and hence preserve some of the resources. Everyone can be a conservationist by using less gasoline and by using what is needed more efficiently. What people often

What people often do not understand is the conservationist role that chemical information plays.

do not understand is the conservationist role that chemical information plays. If every barrel of crude oil pumped from the ground now yielded no more engine power than a barrel of crude oil did in the early years of this century, petroleum resources would be consumed almost ten times as fast as they actually are.

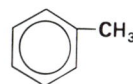

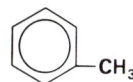

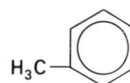

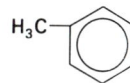

all the same!

This saving is possible because we get more and better gasoline from a barrel of crude oil. Better refining processes come from knowing more about the structure and properties of the hydrocarbon molecules. The concepts of homologous series and isomerism are key ideas in this understanding.

The crude oil that goes to a refinery is a complex mixture of all the types of hydrocarbons we have discussed: continuous and branched chains, paraffins and unsaturated series, cycloparaffins and aromatics. In addition, many members of homologous series of each of these types are present. In all homologous series, the smaller molecules, the ones with fewer carbon atoms, have lower boiling temperatures. Figure 7.7 shows how boiling temperature varies with the number of carbon atoms in continuous-chain paraffin molecules. The relationship represented by Figure 7.7 is typical of all homologous series.

One way to accomplish at least a partial separation of a mixture such as crude oil is by *distillation*. The complex mixture can be separated into *fractions,* each of which contains a limited mixture of compounds. Figure 7.8 is a diagram of a laboratory apparatus used to accomplish this. The liquid in the pot is heated. When the temperature of the mixture is still low, the substances with low boiling temperatures change to gases and bubble out of the mixture. These are the molecules with few carbon atoms in them. As the temperature rises, successively larger molecules boil from the liquid mixture. The vapors boiled from the mixture in the pot rise through the fractionating column. The fractionating column is a tube packed with glass beads or some other material with a large amount of surface. There the gas molecules partially recondense to liquids on the surface of the beads and drip

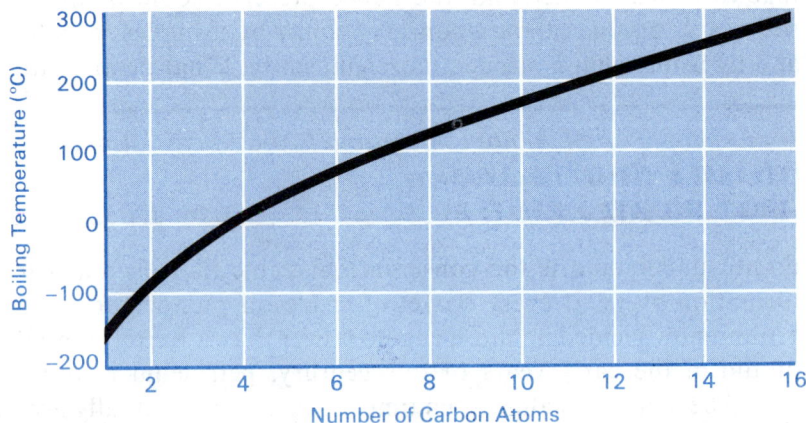

Figure 7.7 Boiling temperatures of continuous-chain hydrocarbons of the paraffin series.

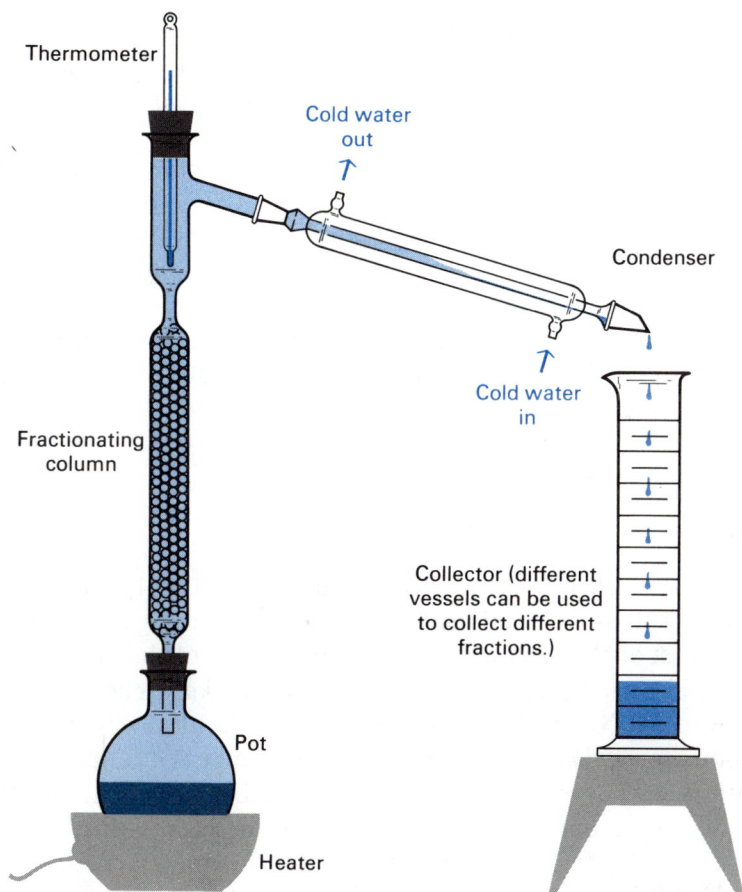

Figure 7.8 Laboratory apparatus for the fractional distillation of a mixture of liquids.

back down, to be bathed by more rising vapors. Gradually, the substances that become gas at the lower temperatures work their way up through the column and consequently leave first. They go past the thermometer into the condenser. The condenser is a glass tube that is kept cold by a jacket containing cold water. In the condenser, the distilling vapors condense to liquids and drip into the collector. As the components with low boiling temperatures are removed, the temperature of the pot and the column rises, because substances with higher boiling temperatures remain. Gradually, these also rise as vapors, enter the condenser, become liquids, and drip into the collector. If a series of vessels is used to collect each fraction of the vapors boiling over a limited range of temperatures, the original mixture becomes separated into fractions.

The apparatus used for fractional distillation in a petroleum refinery is a much larger unit made of steel pipes and columns arranged for continuous operation. But the principle is the same as that illustrated by the laboratory apparatus. It is seldom practical to try to separate out individual hydrocarbons by this method. But each fraction collected over a range of temperature —for example, from 100°C to 180°C—contains molecules of nearly the same size. If you refer to Figure 7.7, you see that molecules of the paraffin series with 7 to 10 carbon atoms boil in this range of temperature. Hydrocarbons in this boiling range make the best gasoline. Smaller ones tend to evaporate from the automobile gasoline tank, at least in hot summer weather; larger ones do not vaporize readily enough to mix properly with air in the engine cylinders.

INCREASING THE YIELD OF GASOLINE FROM CRUDE OIL

When a typical crude oil is fractionally distilled, only about 15% to 20% of the original substance shows up in the gasoline fraction. Most of the mixture of crude oil consists of the larger molecules. For example, the fraction suitable for kerosene boils in a range of 180°C to 250°C and contains molecules with 9 to 14 carbon atoms. Diesel fuel, boiling in the range 250°C to 400°C, contains still larger molecules. Lubricating oils are further up the line, both in boiling temperature and molecular size. These higher-boiling fractions from crude oil have important uses, but the demand for them is not nearly as great as the demand for gasoline.

This imbalance between the amount of the gasoline fraction in crude oil and commercial demand for gasoline was tackled as a chemical problem. The first process to be developed was "cracking" larger molecules into smaller ones. The first discovery was that this cracking can be accomplished by heat and pressure. A typical reaction is

$$C_{16}H_{34} \rightarrow C_8H_{18} + C_8H_{16}$$

In the 1930's, catalysts for this type of reaction were discovered, which made it possible to convert a wide range of large hydrocarbon molecules into those appropriate for gasoline. This process is widely used in petroleum refineries. The most prominent structures in any refinery are the tall "cat crackers," the maze of pipes, valves, and tanks in which *catalytic cracking* reactions are going on.

A "cat cracker" where heavy hydrocarbon molecules are made into lighter ones suitable for gasoline.

Also in the 1930's, a reverse process was developed to build larger molecules out of smaller ones. Some crude oil has relatively large amounts of C_4 hydrocarbons. Use of proper conditions of temperature, pressure, and catalysts converts two C_4 molecules into one C_8 molecule

$$C_4H_{10} + C_4H_8 \rightarrow C_8H_{18}$$

By using these two types of reactions, cracking and building, more valuable gasoline can be taken from crude oil. But it is not possible to make 100% of any sample of crude oil into gasoline. Even if this were scientifically possible, it would not be feasible for economic reasons. But knowledge of these chemical reactions, how to go down or up homologous series, has played a decisive conservationist role by making it possible to convert a larger portion of crude oil into gasoline.

IMPROVING THE QUALITY OF GASOLINE

The other part of a discussion of how better gasoline is made takes isomerism as its central theme. By better gasoline, we mean gasoline that can burn more smoothly with less waste and that allows the engine to operate most efficiently. The search for such a gasoline also is a conservation measure.

When an automobile engine "knocks," power is being lost. The "ping" that you may hear when pulling up a steep grade means that the mixture of gasoline vapor and air in the cylinders is igniting before it should and is burning unevenly. More than engine design and adjustment is involved in the problem. Knocking also depends on the structure of the molecules in the fuel. The

The greater the proportion of branched-chain molecules in the gasoline, the less it will tend to knock.

greater the proportion of branched-chain molecules in the gasoline, the less it will tend to knock.

The quality of the gasoline you buy for your automobile is expressed on a scale of *octane numbers*. The basis for this rating is as follows. A gasoline being rated is used to operate a test engine. The gasoline's performance is compared to the performance in the same engine of standard fuels containing varying amounts of only two hydrocarbons. One of these two is a continuous-chain paraffin, C_7H_{16}, called normal heptane (*hept* means "seven"). This substance knocks much more than ordinary gasoline. The other ingredient of these special mixtures is a branched-chain paraffin, C_8H_{18}, called *isooctane* (*oct* means "eight").

normal heptane

(knocks badly)

isooctane

(knocks very little)

Normal heptane

Isooctane

When a gasoline behaves the same in the test engine as a mixture of 90% isooctane and 10% normal heptane does, the sample is given a rating of 90 octane number. Thus the octane number of a particular gasoline is the percentage of isooctane in this synthetic blend that matches the performance of that particular gasoline.

The octane number, the antiknock quality of a gasoline, can be improved by adding a small amount of the compound lead tetraethyl, $Pb(C_2H_5)_4$ (*Tetra* means "four"; *ethyl* is C_2H_5, derived from ethane, C_2H_6; *lead* is Pb). Just how this compound works in the engine is not entirely understood. Apparently, when it decomposes in the hot cylinder, the C_2H_5 groups are set free and are involved in the combustion reactions of the other hydrocarbons to smooth out the burning process.

In recent years, a great deal of concern has developed over the consequences of widespread use of lead tetraethyl. The lead atoms form gaseous compounds that are emitted into the atmosphere along with the engine exhaust. The ultimate fate of this lead is the cause for concern. All heavy metals, such as lead, are poisonous to living organisms. Lead is known to interfere with the body's manufacture of hemoglobin. At the present time, evidence suggests that most of the lead goes from the atmosphere into stable compounds in the soil. What remains in the atmosphere has not increased beyond the limits of human bodies' normal defenses. But the evidence also is strong that levels of lead contamination in soil and atmosphere have been increasing with alarming rapidity in the years since 1940, when $Pb(C_2H_5)_4$ began to be widely used in gasoline. Lead in the soil may find its way into plants that are used for food by humans. Ultimately, some danger point is sure to be reached. We must stop adding lead to the atmosphere before then. Legal restrictions are now in force to reduce drastically the amount of lead tetraethyl in gasoline.

Another reason for keeping lead tetraethyl out of gasoline is the design of engine exhaust systems in the newer automobiles. These devices, called catalytic converters, contain catalysts for reactions that can remove smog-producing molecules from the exhaust gases. Lead spoils the action of these catalysts, so lead compounds cannot be present in the gasoline that is burned. To keep lead tetraethyl out of gasoline, automobile engines must be designed to run efficiently on lower-octane-number fuels. The composition of gasoline also must be altered to contain more of the hydrocarbon isomers with good antiknock qualities; then lead tetraethyl will not be necessary in gasoline.

Summary

Atoms of elements in families 4A, 5A, 6A, and 7A, as well as hydrogen, are able to share pairs of electrons with other atoms. The bond formed by such sharing action is called a covalent bond. Some atoms, especially those of period 2, form multiple covalent bonds.

The structural shape of molecules has an influence on the properties of the substance. This is particularly notable in the case of water. The bent shape of the H_2O molecule makes it a dipole, with uneven distribution of electric charges. Consequently, H_2O molecules have attractions for one another; this gives water its relatively high boiling temperature.

Carbon, the element essential to life, exhibits a unique tendency to form chains and rings, with the C atoms bonded together. Chains can be in a line or be branched. This makes it important to use structural formulas to identify molecules. Two molecules with the same numbers of the same kinds of atoms but with different structural formulas are called isomers. A series of molecules with similar structural formulas, differing by a —CH_2— unit, is called a homologous series.

Hydrocarbons—compounds of hydrogen and carbon atoms—can undergo substitution reactions in which some other atom takes the place of a hydrogen atom. Hydrocarbons with a C=C, a double bond, can react with other substances by addition reactions.

Crude oil is the source of gasoline. The naturally occurring gasoline in a crude oil can be separated out by fractional distillation. It is also possible to convert large hydrocarbon molecules, such as those normally in oil or diesel fuel, into molecules suitable for gasoline by the process of cracking. The quality of gasoline is improved by the presence of branched-chain hydrocarbons. The compound lead tetraethyl also can be added to improve gasoline quality. The ability to make larger amounts of gasoline and more powerful gasoline from a given amount of crude oil is one way chemists help conserve energy.

Glossary

The number in parentheses indicates the text page where you can find the term defined in context.

addition reaction   a chemical reaction in which two molecules combine to become one molecule, usually describes the reaction of a molecule with a carbon-carbon double or triple bond (204)

aromatic hydrocarbon   a carbon and hydrogen compound containing one or more benzene-like rings (205)

catalytic cracking   an industrial process in which heat, pressure, and catalysts are used to convert long-chain hydrocarbons from crude oil into shorter ones suitable for use as gasoline (210)

chemical bond   a condition in which two atoms remain close to each other because this linkage offers an energy state lower than those of the separate atoms (188)

combining capacity   the ability (but not the tendency) of an atom to form bonds with other atoms (188)

covalence   the number of single bonds that can be made (electron pairs that can be shared) by an atom (188)

covalent bond   a chemical bond formed by the sharing of pairs of electrons by two atoms (188)

cycloparaffin   a saturated hydrocarbon molecule whose carbon atoms form a ring (204)

diatomic molecule   a molecule containing two atoms (192)

dipole   a body within which the positive and negative electric charges are unevenly distributed (194)

distillation   a method by which the components of a mixture are separated out according to their boiling points (208)

homologous series   an ordered group of compounds in which each member is larger than the preceding member by a constant CH_2 structural unit (200) *hydrocarbon* a compound made up only of carbon and hydrogen atoms (197)

isomer one of two or more molecules that have the same chemical formula but differ in the spatial arrangement of their atoms (200)

linear molecule a molecule all of whose atoms are arranged in a straight line (203)

multiple bond the sharing of more than one pair of electrons by two atoms (192)

octane number a performance rating for gasoline, based on its behavior compared with those of various standard mixtures of isooctane and normal heptane (211)

organic chemistry the study of carbon-containing compounds (195)

paraffin series the homologous series of saturated hydrocarbons, general formula C_nH_{2n+2} (201)

planar molecule a molecule all of whose atoms lie in one plane (203)

saturated hydrocarbon a carbon- and hydrogen-containing molecule that has no multiple bonds (203)

structural formula a diagram that shows the spatial sequence in which the atoms of a compound are joined to each other (199)

substitution reaction a chemical reaction in which an atom on one molecule exchanges places with an atom on another molecule; there are two reactants and two products (201)

unsaturated hydrocarbon a carbon- and hydrogen-containing molecule that has multiple bonds (203)

Exercises

7.1 How could the phrase "get married and settle down" be interpreted to describe the behavior of electrons in molecules?

7.2 Contrast what happens to electrons when atoms form ionic bonds with what happens to electrons when atoms form covalent bonds.

7.3 Consider the following formulas of hydrogen compounds: CH_4*, NH_3*, H_2O*, HF*.

   a. How many covalent bonds does each starred atom make?

b. Locate each starred atom in the periodic table. Make a generalization that connects the number of an element's family with the number of covalent bonds it forms.

7.4   The element phosphorus (P, atomic number 15) forms the molecule P_4. It has a structure that looks like this:

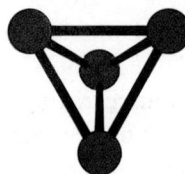

Locate P in the periodic table. Count the number of electrons in the outermost energy level of the atom. Account for the structure of the P_4 molecule by making an electron-dot diagram.

7.5   The element sulfur (S, atomic number 16) forms the molecule S_8. It has a structure that looks like this:

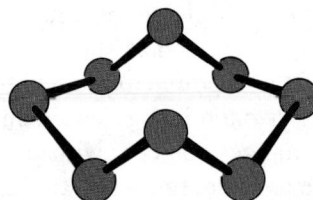

Locate S in the periodic table. Count the number of electrons in the outermost energy level of the S atom. Account for the structure of the S_8 molecule by making an electron-dot diagram.

7.6   Consider the structure of the three molecules CH_4, NH_3, and H_2O.

a. Do you think NH_3 will have some of the dipole character of H_2O? Explain.

b. Use your answer to 7.6*a* to predict the relative strengths of attraction that exist between molecules in liquid CH_4, liquid NH_3, and liquid H_2O.

c. Carry the idea that differences in structure allow you to predict differences in properties one step further. CH_4 boils at $-164°C$; H_2O boils at $+100°C$. Which of the following would you expect to be the boiling temperature of liquid NH_3: $-200°C$, $-33°C$, $+150°C$? Justify your answer.

7.7   How does the electron-dot bookkeeping system of representing the structure of molecules account for the fact that the energy required to separate the F_2 molecule into two F atoms is 36 kcal/mole, while that to separate O_2 into two O is 118 kcal/mole, and that to separate N_2 into two N is 225 kcal/mole?

7.8   Draw the electron-dot diagrams for each of the following molecules. (Only single bonds are involved.)

    a.   Chloroform, one of the first anesthetics $CHCl_3$

    b.   Methyl alcohol, also known as wood CH_3OH
    alcohol

    c.   Arsine, a very poisonous gas. (Hint: Find AsH_3
    arsenic, atomic number 33, in the periodic
    table.)

    d.   Hydrogen sulfide, a foul smelling gas H_2S
    released by rotten eggs. (Locate sulfur in the
    periodic table.)

7.9   Draw the electron-dot diagram for the following molecules. Observe that the atoms in these multiple-bond structures also illustrate the rule that a covalently bonded atom has eight electrons around it. Remember that each single bond represents a pair of electrons.

    a.   Ethylene, a com- C_2H_4
    pound used to make
    plastics

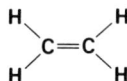

    b.   Formaldehyde, a CH_2O
    preservative for
    biological specimens

    c.   Urea, an animal H_2NCONH_2
    waste product

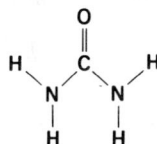

7.10   In each of the following sets, choose the structure that does *not* represent the same molecule as do the other two.

a.

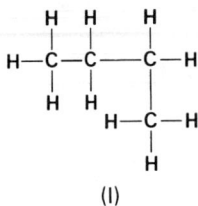

(I)

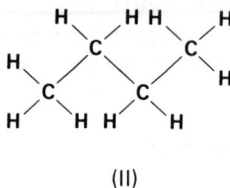

(II)

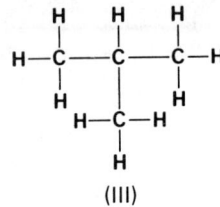

(III)

b.

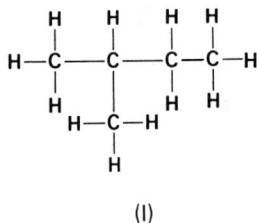

(I)

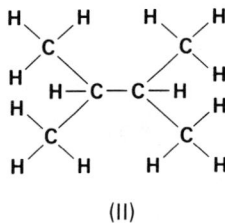

(II)

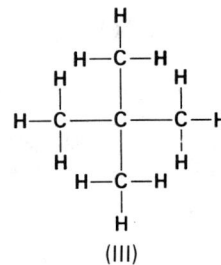

(III)

c.

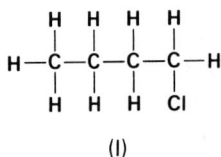

(I)

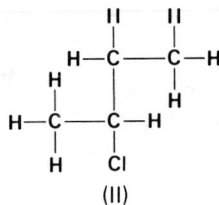

(II)

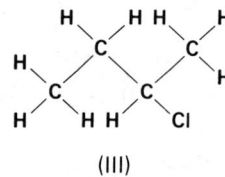

(III)

d.

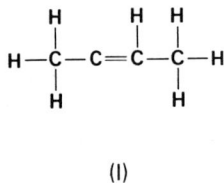

(I)

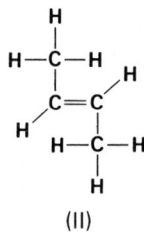

(II)

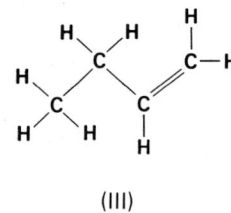

(III)

7.11   Locate the error in each of the following formulas.

a.

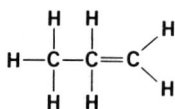

b.

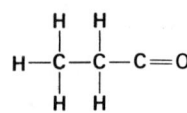

c.

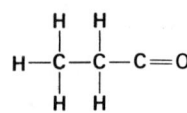

7.12   Identify the two structures among the following that show the two molecules that are isomers.

a.

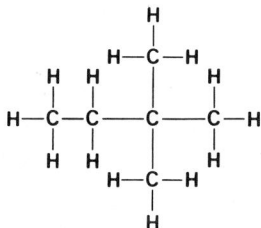

b.

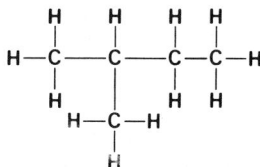

c.

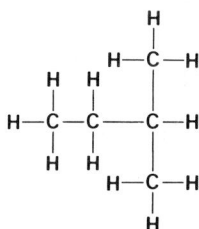

7.13   Draw the structure for the continuous chain molecule C_4H_{10} (normal butane).

  a.   How many *end* H atoms are there?

  b.   Show the structure of the C_5H_{12} molecule you form by removing an end H atom from C_4H_{10} and replacing it by CH_3.

  c.   How many *middle* H atoms are there?

  d.   Show the structure of the C_5H_{12} molecule you form by removing a middle H atom from C_4H_{10} and replacing it by CH_3. Be sure you recognize that no more than one new structure is possible.

7.14   Draw the structure for the branched chain molecule C_4H_{10} (isobutane).

  a.   How many *end* H atoms are there?

  b.   Show the structure of the C_5H_{12} molecule you form by removing an end H atom from isobutane and replacing it by CH_3.

  c.   Compare the structure you wrote for 7.14*b* with that you wrote for 7.13*d*.

  d.   How many middle H atoms are there in isobutane?

e. Show the structure of the C_5H_{12} molecule you form by removing a middle H atom from isobutane and replacing it by CH_3.

f. Count the number of different structures that are possible for a molecule with the formula C_5H_{12}.

7.15 What is the general formula $(C_nH_?)$ of the homologous series that has acetylene, C_2H_2, as its first member? (Hint: Build several consecutive structures by replacing one H with CH_3 and then see what consistency appears in the numbers of H atoms relative to the numbers of C atoms in these formulas.)

7.16 Which of the following is a substitution reaction and which is an addition reaction?

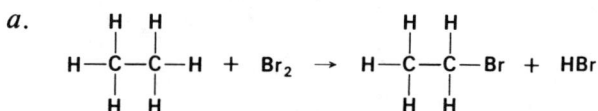

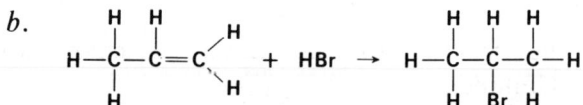

a.

$$H-\overset{\overset{\displaystyle H}{|}}{\underset{\underset{\displaystyle H}{|}}{C}}-\overset{\overset{\displaystyle H}{|}}{\underset{\underset{\displaystyle H}{|}}{C}}-H \ + \ Br_2 \ \rightarrow \ H-\overset{\overset{\displaystyle H}{|}}{\underset{\underset{\displaystyle H}{|}}{C}}-\overset{\overset{\displaystyle H}{|}}{\underset{\underset{\displaystyle H}{|}}{C}}-Br \ + \ HBr$$

b.

$$H-\overset{\overset{\displaystyle H}{|}}{\underset{\underset{\displaystyle H}{|}}{C}}-\overset{\overset{\displaystyle H}{|}}{\underset{}{C}}=C\overset{\nearrow H}{\searrow H} \ + \ HBr \ \rightarrow \ H-\overset{\overset{\displaystyle H}{|}}{\underset{\underset{\displaystyle H}{|}}{C}}-\overset{\overset{\displaystyle H}{|}}{\underset{\underset{\displaystyle Br}{|}}{C}}-\overset{\overset{\displaystyle H}{|}}{\underset{\underset{\displaystyle H}{|}}{C}}-H$$

7.17 Even though members of the paraffin series are considered relatively unreactive, all of them burn by combining with oxygen. Recall the reactions discussed in Chapter 3. Write complete chemical equations for the following.

a. $CH_4 \ + \quad O_2 \rightarrow CO_2 + \quad H_2O$

b. $C_3H_8 \ + \quad O_2 \rightarrow$

c. The reaction that occurs in the flame of a butane cigarette lighter.

7.18 There is some medical evidence to suggest that individuals troubled by hardening of the arteries are benefited by using liquid oils rather than solid fats in their diets. Writers of advertising copy urge them to use "polyunsaturated" cooking fats. Explain what this term means.

7.19 Consider the following substances.

a. Phenol (also called carbolic acid), a strong disinfectant *b.*   Trinitrotoluene (TNT), a powerful explosive

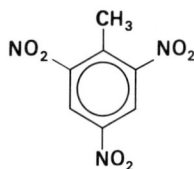

c.   Paradichlorobenzene, moth crystals

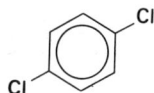

From what simple aromatic hydrocarbon molecule can all of
these substances be derived by substitution reactions?

7.20   Explain how a knowledge of each of the following phenomena in the field of organic chemistry could be responsible for
conserving some of the world's supply of petroleum that is being
made into gasoline.

a.   Homology (the existence of homologous series)

b.   Isomerism (the existence of isomers)

c.   Catalytic cracking

8 Energy and the Energy Crisis

☐ What are the origins of the "energy crisis"?

☐ What is the connection between energy, work, and power?

☐ How does nature limit the conversion of heat into work?

☐ Why is air conditioning so "expensive" in terms of energy?

☐ What is thermal pollution and why is it inevitable?

☐ Why is it difficult to store and transmit energy?

☐ How can coal be used as an energy source without creating environmental hazards?

Energy is everyone's concern. Our awareness of how much the lives of individuals and nations depend on energy resources increases every time we read or hear the daily news. Humans have always used energy to accomplish work. Here we are not concerned with the mental energy we say we use to study or take an examination. Rather, we are dealing with energy and work as defined in science. *The energy in something is its capacity for doing work. Work is done when a force is exerted to lift or push an object through a distance.* Centuries ago, such work was done by muscle power. The source of this energy was the chemical potential energy of the food eaten by humans and animals. But now, two centuries since the steam engine was invented, the picture has changed completely.

At first, simple machines relieved humans and animals from long hours of drudgery and back-breaking labor. We are reminded of this every time we find a motor rated in "horsepower." When James Watt, who invented the steam engine, tried to sell his engines to the operators of coal mines, they wanted to know how many horses an engine could replace. Watt experimented and decided that an "average" horse could lift a weight of 550 pounds through 1 foot in 1 second. These early machines also made possible the manufacture of all sorts of goods. Commerce and industry grew. The wealth bred by machines was poured back into a growing technology. Complex and sophisticated machines have been invented for millions of uses.

The ultimate source of potential energy for machines is fuel. Most of this fuel has been of the *fossil fuel* variety, deposits of coal and petroleum that accumulated in the crust of the earth eons ago. Such fossil fuel, because it originates from ancient

---

Fossil fuel represents stored sunlight.

---

plant life, represents stored sunlight. Our tremendous expenditures of energy have been made out of this rich inheritance from the earth's past.

Until a few years ago, the chief concern of those who looked to the future was the diminishing supplies of raw materials from which manufactured goods are made. We might say they envisioned a time when the law of conservation of mass would catch up to us. Supplies of potential energy seemed limitless. Now we are beginning to realize they are not. The extent and complexity of our technology is causing our inheritance of potential energy to disappear at an alarming rate. The law of conservation of energy is catching up to us. This is why we say we have an energy crisis. Even the food we eat has its cost in terms of fossil fuel. The production of food for one United States citizen is estimated to require the equivalent of 112 gallons of gasoline per year. This estimate includes the fuel used by farm machines, the energy required to make the fertilizer, and so on.

Obviously, the energy crisis, involving as it does all aspects of human living, from the economics of personal budgets to international politics, is more than a scientific problem. Yet knowledge of scientific principles is one basis for understanding the problem and considering possible actions to relieve it. What are the laws of the natural world that describe how energy can be changed from one form to another? Where can we look for stores of potential energy, and how can we unlock these stores most efficiently? Is our use of some energy resources accompanied by more hazard to the environment than our use of other resources? These are some of the questions we will approach in this chapter.

WORK, ENERGY, AND POWER

Just what is energy? We say that science deals with matter and energy and the transformations they undergo. The idea of matter is easy to grasp. Matter is stuff; it takes up space; it has mass. All our descriptions of matter, in terms of electrons, atoms, and molecules, are merely elaborations of a comfortable fundamental idea. But energy is a much more abstract idea. Visible light is the only form of energy we can see. We can feel some forms of energy—such as heat, when we warm our hands at a fire, or the energy of motion, when we catch a ball that has been thrown. But potential energy, either the kind an object has because of its position or the kind represented by chemical bonds in molecules, must be imagined. The key to appreciating and understanding energy is learning how the various forms are related and how transformations can be accomplished.

We can start to understand what energy is by realizing that the essential property of *energy is the ability to do work. Work* is defined:

$$\text{Work} = \frac{\text{force}}{\text{exerted}} \times \frac{\text{distance through}}{\text{which the force moves}}$$

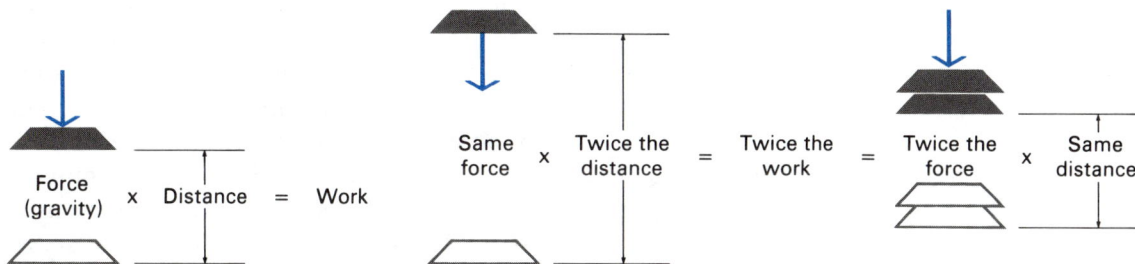

Force (gravity) x Distance = Work

Same force x Twice the distance = Twice the work = Twice the force x Same distance

Notice that this definition also allows us to measure work and therefore energy. We can measure the force we have to exert to lift a brick against the force of gravity. We can also measure the height to which we lift the brick. If we lift two bricks the 'same distance, we do twice as much work. Or if we lift one brick twice as far, we do twice as much work. Every machine exerts a force through a distance. The greater the force or the greater the distance, the more energy is required.

This definition of work says nothing about the time it takes to do work. If we run up a flight of stairs, we do the same amount of work when we walk up slowly. But why are we more tired if we

If we run up a flight of stairs, we do the same amount of work as when we walk up slowly.

run? What we need to answer this question is some definition of the rate of doing work. The rate of doing work is called *power* .

$$\text{Power} = \frac{\text{work}}{\text{time}}$$

Less power is required if we take a longer time to do a certain amount of work. If we walk up the stairs in 60 seconds instead of running up in 10 seconds, we exert only one-sixth as much power. Another example of the use of this definition is the way a power-generating station is rated in terms of the electric energy it can generate in a given time. The usual unit of electric power is the watt; 1 horsepower is 746 watts. Most generating stations are so large that their capacity is rated in kilowatts (1000 watts) or megawatts (1,000,000 watts).

The tremendous demand of modern human societies for work to be done by machines is the cause of the energy crisis we all face. To supply the energy to do work, we turn to fuels. But the energy in a fuel can only be transformed into heat, and then the heat somehow must be harnessed to do the work. So our chief concern in this chapter is with heat and the laws that nature follows in changing heat into work. This branch of science is called *thermodynamics* (*thermo*, "heat"; *dynamics*, "motion"). The harnessing of heat may be accomplished in one step, as with a steam engine or gasoline engine. Or one of these engines may be used to generate electricity, and the electricity then may be used to run a motor to do work. The overall conversion is the same: heat into work.

THE CONSERVATION OF ENERGY

The cornerstone on which we must build any discussion of the conversion of heat into work (or the transformation of any form of energy into another) is the **first law of thermodynamics**, the **law of conservation of energy.**

Energy can neither be created nor destroyed.

Here is the same kind of "you-can't-get-something-for-nothing" idea with which we began our scientific description of matter, the law of conservation of matter. Nature is equally consistent about energy. The bookkeeping system at nature's energy bank leaves no room for embezzlement.

This idea is easy to grasp, even to take for granted. For example, the principle of the conservation of energy is really what is behind our recognition that the reverse of an exothermic reaction

The bookkeeping system at nature's energy bank leaves no room for embezzlement.

must be an endothermic one. If energy is given off when hydrogen combines with oxygen to produce water, then energy must be added to water in order to make it decompose into hydrogen and oxygen. Similar reasoning helps us realize that the light emitted by excited atoms must be equal to the energy they absorbed in the first place.

The conservation principle is also basic to understanding the idea of efficiency. *The efficiency of any process is the ratio of results produced to effort expended.*

$$\text{Efficiency} = \frac{\text{results produced}}{\text{effort expended}} = \frac{\text{work done}}{\text{energy put in}}$$

For example, you sharpen ice-skate blades so that the effort you put into moving your body will take you farther across the ice. If you wear dull skates, more of your effort is used to overcome friction. This energy has not been destroyed, it merely is converted into heat that uselessly scratches and melts more ice than if your skates are sharp. When electricity is used to run a motor, the motor temperature rises. The heat that causes the rise in temperature is not helping the motor to do useful work. In these instances, we say that some of the energy put into the process has been dissipated or lost from the process rather than destroyed.

Obviously, our concern for conserving our fuel resources must include thinking about efficiency. We can conserve mechanical work by cutting down the amount that is converted back into heat. We can put oil or grease on moving parts to reduce friction and hence reduce the dissipation of some of the mechanical energy in the form of useless heat. But what about the heat-to-work transformation that must be accomplished first? Does nature impose limitations on the efficiency of this process? If so, how can we change heat into work most efficiently? To answer these questions, we need to examine heat in more detail.

THE SPONTANEOUS FLOW OF HEAT

Heat is a form of energy to which our senses respond. More accurately, we should say that our senses respond to the *flow* of heat. An infant learns very early the discomfort of losing heat to a cold environment. The baby's cries cease with the flow of heat into its body when embraced by a warm adult body. Heat is more properly called *thermal energy,* because the direction of its flow is always determined by the temperature. *Heat always flows from a body at a higher temperature to one at a lower temperature.* Remember, heat and temperature are different things. A sip of hot coffee holds much less heat than your whole body. Yet you feel the heat flow from a drop of hot coffee to your tongue because the temperature of the coffee is higher.

If we stop to think about it, this simple statement about heat flow describes much of our experience with heat. It is a law of

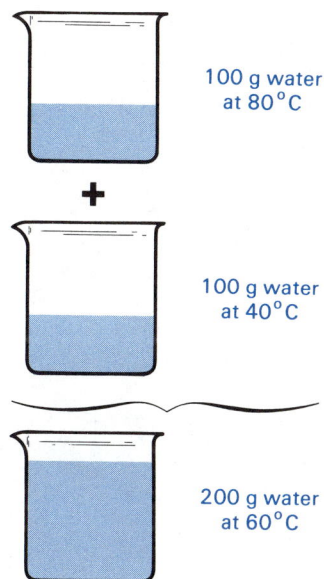

100 g water
at 80°C

+

100 g water
at 40°C

200 g water
at 60°C

nature. Scientists have named it the **second law of thermodynamics:**

> Heat spontaneously flows from a body at a higher temperature to a body at lower temperature.

Calling such an obvious generalization by such a high-sounding name may appear to be a pompous elaboration. However, as our discussion develops, you will begin to recognize that this second law of thermodynamics has very profound and far-reaching implications.

The first of these implications comes from the word *spontaneously*. The natural world tends to distribute heat evenly. This "share-the-wealth" idea may at first be appealing until we realize it means that nature tends to lock thermal energy (heat) away so that it is not available for conversion into any other form. When heat flows out of a hot body to the surroundings, the surroundings warm up and the body cools. When both the body and its surroundings attain the same temperature, heat is no longer transferred. We cannot get at the heat to use it unless we move the whole operation to some new surroundings with a still-lower temperature. When we do that, the temperature of the new place rises as the heat is stored in it. And so it goes. We can only imagine some time in the infinite future when all the heat in the universe will be tucked away in all the matter of the universe. Then all the matter of the universe will be at some ultimate constant equilibrium temperature. If every object in the universe is at the same temperature, no flow of heat can occur. If you have a philosophical turn of mind, you can envision what some thinkers refer to as the "heat death" of the universe.

REFRIGERATORS AND AIR CONDITIONERS

Another implication of *the second law of thermodynamics* is that, if we want to make heat flow "uphill" in the temperature sense (from a cold body to a warmer one), nature requires us to pay for this reversal by expending extra energy. This requirement is why you have to buy electrical energy to run a household refrigerator. You may want to make food that is cooler than room temperature even cooler. Or you may want to remove heat from cold water to freeze it into ice cubes. The refrigerator does this job for you and dumps the heat from its cool interior into the warm room. (You can demonstrate this to yourself by feeling the air in the space near the back of the refrigerator.) But to reverse nature's spontaneous tendency this way, you must supply extra energy into the motors, compressors, and other machinery of the refrigerator.

The second law of thermodynamics also indicates why the widespread use of air conditioning creates such a drain on our fuel resources (potential energy). In many densely populated areas, the domestic and commercial demands for energy in the summer equal or exceed the needs in winter. An air conditioner is a refrigeration device. It keeps a room cool by removing some of the heat from the air of the room and dumping it outside into the already-warm environment. This process goes against nature's spontaneous pattern of heat flow, so we have to pay for it in extra electrical energy to run the air conditioner. Then, as the inside gets cooler and the outside warmer, even more heat flows naturally into the room. The only way we can beat Mother Nature at her own game is to run the air conditioner even more. The time may come when the cost of widespread air conditioning, in terms of using up fuel resources, will have to be balanced against factors other than human comfort. Chief among these factors will be the need for energy for other purposes.

THERMAL ENERGY AND MOLECULAR MOTION

We can use the model of gases developed in Chapter 4 to illustrate how any substance holds thermal energy. Key postulates of the kinetic molecular theory are that gas molecules are in ceaseless random motion and that their average kinetic energy is proportional to the Kelvin temperature. The kinetic molecular theory implies that the only way a gas can absorb work is to increase the random motion of its molecules. This increase in molecular motion increases the average kinetic energy of the gas molecules, so the temperature rises. Thus heat, or thermal energy is related to the amount of random molecular motion.

If you have ever pumped up a bicycle tire with a hand pump, you have experienced the transformation of work into heat. You certainly know that work was done on the air. You applied force to move the pump piston through a distance. The air molecules absorbed the energy and warmed up, so that the hose connecting the pump to the tire felt hot. However, if you fill your tire from a tank of compressed air, the hose does not warm up. The heat from compressing the air in the tank with a pump has long since left the air and the tank to warm up the surrounding atmosphere. So the compressed air in the tank is at the same temperature as the bicycle tire and the rest of the surroundings.

The random motion of molecules of solids also increases in response to the absorption of thermal energy. The molecules of a solid are not free to break away from their fixed positions, but they increase their vibrations about those spots. One consequence of this increase in vibrations is the increase in volume that occurs with a rise in temperature. For example, a strip of metal expands to greater length when the temperature rises. The steel

One end of a steel bridge is on rollers to allow for expansion as temperature rises.

The steel beams of a bridge are longer in the hot summer than they are in the cold winter.

beams of a bridge or building are longer in the hot summer than they are in the cold winter. The design of either the bridge or the building must allow for this expansion or dangerous strains and stresses develop. In the thermostat that controls the on–off switch of a house furnace, a metal strip expands to close an electrical contact when the room temperature reaches a predetermined value.

Molecules in liquids, too, increase their random motion by rolling over and jostling one another when the temperature rises. This increase in random motion makes the volume of the liquid increase. An illustration is the expansion or contraction of a column of mercury in a thermometer in response to changes in temperature. The thermometer absorbs or releases thermal energy to its surroundings until the temperature of the liquid mercury is equal to that of its surroundings.

1.00 liter at 0°C

1.04 liter at 100°C

Liquid water expands little with rise in temperature.

HEAT INTO WORK

The expansion of solids and liquids in response to the absorption of thermal energy is seldom large. But the increase in volume when a gas absorbs thermal energy is very much larger. The volume of a sample of liquid water increases by only 4% when the temperature rises from 0° C to 100° C. The volume of a sample of gas increases by approximately 36% over the same temperature interval if it is confined at a constant pressure. If the gas has been placed in a cylinder with a movable piston, the expanding volume of the gas pushes the piston back in the cylinder, as suggested by Figure 8.1. Thus whatever force holds the piston in place has

1.00 liter at 0°C

1.36 liter at 100°C

Gases expand much more with rise in temperature.

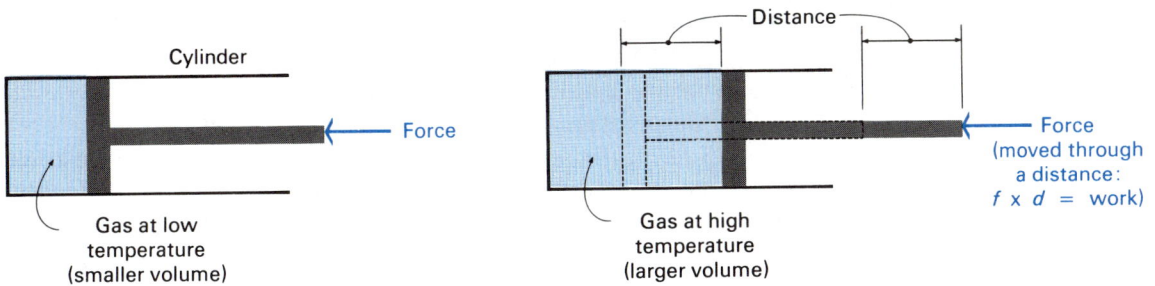

Figure 8.1   An expanding gas performs work.

been moved through a distance; work has been done, because work = force × distance. The gas has been the medium through which energy, in the form of heat, has been changed into work.

This qualitative view of the conversion process heat → work does not answer the fundamental quantitative question of how much work of pushing the piston can be obtained by adding a given amount of heat to the gas. This question of the efficiency of the process can better be approached by considering the essential steps in the operation of an engine.

The operation of a steam engine is typical of any device that continuously converts heat into work. Let us consider the essential steps of its operation, as suggested by Figure 8.2. At a high temperature, the expanding steam forces the piston back. The piston is fastened to a drive shaft or some other means of harnessing the work. However, if the piston is to do its job again, and the engine is to continue to do work, the piston must return to its starting position. To do this, the steam in the cylinder must be allowed to escape through an exhaust valve. Then another charge of hot steam must be admitted to start another stroke.

If we focus on the gas involved in one stroke of the engine, we realize that by no means can all the thermal energy of the gas molecules be used to push the piston. Only those molecules going

Figure 8.2   The operation of a simple steam engine.

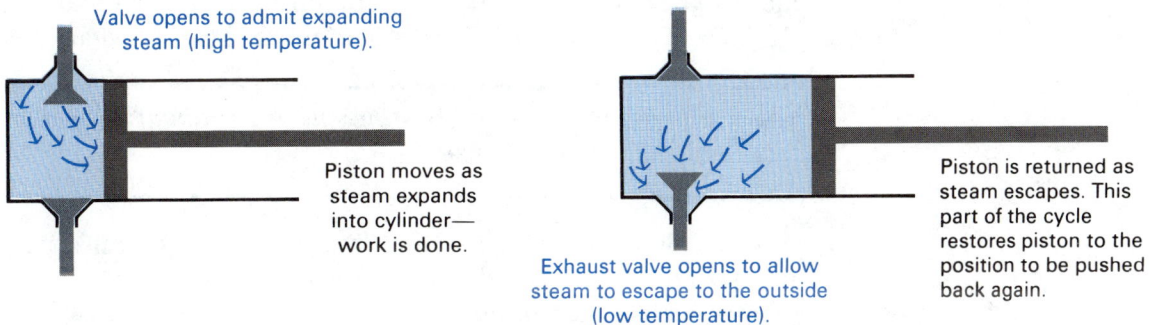

in the direction that the piston moves can help push it. Moreover, if the gas is to escape through the valve, the temperature of the air outside the cylinder must be lower than the gas inside. If the outside temperature is equal to the steam temperature, no steam can leave the cylinder, and the piston cannot return for another stroke. The engine cannot change heat into work. The greater the difference between the inside and outside temperatures, the easier it is for the steam to escape and for the piston to return. In every case, the steam must carry *some* of its excess thermal energy out of the engine to allow the cycle of one stroke to be completed and to prepare the engine for the next. This thermal energy when carried outside the engine warms up the environment.

EFFICIENCY OF THE PROCESS: HEAT INTO WORK

Our analysis brings us to an important general statement about the way nature limits the conversion of heat into work. *It is never possible to convert heat completely into work.* This is another statement of the *second law of thermodynamics*. This statement means that, try as we will, we can never avoid wasting some heat in an engine. In other words, the efficiency of the conversion heat $\rightarrow$ work can never be 100%. A detailed theoretical analysis of the problem gives a prediction of the maximum efficiency any heat engine can attain.

$$\text{Efficiency} = \frac{\text{work obtained}}{\text{heat put in}} = \frac{T_{\text{absorbed}} - T_{\text{delivered}}}{T_{\text{absorbed}}}$$

Here T_{absorbed} means the high temperature at which the steam absorbs heat and enters the cylinder. $T_{\text{delivered}}$ means the low temperature of the outside environment to which the steam delivers the unused heat. Both of these temperatures are measured on the Kelvin scale. You can also see that, if the temperature outside is equal to the temperature of the steam, no work can be obtained: $T_{\text{absorbed}} - T_{\text{delivered}} = $ zero! This situation is exactly what our

You will find how surprisingly stingy nature is about allowing heat to be converted into work.

previous discussion of the second law of thermodynamics predicts. If no temperature difference exists, no heat flows; consequently, no work can be done.

If you make a calculation, you find how surprisingly stingy nature is about allowing heat to be converted to work. If heat is delivered to a simple gas-expansion engine at 150° C (423° K) and exhausted at 40° C (313° K), the efficiency is $(423 - 313)/423 = 110/423 = 0.26$, or only 26%. This percentage is the best we can possibly do. The actual operating efficiency of the engine will be less than this predicted thermodynamic efficiency. For one thing, friction is sure to cause some of the mechanical energy of moving parts of the engine to be lost. And no insulation is perfect, so some of the heat in the steam will be lost as the steam goes through pipes to the engine.

The important point to recognize is that inevitably some of the heat put into the steam to make the engine do work is transferred to the surroundings. The temperature of the surroundings rises. The molecules of the surroundings increase their random motion. Thus some of the heat originally put into the steam cannot possibly be converted into work done by the engine.

POWER PLANTS AND THERMAL POLLUTION

The term *thermal pollution* is frequently found in news stories dealing with threats to the environment caused by society's demand for energy. Now we can understand why this problem is unavoidable. Most of our electricity is generated in power plants. The heat released by burning fuel boils water into steam and raises the temperature of the steam. The expansion of the steam turns the blades of a turbine. The turning turbine shaft rotates

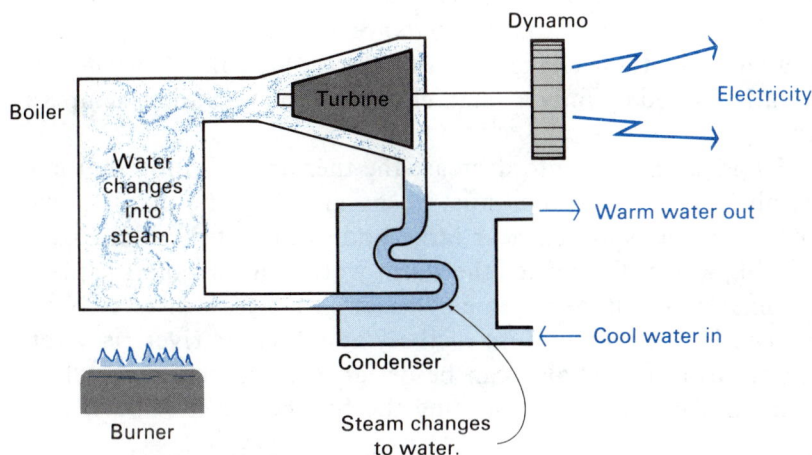

An electrical power plant

the core of a dynamo. A coil of wire fastened to the dynamo core rotates in a magnetic field. This oscillation in the magnetic field influences the electrons in the wire so that they can be drawn from the end of the wire as an electric current. The problem comes from the part of the heat that remains in the exhaust steam. This heat must be "dumped" in order for the turbine to operate. The exhaust steam is cooled so that it becomes liquid water again, and then the water is recycled to the boiler to be reheated into steam.

The exhaust steam is cooled by leading it through condensers or heat exchangers. These condensers are systems of pipes around which water from a nearby lake or river is pumped. Cool water is pumped in; it absorbs heat from the steam and becomes warmer. Then the warm water returns to the lake or river. The net result is that the temperature of the nearby lake or river rises. The heat responsible for this is the part in the steam that the second law of thermodynamics says must be dumped. This dumping of some heat always happens when any conversion of heat into work takes place. If the power plant is a large one and the lake or river is a small one, the resulting temperature rise in the body of water may be considerable.

The consequence of this temperature rise in a lake or river is a biological disturbance. Oxygen is less soluble in hot water than in cold. So less oxygen dissolves into the lake from the air. This affects the amount available for the respiration of aquatic animals. The problem is aggravated by the increased rate of all metabolic reactions at higher temperatures. Therefore, all aquatic life requires *more,* not less, oxygen in warmer temperatures. Not only may this disturbance cause death, but whole life cycles of some species may be altered—for example, in the spawning, fertilizing, and hatching of fish eggs. The balance among species populations is affected by these changes, so that the ecological balance may be greatly altered by only a small change in the prevailing temperature.

Steps can be taken to decrease the thermal pollution that may result from the heat that must be discarded by the plant. Power plants can be situated near large bodies of water as sources of cooling water. Then when the warm water comes from the plant, it mixes with a large volume of water in the lake or river. Consequently, the temperature of the whole lake or river rises very little. Power plants also can be designed to recycle more of the heat—for example, by warming the fuel before it is burned. If

climatic conditions are appropriate, more of the heat can dissipate into the air. Some advanced designs plan for adjacent manufacturing or greenhouse-type agricultural operations. These can
make good use of the moderate-temperature discarded heat.

FUEL TO POWER— THE TOTAL PICTURE

Like all problems that society creates for itself, the question of
suitable long-term planning for the use of energy resources is
both tremendous and complicated. Ultimately, society must
establish some broad policy. So it is important that all of us, as
members of society, be not only concerned but also informed.
Many economic factors will influence formation of a policy. So
will potential effects on the environment. Dominating these con-
siderations are the facts and principles of the natural world that
cannot be changed or avoided. (An example is the way thermal
pollution is a logical result of the second law of thermodynamics.) It is impossible in one chapter to cover all aspects of
society's energy problem. That requires many books. But by
examining some features of the large-scale production of power
to meet society's various demands, we may be able to gain some
perspective on the total view.

IT IS DIFFICULT TO STORE ENERGY

One characteristic of energy that creates a problem is that *the
storage of energy is difficult to accomplish*. It is virtually impossible to store mechanical energy. Some engines are built to
store kinetic energy in heavy rotating flywheels, but the amounts
stored are small and only useful in leveling out the loads. Improved designs may make this type of energy conservation more
feasible in the future.

Electrical energy can be stored in small amounts in a battery
like the one you use for a flashlight or to start your automobile
engine. However, large-scale storage of electricity is neither
feasible nor efficient. As an alternate, a hydroelectric plant may
use some of its output in periods of low demand to run pumps
that will push water back uphill into the reservoirs. This rede-
posited potential energy is then available to make more electricity in periods of high demand. What appears to be an expensive reconversion of electrical energy back into potential energy

is less wasteful than attempting to store the electrical energy. Research to develop lightweight and more efficient batteries is under way so that all-electric cars or trucks with limited cruising range probably will become available. But these too generally will be small-scale storers and converters of energy.

Thermal energy, the primary form used in large-scale energy-converting systems, cannot be stored. Recall that nature's second law of thermodynamics prevents it. The spontaneous flow of heat from a hot body to a cold body is a natural process. Whenever any collection of molecules with low kinetic energies comes in contact with a collection of molecules with high kinetic energies, the random collisions tend to distribute the total kinetic energy more evenly over both sets of molecules.

Insulating materials, such as asbestos, rock wool, or glass fibers, merely slow the flow of thermal energy. The molecules of such *insulators* are tightly held, and because they cannot vibrate readily, they do not absorb much thermal energy. The materials made from them also contain trapped air, which can absorb only small amounts of thermal energy. Free-flowing air or a breeze cools any hot object because a constantly renewed supply of slow-moving molecules encounters the object. The only perfect insulator is a vacuum containing no molecules at all. A perfect vacuum is impossible. Even any vacuum is difficult to maintain on a large scale. However, an important conservation measure is doing the best we can, using insulating materials in the walls and roofs of buildings. Such insulation slows the flow of heat to the outside during the winter and from the outside in during the summer.

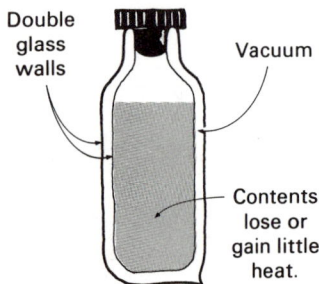

Double glass walls

Vacuum

Contents lose or gain little heat.

A Thermos bottle uses a vacuum to insulate the contents.

IT IS DIFFICULT TO TRANSMIT ENERGY

Another characteristic of energy that creates problems is that *the transmission of energy is difficult to accomplish*. Radiation, or electromagnetic energy, such as light, can be sent over great distances. However, the large-scale transmission of energy by this method is not feasible for power production. Electrical energy is the only form of energy that can be transmitted on a large scale over appreciable distances. A considerable fraction of electrical energy is lost in transmission. The moving electrons in the wires of the transmission lines dissipate some of their energy as heat. Mechanical energy cannot be transmitted, because friction inevitably degrades it into heat.

Thermal energy can be transmitted in the form of hot gases or hot liquids, but not over large distances. Actually, in the transmission of thermal energy, it is matter, the molecules, that is moved. The molecules carry their thermal energy of random motion with them.

A COMPLEX COMPROMISE IS REQUIRED

These limitations on the storage and transmission of energy are a reason why we have used so much of our depletable resources, such as coal, gas, and petroleum, and have used very little of our renewable resources, such as water power, winds, heat in the earth, or sunlight. (In a later chapter, we will discuss some of the problems in using what is by far our greatest renewable energy resource, solar energy.) The tremendous potential energy of Victoria Falls in Africa cannot be used to run air conditioners in Capetown or Cairo, much less in New York or Chicago. Wind power is an almost totally neglected renewable resource. The

The tremendous potential energy of Victoria Falls in Africa cannot be used to run air conditioners in Capetown or Cairo, much less in New York or Chicago.

wind simply does not blow continuously over big cities. But small-scale wind-powered electric generators are being used in rural areas. Geothermal energy, the heat in the outer layers of the earth, comes conveniently through the earth's crust in relatively few places such as geysers and hot springs. In recent years, all these renewable energy sources are receiving increasing attention. Even though they may ultimately contribute small fractions to the total need, such resources undoubtedly will be exploited in the future.

Because we can neither store nor transmit energy efficiently, we are forced to store and transmit fuels, the materials that hold potential energy. Power-generating plants can be built close to where the power is to be used. Consequently, the economics of fuel transportation becomes important. The transportation of solid fuels, such as coal, requires costly energy expenditure. Liquids and gases can be transported more easily through pipes, but the pipes must be permanently in place. Legalities of property ownership and environmental disruption are two of the problems this raises.

Limitations imposed by nature are magnified by what may be termed the human factors of society's demand. The economics of projects limits our possibilities. We have to pay for energy. And we value convenience. For example, we pay for the convenience of private automobiles and individual homes just as we pay for fuel. Some kind of an internal-combustion engine will long be favored for vehicles. It is disturbing to realize that our future gasoline resources may be burning up now in a power-generating plant that uses petroleum fuel. If we ultimately run out of petroleum fuels, we may still prefer to pay the higher cost of having coal converted into gas or alcohol rather than the lower cost of having to shovel coal into a vehicle powered by a steam engine.

Whatever long-term policy society establishes for using energy resources is sure to balance the influence of all these factors. Compromises must be made among conservation, efficiency, economics, convenience, and potential environmental hazards. To illustrate the complexity of balancing these often competing influences, let us consider one recent proposal that, at first glance, appears to be a "way-out" dream. This is the suggestion that hydrogen be used as an alternative fuel.

THE HYDROGEN ECONOMY

Elemental hydrogen, in the form of the H_2 molecule, is not found in nature, so it is not considered a primary fuel, such as petroleum or coal. The production of hydrogen from its compounds, of which water, H_2O, is the most abundant, requires the expenditure of energy. The most convenient and efficient means of obtaining hydrogen is by using electrical energy to break water apart. This process is called *electrolysis*.

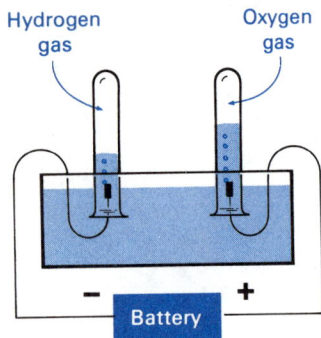

$$2H_2O + \begin{array}{c} \text{electrical} \\ \text{energy} \end{array} \rightarrow 2H_2 + O_2$$

This reaction has the effect of storing the energy added in the form of electricity as potential energy in the hydrogen.

Hydrogen as an energy carrier is highly efficient. This is why hydrogen is chosen as the fuel to power rockets and spacecraft. The molecular mass of hydrogen is small, but the amount of energy released when a molecule reacts is large

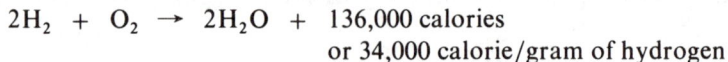

$$2H_2 + O_2 \rightarrow 2H_2O + 136{,}000 \text{ calories}$$
$$\text{or } 34{,}000 \text{ calorie/gram of hydrogen}$$

Hydrogen gas   Oxygen gas

− +
Battery

The electrolysis of water

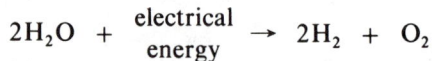

For hydrogen, the ratio of energy produced per mass of fuel burned is almost three times the ratio for gasoline. In the form of a gas, hydrogen can be pumped through pipes easily. Hydrogen gas can be condensed to a liquid if it is cooled below its normal boiling temperature of 20° K (−253° C). Although such a temperature is not easily achieved, the technology of *cryogenics* (maintaining very low temperatures) developed by the space program makes the production, storage, and transportation of liquid hydrogen practical and not very costly. Even though hydrogen is

Burning hydrogen is merely recycling water.

extremely flammable, adequate safety precautions can be taken. Thus, we can envision the use of hydrogen in any situation where the mobility of the fuel is a major consideration.

Hydrogen is an almost "perfect" fuel in terms of potential hazard to the environment. Its only combustion product is water. Because the earth contains such huge amounts of water in equilibrium between atmosphere and oceans, burning hydrogen is merely recycling water.

HYDROGEN–OXYGEN FUEL CELLS

Another important quality of hydrogen as an energy carrier is that its potential energy can be changed directly into electrical energy in a *fuel cell*. These devices are used extensively on spacecraft, instead of batteries, to supply electrical energy. The operation of a fuel cell is essentially the reverse of the electrolysis reaction by which hydrogen is produced from water. The reaction in the fuel cell releases electrical energy; the electrolysis reaction consumes electrical energy.

Figure 8.3 is a simplified diagram of a hydrogen–oxygen fuel cell. Each gas is led into a separate chamber. One wall of each chamber acts as a barrier to the gas. These barriers are made of porous carbon in which catalysts are embedded. As the hydrogen gas goes through its barrier, the molecules give up electrons to become hydrogen ions (H^+). The electrons are led off into the electrical circuit. The circuit is completed by a connection to the barrier on the oxygen side of the cell. The electrons flow from the circuit to the barrier, where they combine with molecules of oxygen gas. The oxygen molecules, along with the electrons, combine with some water molecules to produce hydroxide ions

An assembled fuel cell in which the reaction

$$2H_2 + O_2 \rightarrow 2H_2O + \text{electrical energy}$$

supplies power for an Apollo spacecraft.

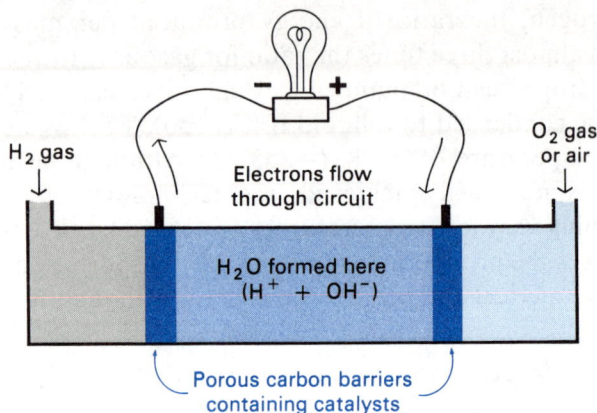

Figure 8.3 Simplified diagram of a hydrogen–oxygen fuel cell.

(OH⁻). We can represent these reactions by the following set of chemical equations:

At the hydrogen barrier:	*electrons go through the circuit*	$2H_2 \rightarrow 4H^+ + 4e^-$
At the oxygen barrier:		$4e^- + O_2 + 2H_2O \rightarrow 4OH^-$
Total reaction:		$2H_2 + O_2 + 2H_2O \rightarrow 4H^+ + 4OH^-$
		$4H_2O$

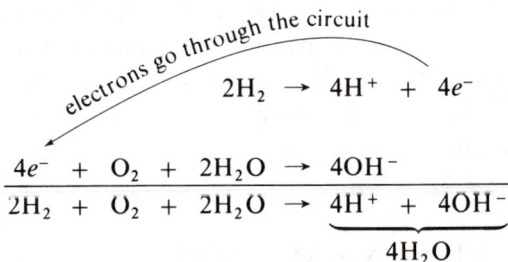

Net overall reaction:	$2H_2 + O_2 \rightarrow 2H_2O +$ electrical energy

In the central chamber of the cell, the H^+ ions and the OH^- ions combine to form water, H_2O. The central chamber is filled with a substance such as potassium hydroxide, KOH, that contains a large number of potassium ions, K^+ and hydroxide ions, OH^-. The water that is formed can be removed from the central chamber.

Notice that the equation for the reaction that occurs in a fuel cell is the same as that for the burning of hydrogen. The important difference is that the energy, instead of being given off as heat, appears as the electrical energy of the flow of electrons in the circuit. The efficiency of the conversion, potential energy $\rightarrow$ electrical energy, is greater than the efficiency of a stepwise conversion, potential energy $\rightarrow$ heat $\rightarrow$ mechanical energy $\rightarrow$ electricity. Fuel cells are not adaptable to the large-scale generation of electricity. Rather, hydrogen can be transmitted to the site of many small-scale fuel cell converters to supply electricity needed for lighting and running motors.

THE USE OF COAL AS A FUEL

A look toward the future assures us of one inescapable conclusion: The fossil fuel of the future will have to be coal. Estimates of total reserves in the United States show about 12 times as much potential energy available in the form of coal as in our petroleum and natural gas supplies. Ultimately, the convenience of using petroleum will be overshadowed by the scarcity of petroleum. Let us examine some of the problems in using coal and what may be done about them.

Coal, because it is a solid fuel, is the most costly to transport. Consequently, in the past most major users located near the supply. For example, the center of the steel industry developed in the United States far from the source of iron ore but close to the coal it requires in huge amounts.

The biggest problem associated with coal as a fuel is the presence of substances other than the carbon that gives coal its fuel value. The most widely used type of coal, bituminous, or "soft," coal, averages about 3% sulfur and 10% ash. When coal burns, the sulfur combines with oxygen to make sulfur dioxide, SO_2. (We mentioned this reaction in Chapter 3 when we discussed the oxides of nonmetals that become atmospheric pollutants.) The sulfur in coal ultimately becomes sulfuric acid by the following series of reactions:

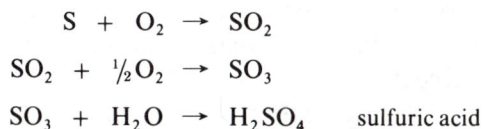

$$S + O_2 \rightarrow SO_2$$

$$SO_2 + \tfrac{1}{2}O_2 \rightarrow SO_3$$

$$SO_3 + H_2O \rightarrow H_2SO_4 \qquad \text{sulfuric acid}$$

Sulfuric acid corrodes metals, makes holes in wood and cloth, reacts with stones, mortar, and cement, and damages living tissue. A criminal with so many potential victims cannot be allowed to go free. Various methods are used to capture either the sulfur in coal before it is burned or the SO_2 in the mixture of smoke and gases in the exhaust stack. These methods have been developed only recently, and many are still experimental. Only in recent years has our concern for eliminating hazards from our environment made us willing to pay the cost of sulfur removal. One promising method "scrubs" the emerging gases with substances such as lime or limestone, which react with SO_2 to form solids and additional CO_2 gas.

$$SO_2 + CaCO_3 \rightarrow CaSO_3 + CO_2$$
$$\qquad\qquad\quad \text{solid} \qquad \text{gas}$$

The solid product itself creates an environmental problem because it must be piled up or put somewhere. A solution to one problem often creates new problems.

The answer is not simply to use coal with naturally low sulfur content. Such high-grade coal, the anthracite that has been removed for years from deep mines (with a considerable cost in human lives), is almost gone.

Another problem created by coal is the almost unavoidable smoke its burning produces. The black soot in smoke is chiefly unburned carbon. Not only is this soot an environmental hazard, but it also represents a loss of energy. Much of this loss can be prevented by properly designing burners into which powdered coal and abundant air are forced. This design creates a hotter flame, which also increases the thermodynamic efficiency of the turbines. (Remember that the second law of thermodynamics predicts the efficiency of an engine is greater if $T_{absorbed}$ is higher.) The ash or noncombustible part of coal is a different problem. Fine particles of ash are swept into the gases emerging from the burner. These particles must be removed before or while the gases rise through the smokestack. This removal is usually accomplished by having the smoke go past electrically charged metal plates, or baffles. The fine particles in the smoke are attracted to the plates, collect on them, and thus are removed. The process can be very effective; the only real drawback is its cost. However, the great improvement in air quality that has resulted from smoke removal in recent years demonstrates that the process is worth the cost.

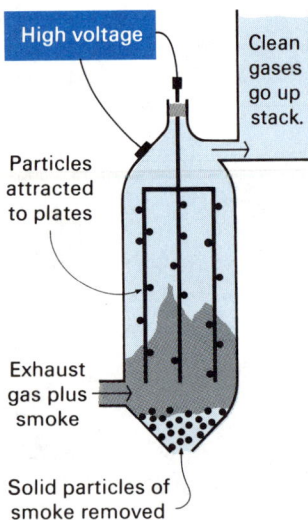

The largest coal reserves remaining in the United States, mostly in the Northern Plains and Rocky Mountain states, are poorer in quality than bituminous coal. This coal contains more moisture, volatile matter, and ash than carbon. Consequently, its heating value is low. Transporting it great distances is therefore more costly in terms of energy value. Using such coal near

High voltage

Clean gases go up stack.

Particles attracted to plates

Exhaust gas plus smoke

Solid particles of smoke removed

The result was such highly visible pollution that even the Apollo astronauts could see the smoke plumes from hundreds of miles out in space.

where it is found is more economical. This economy was the rationale for locating the large coal-burning power-generating station called The Four Corners in northwestern New Mexico. The remote location was used as a justification for not installing equipment to eliminate smoke. The result was such highly visible pollution that even the Apollo astronauts could see the smoke plumes from hundreds of miles out in space. The reaction of the public to this mistake has forced the installation of smoke-removal devices. Similar installations will be demanded in the future wherever this type of coal has to be used.

ALTERNATIVES TO BURNING COAL

One possible solution to problems caused by using low-grade coal may be converting it to combustible gases. This procedure is called coal *gasification*. These gases, rather than the coal itself, are then burned to provide heat. The essential chemical reactions of coal gasification seem deceptively simple. In practice, however, they are difficult to control on a large scale. Nevertheless, intensive research is under way. Some people predict that within a few decades much less coal will be burned directly.

Coal gasification involves heating pulverized coal to a high temperature to drive off volatile materials. Then steam is introduced under high pressure. One reaction that occurs is

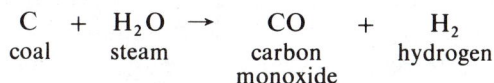

$$\underset{\text{coal}}{C} + \underset{\text{steam}}{H_2O} \rightarrow \underset{\substack{\text{carbon} \\ \text{monoxide}}}{CO} + \underset{\text{hydrogen}}{H_2}$$

Both these product gases are combustible. A variety of other possible gases can be formed. If the conditions are properly adjusted, methane, CH_4, can be formed

$$3H_2 + CO \rightarrow CH_4 + H_2O$$

At the same time that these combustible gases (H_2, CO, and CH_4) are being formed, the sulfur and other impurities in the coal are also forming gaseous products. These undesirable substances can be removed from the gas mixture. Therefore, proponents of coal gasification are justified in their claim that the process may well offer the *clean* energy of the future.

It may even be possible to avoid the undesirable environmental effects of having to remove coal from the ground. The devastation of a landscape by strip mining is one such effect. Proposals have been made to use compressed air to burn coal while it is still underground. The heat generated by the burning would raise the temperature of the various gases so that a further variety of possible reactions would occur. Controlling the amount of air could result in limiting the combustion of the coal to carbon monoxide.

$$C + \tfrac{1}{2}O_2 \rightarrow CO$$

When steam is added, other reactions can occur, such as one to produce hydrogen gas.

$$CO + H_2O \rightarrow CO_2 + H_2$$

If appropriate pipes and valves could be installed, such a system of underground coal gasification could produce mixtures of combustible gases.

THE USE OF PETROLEUM AND NATURAL GAS

Few homes in the United States are heated by furnaces that use coal as fuel. Fuel oil from refined petroleum and natural gas are used almost universally. These fuels are easier to transport through pipes and hoses than coal, which must be hauled, piled up, and shoveled around. More than mere convenience is a factor. Both petroleum and natural gas can be burned to produce little or no smoke and no ashes. Except for undesirable SO_2 produced from the sulfur in some fuel oil, the gases produced by burning these fuels are not harmful to the environment. Fuel oil and natural gas are much "cleaner" fuels than coal.

As long as our resources of petroleum and natural gas appeared to be plentiful, the economic and environmental advantages of using these two fuels also made them attractive for use in power plants. Now, however, with the great demand for petroleum to be used for the transportation of people and freight and for heating, the remaining resources are beginning to appear small. Oil and gas now burned in power plants will not be available for transportation and heating in years to come. Long-term planning for the future strongly suggests that it is a mistake to go on burning oil and gas to make electricity when other more abundant fuels can supply this large-scale need for heat. Again, we find competition among various factors of the total problem. In this case, conservation may ultimately dominate over economics and convenience.

Summary

Work is the form of energy we ultimately want from any source of potential energy. However, when we try to convert the potential energy of any fuel into work by way of heat, nature imposes unavoidable limitations on the conversion. The first law of thermodynamics states that the energy content of the universe is

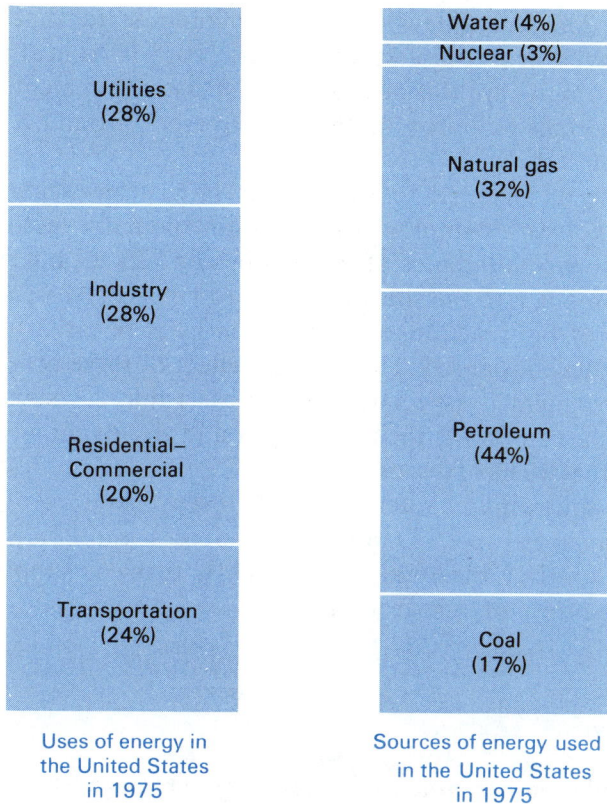

Uses of energy in
the United States
in 1975

Sources of energy used
in the United States
in 1975

**Table 8.1   A comparison of coal, petroleum, and
natural gas as sources of energy**

Fuel	Advantages	Disadvantages
Coal	Most abundant fossil fuel Technology already developed for high thermodynamic efficiency	Produces smoke and ash High sulfur content Higher transportation costs than for liquid or gaseous fuels Environmental problems created by extensive mining
Petroleum and natural gas	Clean fuel—no ashes, little smoke Easier to transport than solid fuel Technology already developed for high thermodynamic efficiency	Some sulfur in fuel oil Resources running low Need to conserve for special uses

constant. The second law of thermodynamics states that we can never convert heat completely into work. Part of the heat must be allowed to flow into the surroundings. This requirement can result in thermal pollution of the environment around a power-generating station.

The use of any fuel to supply our large-scale demands for power involves a complex consideration of many factors. Not only must the abundance of the fuel be considered, but also the cost of moving it to the site where it is to be used. Moreover, the products of the reactions by which energy is released from the fuel must be considered. The possible effect of these products on the environment is especially important. Table 8.1 summarizes and compares some of the advantages and disadvantages of the types of fuel we have discussed.

In the following chapters, we will discuss sources of energy other than fossil fuels. One of these sources is nuclear energy. Another source is the huge and constantly renewed supply of energy in the form of light from the sun.

Glossary

The number in parentheses indicates the text page where you can find the term defined in context.

cryogenics   the technology of achieving and maintaining very low temperatures (239)

efficiency   the ratio of the output of work to the input of energy (227)

electrolysis   the decomposition of a substance by the absorption of electric energy (238)

energy   the ability to do work; energy takes various forms such as light, sound, mechanical, electric, chemical, or nuclear. Each form can be classified as either potential (stored) or kinetic (used in motion). (223)

first law of thermodynamics   the law of conservation of energy, energy is neither created nor destroyed (226)

fossil fuels   the residue of ancient plant life; these fuels represent stored sunlight, converted to potential energy by photosynthesis; coal, natural gas, and petroleum (223)

fuel cell   a device in which a spontaneous chemical reaction occurs in such a way that it releases energy directly in the form of electricity (239) *gasification*   the conversion of coal to a mixture of gases from which undesirable substances can be removed, leaving only combustible gases (243)

insulator   a substance that slows the spontaneous flow of heat from a body at high temperature to one at low temperature (236)

power   the rate at which work can be done, defined as the ratio of work to time (225)

second law of thermodynamics   heat flows spontaneously from a body at higher temperature to one at lower temperature (228)

thermal energy   heat, the ceaseless random motion of molecules (227)

thermal pollution   heat energy released as a by-product of an industrial process, which then raises the temperature of some part of the environment (233)

thermodynamics   the study of the conversion of energy from one form to another, such as heat into work (226)

work   the product of the force exerted and the distance through which it is exerted (225)

Exercises

8.1   Many 150-pound persons are able to jump 2 feet in the air in one-half a second. How does their strength compare with one of James Watt's horses?

8.2   Identify the various forms of energy and the conversions involved in the following:

    a.   The whistle of a teakettle that has been filled with water and placed over a burning gas flame

    b.   A running horse that fed on hay grown in the sunshine

    c.   A burn on your hand from a hot electric light bulb

    d.   A gasoline-powered engine used to run a generator that charges a storage battery

    e.   A hydroelectric power plant generating electricity

8.3   Explain the following in terms of energy transformations and conservation.

    a.   The chemical reaction of iron with acid is exothermic. When a coiled iron spring is compressed and dissolved in acid, the temperature of the resulting solution is higher than when the same reaction occurs with an iron spring that is not compressed.

b. The temperature of water at the bottom of a waterfall is slightly higher than the temperature of the water at the top of the waterfall.

c. A fire can be started by rapidly rotating a hardwood stick in a hole in another piece of wood.

8.4 Environmentalists say that we can no longer speak of "throwing (something) away" because there really is no "away." How does this statement reflect the natural laws governing mass and energy?

8.5 Which of these statements involve the idea of efficiency?

a. A store advertises that when you buy its merchandise, you "get more for your money."

b. One car gets 28 miles per gallon of gasoline, while another of equal mass gets 18 miles per gallon.

c. A garment worker is paid according to the number of pieces finished.

d. By planning ahead, you can do just as many errands as you do now with fewer trips.

e. In buying food, a cook considers the number of people who will eat dinner.

8.6 The formula for kinetic energy (energy of motion) is

$$\text{Kinetic energy} = \tfrac{1}{2}\,\text{mass} \times (\text{velocity})^2$$

Explain why a car traveling at a speed (velocity) of 60 miles per hour will skid four times as far if the brakes are suddenly applied as it will if it has been traveling at a speed of 30 miles per hour.

8.7 Which of the following statements illustrate the principle expressed by the second law of thermodynamics?

a. In winter, the temperature in the house drops when the furnace has been turned off.

b. You cannot see around corners.

c. If you touch a hot stove, you burn your hand.

d. You cannot use the thermal energy stored in the water of the ocean to drive a ship floating on the ocean.

e. A pot of water placed in a freezer does not show a rise in temperature.

f. It is more difficult to lift a 100-pound weight than a 50-pound weight.

8.8 The maximum efficiency of any conversion of heat → work is established by the second law of thermodynamics.

$$\text{Efficiency} = \frac{T_{abs} - T_{del}}{T_{abs}}$$

a. Since T_{del} is usually the established prevailing temperature of the surroundings in which a power generating station operates, what is the only variable that can be altered to increase the efficiency of such plants for converting the energy of fuel into electricity?

b. The overall efficiency of electric power generating stations has just about doubled in the past 50 years. How do you think this has been accomplished?

c. Steam at high temperatures generates great pressure. The boilers, pipes, and all parts of a power plant that handle the steam must be built of metal strong enough to withstand pressures that go as high as hundreds of atmospheres. If you were engaged in metals research, would you feel that your work had implications for conserving the world's energy resources? Explain in light of your answers to 8.8*a* and *b.*

8.9 If you tried to cool off your kitchen in the summer by keeping the refrigerator door open, you would be disappointed. This procedure would make the kitchen even warmer. Explain.

8.10 List at least four ways in which energy could be conserved (and, consequently, less purchased) by people living in an average house.

8.11 The following table gives comparable energy expenditures (in arbitrary units) per passenger mile for transportation devices in the United States.

	Within Cities	Between Cities
buses	37	16
railroads		29
automobile	81	34
airlines		84

Contrast these data with what you know about the transportation preferences of most United States residents. (For example, during the 1960's, railroad passenger traffic decreased by half and airline traffic increased three times.)

8.12   If you see huge plumes of white "smoke" rising from the stacks of a factory on a cool day, do not rush to accuse the factory of being a polluter. What you see is probably not smoke but water droplets condensed from steam produced by the burning of fuel. The larger the proportion of H_2O in the combustion products, the more likely the white plumes will form. Write the chemical equation for the burning of the following fuels.

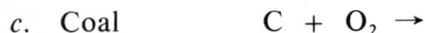

  a.   Natural gas $CH_4 + O_2 \rightarrow$

  b.   Fuel oil $C_{20}H_{42} + O_2 \rightarrow$

  c.   Coal $C + O_2 \rightarrow$

  d.   Which of these fuels shows the highest H_2O-to-CO_2 ratio in the combustion products?

  e.   What kind of fuel probably is being used wherever white plumes rise from the stacks?

8.13   Consider the various factors that complicate the total conversion of fuel into useful energy (such as cost of fuel, cost of transporting fuel, and undesirable environmental consequences). Compare the advantages and disadvantages of the following types of power plants for generating electricity.

  a.   A hydroelectric plant at Niagara Falls

  b.   A geothermal plant north of San Francisco

  c.   A plant using the rise and fall of ocean tides

  d.   An oil-fired plant near New York City

  e.   A coal-fired plant near New York City

  f.   A windmill-driven generator on a farm in the Midwest prairies.

  g.   A natural-gas-fired plant in Texas

8.14   Why is the process of gasification of coal referred to as a means of providing "clean" energy?

8.15   What would be the potential advantages and disadvantages of underground gasification of coal?

8.16   Some who object to plans for coal gasification in western states point to the fact that the west's water resources already cannot support the demands made on them. What is the connection between water resources and coal gasification?

8.17   Why does a power plant have to be designed to supply an amount of energy equal to its peak load, even though it may be required to run at this rate only half the time?

8.18 How does nature's following the second law of thermodynamics make the large-scale storage of energy very difficult to accomplish?

8.19 What are the advantages and disadvantages of hydrogen as a fuel?

8.20 Some scientists have argued that *methyl alcohol* should be used as an energy carrier. (The term *energy carrier* was introduced in the discussion of the hydrogen economy.) Methyl alcohol, CH_3OH, can be made from the gases CO and H_2. Methyl alcohol is a liquid that mixes with gasoline and can be burned in automobile engines.

    a. Why would some experts claim that coal, by way of the gasification process, could serve as a replacement for petroleum?

    b. Complete the equation for the burning of methyl alcohol:

$$CH_3OH \; + \; O_2 \; \rightarrow$$

    c. Would you call methyl alcohol a "clean" fuel in terms of the effect its use would have on the environment?

    d. Why would advocates claim that the use of methyl alcohol would result in an immediate saving of petroleum?

    e. How does the fact that methyl alcohol is a liquid contribute to its advantages as an energy carrier?

8.21 The accompanying figure suggests the world consumption of fossil fuels—past, present, and future. Comment on the shape of the curve.

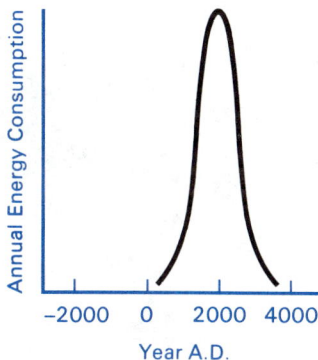

9 Energy from Atomic Nuclei

☐ What conversion takes place when energy is released by atomic nuclei?

☐ Why does a nucleus have a smaller mass than the components from which it is made?

☐ How do nuclear reactions change the elemental identity of an atom?

☐ What are the beneficial and harmful effects of nuclear radiation on living systems?

☐ What are the two conflicting sides of the nuclear power debate?

☐ What is the source of the energy radiated by the sun?

☐ What problems must be solved to make a fusion reactor work?

The first controlled release of the awesome energies locked into the nuclei of atoms was surely one of the crucial events in human history. It represented a big step in understanding and making use of one more natural phenomenon. But its greatest significance is the sheer size of the amounts of energy involved. The release of so much energy so rapidly in so small a spot can mean either potential benefit or potential hazard for humans and other living things. The discovery that the nucleus of one isotope of uranium could be made to break apart and by so doing release a great amount of energy was made at a time when much of the world was at war or preparing for war. Less than 7 years later, in 1945, the discovery led to the fearsome nuclear explosions that brought the war to a close. Since that time, the threat that some humans may wreak similar havoc on their neighbors has continued. Such is the potential danger typical of any human use of technology.

However, the hope that the energy from atomic nuclei could provide a new source of energy to benefit humankind led to research in that direction. Developments begun in the late 1940's led first to nuclear power plants used to drive naval vessels and later to large installations capable of generating megawatts of electricity. In spite of the great sums spent and the efforts invested, at present only about 3% of the power needs of the United

At present, only about 3% of the power needs of the United States are met by nuclear energy.

States are met by nuclear energy. Nevertheless, nuclear energy has become a major supplement to traditional sources of energy. Many people consider it to be an essential means of relieving the drain on fossil fuels in the future. Some foresee the need to have nuclear energy meet at least half our energy needs by the middle of the next century.

Let us examine first just what nuclear energy is and how it can be unlocked in a controlled manner. Then let us discuss the operation of a nuclear power plant called a nuclear reactor. The

The vessel containing the core of a nuclear reactor is lowered into its location behind a heavy protective shield.

greatest potential hazard associated with nuclear reactors is not in their operation, but in the disposal of the substances they produce as the nuclei break apart. An alternative kind of nuclear reaction that builds large nuclei from smaller ones may be used someday to generate power. How this works and its feasibility as a source of industrial power also will be discussed.

THE CONVERSION OF MASS INTO ENERGY

At the beginning of the twentieth century, Albert Einstein predicted that mass can be transformed into energy. He calculated that 1 gram of matter is equivalent to 2.2×10^{13} calories (22 million million calories) of energy. This energy release is many, many times that normally encountered with chemical reactions. (For example, the burning of 1 gram of hydrogen with oxygen releases only 34,000 calories of energy.) Fortunately, most of the large chunks of matter we encounter on the earth are stable and do not explode spontaneously into energy.

However, individual atomic nuclei *can* change spontaneously into different nuclei or can be changed by high-energy collisions

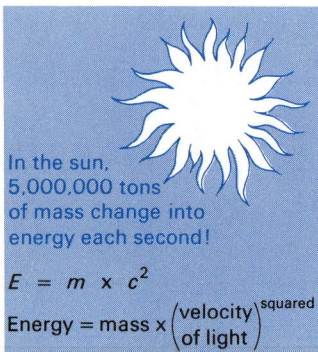

In the sun, 5,000,000 tons of mass change into energy each second!

$$E = m \times c^2$$

Energy = mass x $\left(\begin{array}{c}\text{velocity}\\\text{of light}\end{array}\right)^{\text{squared}}$

Most of the large chunks of matter we encounter on the earth are stable, and do not explode spontaneously into energy.

---

of atomic particles in atom-smashing machines. In these cases, Einstein's prediction is correct. For example, atoms of some heavy elements, such as uranium (atomic number 92) or radium (atomic number 88), are *radioactive*. They spontaneously change into other atoms by emitting alpha, beta, or gamma rays. When all the masses and energies of the particles involved in a radioactive change are measured, calculations show that the mass that disappears equals the energy that is released. Although the changes of mass are very small, the typical energy release in a radioactive event is about 100,000 times the energy that would be released if the atom were to undergo a typical chemical reaction.

It should not be surprising to find that a mass $\rightarrow$ energy conversion is involved when one atomic nucleus changes into another. The protons and neutrons in a nucleus must be held together by very strong forces. Any readjustment of nuclear particles must involve the readjustment of great energies. The large energy changes then must come from the conversion of the very small amount of mass either taken from or given to the individual particles in a nucleus.

THE MASS DEFECT OF NUCLEAR PARTICLES

We can now see a reason for a statement made in Chapter 5 when we were discussing the atomic mass of isotopes. The mass of the individual protons and neutrons is slightly different in different atomic nuclei. The actual mass of a nucleus is always less than the mass we would calculate by adding up the numbers for the masses of the protons and neutrons if they were free and not combined in a nucleus. We can use the calculation for the helium nucleus as an example. The carefully determined atomic mass of a free proton is 1.007277. The atomic mass of a free neutron is 1.008665. The helium nucleus, made up of two protons plus two neutrons, then *should* have an atomic mass of $(2 \times 1.007277) + (2 \times 1.008665) = 4.031884$. Actually, the atomic mass of a helium nucleus is 4.001506. The difference, 0.030378 units of atomic mass, represents energy that was released when the helium nucleus was formed.

The calculated loss in the mass of a particle when it is in a nucleus is called its *mass defect*. Figure 9.1 suggests the way in which the mass of a nuclear particle changes in nuclei of various mass number (the mass number is the number of protons plus

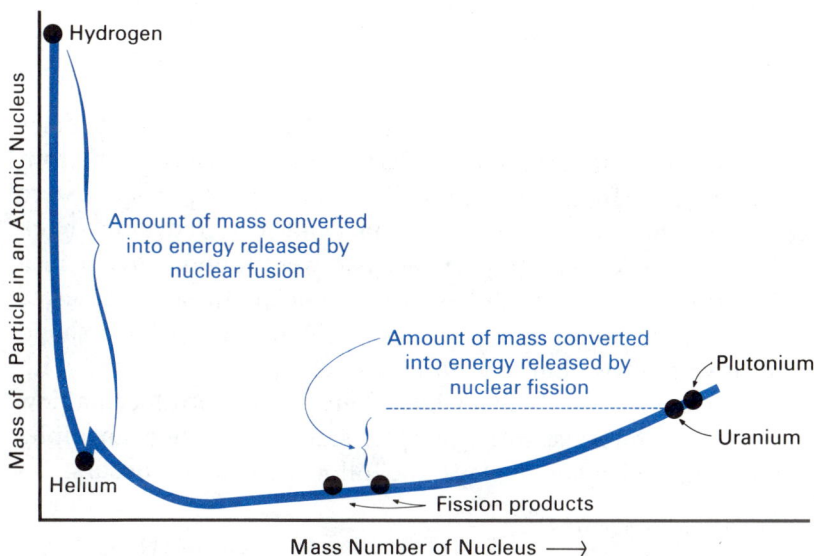

Figure 9.1 The variation of the mass of a nuclear particle as the mass number of nuclei increases.

neutrons). The greater the mass defect (the loss of mass per particle), the more energy is released when that nucleus is made. Consequently, we can anticipate that the greater the mass defect, the more stable is the nucleus. The nucleus, having lost more energy, is in a position of lower potential energy and hence is more stable. The most stable nuclei, those whose particles show the greatest mass defect, are those with mass numbers of about 50 to 60.

Two features shown in Figure 9.1 are important for our future discussion. Later in the chapter we will describe the phenomenon of nuclear fusion. This term is used to identify reactions in which very small nuclei, such as those of the isotopes of hydrogen, are forced together (fused) into a larger nucleus, such as an isotope of helium. When this occurs, a great deal of mass becomes converted into energy, because each nuclear particle needs less mass in the product nucleus. This lost mass is the source of the energy released in nuclear fusion.

At the other side of the diagram, we see the effect of breaking a large nucleus, such as that of uranium or plutonium, into two smaller ones. This type of reaction is called *nuclear fission*. The nucleus undergoes fission as a living cell does when it divides into two daughter cells. The energy released by the process of nuclear fission comes from some of the mass held by each nuclear particle in the uranium nucleus. All this mass is not needed by the particles in the more stable new nuclei, the fission products. Consequently, the unneeded mass is released as energy.

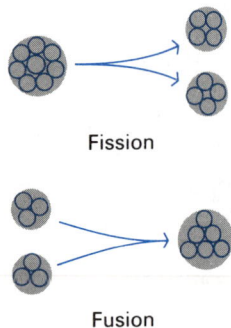

Fission

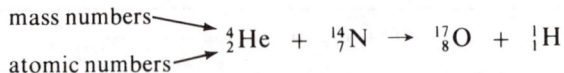

Fusion

NUCLEAR REACTIONS

You recall that Lord Rutherford used a beam of alpha particles (He^{++} nuclei) in the experiments that led to his idea that an atom has a nucleus. In these experiments, the alpha particles were repelled by the nuclei of the gold atoms and scattered, some back toward the source. Alpha particles also were found to react in a different way with some other atomic nuclei. In some cases, a positively charged alpha particle penetrated the repulsive shield of positive electric charge around a nucleus and combined with the particles already in the nucleus. This process produced a new and different nucleus. For example, sometimes when an alpha particle hits a nitrogen nucleus, the following reaction occurs:

$$\text{mass numbers} \longrightarrow \quad \text{atomic numbers} \longrightarrow \quad {}^{4}_{2}He + {}^{14}_{7}N \rightarrow {}^{17}_{8}O + {}^{1}_{1}H$$

A proton (H nucleus) is set free, and a nucleus of the oxygen isotope of mass number 17 is formed. Notice the convention we use in writing equations for such nuclear reactions. The mass number of each nucleus is written as a superscript and the atomic number as a subscript in front of the symbol for the nucleus. You can simply add up the mass numbers on one side of the equation to equal the sum on the other side: $4 + 14 = 17 + 1$. The same is true of the atomic numbers: $2 + 7 = 8 + 1$.

When a beam of alpha particles is directed at a target of beryllium atoms, a different nuclear reaction occurs.

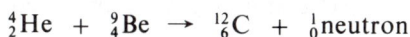

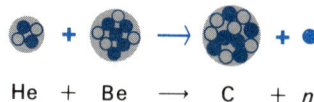

$$^4_2\text{He} + ^9_4\text{Be} \rightarrow ^{12}_6\text{C} + ^1_0\text{neutron}$$

He + Be → C + n

In this case, neutrons are set free and a nucleus of a carbon atom is produced.

Neutrons, too, can be used as bullets to be shot at atomic nuclei. In fact, a neutron is better able to penetrate an atomic nucleus than is an alpha particle or other particle that bears a positive electric charge. The positive charge of the nucleus tends to repel any invading positively charged particle. This repulsion has to be overcome by the high kinetic energy of the charged particles in the beam. However, the neutron, bearing no electric charge, meets no shield of repulsion and hence enters the nucleus more readily.

An important experiment was performed in about 1940 in which neutrons were used to bombard uranium (atomic number 92) nuclei. The experiment was particularly interesting because of what happened to the uranium nucleus after it absorbed the neutron. It kicked out a beta particle (an electron from its nucleus) and became the nucleus of a new element with atomic number 93. Then this nucleus kicked out another beta particle to become the nucleus of another new element with atomic number 94. We can represent this series of reactions with the following sequence of equations:

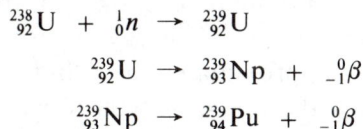

$$^{238}_{92}\text{U} + ^1_0 n \rightarrow ^{239}_{92}\text{U}$$

$$^{239}_{92}\text{U} \rightarrow ^{239}_{93}\text{Np} + ^0_{-1}\beta$$

$$^{239}_{93}\text{Np} \rightarrow ^{239}_{94}\text{Pu} + ^0_{-1}\beta$$

Here we use the symbol $^0_{-1}\beta$ to represent a beta particle. Its charge is -1, but its mass, like the mass of any other electron, is so small that it can be considered as zero in our nuclear arithmetic.

The new elements formed by these reactions were given the names neptunium (atomic number 93) and plutonium (atomic number 94). When uranium was discovered in 1841, it was the element with the heaviest known atomic mass. Therefore it became the last element on the periodic table. So it was named for the planet Uranus, then thought to be the last planet in the solar system. Years later, the planets Neptune and Pluto were discovered. Accordingly, when the physicists and chemists caught up to the astronomers and made two more elements in 1940, they named the elements for the additional planets.

Since 1940, the periodic table has been extended all the way to element 106. A variety of nuclear reactions have been used to produce these synthetic elements. All have used some kind of a bombardment of one nucleus with some particle that enters the nucleus and changes it into a new one.

In 1974 Professor Glenn Seaborg and associates at the University of California created nuclei of element 106 by bombarding nuclei of californium-249 with nuclei of oxygen-18.

$$^{249}_{98}Cf + ^{18}_{8}O \rightarrow 4\,^{1}_{0}n + ^{263}_{106}(?)$$

NUCLEAR FISSION—
A CHAIN REACTION

In 1939, two German scientists, Otto Hahn and Fritz Strassman, made an accidental discovery while bombarding uranium with neutrons. After one experiment, they found atoms of barium (atomic number 56) in the sample of uranium. Barium nuclei are only about half the size of uranium nuclei. Some of the bombardments also released much greater spurts of energy than had been observed in other previous experiments. The mystery was solved by Lise Meitner and Otto Frisch, then fugitives from Nazi Germany working in Sweden. Their observations and calculations indicated that an isotope of uranium of mass number 235 was responsible. When the neutron enters the nucleus of this isotope, the resulting nucleus splits into two fragments, each about half the size of the original uranium nucleus. We say "about" half the size, because it is now known that almost 100 different pairs of fragments can be formed. Moreover, two or three additional neutrons are set free. And the energy released, calculated on the basis of a mole of uranium reacting, is a tremendous 5 billion kilocalories. The amount of energy released by such a nuclear fission reaction is approximately a million times the amount of energy released by the range of typical exothermic chemical reactions. Rather than trying to write equations for the complexity of products produced by a nuclear fission reaction, we represent it diagramatically in Figure 9.2.

A very important special feature of the nuclear fission reaction is that, in addition to releasing a great amount of energy, the capture of one neutron causes the release of two or three new

Burning 1 pound of coal releases 3000 kcal of energy.

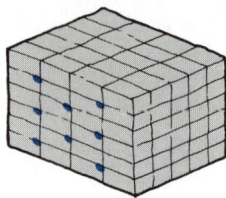

The fission of 1 pound of uranium releases 8,000,000,000 kcal of energy.

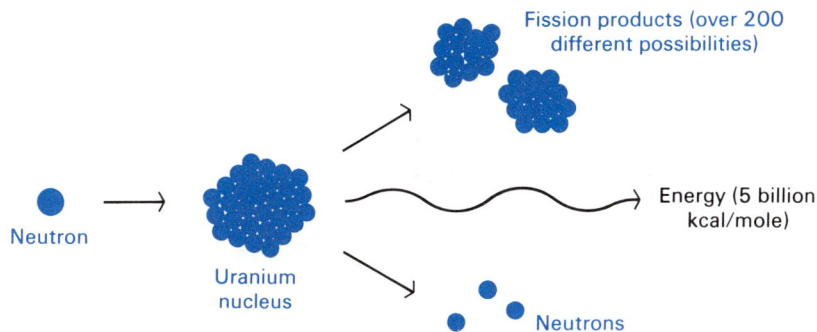

Figure 9.2   The nuclear fission reaction.

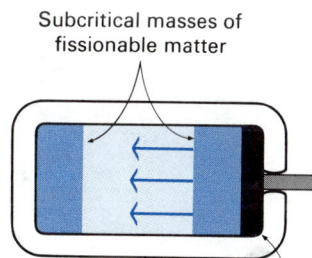

Nuclear fission chain reaction

neutrons. If another uranium nucleus is nearby, it may be hit by one of these neutrons, and another fission reaction occurs. This circumstance produces a *chain reaction*. It is like the game of standing dominoes on end in a line. When one is tapped, it falls to hit the next, and so on. One tap knocks down the whole line.

If the volume of the uranium sample in which a chain reaction starts is small, most of the neutrons may escape without hitting another uranium nucleus, because they are moving so fast. However, if the volume is larger than a particular size, called the *critical mass,* one or more neutrons from almost every fission will cause another fission to occur. Because each fission occurs in about a billionth of a second, the chain reaction can become an explosion. This is the way the nuclear fission reaction was used in the "atomic bombs" of World War II or now is used in nuclear explosion devices. Two subcritical masses of fissionable nuclei in which a chain reaction has been started are suddenly joined together to make a supercritical mass (that is, larger than the critical mass). An explosion results. Because the energy release is so great, often equal to that in the explosion of thousands or millions of tons of TNT, nuclear explosions are rated in kilotons (1000 tons) or megatons (1,000,000 tons).

NUCLEAR FISSION REACTORS

Fortunately, the nuclear fission chain reaction can be controlled or slowed down so that it need not be an explosion. For one thing, a few of the neutrons are not emitted instantaneously but are delayed a fraction of a second. It is also possible to assemble the uranium in the "core" of a reactor in such a way that metal rods can be moved in and out of the core. These rods, called control rods, are made of an element such as cadmium (atomic number 48) or boron (atomic number 5). The nuclei of these atoms

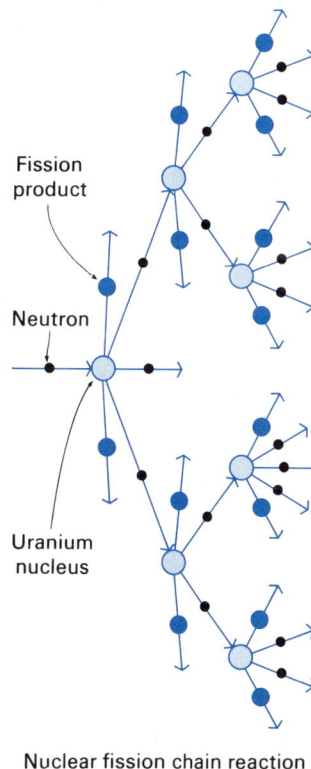

How a nuclear fission bomb is exploded

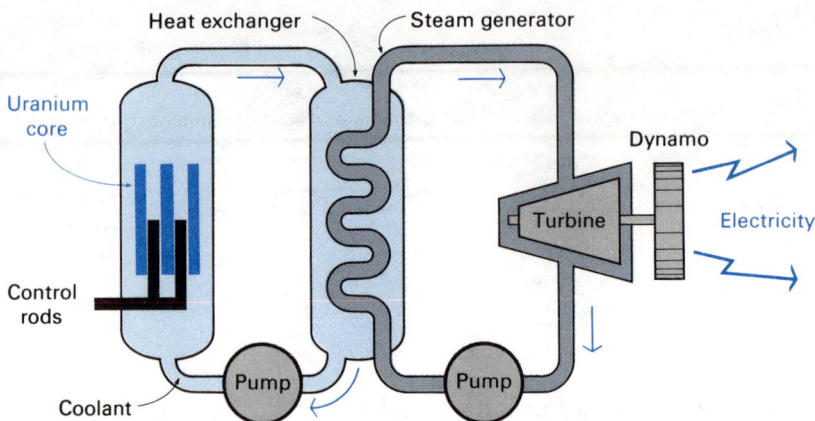

Figure 9.3   A nuclear fission reactor to generate electric power.

absorb neutrons without undergoing fission. Thus if some neutrons are removed, the chain reaction stops. The fact that at least a few neutrons are emitted more slowly allows enough time to move the control rods into the core to keep just the right balance between neutrons lost and those continuing the chain reaction.

Figure 9.3 shows the essential features of a nuclear fission reactor. A coolant fluid is circulated around the core to absorb the heat generated there by the fission. Various coolants can be used, for example, helium gas, water, or molten metals. The coolant fluid is pumped to a heat exchanger, where the heat changes water into steam to drive a turbine, as in any power-generating plant.

The uranium isotope of mass number 235 makes up only 0.7% of natural uranium. The isotope of mass number 238 is the abundant one. Uranium-238 does not undergo fission. Consequently, uranium for use in reactors is treated to remove some of the uranium-238. Manufacture of this "enriched" uranium (uranium with a higher than natural proportion of uranium-235) is costly. It is accomplished on a large scale only at installations such as those built by the United States Atomic Energy Commission at Oak Ridge, Tennessee.

BREEDER REACTORS

The uranium isotope of mass number 238 undergoes a different type of nuclear reaction in the core of a reactor. This is the reaction in which a uranium-238 nucleus absorbs a neutron and eventually changes into a nucleus of plutonium, atomic number 94, with a mass number of 239. Plutonium-239 also is capable of undergoing a nuclear fission reaction. This makes it a valuable commodity, because it represents a new source of "nuclear fuel." Moreover, it can be manufactured in a reactor at the same time the reactor is being used to generate power. A reactor designed to do this is called a *breeder reactor*. Figure 9.4 suggests how this works. Because only one of the two or three neutrons released in fission is needed to keep the chain going, one or more of the other neutrons can be used to convert nonfissionable uranium-238 into fissionable plutonium-239.

The idea of releasing energy and, at the same time, making new fuel is very attractive in terms of conserving energy resources. Up to now all the plutonium that has been manufactured has been used either in nuclear explosion devices or stockpiled for that purpose. But reactors designed to use plutonium as fuel for power are being developed. One difficulty with plutonium is that very few neutrons from its fission are delayed. They all come out instantaneously, so the control of a plutonium reactor is a very delicate process. If breeder reactors and plutonium reactors take their place along with conventional uranium reactors, the total available resources of nuclear power are very great. United States' reserves of fissionable nuclei have potential energy about equal to that of our coal reserves.

A great variety of nuclear fission reactors is in operation. Some reactors supply power to submarines and aircraft carriers. In 1975, approximately 100 reactors in the United States and other countries were supplying power to utility companies. Many more are in various stages of planning. Advocates of nuclear power point out that it is "clean" energy, in the sense that no smoke or

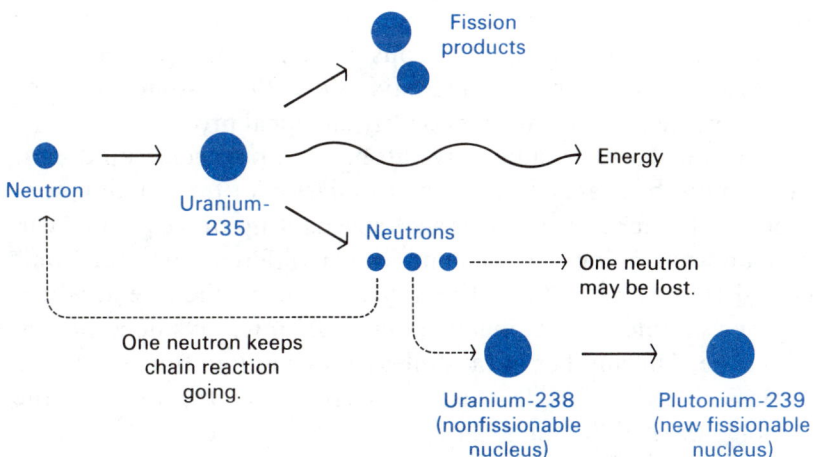

Figure 9.4   Nuclear reactions in a breeder reactor.

other atmospheric contaminants, such as those from burning coal, are produced. The great concentration of potential energy in uranium means lower fuel transportation costs and less disruption of the environment by mining. But the cost of building a nuclear plant is far greater than that of a conventional fuel converter. Moreover, some very serious potential hazards accompany the use of fission reactors.

Chief among these hazards is the fact that all of the fission product nuclei are radioactive. They tend to emit great numbers of beta and gamma rays. This emission of large amounts of energy can be dangerous to any living thing that comes in direct contact with the fission products. To understand more fully why fission products create this danger, let us discuss the typical behavior of a sample of radioactive material.

RADIOACTIVE DECAY

Two features about the phenomenon of radioactivity impressed scientists when it was discovered, at about the beginning of the twentieth century. The first was that each atomic event (each single change in a nucleus) releases a very large bundle of energy.

The energy released is an incredible 5 billion kilocalories.

This energy may be in the form of the kinetic energy of an alpha particle or a beta particle leaving the radioactive nucleus. Or in the case of the third type of rays, the gamma rays, the bundle of energy is a quantum of very energetic radiation like light. The other feature was the spontaneous nature of the process. The emission of rays is a characteristic property of some kinds of atomic nuclei, just as the mass or any chemical property is.

The rate at which radioactive atoms of a particular kind emit rays cannot be changed by any manipulation of the reaction conditions. Neither heat nor pressure speeds it up or slows it down. Nor does changing the element into a different form or compound alter its rate of emitting rays. However, the rate at which any one sample of a radioactive material emits rays does change with time. The number of rays released in a unit of time decreases as time goes on. This decrease is referred to as the *decay* of the sample. Each different radioactive material decays at its own characteristic rate.

There is no way of predicting when any one particular nucleus will undergo decay. However, when large numbers of the same kind of nucleus are in a sample, the statistical probability of many behaving in a predictable way is quite reliable. An analogy is the way a life insurance company predicts statistical probabilities of life expectancy. Any one individual person may live a longer or shorter time than the statistical average. Yet the predictions are valid for large numbers of people of any one age.

The decay of all radioactive samples follows a strikingly similar pattern. In every case, it always takes a characteristic period of time for half of the radioactive nuclei to decay. This time is called the *half-life* of that particular nucleus. The form of this regularity is shown in Figure 9.5. If the time for half of the nuclei to decay is 10 minutes (one half-life), then at the end of 20 minutes (two half-lives), only $\frac{1}{2} \times \frac{1}{2} = \frac{1}{4}$ of the original number will remain. At the end of 30 minutes, $\frac{1}{2} \times \frac{1}{2} \times \frac{1}{2} = \frac{1}{8}$ of the original nuclei will remain. At each stage of this calculation, you notice that some appreciable fraction remains. In principle, the decay is never completely finished. When radioactivity measurements are made in laboratory experiments, a decaying sample usually is considered to be "all gone" after about ten half-lives have passed. The fraction remaining then is $(\frac{1}{2})^{10}$, or $\frac{1}{2} \times \frac{1}{2}$ ten times over. The numerical value of this fraction is 0.000977, or about one-thousandth of the original sample. However, when the problem of potential danger to a living organism is considered, a longer period of decay is required. A sample is not considered

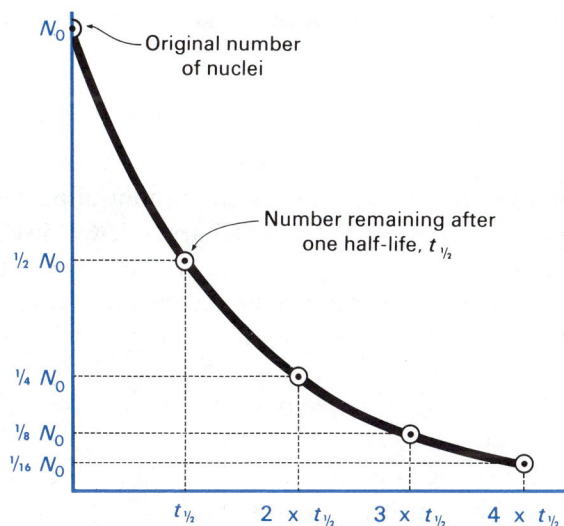

Figure 9.5 Characteristics of radioactive decay.

biologically "safe" until 20 half-lives have passed. After this time, a radioactive sample has decayed until only one-millionth of the original remains.

THE BIOLOGICAL EFFECTS OF RADIATION

The dangerous effects on living cells of the high-energy rays coming from a radioactive nucleus can be understood in terms of what we have already discussed about the structure of molecules. Many complicated molecules are involved in the chemical reactions that take place in living cells. These molecules are held together chiefly by pairs of electrons in covalent bonds between atoms. Some of these molecules are very delicately balanced in terms of the energy required to alter their structure. When a high-energy electron in the form of a beta ray enters a living cell, the energy of the beta ray may knock other electrons out of a molecule. When a negatively charged electron is removed from a neutral molecule, the remaining part of the molecule has a positive charge. The molecule, therefore, becomes an ion. A gamma ray, because it is a very high-energy quantum of light, also may knock electrons out of molecules. Consequently, beta and gamma rays are referred to as *ionizing radiation*. Ions undergo chemical reactions quite different from those of the molecules from which they are made. So the result of having ionizing radiation enter a living cell is a disruption of the cell's normal chemical reactions.

Other kinds of molecular disruptions also can occur. For example, a water molecule can be broken apart into two fragments.

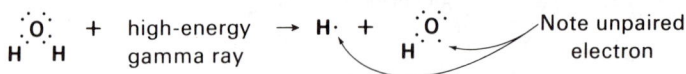

We have written these fragments each with an unpaired electron dot to emphasize what to expect. In Chapter 7, we discussed how electrons tend to form pairs to make stable bonds. The single electron on a fragment of a molecule makes these fragments very reactive. Two OH fragments can join together.

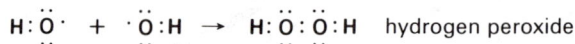

The product, hydrogen peroxide, is a substance you may recognize as a bleaching agent. A solution of hydrogen peroxide can be used to destroy coloring matter in dark hair (which is dead material) to make it blond. Now imagine what can happen inside a *living* cell if hydrogen peroxide has been put there. Not just bleaching, but all sorts of reactions can occur in the many delicate, complicated molecular structures present in any biochemical system.

The reactions caused by ions or molecular fragments formed inside living cells by high-energy radiation lead to biological damage. If the amount is small, the organism can repair the damage. If many cells are knocked out, the effect is serious. This is the reason that large doses of radiation are much more harmful than repeated small doses.

The molecules involved in a cell's mechanism of heredity are particularly susceptible to disruption by added energy. Changing

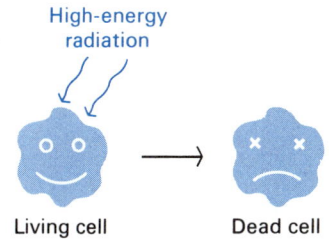

High-energy radiation

Living cell   Dead cell

The molecules involved in a cell's mechanism of heredity are particularly susceptible to disruption by added energy.

---

a few bonds in these molecules can result in sending different information or misinformation to new cells being formed. Thus radiation can cause serious damage over generations.

Two long-lived radioactive isotopes among the fission products are particularly dangerous to humans. One of these is an isotope of strontium (atomic number 38) with a half-life of nearly 29 years. Because strontium is located in family 2A of the periodic table, its chemical properties are similar to those of calcium, also in family 2A. Calcium is a chief constituent of bone, so strontium taken in by a human through food or water would be expected to lodge, along with the calcium, in bones. There, the radioactive energy of the decaying strontium could cause cell damage. The other villain is an isotope of cesium (atomic number 55). Its half-life is 30 years. Radioactive cesium taken into the body can be expected to react chemically as sodium and potassium react, because cesium is in the same family of the periodic table. All body fluids contain sodium and potassium, so radioactive cesium might cause cell damage in any part of the body.

The plutonium manufactured in a fission reactor is also a biological hazard. Plutonium atoms are readily taken into living organisms that may be exposed to them. Plutonium nuclei are intensely radioactive and have a half-life of 24,000 years. Consequently, an environment in which plutonium was spread around would be contaminated and dangerous to living things for hundreds of thousands of years.

A BENEFICIAL, MEDICAL
USE OF RADIATION

The rays from radioactive nuclei also have been used for lifesaving, beneficial purposes. One important use is in radiation therapy. Radiation can stop or retard the growth of cancer in an afflicted person. When gamma rays are used to treat cancer, the aim is to destroy the cancer cells. This is why the radiation is focused or concentrated as closely as possible on the diseased organ. Cancer cells are usually larger and reproduce more rapidly than normal cells. Both of these properties make cancer cells more susceptible to radiation damage. But healthy, normal cells are damaged also, so radiation treatment must be carefully controlled.

One way to concentrate the radiation where it is needed is to introduce a radioactive isotope of some element that the body uses in a particular way. For example, if food taken into the human body contains iodine, the body concentrates it in the thyroid gland. So a cancer of the thyroid can be irradiated right on the spot by any radioactive iodine that is in a person's food. An afflicted person can drink an "atomic cocktail" containing sodium iodide dissolved in water. The iodine of the sodium iodide contains some radioactive iodine isotopes. These radioactive atoms then become concentrated in the diseased thyroid and emit their radiation where it will do the most good.

THE POTENTIAL DANGERS
OF FISSION REACTORS

All the atomic nuclei produced by nuclear fission are radioactive. These fission products cannot be allowed to accumulate in a reactor core, because some of them very readily absorb neutrons. Neutrons needed to keep the chain reaction going are thereby lost. So the reactor must be shut down periodically to replace the fuel. The spent fuel is too valuable to discard, so the fission products must be removed by chemical treatment and the unused fuel recycled. Any valuable plutonium that has been manufactured in the reactor also must be removed. This operation is done by remote handling behind massive concrete and metal shielding. During this treatment time, rapidly decaying fission products (those with half-lives of seconds, minutes, or hours) become harmless, stable nuclei. Others decay slowly; it is these that are potentially dangerous. Some have half-lives of many years. Because we cannot alter the rate at which radioactive atoms decay, we must accept the fact that dangerous fission products cannot be forced to get their decay over with quickly.

Because of the potential danger from radiation and because the radiation from fission products with long half-lives continues for many years, fission products must be stored safely away from any living thing. They cannot be stored in leaky containers that would allow them to get into underground water supplies or oceans. One suggested location for safe storage is the caverns of abandoned salt mines. The argument is that there must be no water in the ground near the mines or else the salt would have dissolved away long ago. However, other evidence suggests that the caverns may not be geologically stable over the centuries that would be required. At present, no permanent solution to the perplexing problem of fission product storage has been found. As more fission reactors are put into operation, the problem of safe storage of the fission product wastes will become even worse.

The plutonium produced by reactors represents a potential hazard in two ways. One is its long-lived radioactivity, which can endanger living organisms. The other is the fact that plutonium can be used to make a nuclear bomb. Obviously, the expanding supply of plutonium must be kept out of the hands of irresponsible groups. Consequently, much stricter security measures than those that have been in effect must be instituted to safeguard the transportation and storage of plutonium.

Some people fear that an accident in a nuclear reactor might produce an explosion comparable to an atomic bomb. This fear is not justified. A reactor cannot possibly blow up as a bomb

A reactor cannot possibly blow up as a bomb would, spewing radioactive contaminants all over the neighborhood.

would, spewing radioactive contaminants all over the neighborhood. The design of a reactor is entirely different from that of an atomic bomb. The fissionable material in a reactor cannot come together to make a supercritical mass as in an exploding bomb.

What conceivably can happen in a reactor is that a blockage or leak could develop in its coolant-circulation system. Then heat would rapidly build up in the core. The heat could become so intense that the containing structure would melt. However, all reactors are designed with second-level containment systems built in behind the radiation shield. Such an accident would require costly and time-consuming repairs, but it would not result in widespread contamination of the surroundings.

One other design feature of reactors is receiving increasing attention. The intense radiation in the core and cooling loops may cause pinhole leaks or fine cracks to develop in the piping. These would allow the cooling water or gas to become contaminated with radioactive fission products. Elaborate and continuous inspection systems must be used to avoid this danger. Properly operating, a reactor allows almost no radioactive contamination to escape.

THE POTENTIAL BENEFITS OF FISSION REACTORS

Nuclear fission reactors hold very real promise as a means of meeting future energy needs. At least it is clear that this existing technology can supply energy until other sources, such as solar energy, can be exploited on a large scale. A power plant generating electricity from nuclear fuel has many advantages. It is compact, the costs of transporting its fuel are limited, and it produces no smoke, sulfur dioxide, or ashes to contaminate its environment. A very important feature is the fact that using nuclear energy can conserve our vital reserves of fossil fuels for other important uses.

The uranium used to supply the energy for a reactor is a natural resource in limited supply. However, the development of breeder reactors can produce plutonium as nuclear fuel. Breeder reactors can provide a new source of potential energy at the same time that potential energy is being converted into electric power. This means that, when the time comes that breeder and plutonium reactors are operating, we can produce power without rapidly exhausting our energy reserves. Such a conservation policy has obvious long-term advantages.

The potential dangers from fission reactors cannot be overlooked. Certainly, a public awareness of these is essential. But it is also essential that this awareness does not generate a fear that totally blocks balanced consideration. The nuclear power industry now has a safety record far superior to that of any other industry. This can never be allowed to result in complacency or to stand in the way of constant research to assure safe operation without accident. The possibility of accidents always exists, and we must be sure that steps are taken to minimize such a possibility. The same is true in our private lives; for example, we constantly risk our lives in vehicles for transportation, and we defy the statistics demonstrating that most home accidents occur in the bathtub! Risk can never be totally avoided. However, sensible awareness and careful planning can decrease any risk. The rewards are worth the planning.

NUCLEAR FUSION

Another type of nuclear reaction holds the key to unlocking tremendous resources of potential energy. This reaction is known as *nuclear fusion*. If the nuclei of various isotopes of the lightest element, hydrogen, are forced to join together (fuse) into nuclei of the next largest element, helium, mass is converted into energy. This is suggested on the left side of Figure 9.1.

Scientists believe reactions of this type are responsible for the energy released by stars. The star closest to us is the sun. Calculations based on the miniscule fraction of the sun's total radiation that reaches the earth suggest that the sun converts about 5

The sun converts about 5 million tons of mass into energy each second.

---

million tons of mass into energy each second. Yet it still has millions of years to go. The sun contains mostly hydrogen and helium. The temperature of the sun's core, where its energy is released, is estimated to be 10 million degrees. (When an estimate of a temperature as high as this is made, it makes little difference whether the degrees are Fahrenheit, Celsius, or Kelvin!) The nuclear fusion reaction proposed as the source of the sun's energy is represented as follows:

$$4\text{H nuclei} + 2\text{ electrons} \rightarrow \text{He nucleus} + \text{energy}$$

This looks like a simple process until we realize what has been accomplished.

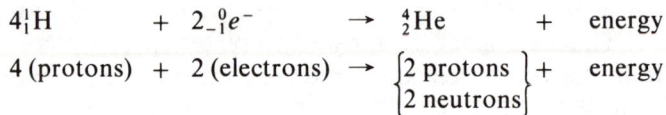

$$4{}^1_1\text{H} \quad + \quad 2{}^{\ 0}_{-1}e^- \quad \longrightarrow \quad {}^4_2\text{He} \quad + \quad \text{energy}$$

$$4\text{ (protons)} + 2\text{ (electrons)} \longrightarrow \begin{Bmatrix} 2\text{ protons} \\ 2\text{ neutrons} \end{Bmatrix} + \text{ energy}$$

Four protons have somehow overcome the huge repulsive forces of their positive charges, and two of the four have been changed into neutrons. Only protons moving with tremendous kinetic energies could do this. This need for tremendous energy of motion is why the temperature must be so high. Only at very high temperatures are nuclei stripped of electrons. Matter in this state is called *plasma*. In the plasma state, positively charged nuclei and negatively charged electrons are in chaotic motion. All sorts of collisions occur.

Establishing the necessary condition, intense heat, for fusion reactions to occur is the great problem in making a controlled fusion reaction work on earth. The H-bomb, the nuclear weapon exploded over Bikini in 1952, achieved this condition. The necessary high temperature for fusion to occur was attained by placing the reacting hydrogen isotopes inside a fission bomb. But creating such a powerful explosion as this is no way to release energy to run a power plant. Controlled nuclear fusion reactions must be contained in some device that can allow the tremendous energy, even if it comes in bursts, to be used as heat. No metals or any other form of matter can hold together at such high temperatures.

Research scientists, chiefly in the United States, Great Britain, and the Soviet Union, are developing devices in which the plasma is held together by very intense magnetic fields. The problem is holding the reactive nuclei close enough together for a long enough time for a reaction to occur. Such progress has been made that scientists are confident this problem can be solved. But other extremely difficult problems remain. Energy will be released in very intense bursts. Each burst of energy must be absorbed by some coolant that will carry it from the core of the reactor where the fusion occurs to where it can generate steam or another hot gas to run a turbine. Solutions to these engineering hurdles are still far in the future.

Various possible fusion reactions are under consideration. The reaction that promises the best control involves bringing the nuclei of two different hydrogen isotopes together to make a

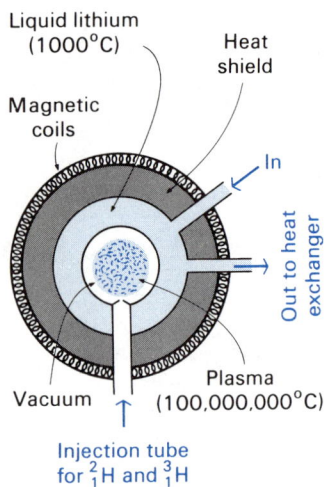

Liquid lithium (1000°C)

Heat shield

Magnetic coils

In

Out to heat exchanger

Vacuum

Plasma (100,000,000°C)

Injection tube for ${}^2_1\text{H}$ and ${}^3_1\text{H}$

Diagram of a proposed fusion reactor

helium nucleus. One of these hydrogen isotopes is called *deuterium*. Its nucleus contains one neutron and one proton. The other isotope of hydrogen is called *tritium*. Its nucleus contains two neutrons and one proton. The reaction can be represented as follows:

$$^2_1H \quad + \quad ^3_1H \quad \rightarrow \quad ^1_0neutron \quad + \quad ^4_2He \quad + \quad energy$$

$$\begin{matrix} deuterium \\ nucleus \end{matrix} \quad + \quad \begin{matrix} tritium \\ nucleus \end{matrix} \quad \rightarrow \quad neutron \quad + \quad \begin{matrix} helium \\ nucleus \end{matrix} \quad + \quad energy$$

$$\left\{\begin{matrix} 1\ proton \\ 1\ neutron \end{matrix}\right\} + \left\{\begin{matrix} 1\ proton \\ 2\ neutrons \end{matrix}\right\} \rightarrow 1\ neutron + \left\{\begin{matrix} 2\ protons \\ 2\ neutrons \end{matrix}\right\} + energy$$

The potential advantages of fusion power are a strong stimulus to its development. The energy yield from this fusion of nuclei is ten million times the energy that would be released by burning the same mass of hydrogen with oxygen. The product of the reaction, a helium nucleus, is very stable. It does not decay by a radioactive process as the nuclei produced from fission reactions do. Thus all the long-term hazards involved in handling and storing the products produced by a nuclear fission reactor are avoided. The fuel for a fusion reactor will be cheap and in abundant supply. Deuterium is found with ordinary hydrogen atoms in water. The huge amounts of water on the earth thus represent an almost limitless supply of deuterium. The other nucleus, tritium, is made right in the reactor from the element lithium (atomic number 3). Lithium is a metal and will be used, melted into liquid form, as the coolant that carries the heat away from the reaction zone.

Development of practical fusion reactors to deliver heat to power plants is still many years away. The advantage of having a great abundance of starting materials and the advantage of producing products that cause minimal environmental harm or hazard to health are very attractive. These two advantages may well grow in significance in the future, so the cost and effort of developmental research are certainly justified.

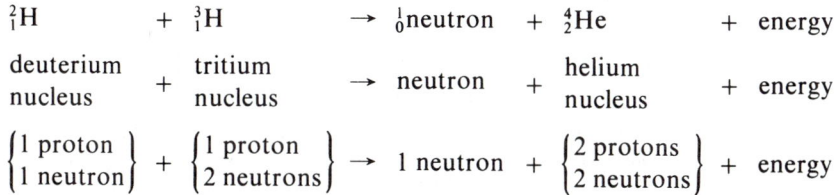

One ^2_1H for every 5000 ^1_1H in water

The oceans of the world contain 7×10^{13} tons of ^2_1H

This much deuterium used in a fusion reactor would release 5×10^{27} kcal of energy

This amount of energy would meet human needs for 4×10^9 years

Summary

Mass is converted into energy when an atomic nucleus is altered in a reaction. Protons and neutrons have less mass when combined into nuclei than they do as free particles. This mass defect can be released by two types of nuclear reactions, nuclear fission and nuclear fusion.

The fission of a uranium-235 nucleus into two fragments can be initiated by bombardment with a neutron. The nucleus splits into two nearly equal fragments (fission products). Several neutrons also are released. These neutrons can be used to trigger a chain reaction in a sample of uranium by colliding with other uranium nuclei. The tremendous energy released by nuclear fission can be used explosively in a bomb or, in a more controlled fashion, in a reactor to generate power. The nuclei produced by fission are highly radioactive. Some decay slowly. The energy of the radiation they emit is potentially harmful to living cells. The safe storage of fission products away from living things is the greatest safety problem associated with fission reactors.

Uranium-238 nuclei can be converted into plutonium-239 nuclei in a reactor. Plutonium-239 also can undergo fission to release energy. A breeder reactor generates plutonium fuel at the same time that it uses uranium fuel to produce power. When breeder reactors and plutonium reactors begin to operate, we will achieve the goal of producing power and fuel at the same time, thus conserving other supplies of potential energy.

In the nuclear fusion reaction, isotopes of hydrogen can be fused together to form helium nuclei. This reaction occurs in

Table 9.1   A summary of advantages and disadvantages of nuclear energy as potential energy for power

Reaction	Advantages	Disadvantages
Nuclear fission	Abundant resources, especially when breeder reactors are used	Radiation hazards, especially from long-lived fission products
	No smoke, ashes, or SO_2	
	Very concentrated fuel, low transportation costs, little disruption of environment by mining	Much research still needed to develop assurance of safety
		Costly research and development still needed for breeder and plutonium reactors
Nuclear fusion	Almost limitless potential fuel	Not yet proved feasible
	No hazard from radioactive products	Much research and development still needed to solve problems of containment and heat removal
		Very costly

stars to release tremendous amounts of energy. Research is under way to accomplish the control of this type of reaction on earth. If this can be done successfully, we will have available an almost limitless store of potential energy for power.

Table 9.1 summarizes the advantages and disadvantages of nuclear reactors as suppliers of potential energy for power production.

Glossary

The number in parentheses indicates the text page where you can find the term defined in context.

breeder reactor   a nuclear reactor that manufactures fissionable nuclei at the same time that it generates power from the fission of other nuclei (261)

chain reaction   a series of identical reactions in which the product of one reaction initiates the next reaction (259)

critical mass   the mass of fissionable material that is just large enough to trap sufficient neutrons to sustain a chain reaction (259)

deuterium   the isotope of hydrogen whose nucleus contains one neutron and one proton (271)

half-life   the time required for half of any sample of a particular radioactive material to decay (263)

ionizing radiation   rays of fast-moving particles or highly energetic quanta, so called because they may knock electrons from atoms or molecules, forming positive ions (264)

mass defect   the mass lost (converted to energy) by a proton or neutron in forming part of a nucleus; the difference between the mass of a nucleus and the individual masses of its component neutrons and protons (255)

nuclear fission   the splitting of a large atomic nucleus into two smaller nuclei (258)

nuclear fusion   the joining of two small atomic nuclei to form a larger one (269)

plasma   matter at so high a temperature that nuclei and electrons are no longer associated (270)

radioactive decay   the decrease in the amount of radiation released from a radioactive sample over the course of time (262)

radioactivity   the spontaneous decay of one atomic nucleus to become a different nucleus by emission of alpha, beta, or gamma
radiation (254)

reactor   a device in which the rate of a nuclear fission reaction
can be controlled, yielding heat that can be converted to electric power (259)

tritium   the isotope of hydrogen whose nucleus contains two
neutrons and one proton (271)

Exercises

9.1   In most of our previous discussions, we have treated mass
and energy as separate ideas. We discussed a law of conservation
of mass and a law of conservation of energy. Some people suggest lumping these laws into a law of conservation of mass-
energy. How would such a law be stated? What is the reasoning
behind this suggestion?

9.2   *a.*   What three fundamental particles make up an atom?

    b.   Which fundamental particles are in an atomic nucleus?

    c.   Which fundamental particle is outside the atomic
nucleus?

9.3   *a.*   By what process does an atom become a positive ion?

    b.   By what process does an atom become a negative ion?

    c.   Is the nucleus involved in these processes?

9.4   In this symbol, what does each number represent? $^{14}_{7}N$

9.5   Give the number of protons and number of neutrons in the
nucleus of the atom identified as $^{235}_{92}U$.

9.6   Consider the following symbols.

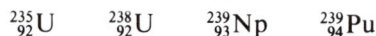

$$^{235}_{92}U \qquad ^{238}_{92}U \qquad ^{239}_{93}Np \qquad ^{239}_{94}Pu$$

    a.   Which two nuclei have the same mass numbers?

    b.   Which two nuclei are isotopes?

c. Which two nuclei contain the same number of protons?

d. Which two nuclei contain the same number of neutrons?

9.7   *a.* What very precise measurements must be made to obtain the data needed for the calculation of the mass defect of a nuclear particle?

b. Justify the statement, "The greater the mass defect of the fundamental particles in a nucleus, the more stable is the nucleus."

9.8   Consider the diagram in Figure 9.1.

a. In what general portion of the curve would we find the most stable nuclei?

b. A nucleus of an isotope of the element americium has a mass number of 243. Is this nucleus more likely to undergo fission or fusion?

c. The nucleus $^{238}_{92}$U undergoes slow radioactive decay by a series of steps. A large amount of energy is released along with alpha, beta, and gamma rays in this process. The final stable nucleus produced is $^{206}_{82}$Pb. Estimate where you would locate this nucleus on the curve of Figure 9.1. Suggest where the energy released by the radioactive decay process comes from.

9.9   The symbols for alpha and beta particles produced by the decay of various radioactive nuclei can be represented as follows:

$$\text{alpha: } {}^{4}_{2}\alpha \qquad \text{beta: } {}^{0}_{-1}\beta$$

a. The symbol of what element could be used in place of α?

b. The symbol for what fundamental particle of matter could be used in place of β?

c. Locate the element radium, Ra (atomic number 88), in the periodic table. What is the atomic number and symbol of the nucleus formed when a radium nucleus, $^{226}_{88}$Ra, emits an alpha particle?

d. When $^{239}_{92}$U decays, a beta particle is emitted and $^{239}_{93}$Np is formed. Since the mass number 239 has not changed, but the atomic number 92 has changed to 93, what conversion of one fundamental nuclear particle into another must have occurred?

9.10   Explain how the fission of $^{235}_{92}U$ or $^{239}_{94}Pu$ can become a chain reaction.

9.11   Explain why a fission bomb does not explode as soon as it is manufactured, but can be made to explode later.

9.12   What is the function of control rods in a nuclear reactor?

9.13   Justify the following statement: The potential danger from a nuclear reactor accident is not an atomic-bomb type of explosion, but rather the possible release of fission products into the environment.

9.14   Why is a breeder reactor unique as a means of conserving energy resources?

9.15   Compare the possible environmental hazards from the operation of a nuclear fission reactor with those of a coal-fired power generator. Consider the problems of fuel resources and production, the problems created by the products generated along with the energy released, and the problem of thermal pollution.

9.16   What do you think happens to the fission products of the reaction when a nuclear fission bomb is exploded in the atmosphere? Why do you think the atmospheric testing of such bombs has been stopped by all but a few major nations?

9.17   Would you ever vote in support of a proposal to get rid of nuclear fission products by shooting them off the earth in a space capsule? Justify your answer.

9.18   Can the rate at which radioactive decay occurs be slowed if the temperature of the sample is lowered?

9.19   If a person's environment should become dangerously contaminated with $^{90}_{38}Sr$ (half-life = 29 years), to approximately what level would the danger diminish if the person lived for 30 years? To what fraction of the original would the danger diminish by the time the person's infant children reached the age of 60?

9.20   It has been said that in these difficult times, the half-life of a college president's tenure at one institution is about 5 years. What proportion of college presidents could be expected to hold office for 20 years?

9.21   Explain in general chemical terms why the presence of many radioactive atoms in a living organism may be harmful to that organism.

9.22   Explain the paradox of using radioactive atoms to combat cancer when there is evidence to suggest that the ionizing radiation from radioactive atoms may cause some kinds of cancer.

9.23   What is the greatest problem yet to be solved before controlled nuclear fusion can be a source of power for use by our technological society?

9.24   List the arguments that would be used by those who advocate financial support from the government for research on nuclear fusion reactors.

10

Light Energy and Solar Radiation

- [ ] How does the quantum theory account for the dual nature of light?
- [ ] What is the connection between sunlight and smog?
- [ ] How are forms of light other than the visible kind generated and absorbed?
- [ ] What happens to sunlight in the earth's atmosphere?
- [ ] Is the proportion of carbon dioxide in the atmosphere changing?
- [ ] How does the atmosphere function like the glass in a greenhouse?
- [ ] Is solar energy a practical alternative to fossil fuels?

In Chapter 8, we discussed the transformation of one kind of energy into another but said very little about light. The idea that light is a form of energy was introduced in Chapter 5. There we were concerned with what the light emitted by an excited atom can tell us about the spacing of energy levels in atoms. In Chapter 9, the connection between light and energy was mentioned again when we described the effects on living cells when they are hit by gamma rays, a highly energetic form of light. Now, to make our picture of energy more complete, it is appropriate for us to discuss the subject of light more fully. For example, what about other invisible forms of light, such as ultraviolet light that darkens our skin or infrared light that warms our bodies?

Another aspect of our discussion in this chapter is sunlight. What are the characteristics of this radiation that constantly bathes the earth with energy? We know that plants take in sunlight and store its energy by the process of photosynthesis. But what are some of the other ways in which sunlight interacts with the matter in the earth and its atmosphere? Is there any way we can capture some of the tremendous total energy of solar radiation to do some of the things we now do with the energy from fossil fuels? Here again, we will find ourselves exploring the idea of conservation. The energy crisis forces that consideration into any discussion of energy.

WHAT IS LIGHT?

We can start to answer the question "What is light?" by reviewing some of the ideas we introduced in Chapter 5.

Light has wave-like properties. The colors we see in the spectrum of white, visible light dispersed by a prism are associated with differences in wavelength. Light at the red side of the rainbow has a longer wavelength than does light at the violet side.

Light also has particle-like properties. A beam of light behaves somewhat as if it were a stream of bullets. You cannot see around a corner. Light goes in a straight line unless reflected by a mirror or unless its direction is changed by a prism or a lens. And the light emitted by a collection of excited atoms appears in little spurts of energy, one from each individual atom.

This dual nature of light, its "schizophrenia" of wave-like and particle-like behavior, is accounted for by the quantum theory of light. *Light consists of bundles of energy called quanta or photons.* The word *photon* is another term used to designate a quantum of light energy. The word photon should not be confused with the word proton. Actually, there is a parallel implication in the two terms. A photon of light is a fundamental bundle of energy as a proton is a fundamental bundle of matter. The graininess of light

---

The graininess of light is comparable to the graininess of matter.

---

is comparable to the graininess of matter. Each photon of light has a particular energy, just as each atom has a particular mass.

The amount of energy in a photon is inversely proportional to its wavelength. Each photon of longer-wavelength red light has less energy than a photon of shorter-wavelength violet light has.

We also can tie together some other information about light. Objects can emit, absorb, or reflect light. All these properties of light can be described in terms of the behavior of photons. When you turn on an electric light bulb, you cause a stream of electrons to go through the tungsten-wire filament in the bulb. The atoms of the filament absorb some of this energy of the moving electrons. Electrons in the tungsten atoms jump to higher energy levels. When these electrons fall back to lower energy levels, the atoms emit photons of visible light. The typical incandescent light bulb emits all the various photons that, mixed together, appear as white light.

If you pass a beam of white light through a prism and put a piece of white paper in the spread-out beam, you see a complete spectrum of all colors of the rainbow. But if you put colored paper in the spread-out beam, you see something quite different. If you use red paper, you will see only the red part of the spectrum; all the rest looks black. Green paper will let you see only the green portion; all the rest will look black. So it is with all colored objects. They look black in light of every color other than their own. What you demonstrate with these simple experiments is that light can be reflected or absorbed. Red-colored objects are able to soak up the energy of all photons except those of red light. Red-colored objects cannot accommodate the energy of a red

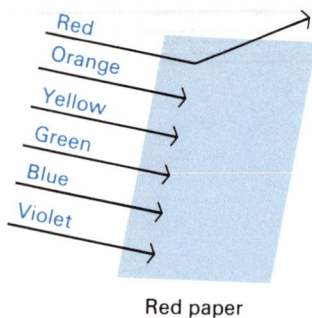
Red paper

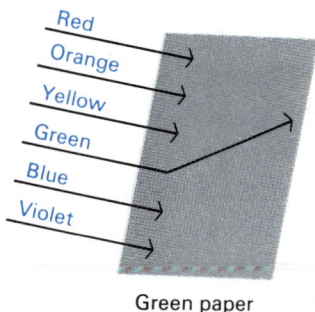
Green paper

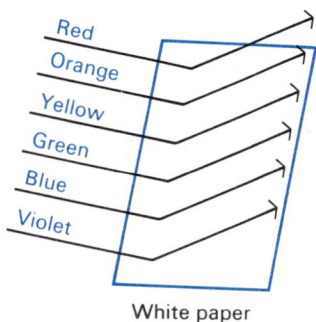
White paper

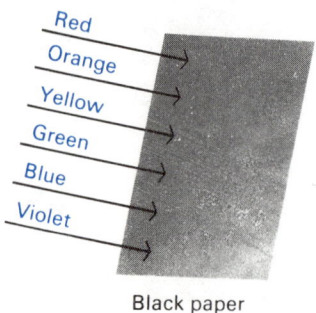
Black paper

photon, so they reject or bounce back (reflect) the red photons. A white object has no way of accommodating the energy of any of the photons in white light. All are reflected. A black-colored object behaves in the opposite way. The energies of all the various photons can be absorbed by it, so none are reflected.

The phenomena of absorption and reflection of light are responsible for your feeling warm when wearing black clothes in the sunshine. Photons of all energies are being absorbed. Some of this energy is converted to heat; the flow of this heat to your body makes you feel warm. If you wear white clothes, you feel cooler. White clothes reflect the sunlight. Few photons are absorbed, little energy is added, so less heat is generated.

When light is absorbed, energy must be absorbed. This fact suggests a way to measure light energy. Each photon or bundle of light carries a quantum of energy. The size of the energy quantum and the number of photons will determine the total energy absorbed. Using money is somewhat like measuring light energy. You can have a total value of $1.00 by taking 100 of the lowest-valued "photons," pennies. You can also have the same total value by taking only 20 of the "photons" with the next-lowest value, nickels, and so on.

THE WAVELENGTH, FREQUENCY, AND VELOCITY OF PHOTONS

Let us use the familiar analogy of waves on the surface of a body of water to point out another important characteristic of all waves. A wave always involves some kind of an oscillation, a regularly repeated motion.

Imagine a cork floating on the surface of water as a wave comes along. Figure 10.1 suggests what you observe. The cork moves up and down perpendicular to the motion of the wave. The cork makes one complete up-and-down *cycle* for the passage of each wave. If we count the number of times the cork completes a cycle in a given time, we have a measure of the *frequency* with which a complete wave passes through the water.

Suppose we have two waves moving across a water surface with the same speed in a particular direction (velocity) but with different wavelengths. Figure 10.2 suggests this contrast. The upper wave moves the same distance (A to B) as the lower wave (C to D) does in the same period of time. So their velocities are equal. But the upper wave has a longer wavelength. Consequently, a cork floating on the surface as the upper wave passes

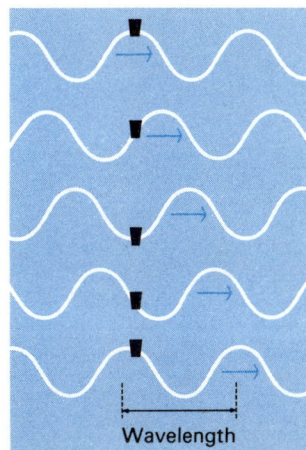

Figure 10.1 The up-and-down cycle of a cork floating on water.

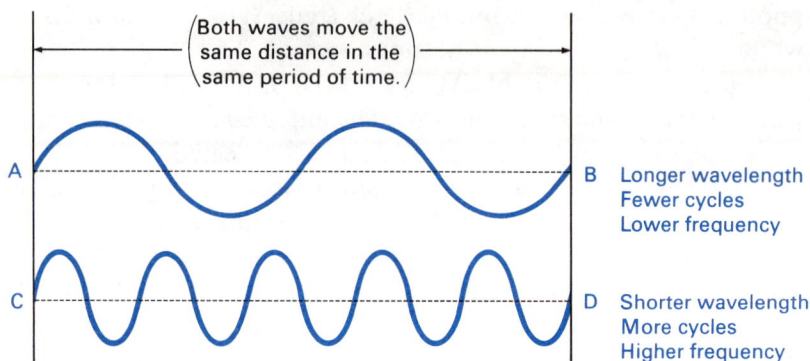

Figure 10.2   A comparison of two waves with the same velocity.

by would make fewer up-and-down oscillations than would a cork moved by the lower wave. The relationship among the wavelength, frequency, and velocity is represented by the formula

$$\text{Wavelength} \times \text{frequency} = \text{velocity}$$

$$\lambda \,(\text{distance}) \times \nu \left(\frac{\text{number}}{\text{time}} \right) = c \left(\frac{\text{distance}}{\text{time}} \right)$$

Light waves behave in the same way as the water waves of our analogy. Here we have introduced the symbols most frequently used to represent these quantities when dealing with photons of light. The symbol λ is the Greek letter lambda (*l* for "length"). The symbol ν is the Greek letter nu (*n* for "number"). The velocity of light is represented by *c*.

When light is measured as wave motion, a remarkable consistency in nature shows up. *The velocity of all kinds of light is constant.* The value of this fundamental constant of the natural world is 300,000,000 meter/second or 186,000 mile/second. In decimal notation, these numbers are written 3.00×10^8 meter/second and 1.86×10^5 mile/second.

The fact that the velocity of all photons of light is constant means that *the wavelength varies inversely with the frequency.* A photon of red light has a longer wavelength than a photon of violet light. Consequently, the frequency of the red photon is the smaller of the two. We have already used the idea that the energy of the red photon is less. We can summarize these relationships as follows for the spectrum of visible light:

$$\text{Violet} \left\{ \begin{array}{l} \longrightarrow \text{ increase in wavelength } (\lambda) \ \longrightarrow \\ \longleftarrow \text{ increase in frequency } (\nu) \ \longrightarrow \\ \longleftarrow \text{ increase in energy } (E) \longrightarrow \end{array} \right\} \text{Red}$$

THE COMPLETE SPECTRUM
OF ALL FORMS OF LIGHT

The kinds of photons we can see with our eyes represent only a very small fraction of all the kinds of light photons that can exist. The complete spectrum extends far beyond both ends of the spectrum of visible light. Figure 10.3 summarizes the relationships among the various kinds of light. The spectrum extends all the way from the very short-wavelength, high-energy gamma rays through X rays and ultraviolet light into the visible range. From the red end of the visible range, it extends on through the infrared waves to radar waves and microwaves into radio waves. These radio-wave photons have the longest wavelengths and the smallest frequencies. Figure 10.3 includes a few numbers obtained by actual measurements to give you some points of reference. The tuning dials of some radios are marked in thousands of kilocycle/second (1000 cycle/second), or in megacycles (1,000,000 cycle/second). The wavelength of this kind of radiation is thus in a range of about 100 meters (approximately the length of a football field). The photons of this kind of radiation have the smallest energy. As we move toward the left in the diagram, we find that the wavelengths become shorter. The energies and frequencies become greater. In the visible range, the wavelengths are just a little shorter than a micrometer, one-millionth of a meter (expressed as 10^{-6} meter). The corresponding frequencies are in the range of 10^{14} cycle/second (100 million million cycle/second). The energy of these photons of visible light is thus 100 million times the energy of the photons of radio waves.

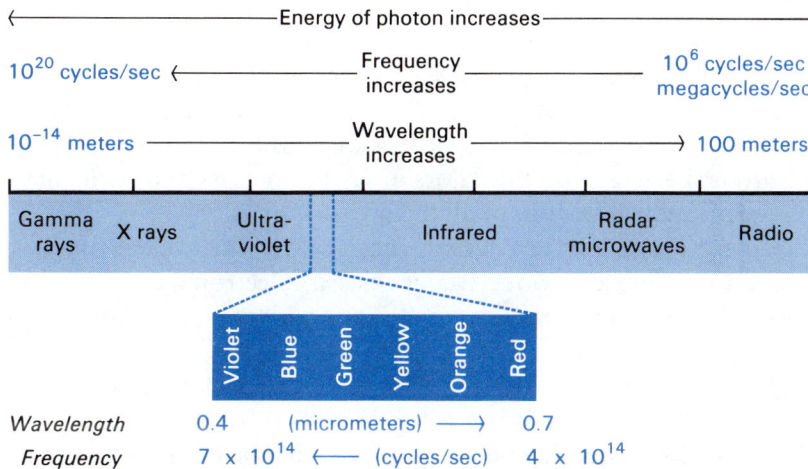

Figure 10.3 The complete electromagnetic spectrum.

Ultraviolet light is made up of photons with shorter wavelengths, in the range of 10^{-8} meters (0.00000001, or one hundred-millionth, of a meter). Consequently, the energy of an ultraviolet photon is about 10 to 100 times the energy of a photon of visible light. If we continue toward the left in the diagram, we find that X rays and gamma rays are made up of photons with even shorter wavelengths and correspondingly greater energies.

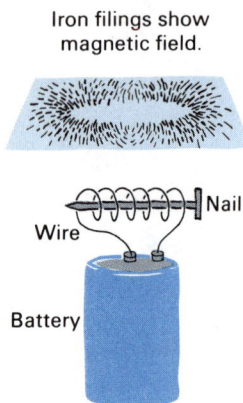

LIGHT IS ELECTROMAGNETIC RADIATION

Iron filings show magnetic field.

Nail

Wire

Battery

A more accurate term for light is *electromagnetic radiation* Nature always connects electricity and magnetism. You can make a nail into a magnet by wrapping a wire around it and connecting the ends of the wire to a battery so that electrons flow

Whenever electrons move, they generate a magnetic field.

through the wire. Whenever electrons move, they generate a magnetic field.

If the movement of the electrons follows a pattern of regular back-and-forth oscillations, then the magnetic field they generate will oscillate also. *A simultaneous oscillation of an electric and a magnetic field in this way generates light.* The light radiates from the spot where the electrons are moving. This is why light is called electromagnetic radiation. You can now recognize why light has a frequency, a wave-like characteristic. The frequency of the oscillations with which the electrons move corresponds to the frequency of the photons that are generated.

Electrons in various kinds of matter are able to oscillate with a wide variety of frequencies. Consequently, electromagnetic energy (in the form of photons) generated by these oscillations has a correspondingly wide variety of possible frequencies. The photons that make up the colors of visible light have only a small range of frequencies out of all the possible ones.

Figure 10.4 summarizes the processes by which electrons can move to generate various frequencies of electromagnetic radiation. The full spectrum is covered. The energy of the various photons increases from top to bottom, just as it does in going from right to left in Figure 10.3. You find included in this summary the type of energy absorption and emission we talked about in Chapter 5. When electrons jump from one energy level to another in

(Increasing wavelength)

(Increasing frequency and energy)

Radio or radar waves

Electrons of an electric current oscillate back and forth in wires of an antenna.

Infrared radiation

Electrons in molecules move as molecules rotate or as the atoms in molecules vibrate back and forth.

Visible or ultraviolet radiation

Electrons in the outermost energy levels of atoms or molecules jump to different energy levels.

X rays

Tightly held electrons in underlying energy levels of atoms jump to different energy levels.

Gamma rays

Electrically charged fundamental particles in atomic nuclei shift to different energy levels.

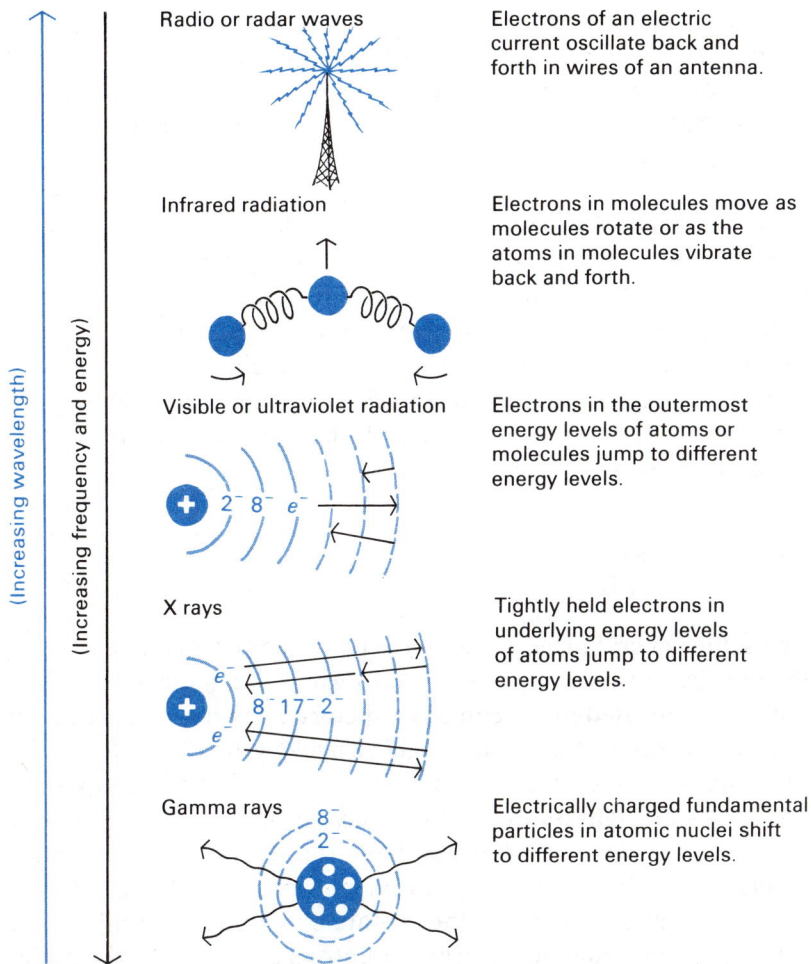

Figure 10.4   The kinds of electron movement involved with various types of radiation.

atoms or molecules, the size of those jumps usually is such that photons of visible or ultraviolet light correspond to the appropriate energy.

THE ABSORPTION OF ELECTROMAGNETIC RADIATION

We have said that electromagnetic radiation is generated when electrons move in some pattern of regular, repeated oscillations. The opposite is also true; the absorption of electromagnetic radiation causes electrons to oscillate. The frequency of the oscillations is the crucial point in understanding how this exchange of energy can possibly occur. A simple illustration of this kind of

energy exchange occurs every time you push a child on a playground swing. Once the child is swinging, you must time your pushes of extra energy to make the swing go higher. Your pushes have to be timed with a frequency to agree with the natural oscillation frequency of the swing.

In a similar manner, if you want your radio or television set to receive (absorb) energy in the form of long-wavelength radiation generated by a particular broadcasting station, you must tune your set. You adjust the electronic circuits so that electrons in the antenna of your set can oscillate at the particular frequency of the photons of radiation you want to receive. Then their motion is timed to accept the extra push of the electromagnetic photons, just as the swing with the child can accept your pushes to make it go higher.

HOW MOLECULES ABSORB INFRARED LIGHT

The same general principle we have just been discussing accounts for the absorption of *infrared radiation* by some molecules. Infrared light is located in the complete spectrum (Figure 10.3) next to the visible range but with longer wavelengths. Therefore, the photons of infrared radiation are smaller bundles of energy than are the visible light photons. So we have to imagine some kind of oscillation of electrons that can have the appropriate frequency to match the frequency of infrared photons. The kinds of motions molecules make when they rotate or vibrate internally have this matching frequency. Infrared photons can "shove" the molecules into these motions just as your properly timed shoves make the child move on the playground swing.

We will illustrate this phenomenon by describing how the carbon dioxide molecule, CO_2, absorbs energy to set it wiggling or vibrating internally. We choose this example deliberately, because later in the chapter we will discuss how important the absorption of infrared photons by CO_2 is to the atmosphere of the earth. The temperature of the earth's surface and, consequently, the earth's climate is kept balanced largely by the ability of CO_2 in the atmosphere to absorb or emit infrared radiation.

Imagine a model of the CO_2 molecule made from three wooden balls that represent atoms joined by springs that represent chemical bonds. If you throw this model into the air, you would see that the wooden balls move in various wiggling fashions relative to one another. The springs alternately push and pull the mass of

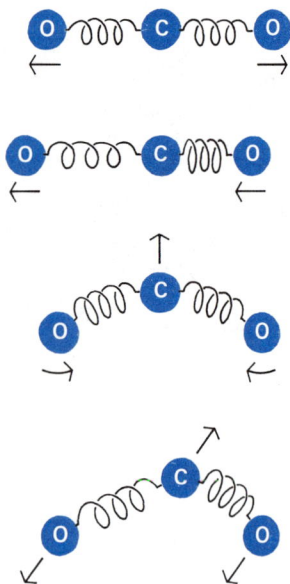

Figure 10.5 Fundamental frequencies of a CO_2 molecule.

the wooden balls so that the whole structure vibrates. A detailed analysis of all the possible motions resolves into the four fundamental motions represented in Figure 10.5. If we made several models with wooden balls of the same mass and springs of the same strength, they all would vibrate in the same way with the same frequencies.

So it is with every CO_2 molecule. Atoms of the same mass have electrons arranged the same way to bond them together. Every CO_2 molecule can vibrate with only these four fundamental frequencies. The movement of the atoms moves the electrons; the result is electric charges oscillating in space. Any CO_2 molecule thus is "tuned" to absorb electromagnetic energy that has the same four frequencies. The photons that have these frequencies are in the infrared range of the spectrum. If a mixture of infrared light of many frequencies hits a CO_2 molecule, it absorbs only those photons that have the right frequency to set it vibrating. This process is just like your tuned radio, which selects waves of one frequency out of all the possible broadcasts in the air.

THE ABSORPTION OF ULTRAVIOLET LIGHT

Let us now turn to the shorter-wavelength side of the spectrum which begins on the violet side of the visible rainbow. The photons of *ultraviolet light* are bundles carrying more energy than those of visible light. Consequently, atoms and molecules make much bigger changes when they absorb these photons. The absorbed energy is often enough to rip electrons off or at least cause molecules to break apart. The fragments so formed are very reactive. In a later section of the chapter, we will illustrate the kinds of chemical reactions that can occur among these fragments when we discuss the ultraviolet portion of the sunlight's total spectrum.

Another phenomenon that involves ultraviolet radiation is called *fluorescence*. Fluorescence is the process of absorbing the larger-energy photons of ultraviolet light and then emitting smaller-energy photons of visible light. Fluorescent lights involve this process. Fluorescence also is responsible for the interesting effects that can be produced by what is called "black light."

Fluorescent lamps are efficient devices for converting electricity into visible electromagnetic energy. If you touch a glowing fluorescent bulb, it feels cool, whereas a glowing tungsten-filament bulb (a conventional light bulb) feels hot. You remember

Atoms share electron pair in a molecule.

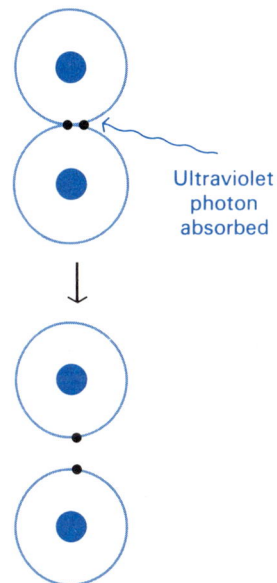

Ultraviolet photon absorbed

Separate atoms (molecule breaks up)

Atom absorbs ultraviolet photon by electron jumping to very high energy level.

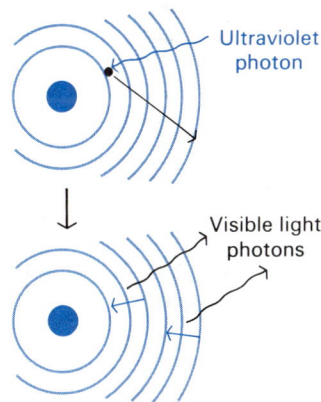

Ultraviolet photon

Visible light photons

Electron falls back from the high-energy level in smaller jumps. Each jump emits a photon of visible light (atom fluoresces).

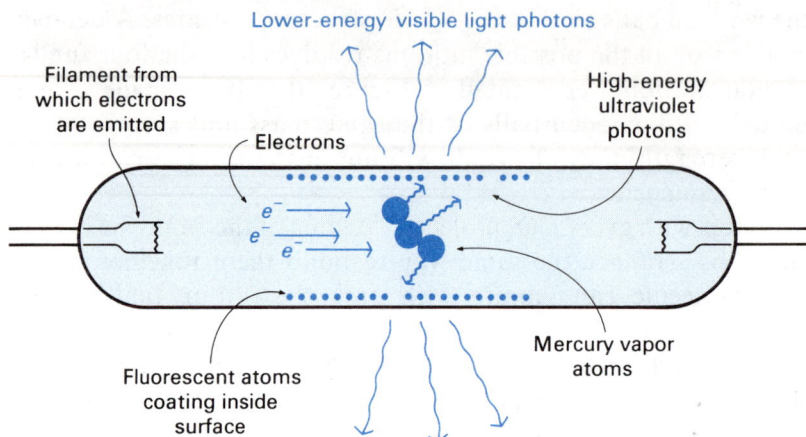

Figure 10.6   The essential features of a fluorescent light tube.

from the discussion in Chapter 8 that any hot object loses its thermal energy to cooler surroundings. So any conversion of electrical energy into light energy that takes place at a low temperature is bound to have a greater overall efficiency. Let us see how fluorescent lights work.

Figure 10.6 shows the essential features of a fluorescent light tube. The glass tube is filled with mercury vapor at a low pressure. When the current is turned on, electrons collide with the gaseous mercury atoms. The mercury atoms become excited by having their electrons jump to very high energy levels. When these mercury electrons return to stable, lower energy levels, high-energy photons of ultraviolet light are emitted. These photons strike atoms in the coating that has been applied to the inside of the glass tube. Atoms in the coating absorb the high-energy ultraviolet photons by exciting their electrons to higher energy levels. When these electrons in turn fall back down to stable energy levels in their atoms, photons of light in the visible range are emitted. Different colors of visible light can be obtained by mixing different kinds of atoms in the tube coatings. The overall process can be represented as follows:

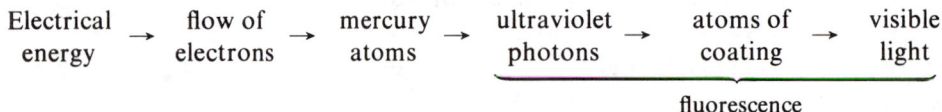

$$\begin{array}{ccccccccc} \text{Electrical} \\ \text{energy} \end{array} \rightarrow \begin{array}{c} \text{flow of} \\ \text{electrons} \end{array} \rightarrow \begin{array}{c} \text{mercury} \\ \text{atoms} \end{array} \rightarrow \begin{array}{c} \text{ultraviolet} \\ \text{photons} \end{array} \rightarrow \begin{array}{c} \text{atoms of} \\ \text{coating} \end{array} \rightarrow \begin{array}{c} \text{visible} \\ \text{light} \end{array}$$

fluorescence

Many molecules, as well as atoms, are able to fluoresce. The mechanism is the same. The molecules absorb the energy of the ultraviolet photon by having electrons jump to a variety of high molecular energy levels. Then, when the electrons fall back to stable energy levels, they do so in perhaps several jumps of a smaller size. These smaller energy jumps also release photons that are in the range of visible light.

BLACK LIGHT AND FLUORESCENCE

The term "black light" is sometimes used to describe ultraviolet light. A lamp to produce ultraviolet light contains mercury vapor, as does a fluorescent tube. But the bulb is made of quartz instead of glass. Quartz is transparent to ultraviolet light, whereas glass absorbs it and thus will not let it through. Our visual senses do not respond to the larger energies of ultraviolet photons; hence, we cannot see them, and therefore, we refer to the light as "black." Moreover, many colored objects totally absorb ultraviolet light and reflect none, so they appear black.

Long wavelength, smaller-energy visible photons are transmitted by glass.

Glass

Short wavelength, high-energy ultraviolet photons are stopped by glass.

It is dangerous to look directly at a source of ultraviolet light. Even though you cannot see them, the high-energy photons are being absorbed by molecules in your eyes. This absorbed energy causes chemical reactions to occur, somewhat like those that take place when an egg cooks. These are irreversible (you cannot uncook an egg!), so permanent damage to your eyes occurs. Your eyes can be protected by wearing glasses, because ultraviolet light cannot go through glass.

Black light is often used, along with the phenomenon of fluorescence, to accomplish very striking decorative effects. Some dye molecules that give an object one color when reflecting visible light fluoresce with an entirely different color of visible light. Other molecules may be totally transparent to visible light yet fluoresce brilliantly when irradiated with ultraviolet light. "Invisible" inks are made of such molecules. Markings made with them are clearly visible in black light. An object planted to catch a thief can be lightly dusted with a powder that will fluoresce. The thief is revealed by the glow of particles of the powder on his or her fingers when they are exposed to a source of ultraviolet light.

THE SPECTRUM OF SUNLIGHT

The most abundant ever-renewed form of energy on earth is the sunlight that constantly bathes our planet. This source of energy has been coming to the earth over the 4.5 billion years of the earth's history. Although we can only speculate about the early years of that history, we must recognize that somehow the energy

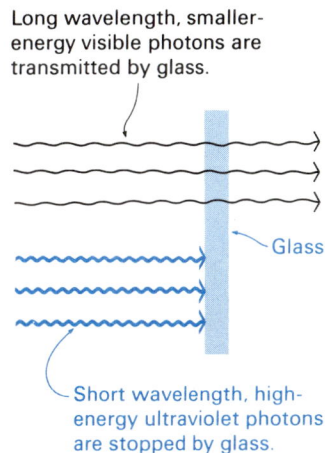

of sunlight altered the earth's surface and provided the energy needed for the beginning of life. Plants, by the process of photosynthesis, store the energy of sunlight as potential energy. Other living things use plants for food and thus make use of the potential energy. Most of the power our society uses to do its work traces back through fossil fuels to sunlight.

Later in the chapter, we will discuss how we may be able to begin to use sunlight directly to meet some of these power needs. But first, let us describe sunlight more fully. We also should examine how some of the products of our complicated technological society affect the sunlight that reaches us through the blanket of atmosphere that surrounds the earth's surface.

Sunlight, more accurately called *total solar radiation,* spreads over a wide range of the whole spectrum. Recall from the discussion in Chapter 9 that the origin of the sun's radiated energy is a nuclear fusion reaction. Protons fuse into helium nuclei in the core of the sun, estimated to be at a temperature of 10,000,000 degrees. Most of this energy is in the form of gamma-ray photons, each a tremendously large bundle of energy. As these gamma rays pass through the matter of the sun, which is about 500,000 kilometers thick, they interact with the atoms of the sun's matter. These excited atoms then emit photons of smaller energy and longer wavelength than the gamma rays. The light that leaves the surface of the sun is chiefly a mixture of ultraviolet, visible, and infrared radiation. Small amounts of gamma-ray and X-ray radiation of the short-wavelength spectrum and some microwave and radio-wave parts of the long-wavelength spectrum also are included.

Figure 10.7 suggests the distribution of the energy of sunlight across the spectrum. Two curves are shown. The upper one suggests the spectral distribution of energy arriving at the outer limits of the earth's atmosphere. This curve is nearly the same as that observed when an object is heated to a temperature of about 5500° C in an oven. The distribution of the light from such a "white-hot" object gives the same kind of a curve. This measurement is a way of estimating the temperature of the sun's surface.

The lower curve of Figure 10.7 suggests what is observed when the energy of sunlight is measured at the surface of the earth on a cloudless day. You can note some important features of the comparison between these two curves.

1. The total amount of energy reaching the earth's surface is much less than the amount that arrives at the outside of the earth's atmosphere. Only about half the total gets through.

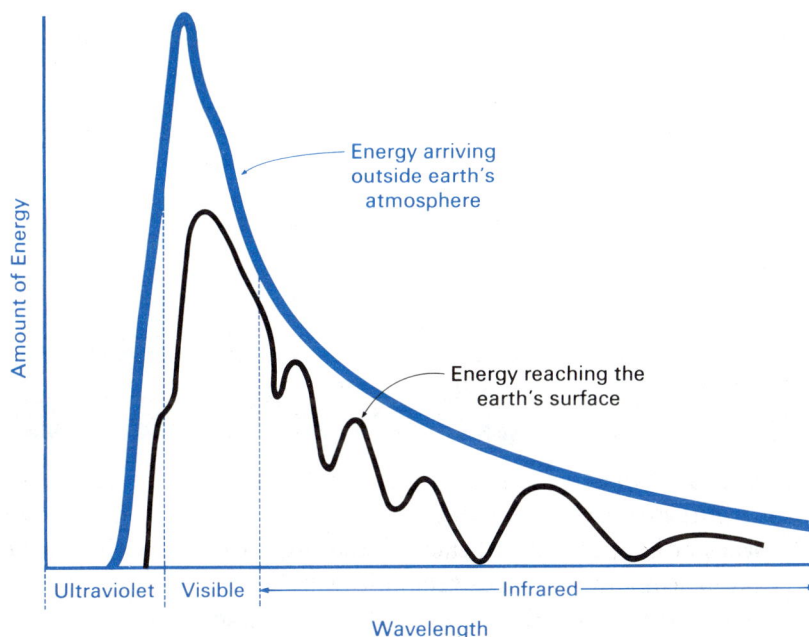

Figure 10.7   The spectrum of solar radiation.

2. The most abundant kind of light in sunlight is in the visible range of the spectrum. The color of light corresponding to the peaks of both curves is blue-green.

3. A large fraction of the ultraviolet light coming from the sun is screened out by the earth's atmosphere. We will describe how this happens and also why it is fortunate for living things that it does occur.

4. Many peaks and valleys appear in the curve showing the infrared radiation that reaches the earth's surface. This irregularity means that photons of some wavelengths or energies get through; others are absorbed. Such selective absorption suggests that certain molecules are responsible. (We have described how the internal vibrations of molecules are "tuned" to absorb only certain energies.)

THE SCATTERING AND ABSORPTION OF VISIBLE AND INFRARED SUNLIGHT

If much of the visible and infrared solar radiation never reaches the earth's surface, what happens to it? Most of the loss comes from the *scattering* of light by water droplets and fine particles of dust or smoke. The scattered light ends up being reflected back into space. If a lower curve were drawn on Figure 10.7 for a day when thick clouds were in the atmosphere, it would show much

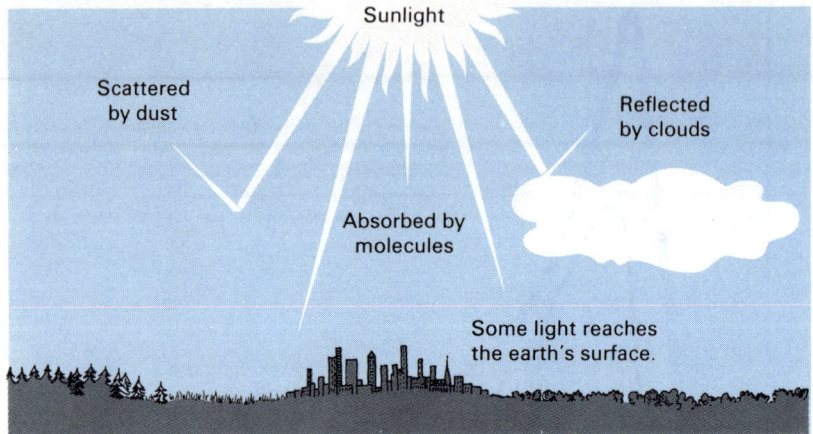

less of the infrared light getting through. Most of the clouds and dust in the atmosphere come from natural causes. So does some of the smoke. But most of the smoke results from the activities of humans using the complex technology of modern society. Smoke particles are responsible for the reflection of much of the infrared

Smoke particles are responsible for the reflection of much of the infrared radiation that is lost.

radiation that is lost. The part of the infrared radiation that does get through the atmosphere heats the earth's surface. So it is clear that, if humans keep thoughtlessly adding to the store of fine particles of solids (smoke) in the atmosphere, enough additional infrared radiation might be lost to lower the temperature of the earth's surface.

Another part of the total radiation loss is caused by the absorption of light energy by various molecules in the atmosphere. Although the air surrounding us seems to be a very "dilute" form of matter and although we know it becomes still thinner within a few miles above the earth's surface, we have to realize that the huge size of the planet means that the total mass of the atmosphere is very large. Estimates suggest that it weighs 6×10^{15} tons (6,000,000,000,000,000 tons). That represents a lot of molecules! Most of these molecules, such as O_2, N_2, H_2O, and CO_2, do not absorb visible light. They are transparent to photons of this range of energy. H_2O, condensed in the form of water droplets piled together in clouds, does stop visible light. However, as we have already learned, molecules such as CO_2 and H_2O are capable of vibrating with internal wiggling, oscillating motions.

The frequencies of these wiggles are comparable to the frequencies of infrared light, so these molecules absorb some of these photons. The result is that some of the infrared radiation is filtered out by the time the sunlight reaches the earth's surface.

THE ABSORPTION OF ULTRAVIOLET SUNLIGHT

On the other side of the solar spectrum, in the ultraviolet range, you note that only a small fraction of the sun's radiation gets through. A great deal is absorbed. The molecules in the atmosphere are not transparent to ultraviolet light. The reason is that the energy of the ultraviolet photons is equal to some of the big energy jumps that electrons can make in the molecules of the atmosphere. If the electrons are knocked completely out of the molecule, ions are formed. Recall that the absorption of gamma rays also forms ions. The same thing happens when X rays are absorbed. This process of ionization occurs in the upper regions of the atmosphere. The part of the atmosphere above 30 miles is called the *ionosphere*.

A different kind of process takes place in the layer of atmosphere known as the *stratosphere*, from about 7 to 30 miles above the surface. Here, the molecules that have been excited or broken apart by the absorption of ultraviolet photons react with other molecules. A chemical reaction such as this, started by the absorption of light, is called a *photochemical reaction*. Many different photochemical reactions can occur. We are going to concentrate here on one important example, the formation of ozone.

THE FORMATION OF OZONE

Ozone is a gas with the molecular formula O_3. It can be formed from ordinary oxygen gas, O_2, by the absorption of light or electrical energy. Ozone has a very sharp, irritating odor. If you have ever been near any short circuit of a heavy electric current that produced an arc or sparks, you probably smelled ozone. The ozone was formed from the O_2 molecules in the air by the electric arc. Special lamps that emit ultraviolet light capable of making ozone from oxygen are used to purify air and kill air-borne bacteria. Some food-storage facilities, or places where food is prepared, use such lamps. Ozone also reacts with molecules responsible for offensive odors. Sometimes ozone-making lamps are installed for that purpose in such locations as zoos or public rest rooms. Ozone reacts very readily with various organic compounds, especially those of biochemical significance. Consequently, ozone in any large amount is considered toxic to plant and animal life.

When an oxygen molecule absorbs an ultraviolet photon, it may break apart into two oxygen atoms.

$$O_2 \ + \ \text{ultraviolet photon} \ \rightarrow \ O \ + \ O$$

Remember that the oxygen atoms in the O_2 molecule are held together by shared pairs of electrons. The absorbed energy disrupts this stable arrangement. Each of the oxygen atoms takes back the electrons it was sharing with the other oxygen atom. Consequently, the oxygen atoms so formed are very reactive, because they have unpaired electrons that they tend to share with any other atom they encounter. You may recognize an analogy in the statistic that shows a majority of people who have been divorced get married again within a few years. One of the reactions such "divorced" oxygen atoms undergo is a combination with an O_2 molecule.

$$O_2 \ + \ O \ \rightarrow \ O_3 \qquad \text{ozone}$$

Ozone itself is a very strong absorber of ultraviolet photons of a slightly different energy than that absorbed by the O_2 molecule. When ozone absorbs an ultraviolet photon, the above reaction is reversed.

$$O_3 \ + \ \text{ultraviolet photon} \ \rightarrow \ O_2 \ + \ O$$

The O atom formed may start the cycle again. Over the eons of time that these reactions have been taking place in the atmosphere, an equilibrium has been established. A small but steady amount of ozone exists, chiefly in the stratosphere.

This layer of ozone is a great benefit to the earth, particularly to living things on the surface of the earth. If the ozone layer were not there to screen out some of the ultraviolet radiation, so much radiation would come through to the surface that all sorts of damaging reactions might occur. For example, ultraviolet light reacts with substances in the skin to form molecules that otherwise are not normally present. Among these are molecules known to produce skin cancers. If the ozone layer were not present in the atmosphere, photochemical reactions, such as molecular decompositions, would occur in such numbers and variety that the surface of the earth and the life on it would be entirely different.

POTENTIAL THREATS TO THE OZONE IN THE ATMOSPHERE

As scientists learn more and more about the constituents of the atmosphere and the photochemical reactions that can occur among these molecules, they have become more and more aware of how important it is not to disrupt the balance that exists.

One of the reasons the United States government decided not to support the development of supersonic transport aircraft is this kind of danger. Supersonic aircraft fly in the stratosphere. The exhaust from their engines adds various molecules that are known to react with ozone in the stratosphere. If this were to happen to too great an extent, the total amount of ozone now established by the balance of naturally occurring reactions might be decreased. This decrease ultimately could remove the protection from ultraviolet radiation that life on earth enjoys. Such a consequence would be far too great a price to pay for a limited technological advance in transportation.

A comparable potential danger has recently been recognized because of the presence of Freon molecules in the atmosphere. Freon is a trade name for several different substances whose molecules contain carbon, chlorine, and fluorine. Two typical formulas are CF_2Cl_2 and $CFCl_3$. These molecules are very stable and react with very few other molecules. Consequently, they have been considered to be safe to use as the compressed gas in aerosol spray cans. The Freon under pressure supplies the force to push out the contents of the can as a mist of very small droplets (an aerosol). The Freon molecules also come out of the can and enter the atmosphere. Because they react so little, they remain dispersed in the air and find their way to the upper atmosphere. Think of how many different substances you can find packaged in spray cans. Imagine how much Freon would be added to the atmosphere if Freon were the gas used in every one.

The trouble begins in the upper atmosphere. Freon molecules very readily absorb ultraviolet photons and a chlorine atom breaks off.

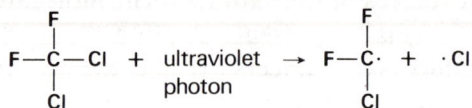

$$\underset{\underset{Cl}{|}}{\overset{\overset{F}{|}}{F-C-Cl}} + \underset{\text{photon}}{\text{ultraviolet}} \rightarrow \underset{\underset{Cl}{|}}{\overset{\overset{F}{|}}{F-C\cdot}} + \cdot Cl$$

The chlorine atom is very reactive, because it has one unpaired electron. If it meets with an ozone molecule, the following two reactions occur:

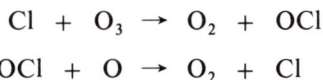

$$Cl + O_3 \rightarrow O_2 + OCl$$

$$OCl + O \rightarrow O_2 + Cl$$

This pair of reactions has the net effect of accomplishing the reaction

$$Cl + O_3 + O \rightarrow 2O_2 + Cl$$

The Cl is used up and remade, so here is a chain reaction. The overall result is to remove ozone from the stratosphere. So we see that putting Freon in the atmosphere may remove the ozone shield against ultraviolet radiation.

It is extremely difficult, if not impossible, to discover whether this speculation reflects what is actually happening. Calculations based on a laboratory-sized sample reaction predict with only limited accuracy what can happen in the huge reaction chamber of the atmosphere. More data are being obtained by new techniques of sampling and analyzing the contents of the atmosphere. Even though sure predictions cannot be made, the realities of the *potential* danger and its *known* consequences are valid reasons for concern and caution about the continued widespread use of Freon-propelled aerosol cans.

PHOTOCHEMICAL SMOG

Beginning in the 1940's, a severe type of pollution began to occur in the air over cities where there was a high concentration of people and automobiles. This pollution is observed as a brown haze that contains many irritating substances. The harm caused by this atmospheric condition ranges from human lung disorders to the loss of billions of dollars worth of crops and other forms of plant life. The obvious cause of this irritating haze, called *smog* (*smoke* + *fog*), must be chemical reactions in the atmosphere. Identifying what is reacting and how it is reacting is not at all

Photochemical smog held near ground by inversion layer

simple. However, scientists have been able to fit together some pieces of the complex puzzle. A key idea in understanding the problem is the role played by photochemical reactions started by the absorption of photons of sunlight. Accordingly, this type of atmospheric pollution is referred to as *photochemical smog*.

Whenever petroleum is burned to obtain heat energy, as in the cylinders of an automobile engine, air is used as a source of the necessary oxygen. However, you recall from our earlier discussions that air also contains nitrogen. (Air is 79% nitrogen.) In the hot region in the engine cylinder where the hydrocarbon fuel burns, some of the oxygen tends also to react with nitrogen to form the compound nitric oxide, NO.

$$\text{Heat} + O_2 + N_2 \rightarrow 2NO \qquad \text{nitric oxide}$$

The higher the temperature, the more this reaction tends to occur. NO molecules thus are added to the engine exhaust and are expelled into the atmosphere. There the NO molecules encounter more oxygen and react very readily.

$$2NO + O_2 \rightarrow 2NO_2 \qquad \text{nitrogen dioxide}$$

Nitrogen dioxide, NO_2, is a brown-colored gas and colors the hazy smog. The NO_2 molecule is capable of strongly absorbing photons of such an energy that, after doing so, the molecule breaks apart.

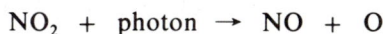

$$NO_2 + \text{photon} \rightarrow NO + O$$

The NO molecules encounter more O_2 to make more NO_2. The O atoms released by the photochemical reaction are capable of reacting with O_2 molecules to make ozone, O_3.

$$O_2 + O \rightarrow O_3$$

NO_2, O, and O_3 all can produce a great variety of reactions with other molecules they encounter, particularly hydrocarbon molecules. If some of the gasoline put into the engine cylinder is not completely burned, the unburned vapors go out into the atmosphere. So also do molecules of gasoline or oil that evaporate from places where some has been spilled or discarded. Such hydrocarbon molecules in the atmosphere are converted by NO_2, O, and O_3 into literally hundreds of different organic compounds that are irritating and harmful to living organisms. Typical of the complex molecules formed is a particularly noxious type called *peroxyacetylnitrate* (PAN).

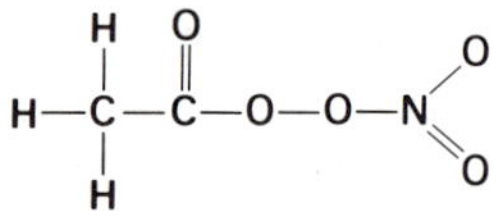

PAN

The essential features of the formation of photochemical smog are represented in the outline diagram of Figure 10-8. The whole complex process is more likely to occur if the mixture of primary and secondary pollutants is kept in place by climatic conditions. In Chapter 4 we discussed the way an inversion layer of warmer air above cooler air creates such conditions. An inversion layer

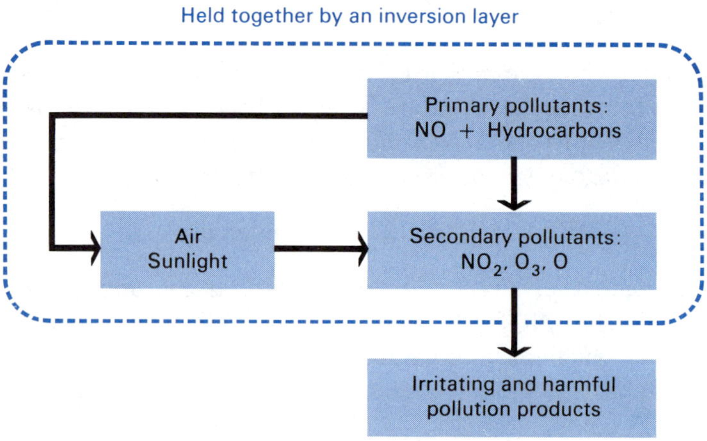

Figure 10.8   The production of photochemical smog.

has the effect of keeping pollutants in a giant cauldron for a long enough time for the complex sequence of reactions to occur.

The elimination of smog obviously involves eliminating both of the primary pollutants (NO and hydrocarbons). We emphasize *both*. One early attempt to control pollution was designed to cut down on hydrocarbon emissions by adding more air in

---

An inversion layer keeps pollutants in a giant cauldron for a long enough time for the complex sequence of reactions to occur.

---

automobile engine cylinders to ensure more-complete combustion of the hydrocarbons. But this had the effect of producing more NO, and the overall result was more smog. An important development of the mid-1970's is a device in which the NO in the exhaust gases is passed over a catalyst at moderate (not high) temperatures. This favors the reversal of the reaction by which NO was formed.

$$2NO \rightarrow N_2 + O_2 + \text{heat}$$

THE GREENHOUSE EFFECT

Another important global phenomenon involves the processes by which the earth's atmosphere controls the temperature of the earth. If there were no atmosphere, as for example on the moon, we would all cook in the daytime and freeze at night. The temperature of the surface of the earth is determined primarily by how much of the total solar radiation is converted into thermal energy and by how much of that thermal energy is lost as radiation back into space.

Thermal energy, the random vibrating and moving of molecules, involves energy bundles corresponding to the energy of infrared photons. So it is convenient for us to think of heat as infrared radiation. Figure 10.7 suggests that the bulk of the energy of solar radiation is in the form of visible light. However, on the surface of the earth, many processes convert this into other forms, including the infrared radiation associated with heat. Part of the conversions that occur in photosynthesis release heat. The temperature of the earth's surface would rise if some of this thermal energy were not reradiated back into space. An overall

balance must be maintained between absorption of solar radiation, chiefly photons of visible light, and reemission by the earth,
chiefly photons of infrared light. For the earth's radiation actually to leave the earth and go off into space, it must pass
through the atmosphere. This passage is the step with which we
are concerned; the atmosphere acts as do the glass walls and roof
of a greenhouse. The natural phenomenon we are describing is
called the *greenhouse effect*.

In a greenhouse, the glass transmits (allows in) the incoming
visible light. The plants and other objects in the greenhouse absorb the visible light, but they give off predominantly infrared
radiation. These lower-energy photons of infrared light are
trapped in the greenhouse because of the glass roof and walls.
Some photons are reflected by the glass; others are absorbed and
reradiated by the glass. So the temperature inside the structure
rises, often to the point that heaters can be turned off. If the glass
were not there, the infrared radiation would escape and the temperature would drop, requiring the heaters to be on all the time.

How does the atmosphere do for the earth what the glass does
for the greenhouse? It is the presence of carbon dioxide, water,
and, to a lesser extent, ozone in the atmosphere that performs
this important function. These molecules are so constructed that
they can vibrate with frequencies corresponding to those of some
infrared photons. Thus we can properly say that the vibrations
of a few simple molecules in apparently just the right concentration in the atmosphere maintain the temperature balance of the
earth.

Figure 10.9 shows, in a qualitative way, how the distribution
of the earth's radiated infrared light compares with the absorption spectra of the molecules we have identified. The lower part
of the diagram is a curve showing how the total energy distribution varies with wavelength of the infrared photons. (Also included is a dotted curve suggesting how the distribution of solar
radiation, shown in Figure 10.7, would look on the wavelength
scale used for this diagram.) The upper part of Figure 10.9 shows
the absorption spectra of CO_2, H_2O, and O_3. The shaded portions of the diagram suggest that the molecules absorb photons
of those wavelengths. The unshaded portions suggest the wavelengths of the photons that will not be absorbed by these molecules. You can see from the locations of the wavelength regions
of absorbed photons that these molecules are capable of soaking
up various parts of the infrared radiation that comes from the
earth's surface.

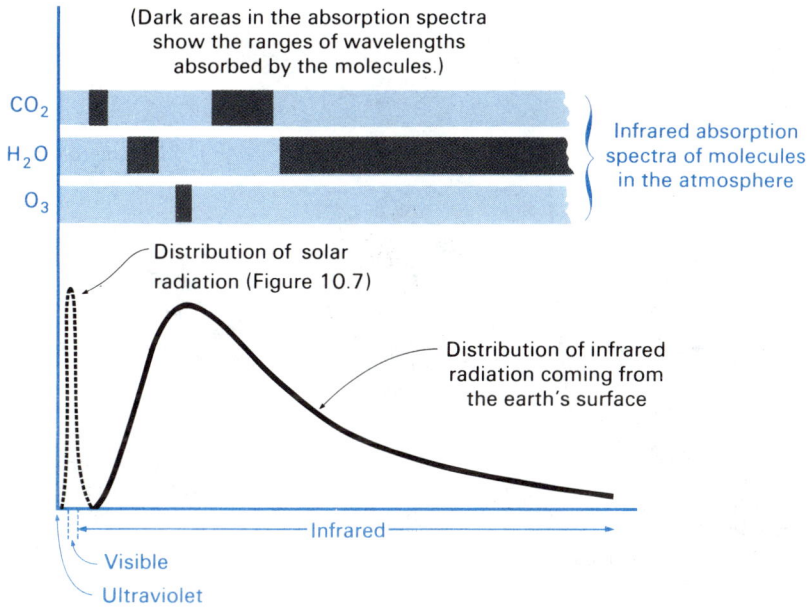

(Dark areas in the absorption spectra show the ranges of wavelengths absorbed by the molecules.)

CO_2

H_2O

O_3

Infrared absorption spectra of molecules in the atmosphere

Distribution of solar radiation (Figure 10.7)

Distribution of infrared radiation coming from the earth's surface

Infrared

Visible

Ultraviolet

Figure 10.9   Comparison of the infrared radiation from the earth with the absorption characteristics of CO_2, H_2O, and O_3.

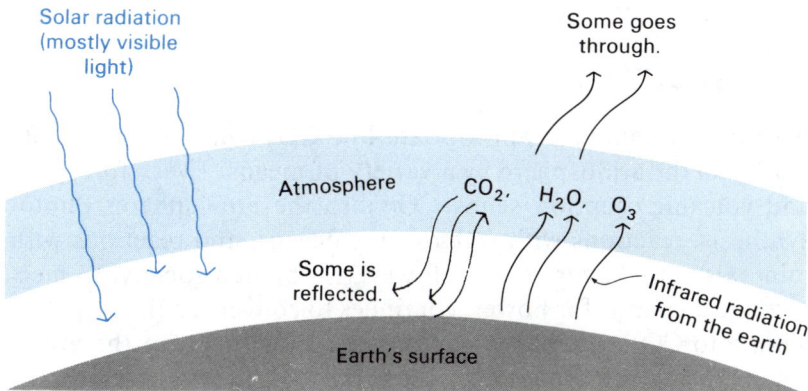

Solar radiation (mostly visible light)

Some goes through.

Atmosphere

CO_2, H_2O, O_3

Some is reflected.

Infrared radiation from the earth

Earth's surface

Figure 10.10   The greenhouse effect.

Figure 10.9 should not be read to imply that *all* the energy the earth radiates in the form of infrared light *is* absorbed by the atmosphere. There are far too few molecules of CO_2, H_2O, and O_3 in the atmosphere to do that. Rather, what is demonstrated is how the part that is trapped becomes absorbed. Figure 10.10 suggests how the present concentrations of the infrared absorbers, CO_2, H_2O, and O_3, are just right to keep the proper balance. Some infrared radiation goes through the atmosphere into space. Another part of it is reradiated back toward the earth by the atmosphere. The whole process is balanced to maintain a very nearly constant average temperature of the earth.

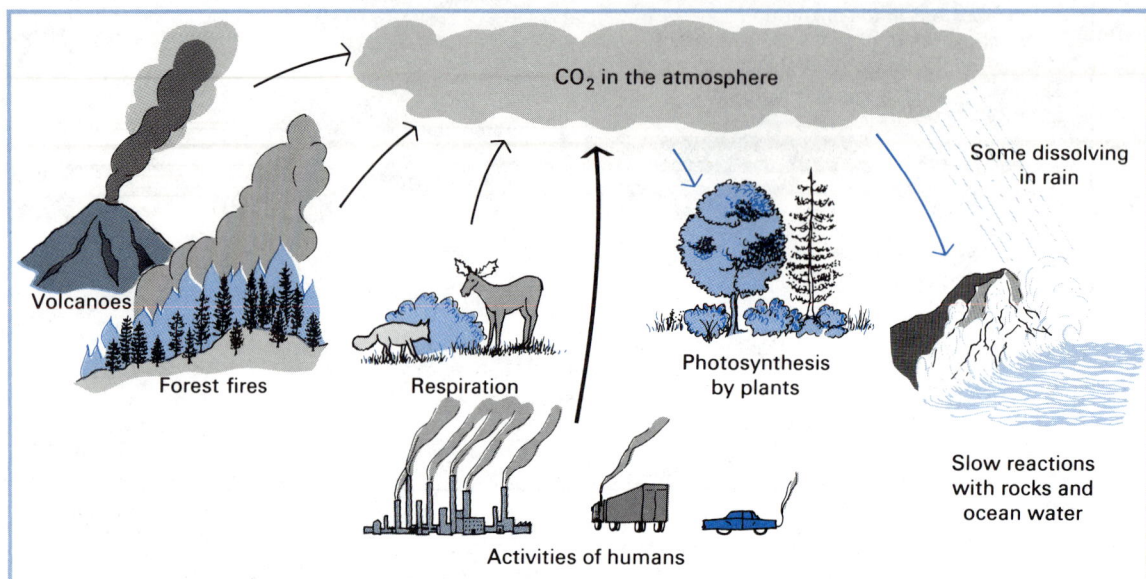

CO₂ in the atmosphere

Volcanoes

Forest fires

Respiration

Some dissolving in rain

Photosynthesis by plants

Slow reactions with rocks and ocean water

Activities of humans

IMPLICATIONS OF THE GREENHOUSE EFFECT

Nature maintains the appropriate low CO_2 concentration (only 0.03%) in the atmosphere by a variety of means. Every forest fire and volcanic eruption sends CO_2 into the atmosphere. Photosynthesis, reactions with water of the oceans, and reactions with minerals in rocks remove it. However, if human society, to meet its huge demands for power, continues to convert coal and petroleum into CO_2, we may be adding too much CO_2 to the atmosphere for nature to maintain the proper balance. Therefore, the greenhouse effect holds dangerous implications for the future. More CO_2 in the atmosphere means that more of the earth's infrared radiation will be trapped. This effect would raise the temperature of the earth. Even a small permanent increase in global temperature would have serious harmful effects on all ecological balances. For example, more of the permanent snow on mountains and ice in glaciers would melt. This extra water would raise the level of the oceans.

Speculation about the effect that increasing the CO_2 content of the atmosphere would have on global temperature can hardly be checked by direct experimentation. However, a mathematical model that accounts for the various interrelated reactions can be put together. These complex mathematical equations, together with known data, have been fed to a computer. When the computer was thus programmed, it gave an answer that matches the facts; the model seems to be a good one. Then the computer was fed the imaginary data for an atmosphere containing twice as much CO_2 as is now present. (This level means 0.06% CO_2 instead of 0.03%, still a small fraction of the total atmosphere.) The computer replied that the additional CO_2 would raise the average temperature of the earth by about 4°F.

You may well be asking if anything is known about the effect of all the extra CO_2 humans have put into the atmosphere from fuels burned in the past century. Unfortunately, we do not have reliable data because the potential problem has not been recognized until recently. Only since about 1940 have accurate measurements of global temperature been made. The interesting fact is that, since that date, the global temperature has *dropped* nearly half a degree! This fact is contrary to the effect we would expect extra CO_2 in the atmosphere to have. Probably this decrease is caused by the extra smoke particles that have gone into the atmosphere to scatter some of the incoming solar radiation. If this speculation is correct and we take steps to cut down smoke emissions, we may be removing this compensating influence. Then the greenhouse effect of the extra CO_2 may become dominant. So we see how all the factors make the picture very complicated. Obviously, the more we learn about the way the natural world works, the more we must plan to bring human activities into line. Here is one more illustration of what is meant by saying that scientific information must be the basis on which human society develops a wise plan for the future.

SOLAR RADIATION—OUR
LARGEST ENERGY RESOURCE

Solar radiation certainly is the most abundant total energy resource available, because it is constantly being renewed. If we could make use of the sunshine that strikes only 0.5% of the land in the United States, we would meet even the large power demands projected for the year 2000. The figure 0.5% may impress you as a small fraction. However, it represents a large actual area. The total land area of the United States is about 3,000,000 square miles. Thus 0.5% of this figure is 15,000 square miles, about one-tenth of the area of the state of California or about the combined area of the states of New Jersey and New Hampshire. If we were to try to use solar radiation to replace *all* our other sources of energy, we would have to cover 15,000 square miles of land area with devices to collect and convert the sunlight. A change in the landscape of such magnitude would very probably raise all sorts of environmental problems.

When we put the facts of the abundance of solar energy beside those that indicate future exhaustion of fossil fuels, or the dangers and uncertainties of nuclear fuels, we must agree with those who advocate supporting the research and development needed to

If we could make use of the sunshine that strikes only 0.5% of the land in the United States, we would meet even the large power demands of projects for the year 2000.

solve the problems of using solar energy. Solar radiation as an energy resource is sure to be important in the long run.

The chief problem with using solar energy is that it comes to us both diffusely and intermittently. By *diffusely* we mean a little at a time on any reasonably-sized area, such as the roof of a building. By *intermittently* we mean the obvious fact that the sun shines on any one spot in the daytime only. The characteristics of this energy source collide head-on with the characteristics of energy itself that we discussed in Chapter 8. The storage and transmission of energy is difficult to accomplish. The overall inefficiency that nature imposes on any system of storing or transmitting energy is bound to result in higher economic costs for using solar radiation to meet our large-scale power demands. And here we have the real reason why solar energy has been used very little in the past.

Scientific and engineering research can improve the efficiency of systems only up to the limits nature allows. When the fuels are gone that have allowed us cheap and convenient power, we will be forced to accept what appears to be the greater cost of using solar energy. Undoubtedly, as more solar energy converters are developed and put into use, the projected costs can be expected to decrease. However, it is important to realize that, although sunlight is "free," energy in the form of usable large-scale power from sunlight by no means will be free. Even at present, less than half the cost of electricity you buy from a utility

Less than half the cost of electricity you buy from a utility company is the cost of the fuel.

company is the cost of the fuel. The rest is the cost of building and operating the plant and transmission facilities.

SMALL-SCALE INSTALLATIONS TO USE SOLAR ENERGY

Probably the first widespread use of solar energy will be in small-scale installations such as individual houses. This small-scale use may not appear to be much relief for the energy crisis. But we must recognize that the total energy used to heat, light, and live in our homes is 20% of all the energy consumed in the United States. Some experimental solar-powered houses have been operating for years. At the present time, many improved designs are being developed. Some states and municipalities are requiring that designs for new homes include plans to make some use of solar energy to relieve the demands on public utilities. We could say that the sun is already being used to meet a part of society's energy needs.

Glass-covered panels can be built to trap solar radiation by using the greenhouse effect. A black absorbing surface of metal is placed under specially-made glass sheets that transmit visible photons but absorb and reflect infrared photons. The light absorbed by the black metal is changed into the thermal energy of hot air, hot water, or some other hot fluid that then is pumped around to heat the house. Some of the heat produced by such panels when the sun is shining can be stored in chemical substances that melt or vaporize readily. When the sunshine stops, these chemicals can be allowed to change back to a solid form and by doing so give up the heat they previously absorbed. The overall efficiency is low, but the abundance of the energy supply makes the process practical.

An important additional conversion is possible. Sunlight can be converted directly into electrical energy. Devices that do so are called *photovoltaic cells* (light → electricity). You probably have seen photographs of space vehicles or satellites that have panels of such cells. The cells are arranged to capture sunlight and convert it into electricity to power electronic systems. Some compounds, such as cadmium sulfide, CdS, have the property of readily giving up electrons when they absorb photons. So, too, will layers of elements like silicon or germanium when small amounts of other elements have been added to them. These electrons can be collected and pushed into wires just as the electrons from a battery or a fuel cell are.

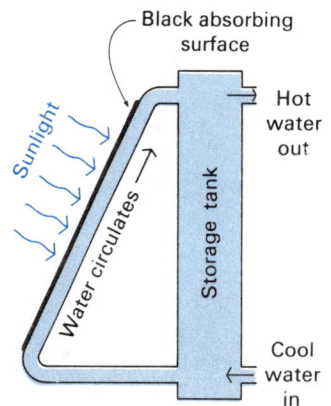

A simple solar water heater

House designed to use solar energy.

Photovoltaic cells generate only about 0.5 volt, so many must be connected together to supply the common household current of 110 volts. This can be done, and with a capacity to more than meet the needs of a house, while the sun is shining. So then comes the problem of storing the electricity for use during the dark hours. Storage batteries can be used, but the efficiency is very low. Probably the best plan is to tie a house equipped with solar cells into a conventional electric utility system. During the daylight hours, the solar cells will feed the excess electricity beyond what is needed by the house into the system to supply some of the total demand on the system. Then, when the solar cells are not generating electricity during the night, the system can feed back into the house.

CAN SOLAR ENERGY BE USED FOR LARGE-SCALE POWER PLANTS?

The large-scale conversion of solar radiation into heat to run a central power station poses additional problems. The first of these is the sheer size of the necessary collection device. A typical power-generating plant requires energy equivalent to the sunlight that arrives on 5 to 10 square miles in a day. Only in arid uninhabited regions will it be feasible to devote this much land area to this purpose.

A problem even more difficult to solve is that posed by the second law of thermodynamics. When heat is converted to work, the efficiency of the process depends on how high the temperature is. For example, the amount of heat in hot water below 100°C can do much less work than the same amount of heat in steam at 350°C. And the same amount of heat in a gas running a turbine at 600°C can be converted into even more work. This demand for high temperatures can be met by solar radiation only if it is concentrated in some way, and then only to a limited extent. Some designs propose that mirrors or lenses be used to concentrate sunlight onto small areas. The principle is the same you may have used to ignite a spot on a piece of paper by focusing the sun's rays with a magnifying glass.

Pipes covered with efficient radiation-absorbing materials would be used to convey the concentrated energy into moderately high-temperature steam. The maximum temperatures would still probably be less than 350°C, much lower than those possible in the present conventional plants burning coal or petroleum.

The use of solar radiation as a source of power would not have the harmful environmental effects that using other fuels has. There would be no smoke, ash, SO_2, or NO as produced by burning coal or petroleum. No strip mining would devastate the landscape. No dangerous radioactive atoms would be formed, as is the case with the products from fission reactors.

However, using solar radiation would have other consequences, some of which would pose environmental problems. One of these is the increased thermal pollution caused by operating at lower thermodynamic efficiencies. Remember, the thermodynamic efficiency is low when the temperature of the steam is low. Lower thermodynamic efficiency means that a larger fraction of the total heat would end up as thermal pollution. More heat would have to be discarded to produce the same power.

The large-scale use of solar radiation for power production also may pose the problem of global temperature balance. If we capture and use a fraction of the energy we receive from the sun, will the amount radiated back into space be the same? Even if the global balance would be little altered by our using such a small amount of the total radiation received by the whole planet, certainly there would be local changes. The weather near a large solar collector would be affected by absorbing energy that previously was reflected.

Summary

Light comes in separate bundles or quanta of energy called photons. These photons have the wave-like properties of wavelength and frequency. The energy of a quantum is proportional to its frequency and is inversely proportional to its wavelength. Light is also called electromagnetic radiation, because it is generated by the oscillation of electrons. The complete spectrum of electromagnetic radiation extends from the short wavelength gamma rays through X rays to ultraviolet on one side of the visible range. Beyond the red side of the visible spectrum, it extends through infrared rays to microwaves and radio waves.

Light absorption in the infrared range involves the vibration of atoms within molecules. The absorption of the more energetic ultraviolet photons involves the excitation of electrons. When this occurs in molecules, bonds may be broken, so that in some cases chemical reactions are initiated.

The spectrum of sunlight reaching the earth's surface shows the effect of scattering, reflection, and absorption of photons by substances in the earth's atmosphere. One consequence of the absorption of ultraviolet light is the existence of a layer of ozone molecules, O_3, formed from oxygen, O_2, in the upper atmosphere. This ozone layer protects living things by screening out some of the ultraviolet light that otherwise would be harmful. It is important that contaminants that would remove the ozone not be added to the atmosphere.

The existence of CO_2, H_2O, and O_3 in the atmosphere results in the phenomenon known as the greenhouse effect. These molecules tend to absorb and reradiate infrared photons emitted by the earth, so the temperature of the earth is held at a constant level. Again, it is important that human activities do not greatly alter the amounts of these molecules in the atmosphere. If this were to happen, the balance of radiation and reradiation of infrared light might be destroyed, so that the earth's temperature would be altered.

Solar radiation represents a tremendous source of energy in place of burning up our dwindling resources of fossil fuels. Small-scale use of solar energy for such purposes as heating buildings is now feasible. Large-scale use for power production has inherent problems that make it less feasible at present. However, solar energy ultimately may have to be used in place of present conventional fuel-fired power-generating stations.

Glossary

The number in parentheses indicates the text page where you can find the term defined in context.

cycle   one complete event in a series of repeated identical events; specifically, one complete pattern in the movement of a wave (281)

electromagnetic radiation   light, the result of oscillation of electrons, which generates oscillating electric and magnetic fields (284)

fluorescence   the process by which an atom or molecule absorbs a photon of higher-energy light and then emits a photon of lower-energy light (287)

frequency   the number of cycles of a wave motion which pass a specific point in a given time (282)

greenhouse effect   the trapping and reflecting of some of the earth's radiated infrared light by carbon dioxide, water, and ozone molecules in the atmosphere (300)

infrared radiation   light of longer wavelength (lower energy) than that of visible light, associated with the rotation and vibration of molecules (286)

ionosphere   the region of the atmosphere that extends 30 miles or more above the earth's surface, in which high-energy radiation is absorbed to produce ions (293)

photochemical reaction   a chemical reaction that is initiated by the absorption of light (293)

photochemical smog   a form of atmospheric pollution that results from the photochemical reactions of oxides of nitrogen, oxygen atoms, and ozone with hydrocarbon molecules (297)

photon   a fundamental package (quantum) of light energy   (280)

photovoltaic cell   a device that converts absorbed light energy directly into electric energy (305)

scattering   the reflection of incoming sunlight back into space by small particles and water droplets in the atmosphere (291)

stratosphere   the layer of the atmosphere that extends from about 7 to 30 miles above the earth's surface (293)

total solar radiation   the complete spectrum of light emitted by the sun (290)

ultraviolet light   light of shorter wavelength (higher energy) than that of visible light, associated with the jumping of electrons between widely spaced energy levels in atoms or molecules (287)

velocity   the rate at which an object or a wave moves through space or a medium (282)

Exercises

10.1   A dancer is wearing a red costume with green trimmings. During a performance, the spotlight on the dancer is changed from white to red and then to green.

       a.   How does the costume appear in each colored light?

       b.   What is happening to the photons of light in each case?

10.2   The chlorophyll in the leaves of plants gives them a green color. Chlorophyll also is the molecule that makes it possible for the plant to accomplish the chemical reactions of photosynthesis. If you illuminated a plant with only green-colored light, do you think it would grow? Justify your answer.

10.3   The native clothing of desert peoples usually is white or light colored; and in temperate zones, it has been traditional to wear light colors in summer and dark ones in winter. Why do you think these color choices are made?

10.4   Children going barefoot on a summer day when the sun is shining soon learn that although they can walk on a concrete sidewalk, it is almost impossible to cross an asphalt street without pain. Why?

10.5   The parts of the complete electromagnetic spectrum in alphabetical order are: gamma rays, infrared, radio, ultraviolet, visible, and X rays.

       a.   Arrange the parts of the spectrum in order of *increasing wavelength.*

       b.   Arrange the parts of the spectrum in order of *increasing frequency* of wave motion.

       c.   Arrange the parts of the spectrum in order of *increasing energy* of the photons.

10.6   UHF (ultra high frequency) television broadcasters use higher frequencies than do VHF (very high frequency) broadcasters.

       a.   Who uses the higher energy radiation?

       b.   Who uses the radiation of longer wavelength?

10.7   Which of the following involve oscillation?

       a.   A working grandfather clock

       b.   A normally functioning heart

       c.   Water passing over a dam

    d. Flight of a bird

    e. Steam coming from a whistling teakettle

10.8 Explain what is meant by saying that a molecule is "tuned" to absorb a photon of infrared light that has a particular frequency?

10.9 Do photons of large energy move faster than photons of small energy? Justify your answer.

10.10 Someone tells you (incorrectly) that fluorescence is the absorption of visible light followed by the emission of ultraviolet light. You reply that this statement contradicts the first law of thermodynamics (conservation of energy). Explain why you can give this answer. (Recall the relative energies of ultraviolet and visible photons.)

10.11 Fabrics can be treated with substances that are called "optical brighteners." White fabrics so treated appear whiter than untreated fabrics. The phenomenon of fluorescence is involved in the effect.

    a. Explain how the molecules of the brightener accomplish the effect of making the fabric look whiter.

    b. Compare the energies of the photons the brightener absorbs with the energies of the photons it emits.

10.12 If you choose to have your skin tanned by sunlight, you must do so outdoors in direct sunlight. Why is the sunlight that comes through window glass ineffective for making your skin darker?

10.13 Why might a dermatologist tell you that a beautiful tan is not worth risking the necessary exposure of your skin to the sun?

10.14 A very good plan of preventive medicine for you is to have a diagnostic chest X-ray examination once a year. Abnormalities or potential trouble spots in your lungs can be detected. However, having a chest X-ray picture taken every week is not a good plan. Explain why.

10.15 Why is the upper region of the earth's atmosphere above a height of 30 miles called the ionosphere?

10.16 Each of the following substances in the earth's atmosphere influences the sunlight that reaches the earth's surface. Tell what each does.

    a. Ozone   *b.* Smoke   *c.* Carbon dioxide

10.17   When the temperature of the night air is near 0°C, a puddle of water is more likely to freeze if the sky is clear than if it is cloudy. Why?

10.18   Why would decreasing the amount of ozone in the upper atmosphere have a bad effect on organisms living on the surface of the earth?

10.19   What do the supersonic transport Concorde and Freon aerosol cans have in common?

10.20   The following diagram shows how the concentration of hydrocarbons, NO, NO_2, and O_3 varies in the air over a city during a day of particularly bad smog.

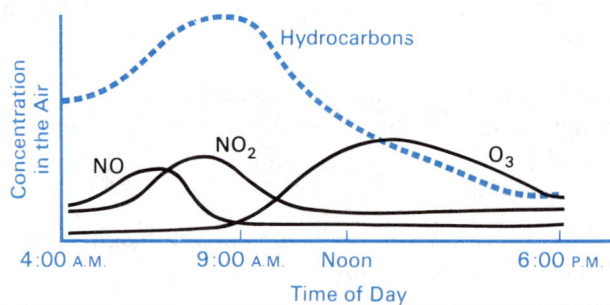

a.   In terms of the chemical reactions by which NO, NO_2, and O_3 are formed, account for the times during the day at which the maximum amount of each is found.

b.   Note that the curve for hydrocarbons has a maximum at the time of heavy morning commuting traffic. Why does the concentration of hydrocarbons drop after that hour?

c.   Draw onto the above diagram a curve you would expect to show the concentration of the molecules in smog that cause eye irritation and lung damage.

10.21   *a.*   Describe the greenhouse effect of the earth's atmosphere.

b.   Why is only some of the infrared radiation from the earth trapped by the greenhouse effect?

c.   How may human activities alter the conditions responsible for the greenhouse effect?

10.22   Why would you save energy by planting a large deciduous tree near the south side of a house? (A deciduous tree has leaves in summer but sheds them at the end of the growing season.)

10.23   Describe in general terms how solar energy can be used to supply heat for a house.

10.24   What is the connection between the second law of thermodynamics and the problems of large-scale use of solar energy as a source of power?

10.25   Why can we never expect solar energy to supply "free" power for industry?

11

The Liquid Form of Matter

☐ How must we modify the kinetic molecular theory of gases to explain the behavior of liquids?

☐ Why is evaporation a cooling process?

☐ Why does the boiling of water occur at temperatures lower than 100°C at higher altitudes?

☐ What factors affect the air's ability to hold moisture?

☐ What special properties does the surface of a liquid have?

☐ Why are the intermolecular forces in water so strong?

☐ In what ways are living things dependent on the unique properties of water?

V ery early in our discussion of how a scientist describes matter, we found that the forms matter takes can serve as the basis for a scheme of classification. A sample of matter may be classified as a gas, a liquid, or a solid. We have become familiar also with the description of these forms of matter in terms of how close the molecules of a sample are to one another and in terms of how tightly they hold onto one another. In Chapter 4, we took a closer look at the behavior of gases. The broad general principles of how gas molecules behave have been very useful in explaining our experiences with gases. Such natural phenomena as how the atmosphere functions, how expanding gases do work for us, and why nature makes it so difficult to store heat all fit together when we use the ideas of molecules and their random chaotic motion in the gaseous form.

Much of our experience in daily living involves liquids, as well as gases. Water is the most obvious example. We encounter water flowing in rivers and stored in lakes and oceans. Something goes on in the atmosphere to pull together the invisible gaseous water molecules into the liquid rain we can see and feel. The same process takes place in an air conditioner to make liquid water drip from the back of it. We encounter many other liquids, too, from gasoline and oil to nail-polish remover. What about the evaporation or boiling of liquids? Why does water in an open pan boil at

---

Why does water in an open pan boil at lower than normal temperature on the top of a mountain, yet at higher than normal temperature in a pressure cooker?

---

a lower than normal temperature on the top of a mountain yet at a higher than normal temperature in a pressure cooker? In this chapter, we will discuss some of the general principles that allow us to tie together all these facts of our experience with liquids. A modified theory of molecules in motion will help us connect and explain these facts.

WHAT HAPPENS WHEN A
GAS BECOMES A LIQUID?

We have already discovered some remarkable consistencies about gases. All gases, regardless of the formulas of the molecules, follow Boyle's and Charles' laws. This fact was the clue that scientists followed in the early nineteenth century to develop the whole grand idea of molecules as the tiny individual units in which matter exists. Avogadro's idea was that equal volumes (at the same temperature and pressure) of any kind of a gas contain the same number of molecules. As we thought about Avogadro's suggestion, we came to the conclusion that the molecules of a gas must be very far apart and must exert no attractive forces on one another. Gas molecules fill whatever volume they are in, not as we would fill a crate with oranges, but rather as we would fill

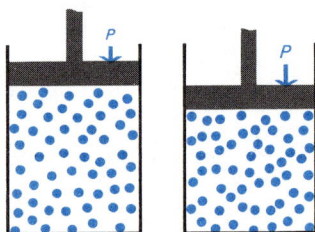

At moderate pressures (*P*,*P*), molecules of a gas follow Boyle's law.

Gas molecules fill whatever volume they are in, not as we would fill a crate with oranges, but rather as we would fill a room with mosquitoes.

a room with mosquitoes. Gas molecules are constantly flying about, bouncing off one another and the walls of the container like super rubber balls.

The kinds of molecules the kinetic molecular theory invented (see Chapter 4) are fine for explaining the properties of gases, but they cannot possibly be the molecules of any but an imaginary *ideal gas*. If we examine real gases, the actual kind we find in the natural world, we find that there are always some conditions of temperature and pressure under which they do not behave according to the predictions of the theory.

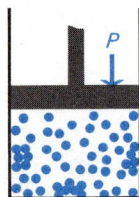

At higher pressure (*P*), drops of liquid begin to form.

Figure 11.1 shows the general pattern of behavior followed by all real gases. The curves connect the data we obtain when we measure both the pressure and the volume of a gas sample. At a constant high temperature, the gas behaves as Boyle's law predicts; the volume varies inversely with the pressure. However, if the measurements are made on the same gas sample at a lower temperature, its behavior is quite different. Only over a range of low pressures (the lower part of the diagram) does the volume vary inversely with the pressure. This range is shown by the portion of the lower curve between points *A* and *B*. As the pressure increases, the lower curve begins to flatten out. At a particular pressure, the volume suddenly changes. Note how the curve is horizontal at that particular pressure. This horizontal section is

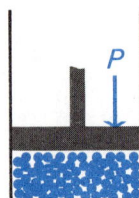

At very high pressure (*P*), all the molecules stick together as the liquid forms.

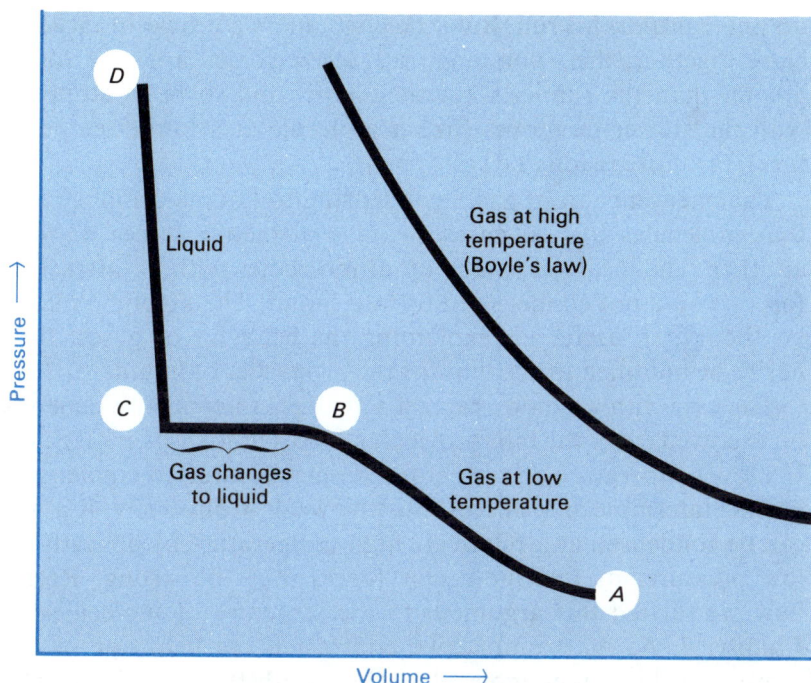

Figure 11.1 The contrast between the pressure–volume behavior of a gas at high and low temperatures.

the portion of the curve between points *B* and *C*. If we were to look at the gas sample at this point, we would see droplets of liquid forming in it. Finally, all the gas turns into liquid. Then if we increase the pressure by a large amount, the volume changes very little; the curve rises very steeply. This rise is the portion of the curve between points *C* and *D*.

We do not have to stretch our imagination far beyond common sense to explain what has happened at the lower temperature. The rapid decrease in volume of the sample when liquid begins to form means that the free-moving molecules of the gas are beginning to stick to one another when they collide. As this happens to more and more of the molecules, the gas *condenses* to a liquid. In the liquid form, the molecules occupy much less volume than they do bouncing around as gas molecules. The molecules aggregate (stick together) because of the *intermolecular forces* of attraction. (Intermolecular means "between the molecules.") These forces are not strong enough to influence fast moving molecules (at higher temperatures) or when they are at a great distance from one another (at lower pressures and larger volumes). An analogy is what often happens on a football field: A fast broken-field runner has enough kinetic energy to overcome the "attractive force" of a single tackler. He can bounce off and continue his run. But if he encounters a crowd of tacklers close together, their combined attractive forces are sure to be greater than the runner's kinetic energy, and so he is stopped. And the slower he moves (like a molecule at a lower temperature), the more readily he is stopped.

The phenomenon of gases condensing to liquids demonstrates that molecules of real gases *do* have attractive forces on one another. The ideal gas, made up of molecules with *no* attractive forces, could not condense to form a liquid. The kinetic molecular theory, so useful for explaining the behavior of gases, thus has to be modified to account for the properties of liquids.

The ease with which we can get a real gas to condense depends on the strength of the intermolecular forces among its molecules. If a low temperature or a high pressure is required to condense a gas, its intermolecular forces must be weak. Conversely, if a gas can be condensed at a relatively high temperature by a relatively low pressure, its intermolecular forces must be strong. Recall how we related this argument to the structure of molecules in Chapter 7. As an example, we contrasted the behavior of the symmetrical methane molecules, CH_4, with the behavior of the potato-shaped water molecules, H_2O. The structure of H_2O molecules is responsible for their strong intermolecular forces; under 1 atmosphere pressure, H_2O gas condenses to H_2O liquid at 100°C. The intermolecular forces among symmetrical CH_4 molecules are weak; CH_4 condenses to a liquid under 1 atmosphere pressure at -161°C.

If we put pressure on a gas sample, we force the molecules closer together. Then the attractive forces can be effective even if the molecules are moving more rapidly. This means that gases under higher pressures can condense to liquids at higher temperatures. (Remember, the higher the temperature, the faster the molecules are moving.) Under 40 atmospheres pressure, water, H_2O, changes from gas to liquid at 251°C. Under the same 40 atmospheres pressure, methane, CH_4, changes from gas to liquid at -86°C.

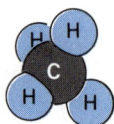

Methane: symmetrical; weak intermolecular forces; boils at −161°C

Water: unsymmetrical dipole; strong intermolecular forces; boils at 100°C

LIQUIDS ARE ALMOST INCOMPRESSIBLE

The form of the lower curve of Figure 11.1 between points *C* and *D* tells us something very characteristic about liquids. Liquids are almost *incompressible*. A very great increase in pressure is required to make their volume appreciably smaller. This property has one very practical application. Liquids can be used as "hydraulic fluids" in devices that transmit pressure from one location to another location. The brake system in automobiles is constructed this way. The brake pedal you push with your foot puts pressure on a liquid completely filling a tube connected to the brake drums on the wheels. The pressure is transmitted through the liquid to push a plate against the side of the rotating wheels. The friction between the plate and the wheel slows down the wheel. If the liquid were not almost incompressible, the pressure of your foot would be absorbed by a change in the volume of the liquid and not be transmitted to the brake. Then you would have to push the brake pedal much farther to slow or stop your car.

The very small change in the volume of a liquid in response to large changes in pressure is very different from the behavior of gases. The incompressibility of liquids is evidence for the idea that the molecules of liquids are in close contact with one another. If you have a balloon filled with water, you cannot squeeze it into a smaller volume, as you can a balloon filled with air. The molecules of a liquid can be shoved over one another, as you can push marbles around in a bag. If the bag is full of marbles, there is no extra space except that left between the spheres where they touch each other. So it is with molecules in the condensed form of matter we call a liquid.

THE EVAPORATION OF LIQUIDS

When a liquid is left in an open container, it will eventually seem to disappear. The molecules change from the piled-together, visible aggregation of the liquid form into the far-apart, invisible, spread-out condition of the gaseous form. This spontaneous change of a liquid into a gas is called *evaporation*. Evaporation is a familiar phenomenon. A puddle of water evaporates from the sidewalk after the rain stops. The evaporation is faster if a breeze is blowing. The evaporation also is faster if the sidewalk is warmed by sunshine than if the day is cold and overcast.

Some liquids evaporate more readily than others. If the cap is left off a bottle of liquid nail-polish remover, the contents probably will be gone by the next day. Nail-polish remover is chiefly a compound of carbon, oxygen, and hydrogen called acetone, C_3H_6O. Ether, $C_4H_{10}O$, also a compound of carbon, oxygen, and hydrogen, is another liquid that must always be stored in a

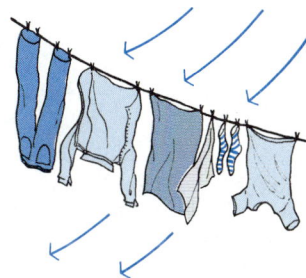

tightly closed bottle. Ether evaporates very readily, as anyone can attest who has smelled it in a room where it has been used. Not only is ether vapor potentially harmful because it puts you to sleep, but when mixed with oxygen of the air, it forms a very combustible mixture. No smoking around ether!

Why are some molecules of a liquid able to overcome the restraining intermolecular forces and go off in the gaseous form?

Does evaporation stop when a liquid is placed in a closed container?

Does evaporation stop when a liquid is placed in a closed container or a cap is put on a partially filled bottle? Answers to these questions involve important generalizations about the properties of molecules in liquids.

Suppose you place an open container of a liquid under a large glass jar. Figure 11.2 suggests the setup. After a time, you observe that the liquid level has fallen. Beyond a certain time, you find that the level remains constant. The initial decrease in the amount of liquid is caused by the escape of molecules into the enclosed air space. As more and more molecules of the evaporating substance fly about in the air space, an increasing number of them bounce back to hit the surface of the liquid. There they are recaptured by the liquid. Finally a state of *dynamic equilibrium* is reached. In this circumstance, just as many molecules are recaptured as are leaving the liquid during any period of time. Evaporation has not stopped; rather it is balanced by the opposite process of condensation. The word dynamic implies movement; dynamic equilibrium means that no net overall change takes

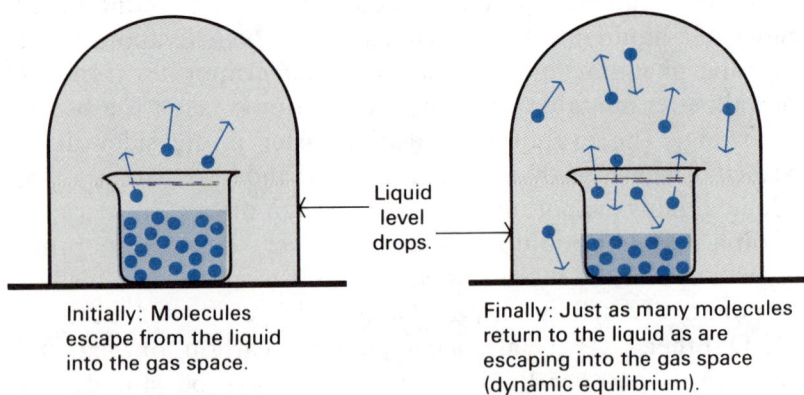

Figure 11.2   Evaporation and condensation are balanced in a closed container.

Initially: Molecules escape from the liquid into the gas space.

Liquid level drops.

Finally: Just as many molecules return to the liquid as are escaping into the gas space (dynamic equilibrium).

place because two opposing movements are in balance. What we have is a situation in which the two opposing processes move with equal rates. It is like keeping to a financial budget so that the rate of expenditures matches the rate of income; thus the bank balance remains the same over a period of time.

This demonstration also suggests the reason why the puddle evaporates faster when a breeze is blowing. The breeze moves the molecules of the gas away from the liquid surface. No dynamic equilibrium can be established because the molecules that have been blown away cannot return to the liquid so readily. So the evaporation continues without the offsetting condensation.

THE VAPOR PRESSURE OF A LIQUID

Let us do a slightly more sophisticated experiment with a liquid and a mercury barometer. (Recall from Chapter 4 how a glass tube filled with mercury can serve as a barometer.) Figure 11.3 is a diagram of the apparatus. A vacuum exists above the mercury in the barometer tube, because the mercury falls to a level that corresponds to the atmospheric pressure. Mercury is a liquid that evaporates very, very little at ordinary room temperature, so the space above the mercury contains almost no molecules. The right-hand part of the diagram shows our experiment. We take some liquid in an eyedropper and squirt a few drops under the mercury at the bottom of the barometer tube. Any ordinary liquid we choose has a density less than the high density of mercury (13.6 g/ml), so the drops of liquid float up to the surface of the mercury inside the tube. There some of the liquid quickly evaporates. The evaporation–condensation dynamic equilibrium is rapidly established. If we do the experiment carefully, there is

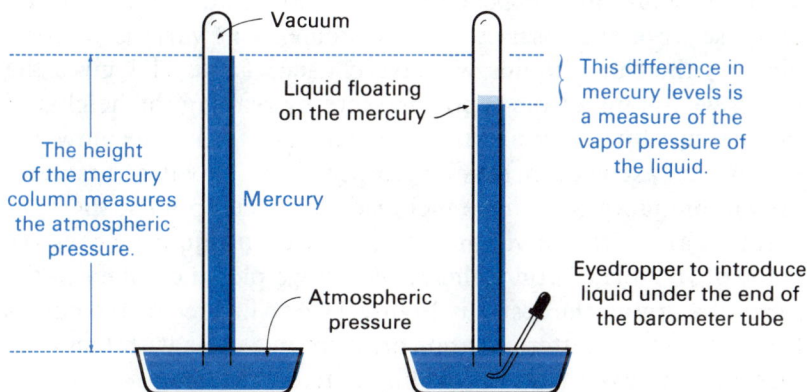

Vacuum

Liquid floating on the mercury

This difference in mercury levels is a measure of the vapor pressure of the liquid.

The height of the mercury column measures the atmospheric pressure.

Mercury

Atmospheric pressure

Eyedropper to introduce liquid under the end of the barometer tube

Figure 11.3 Apparatus used to measure the vapor pressure of a liquid.

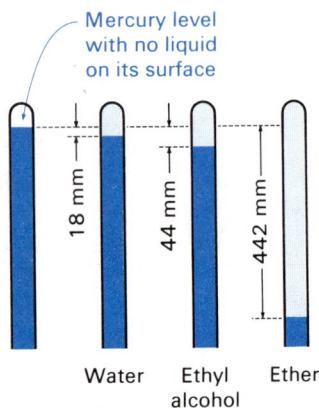

just enough liquid left after some evaporates to form a small droplet on top of the mercury column. The weight of this small amount of liquid pushes the mercury column down by a negligible amount. The only molecules in the space above the mercury column came from the liquid. They exert a pressure in the space that previously was a vacuum. We can measure this pressure by measuring how far down the mercury column is pushed. This equilibrium pressure of the gas over a liquid is called the *vapor pressure* of the liquid.

Mercury level with no liquid on its surface

18 mm

44 mm

442 mm

Water   Ethyl alcohol   Ether

If we use this apparatus to measure the vapor pressure of various liquids, we find that vapor pressure is a characteristic property of specific substances. For example, if we measure it for any sample of water at 20° C, the mercury level is pushed down 18 millimeters. Recall that the term *torr* means "millimeter of mercury." So the vapor pressure of H_2O at 20° C is 18 torr. If we use pure ethyl alcohol (the kind of alcohol in wine or whiskey, formed by fermentation of grains or sugars), we find the vapor pressure of any sample to be 44 torr at 20° C. If ether is the liquid we use, the vapor pressure turns out to be 442 torr at 20° C. Obviously, the vapor pressure of a liquid depends on what molecules are in the liquid. These measurements also give us some indication of the strength of the intermolecular attractive forces in the various liquids. Because more molecules of ether than of water or ethyl alcohol can escape to create a greater gas pressure, the forces holding ether molecules back must be the least strong. The intermolecular forces of water molecules are the strongest. Those of ethyl alcohol molecules are in between.

VAPOR PRESSURE INCREASES WITH INCREASING TEMPERATURE

Vapor pressure also depends on the temperature of the liquid. Suppose we make a series of measurements at various temperatures on the three liquids we have chosen. Table 11.1 gives the values we obtain for the vapor pressure by reading the heights of the mercury to the nearest millimeter. You will note how the values increase with increasing temperature—slowly at first, and then in big jumps. We have included, as special points, the temperature at which the vapor pressure of each liquid reaches the value of 760 torr (1 atmosphere). When we plot these data in the form of a graph, the result is Figure 11.4. All three of the curves follow the same pattern: Vapor pressure increases with temperature, and increases faster at the higher temperatures.

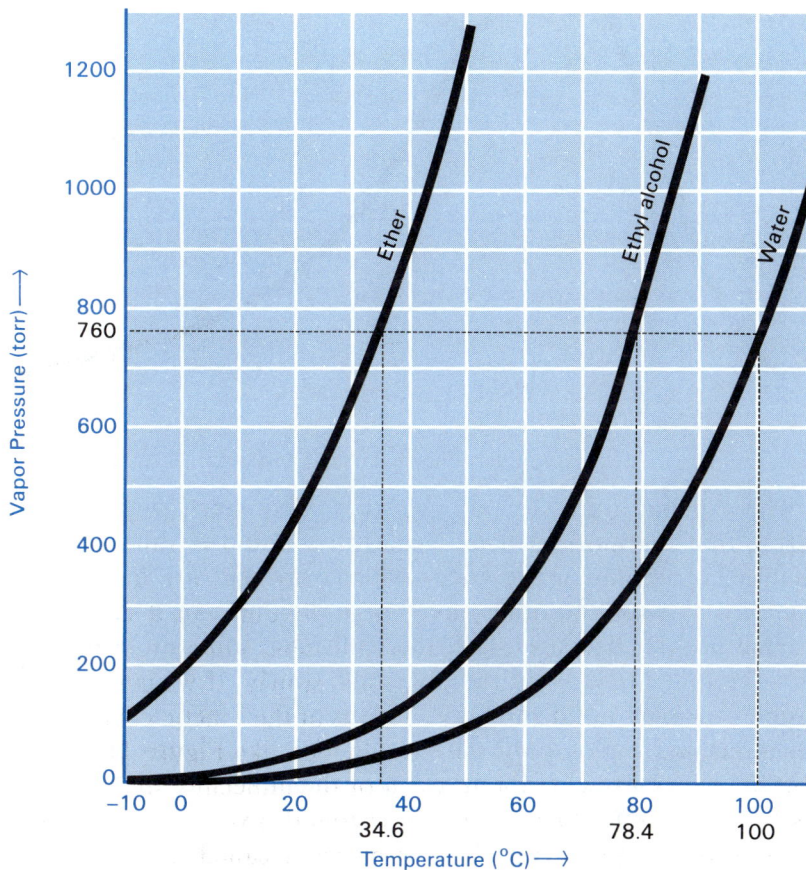

Figure 11.4   The variation of vapor pressure with temperature for ether, ethyl alcohol, and water.

Table 11.1   Approximate values for the vapor pressure of various liquids at various temperatures

Temperature (°C)	Vapor Pressure (torr)		
	Water	Ethyl Alcohol	Ether
0.0	5	12	185
20.0	18	44	442
34.6	—	—	760
40.0	55	135	921
60.0	153	353	1729
78.4	—	760	—
80.0	355	813	—
100.0	760	—	—

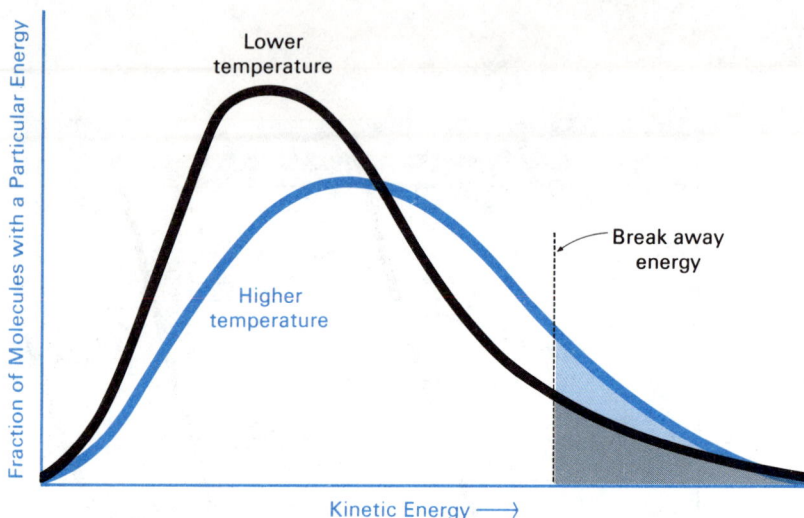

Figure 11.5   Distribution of kinetic energies of molecules in a liquid.

The explanation of this temperature effect starts with recognizing that the motion of molecules in a liquid is random and chaotic. Molecules move every which way and are constantly bumping into one another, like the individuals in a crowd of excited people. Because of all the collisions, some molecules at any one time are moving rapidly, some slowly. If we could look inside a drop of liquid and take a census of the kinetic energies of the molecules, a plot of the data would look like Figure 11.5. The vertical axis represents the fraction of the molecules that have a particular kinetic energy. The horizontal axis represents the amount of kinetic energy. Such a curve is called a *distribution curve*. Any point on the curve directly above some particular energy value tells what fraction of all the molecules have that energy. An analogy would be to plot the age distribution of a population. A point on the curve would show what fraction of the total population had that particular age.

The distribution curve of Figure 11.5 shows that some molecules have each of a wide range of energies. A few are moving slowly, a few very fast, but most are somewhere near the average, which is close to the peak of the curve. Note also in Figure 11.5 that the curve for a higher temperature shows the distribution shifting toward more molecules with higher energies.

The important feature of Figure 11.5 is the shaded areas under the curves, beyond what is labeled the "breakaway energy." The breakaway energy is the amount of kinetic energy a molecule needs to break away from its neighbors, to overcome the inter-

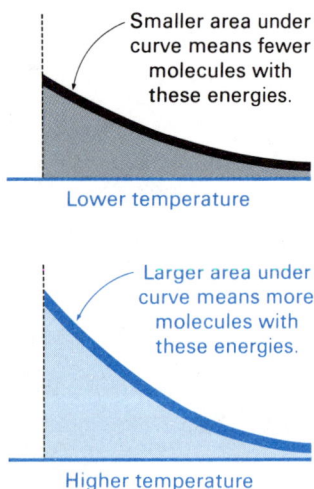

molecular attractive forces. The areas under the two curves to the right of this point represent the number of molecules that have at least this energy or more. Notice that the area under the curve for the higher temperature is much larger than the area under the lower-temperature curve. The larger the area, the more molecules it represents. So the comparison of the two areas means that at the higher temperature more molecules have enough kinetic energy to break away. This difference explains the increase in vapor pressure as the temperature of the liquid increases.

EVAPORATION IS A COOLING PROCESS

The cooling effect of evaporation is a very common experience. When you emerge from a swimming pool, you tend to shiver until you dry off. When you dry your hands with an air blower, the blast of air feels cool until your hands are dry. A drop of alcohol or gasoline on your skin feels cold until it has all evaporated. Figure 11.5 helps us understand why evaporation is a cooling process. Because it is always the molecules with high kinetic energies that leave the liquid, the average energy of those remaining is always less. Because the temperature is proportional to the average kinetic energy, the temperature drops as the faster-moving molecules break away. An analogy is to measure the weight of every member of a class and then calculate the average. Then tell everyone over a certain weight to leave. The average weight of the remaining class members is less.

As the temperature of the evaporating liquid falls, heat flows from your warmer body into the remaining cooler liquid. This flow of heat is what makes you feel cool. The flow of heat into the liquid adds energy to the remaining molecules, so more of them speed up and evaporate. When the liquid is all gone, no more heat flows from your body and you no longer feel cool.

BOILING—HIGH ALTITUDES
AND PRESSURE COOKERS

If you were to defy the old saying and watch a pot of water as you heat it on a stove, you would find not only that it *does* boil, but you would also make some other interesting observations. If you put a thermometer in the water and take temperature readings at regular time intervals after you start to heat the water steadily, you get data that could be plotted as in Figure 11.6. The water temperature rises steadily until the water boils and then remains

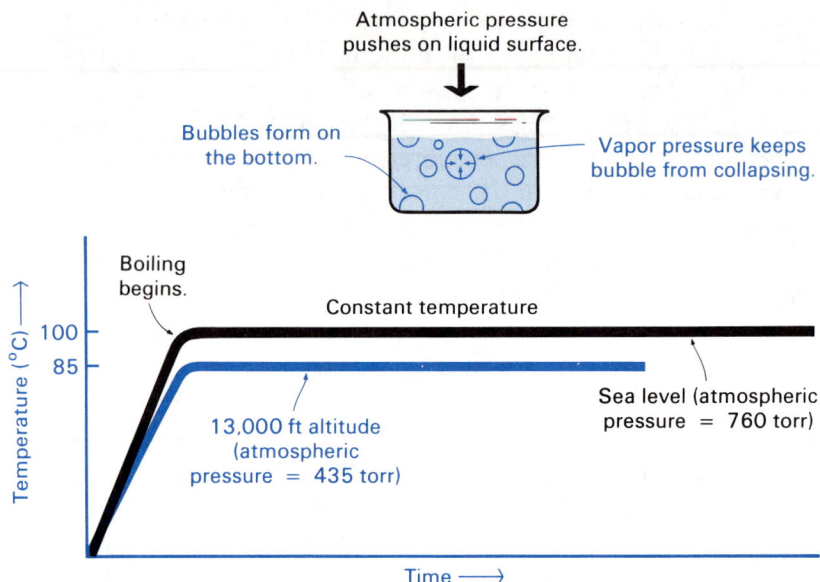

Figure 11.6   Boiling time in relation to temperature (a watched pot *does* boil, given enough time).

constant. If you do the same at a sea-level location and again at a high altitude, you get data for the two curves like the data shown.

One thing this experiment should tell you about cooking is that, once boiling starts, you cannot cook anything in that water faster by having the water boil more vigorously. The rates of the chemical reactions involved in the cooking depend on the temperature. Vigorous boiling occurs at the same temperature as

Vigorous boiling occurs at the same temperature as does mild boiling.

does mild boiling. All you are accomplishing with vigorous boiling is wasting heat energy by evaporating extra water.

Another implication of this experiment is that all cooking done in boiling water takes longer at higher altitudes. The maximum temperature you can reach with boiling water is lower; hence, the chemical reactions of cooking proceed at a slower rate.

The reason the temperature remains constant once boiling starts is clear when we recognize just what happens when a liquid such as water boils. Boiling involves the formation of bubbles of steam on the bottom and throughout the liquid. The first bubbles you see are not bubbles of steam. They are bubbles of air dissolved in the water that are being released as the water warms up. These bubbles appear long before boiling starts and should not be confused with the steam bubbles formed in boiling. The liquid water evaporates into steam bubbles at a pressure equal to its vapor pressure. Each bubble can form only if the vapor pressure inside the bubble is equal to the outside pressure that tends to collapse the bubble. This outside balancing pressure is the pressure of the atmosphere pushing down on the surface of the liquid. No bubbles form until the vapor pressure is high enough to match this. The temperature cannot go higher, because a higher temperature requires a vapor pressure greater than the opposing atmospheric pressure.

At higher altitudes, above sea level, the atmospheric pressure is lower. This means that water attains a vapor pressure equal to the lower opposing pressure when the temperature is less than 100° C. Consequently, the data in Figure 11.6 taken at a higher altitude show a horizontal line at a lower temperature.

The definition of the *boiling temperature* is the temperature at which the vapor pressure of a liquid is equal to the opposing pressure. Usually the opposing pressure is set at 1 atmosphere (760 torr). In Figure 11.4, we have identified the temperature at which ether has a vapor pressure of 1 atmosphere. You will find this temperature, 34.6°C, listed in tables of data as the *normal* boiling temperature of ether. If you were at an altitude where the barometric pressure is 650 torr, ether would boil at 30° C, and water, at about 96° C.

We note from Table 11.1 that the normal boiling temperature of ethyl alcohol (the temperature where vapor pressure is 760 torr) is listed as 78.4° C. We can always estimate a qualitative comparison of the vapor pressures of two liquids by comparing their normal boiling points. The lower the boiling temperature, the higher the vapor pressure of the liquid at some chosen lower temperature. For example, at 20° C the vapor pressure of the lower-boiling ether is 442 torr; the vapor pressure of the higher-boiling ethyl alcohol is 44 torr. We say that the ether is the more *volatile* of the two liquids, because it has a higher vapor pressure at any chosen temperature.

If we examine Figure 11.4 further, we can see why the temperature in a pressure cooker rises above the normal boiling point of water, 100° C. When the lid of the pressure cooker is tightly sealed, the pressure caused by the steam inside can build up. For example, when the gauge reads 2 lb, it means that the pressure inside the cooker is [14.7 (1 atmosphere) + 2] pound/square inch. This condition corresponds to a pressure of about 863 torr. If you look at the curve of Figure 11.4 for the vapor pressure of water, you can estimate the temperature at which water has a

Open pan
Temperature 100°C

Pressure cooker
Temperature 103.5°C

vapor pressure of 863 torr. The temperature at which liquid water has this vapor pressure is about 103.5° C. So this is the temperature at which the pressure equilibrium within the cooker is maintained. The cooking advantage of the higher temperature comes from speeding up the chemical reactions.

HUMIDITY AND THE DEW POINT

"It's not the heat but the humidity that gets you" is a common complaint. What is humidity, and why does it make people uncomfortable?

The atmosphere near the surface of the earth always contains some molecules of gaseous water. Water evaporates from oceans, lakes, and rivers. Water molecules are added to the air when any hydrogen-containing compound is burned. And all breathing animals exhale water vapor. With so many sources and such a variety of conditions, it is no wonder that the gaseous water content of the air changes from place to place and also with the time of day in any one place. The amount of gaseous water in the air can be expressed as the *partial pressure* of water in gaseous form. For example, a sample of air is analyzed, and 2% of all the molecules in the air are found to be water molecules. So if the total pressure of the atmosphere is 760 torr, 2% of 760 torr = 0.02 × 760 torr = 15.2 torr. This is the partial pressure of H_2O in the air when that sample is taken. (Note that the partial pressure is not the same as the vapor pressure.)

A more familiar expression for the moisture content of the air is the *humidity.* On a rainy or foggy day, the humidity may be 100%. This means that the air holds as much moisture as it possibly can hold in the gaseous form. When the humidity is 100%, the partial pressure of gaseous water is equal to the vapor pressure of liquid water. A dynamic equilibrium exists, as it did in the case described in the right-hand part of Figure 11.2. When the air contains this much moisture, it is said to be *saturated;* the air can hold no more molecules of gaseous water without their condensing to liquid water.

What does a reading of 50% humidity mean? A sample of this air contains only half (50%) as much water vapor as it would if it were saturated. The definition of humidity is

$$\% \text{ humidity } = 100\% \times \frac{\text{partial pressure of } H_2O \text{ (gas) in air}}{\substack{\text{vapor pressure of } H_2O \text{ (liquid) at} \\ \text{air temperature}}}$$

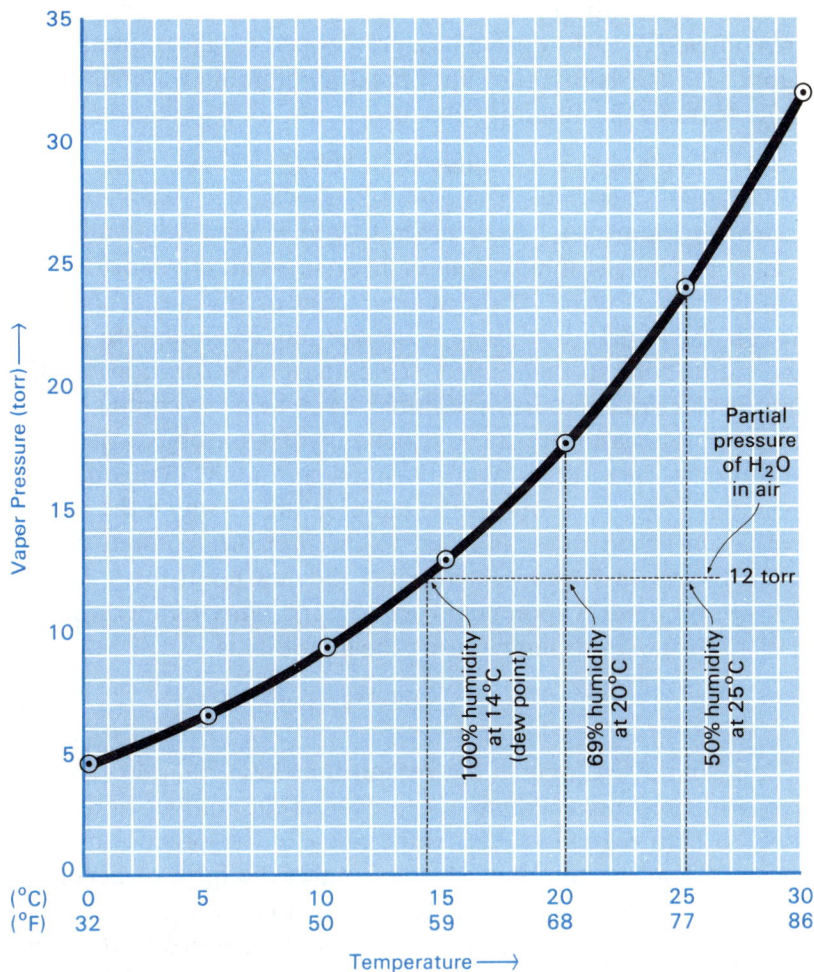

Figure 11.7 The vapor pressure of liquid H_2O and typical humidity calculations.

Let us illustrate with some numbers. Figure 11.7 shows the vapor pressure of liquid water, in some detail, over the range of temperatures from freezing (0° C, or 32° F) to that of a moderately warm day (30° C, or 86° F). Suppose the partial pressure of moisture in the air is 12 torr on a day when the temperature is 25° C (77° F). Consult Figure 11.7, and read the number 24 torr for the vapor pressure of water at 25° C. Then the humidity is

$$\% \text{ humidity } = 100\% \times \frac{12 \text{ torr}}{24 \text{ torr}} = 50\%$$

It is important to note that a humidity reading must always be given for a specified temperature. Otherwise, a humidity value means very little. For example, if the same sample of air used in

Dew point: 100% humidity.

the above illustration is cooled to 20° C (68° F), the same moisture content, 12 torr, represents a greater percentage of humidity. Figure 11.7 shows that the vapor pressure of liquid water at 20° C is 17.5 torr. So the humidity now is

$$\% \text{ humidity } = 100\% \times \frac{12 \text{ torr}}{17.5 \text{ torr}} = 69\%$$

Turn to Figure 11.7 once more to understand what another term means that you may read or hear in weather reports. This term is the *dew point*. If the temperature of the air drops low enough, the moisture content of the air may be enough to represent saturation at that lower temperature. Saturation, 100% humidity, means that drops of liquid water begin to condense out of the air. This appearance of drops of water on objects is what we call dew. Dew often forms during a cool night that follows a warm day. The dew point is the temperature at which the vapor pressure of liquid water is equal to the moisture content of the air. In Figure 11.7 we have indicated the dew point for a partial pressure of water vapor equal to 12 torr. This temperature is 14° C (57° F). At this temperature the humidity is

$$100\% \times \frac{12 \text{ torr}}{12 \text{ torr}} = 100\%$$

You can now recognize the reason that high humidity brings discomfort to humans. One of the devices the human body employs to help regulate body temperature is the process of perspiring. Glands in the skin secrete water in the form of perspiration. When this water evaporates, it has a cooling effect on the body because heat is used in the evaporating process. If a perspiring person is in an atmosphere with high humidity, the amount of evaporation that can occur is small. Consequently, this person is denied the comfortable cooling effect that the evaporation of perspiration brings.

The reason that an air conditioner often "sweats" or drips water also is related to the phenomenon we have been discussing. If the warm air is so humid that the air conditioner cools it below the dew point, liquid water forms in the same way that dew forms on a cool night after a warm day. The air conditioner must be designed either to allow this condensed water to drip out or to evaporate it into the air it pushes out of the room.

CLOUD-CAPPED MOUNTAINS

Have you ever wondered why photographs of high mountains often show clouds around the peak, even though the summer

Direction of the wind ⟶

Clouds form when air temperature drops to the dew point.

Air expands and cools as it rises.

Air contracts and warms as it falls.

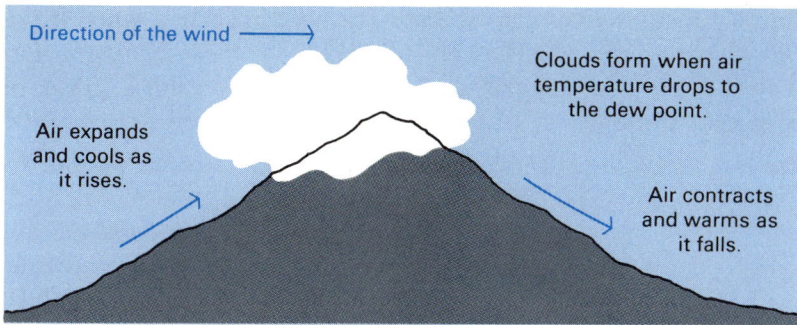

weather over the surrounding country appears to be cloudless? Anyone who has climbed such peaks knows the likelihood that he or she will encounter fog near the top. This phenomenon, too, is related to our present discussion.

When a mass of warm air is moved by a prevailing wind to the base of a mountain, it has to rise over the mountain. Because the air pressure decreases with altitude, the warm air mass expands as it goes up the mountainside. Any such expansion means that the air mass will do work as it pushes back the rest of the atmosphere. The energy to do this work must come from the thermal energy of the air molecules. So the temperature of the air mass drops as it converts thermal energy into the work of expansion. The decrease in temperature for a rapidly moving air mass may be as much as 2° C (3.6° F) for every 1000 feet of climb. The total drop in temperature by the time the air mass reaches the top of the mountain may be enough to reach the dew point for the moisture in the air. Consequently, the gaseous water condenses into liquid water and forms fog or clouds. The clouds are carried along with the rest of the air mass. After the air has passed over the summit and down the other side of the mountain, the whole process is reversed. The clouds disappear. The liquid water droplets evaporate when the temperature of the air mass increases as the air comes down on the far side of the mountain.

HEAT OF VAPORIZATION

If you return to watching that pot of water boil and focus on what is happening to the heat you put into the water, you can make additional important generalizations. The change of liquids to gases always involves the absorption of heat. The first heat you put into the water makes its temperature rise. When boiling begins, the temperature no longer rises, but you keep on adding heat. The heat is absorbed by the process of boiling; the liquid constantly changes to gas. The heat absorbed when a liquid changes into a gas is called the *heat of vaporization*. For water, this heat of vaporization amounts to 540 calorie/gram, or 9.72 kcal/mole.

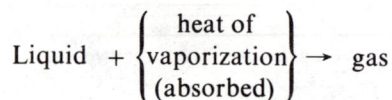

$$\text{Liquid} + \begin{Bmatrix} \text{heat of} \\ \text{vaporization} \\ \text{(absorbed)} \end{Bmatrix} \rightarrow \text{gas}$$

We can visualize what happens to the absorbed heat of vaporization by using our model of molecules held together in the liquid by the intermolecular forces of attraction. In order for a molecule to overcome these restraining forces, it has to work against these forces. Work means energy expended, just as you expend energy to jump off the ground against the force of gravity. So the energy absorbed as heat does the work of moving the molecules from the liquid into the gas form. You also can expect that the stronger the intermolecular forces, the greater the heat of vaporization. Recall how we interpreted the vapor pressures of three liquids in terms of the strengths of their intermolecular forces. The strength of these forces increases in the following order: ether, ethyl alcohol, water. The corresponding heats of vaporization are 4.2 kcal/mole for ether, 9.4 kcal/mole for ethyl alcohol, and 9.7 kcal/mole for water. The model fits the facts.

If heat is absorbed when a liquid changes to a gas, then if a gas changes to a liquid, heat evolves.

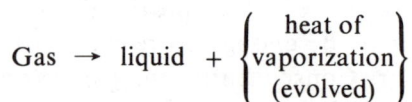

$$\text{Gas} \rightarrow \text{liquid} + \begin{Bmatrix} \text{heat of} \\ \text{vaporization} \\ \text{(evolved)} \end{Bmatrix}$$

If you are aware of the process by which heat evolves, you may avoid painful burns in your kitchen. Never put your finger in a jet of steam! When the molecules of water in the form of a gas (steam) come in contact with your skin, they condense to liquid water and give up heat. Scalding by steam produces more severe burns than those produced by liquid water at the same temperature. The extra heat released as the steam condenses causes more damage to the tissues of the body.

CLOUD FORMATION AND WEATHER

The towering clouds called thunderheads, often seen before severe rainstorms, are formed as a consequence of the process we are discussing. When water molecules condense from gas to liquid raindrops in a cloud, energy is released. This energy raises the temperature of the air in the cloud, so the air rises, carrying the condensed water along with it. The loaded air rolls thousands of feet upward to make the spectacular thunderheads.

The fact that water has a large heat of vaporization and that such large quantities of water exist on the earth's surface is an important factor in determining weather. Tremendous quantities of energy are absorbed in locations where liquid water evaporates. The water in the gaseous form so produced can be carried away by winds. Thus winds transport the energy, because when the water molecules condense to liquid, energy is released in the new location where the condensation occurs. Such a process constantly takes place when storms move from oceans or lakes to land areas.

THE SURFACE TENSION OF LIQUIDS

One of the fascinations of childhood is watching spindly-legged water bugs "skating" over the surface of a pool or quiet stream. The bugs never break the surface of the water. Is this fact caused by a unique ability of the bugs or by a characteristic of the surface? You can get a clue by carefully lowering a steel needle sideways onto a water surface. The needle floats. Yet it sinks if you put it into the water end first. If you take a drinking straw out of a glass of water, a few drops always stay in the straw unless you touch the end against some object. Raindrops never have

Raindrops never have sharp corners.

any sharp corners. All these experiences, and many others, suggest that the surface of a liquid has some special properties different from the properties of the bulk of the liquid. The scientific name for this phenomenon is *surface tension*. The word *tension* implies a stretching because of some pulling force. When you pull on the end loops of a rubber band, you feel the tension in the stretched rubber that tends to snap it back to its original length. The surface tension of a liquid is the tension that tends to pull the surface layers of molecules together.

Our model of molecules with intermolecular attractive forces helps us understand why the surface of a liquid is different in some ways from the rest of the liquid. If we visualize a molecule in the center of a sample of a liquid, we recognize that such a molecule is surrounded uniformly in all directions by other molecules. Figure 11.8 suggests the picture in only two dimensions.

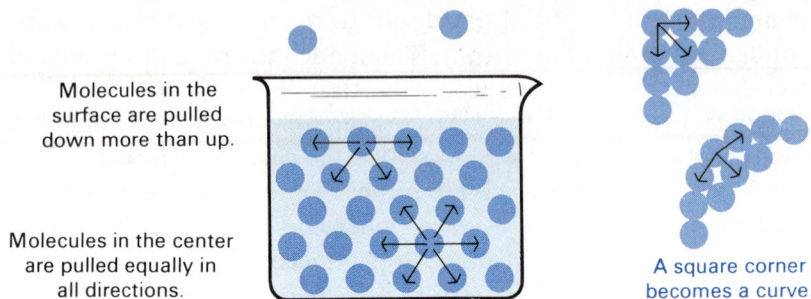

Molecules in the
surface are pulled
down more than up.

Figure 11.8   The direction
of intermolecular forces
among molecules of a liquid.

Molecules in the center
are pulled equally in
all directions.

A square corner
becomes a curve.

The molecule in the center is tugged equally in all directions by
the intermolecular attractions of its neighbors. However, a molecule on the surface of the liquid is not uniformly surrounded.
There are no nearby molecules above it. Any gas molecules that
may have evaporated from the surface are too far away to have
any influence. Consequently, this surface molecule is tugged
more, on the average, toward the center of the liquid. This unbalanced force tends to make the surface molecules act as if they
were packed closer together. This is why the surface layer acts as
a film that resists being broken. The film is capable of supporting
extra weight, such as that of the water bug or floating needle.

SOME CONSEQUENCES OF
LIQUID SURFACE TENSION

The fact that you have never seen a raindrop in the form of a
cube also can be explained by considering the attractive forces
among the molecules. Figure 11.8 suggests this too, in two
dimensions. A molecule on the corner of an imaginary cubical
raindrop would be pulled in an even more unbalanced way
toward the center of the drop than would a molecule located in
the side of the cube. Because the molecules of a liquid can roll
over one another, the corner molecule shifts its position to adjust
the unbalanced force. In the curve of a spherical surface, all the
molecules are tugged equally toward the center. A sphere also is
the three-dimensional figure with the smallest possible surface for
the volume it contains. Thus the tension pulling the surface molecules together produces a form with the least surface whenever
the molecules are free to move in response to this tension.

  This same process explains why the sharp corners of a pat of
cold butter become rounded as the butter warms up. As the temperature rises, the molecules in the butter gain more energy and

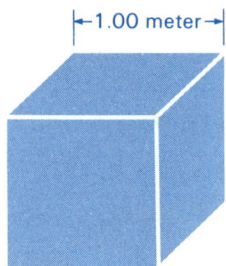

A cube, one meter
on edge, has

Volume = 1.0 (meter)^3 ←———Same———→ Volume = 1.0 (meter)^3

Area = 6.0 (meter)^2 ← More   Less → Area = 4.8 (meter)^2

A sphere, 1.24 meter
in diameter, has

can wiggle around more, as do the molecules of any liquid. Consequently, they roll over one another to adjust the forces on them toward greater uniformity. The corners change to curved surfaces.

If you put a glass or paper tube of small diameter (such as a drinking straw) into a glass of water, you see the water rise in the straw to a level above the liquid surface outside the straw. If you look carefully, you will notice that the surface curves up around the outside of the tube, too. Both of these effects occur because water molecules next to the walls of the tube tend to stick to the walls (the water "wets" the walls). These adhering molecules pull other surface molecules along with them. The result is a vertical film up along the wall of the tube made from the surface molecules of the liquid. On the inside of a small-diameter tube, this vertical film all around the inner surface pulls the liquid up into the tube. When enough liquid is pulled up to make a column with a total weight equal to the forces of surface tension, the liquid stops rising. The stronger the intermolecular attractive forces, the stronger the surface tension. The stronger the surface tension, the higher the liquid will rise in a tube.

Surface tension also is responsible for the fact that a few drops always remain in the bottom of a tube that is being emptied. As

A liquid rises in a glass tube to a greater height when the tube diameter is smaller.

A few drops always remain in the bottom of a tube that is being emptied.

The last few drops remain held in the small-diameter straw.

long as some molecules are wetting the sides of the tube, they tend to hold other molecules of liquid with them. The last drops defy even the pull of gravity and stay in the tube. If you touch the filled end of the tube to some other object, some of the liquid

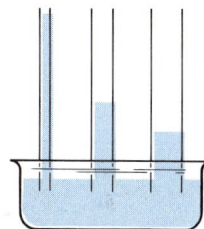

molecules wet that object. These molecules then pull some other molecules of the liquid with them, helped along by the pull of gravity, and the tube empties.

If the diameter of a tube dipped into a liquid is very small, the surface tension of the liquid forces it to rise to a considerable height. If a glass tube with an internal diameter of 0.2 millimeter is used, the large surface tension of water at 20° C forces it up to a height of almost 15 centimeters. Such tubes of small internal diameter are called capillaries. The word *capillary* comes from the Latin word for "hair"; capillaries are hairlike, small-bore tubes. For example, your body contains a network of such small capillary blood vessels.

The phenomenon of liquids rising in small-bore tubes is called *capillarity*. Many materials, such as cloth and paper, are made of fibers arranged so that the spaces between them act as pores or tubes of capillary size. When the end or edge of a piece of cloth is dipped into water, the molecules enter these pores and appear to climb into the cloth by capillarity. The water molecules wet the cloth, and the surface tension of the water causes the liquid to rise into the cloth. A sponge picks up water this way, too. Capillarity also is one of the processes by which water absorbed by the roots of a plant rises into the woody structure of the plant. Another example of capillarity is the way the liquid wax at the base of a candle wick rises to the top of the wick where the flame is.

HYDROGEN BONDING IN WATER

We have used water repeatedly as an example of a familiar liquid to illustrate the general properties of the liquid form of matter. By now you are aware that the strong intermolecular forces among water molecules are responsible for the fact that water is a liquid at normal temperatures. Water has a relatively high boiling point and large heat of vaporization. These properties are in marked contrast to the properties of the hydrogen compounds of oxygen's neighbors in the periodic table. CH_4, NH_3, and HF are all gases at room temperature. So are the *hydrides* of other members of the oxygen family: H_2S, H_2Se, and H_2Te. In Chapter 7, we discussed how the atoms in these molecules are all held together by covalent bonds. Why are the intermolecular forces in water so much stronger? Part of the answer lies in the dipolar structure of the H_2O molecule, which we described in Chapter 7. However, there is more involved than just the electric attraction of one negatively charged dipole for the positive charge of another dipole. Because water is such an important compound in both our bodies and the natural world, let us examine the structure of liquid water in more detail.

Recall the geometric form of the water molecule. The hydrogen atoms are not joined to the oxygen atom in a straight line. The molecule is bent. The H—O—H angle is almost the same as the H—C—H angles in the tetrahedral molecule CH_4. In the H_2O molecule, two pairs of electrons on the oxygen atom are not involved in bonding the hydrogen atoms. When the H_2O molecules are in close contact with one another, as they are in liquid water, a special kind of strong interaction takes place among them. This interaction is shown in Figure 11.9. A hydrogen atom, although sharing a covalent bond of a pair of electrons with one oxygen atom, is attracted by the unused pair of electrons on an oxygen atom in another H_2O molecule. The result is that the hydrogen atoms act almost as a kind of bridge, and link the H_2O molecules together into clusters throughout the bulk of the liquid. This special kind of linking among the molecules is called *hydrogen bonding*

The internal arrangement of the H and O atoms in the H_2O molecule is responsible for its being a dipole.

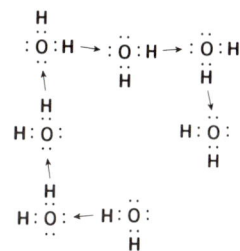

Figure 11.9 Hydrogen bonds between molecules of liquid water.

SOME CONSEQUENCES OF HYDROGEN BONDING IN WATER

The important consequence of hydrogen bonding is the feature about water we have already noted: its very strong intermolecular attractions. This feature is in turn responsible for the various

Hydrogen bonding is responsible for many of the properties of water.

properties of water, such as its relatively low vapor pressure at ordinary temperatures (only 17.5 torr at 20° C, or 68° F), its high boiling point (100° C, or 212° F), and its high heat of vaporization (9720 calorie/mole, or 540 calorie/gram). The low vapor pressure and high boiling point of water mean that humans can live comfortably at ordinary temperatures, because our bodies are 65% water. (Our blood is about 83% water.)

The high heat of vaporization of water also helps us to live more comfortably. Earlier in this chapter, we discussed how the evaporation of any liquid is a cooling process. When perspiration evaporates from our skin, the cooling effect helps us regulate body temperature. As a drop of perspiration cools, it absorbs heat from the body. This heat vaporizes more of the perspiration. Because water has a high heat of vaporization, the evaporation of a small amount of water absorbs a large amount of heat. So the net effect is that the evaporation of a small amount of liquid perspiration cools our body considerably.

The same process, the vaporization of liquid water, is important in regulating climate. Much of the heat in the sun's rays is absorbed on the surface of the earth by the evaporation of water from oceans, lakes, and rivers. The climate near an ocean or a large lake is always modified by this effect; there are fewer abrupt temperature changes along the shore.

We will return to the subject of hydrogen bonding several times in later discussions. Hydrogen bonding is a determining factor in the structure of solid water (ice), as we will discover in the next chapter. It also has an influence in determining what kinds of substances will dissolve in water, as we will discuss in Chapter 13. And hydrogen bonding is responsible for important properties of some big molecules that regulate biological processes, such as muscle action and heredity, as we will see in Chapter 18.

Summary

Much of the discussion in this chapter illustrates the typical manner in which we can observe the properties of a macroscopic (large) sample of matter and interpret these properties in terms of the behavior of microscopic (molecule-sized) units. The ideas of the kinetic molecular theory that satisfactorily interpret the behavior of gases can be modified to account for the properties of liquids, too. The chief difference is the idea of intermolecular attractive forces, which are responsible for the molecules holding onto one another in the liquid form.

Because the molecules of liquids are in close contact with one another, there is little space between them. Thus liquids are almost incompressible. Some molecules have enough kinetic energy to break away from the bulk of the liquid. Consequently, liquids exert a vapor pressure. The vapor pressure of a liquid increases with increasing temperature. When the vapor pressure equals the external pressure on the surface of a liquid, boiling occurs.

The change from the liquid form to the gaseous form involves the absorption of an amount of energy called the heat of vaporization. As a consequence of this phenomenon, evaporation always has a cooling effect on the surroundings.

Surface tension is another property of liquids. Because the attractive forces of the molecules in a liquid surface are unevenly distributed, the surface tends to contract. The spherical shape of drops of liquids is one evidence of this surface tension. Another is the way liquids rise in small-bore tubes, the phenomenon of capillarity.

The relationship between the amount of water vapor in the air and the vapor pressure of liquid water at the prevailing temperature is called the humidity. The temperature at which the humidity is 100% is called the dew point.

Water molecules have a strong attraction for one another because of hydrogen bonding. A hydrogen atom on one H_2O molecule is attracted by the electrons on the oxygen atom of another H_2O molecule. Because of hydrogen bonding, the intermolecular attractions in liquid water are strong. So we find that water has a large surface tension, low vapor pressures at moderate temperatures, and a high heat of vaporization. This last property accounts for the large cooling effect of the evaporation of perspiration from human bodies and for the weather-modifying influence of lakes and oceans.

Glossary

The number in parentheses indicates the text page where you can find the term defined in context.

boiling temperature   the temperature at which the vapor pressure of a liquid is equal to the opposing pressure on the surface of the liquid; the liquid molecules have sufficient kinetic energy to break the intermolecular bonds and enter the gas state (327)

capillarity   the tendency of a liquid to rise, because of surface tension, in a thin tube or between fibers (336) *condensation*   the formation of a liquid from the gaseous state, caused by the collision and attraction of gas molecules which do not have sufficient kinetic energy to break the intermolecular forces (317)

dew point   the temperature at which dew forms because the vapor pressure of liquid water is equal to the gaseous water content of the air (330)

distribution curve   a plot showing the number of individuals in a sample that have each of a series of values (324)

dynamic equilibrium   a process in which two opposing processes are proceeding at equal rates so that there is no net change (320)

evaporation   the escape of molecules from the liquid state into the gaseous state (319)

heat of vaporization   the amount of heat that a certain quantity of a liquid substance must absorb to change to a gas (332)

humidity   the ratio of the partial pressure of water vapor in air to the vapor pressure of liquid water at a given temperature (328)

hydride   a compound in which one other element is combined with hydrogen (336)

hydrogen bonding   a strong intermolecular force caused by the attraction of a covalently bound H atom to a pair of electrons on an atom in another molecule (337)

incompressibility   the resistance of a liquid or solid to be found to occupy a smaller volume due to applied pressure (318)

intermolecular forces   the forces of attraction between molecules (317)

partial pressure   the relative amount of one gas in a mixture of gases (328)

saturation   the condition in which air holds as much gaseous water as it can; the partial pressure of water in the air equals the vapor pressure of liquid water (328)

surface tension   the tension on the surface of a liquid that tends to make the surface area as small as possible, caused by imbalance in the intermolecular forces acting on the surface molecules (333)

vapor pressure   the pressure established by the gas when a liquid is allowed to evaporate until equilibrium is established in a closed container (322)

volatility   the tendency of a liquid to vaporize (327)

Exercises

11.1   *a.* How must the kinetic molecular theory be modified to account for the fact that gases can be condensed to liquids?

b. Explain why we cannot expect an ideal gas to condense to a liquid.

11.2   Consider the lower curve of Figure 11.1. Describe what is happening to the molecules of the substance in each region of the curve.

    a.   Region A–B   *b.*   Region B–C   *c.*   Region C–D

11.3   Why do not all gases condense to liquids at the same temperature?

11.4   If you fill a rectangular can with steam (H_2O gas at $100°$ C) and seal the can, as it cools the can collapses. Explain.

11.5   A naval surface ship can damage a submarine by exploding a depth charge in the water *near* the submarine. (The depth charge does not have to hit the submarine directly.) Explain how the liquid property of incompressibility makes this possible.

11.6   Explain why liquids exert a vapor pressure.

11.7   An open pitcher of orange juice stored in a refrigerator in which there is an uncovered sliced green pepper acquires a definite green-pepper flavor without touching the pepper. What has happened?

11.8   Which of the following cases represents a dynamic equilibrium? Justify your answer.

    a.   People commute from city A to city B in the morning and back at night.

    b.   People commute from city A to city B, and the same number of people commute from B to A.

11.9   Why does the vapor pressure of a liquid increase as the temperature increases?

11.10   Explain how evaporation is a cooling process.

11.11   The water in a rapidly boiling pot is at the same temperature as it is in one where the water is just barely boiling slowly. What is the difference in terms of what is happening to water molecules?

11.12   Denver, Colorado is referred to as "the mile-high city." Why does it take longer to cook vegetables in an open pot in Denver than it does in San Francisco?

11.13   Why does a pressure cooker attain a higher temperature than water boiled in an open pan?

11.14   Why must a humidity reading always be given at a specified temperature in order to have any valid meaning?

11.15   Refer to Figure 11.7.

a.   Estimate the dew point for air that has 50% humidity at 30° C (86° F).

b.   On a certain day, the maximum temperature is 80° F and the humidity is 60%. That night the temperature falls to 55° F. Does dew form?

11.16   What happens to the energy that a liquid absorbs as its heat of vaporization?

11.17   In your refrigerator, a fluid is pumped around a loop of tubing. In one part of the loop, the fluid is in liquid form; it evaporates to a gas. Then the gas is compressed by a pump back into liquid form to be recirculated.

a.   Which process, evaporation or condensation, occurs in the tubing around the trays in which ice cubes are formed?

b.   Which process, evaporation or condensation, occurs in the tubing in the base or back of the refrigerator where heat is dissipated?

11.18   When the air is still, just before a summer thunderstorm, the temperature often rises. After the storm has passed and a breeze is blowing, the temperature drops. Explain how these temperature changes are caused in terms of the condensation and evaporation of water vapor and rain.

11.19   Old kerosene-burning lamps always had a reservoir of liquid fuel into which a cloth wick dipped. The kerosene was burned by a flame on the upper end of the wick. Explain how the kerosene traveled from the liquid reservoir to the flame.

11.20   You are told that the proper method of watering a house plant is to set the flower pot in a dish of water and allow the water to be absorbed through the hole in the bottom of the pot. Even if the dish is shallow, water eventually is distributed throughout the pot. Explain how this happens.

11.21   A liquid that has a high vapor pressure at ordinary temperatures also has a low surface tension at ordinary temperatures. Explain why this is true.

11.22   As the temperature rises, molecules of a liquid move faster and, consequently, are less influenced by intermolecular forces. Would you predict that the surface tension of a liquid is larger or smaller at a higher temperature? Explain.

12

The Solid Form of Matter

- [ ] In what ways is the structure of solids different from that of liquids or gases?
- [ ] Is there any molecular movement at all in a solid?
- [ ] Can a solid ever change directly into a gas?
- [ ] How do the arrangement and kind of particles in a solid affect its properties?
- [ ] How does the arrangement of atoms in metals explain their characteristic properties?
- [ ] Why does ice float on the surface of liquid water?

In our descriptions of the gaseous form of matter, we frequently used the words "free," "chaotic," and "motion." The implications of these terms can be summed up in another word—disorder. The molecules of a gas are not restricted to orderly arrangements. A gas expands to fill uniformly any container in which it is placed. Gases diffuse readily because the molecules, in their chaotic motion, are free to go in all directions.

In liquids, we find much more restriction placed on the molecules. Except for some that evaporate at the surface, the molecules stay within the body of a liquid. So the sample has a surface. Below the surface, a liquid has a shape, determined by the shape of the container. Yet within the liquid, the molecules can move around; diffusion in liquids takes place readily, though at slower rates than in gases. There is still a considerable amount of disorder in the arrangement of molecules in a liquid.

The situation in solids is quite a contrast. In solids, the units making up the solid are very much restricted in their motions. A sample of a solid—like a grain of sand, a steel beam, or a stone—has a definite shape all its own. The units in a solid are arranged

---

Diffusion in solids is often so slow that it cannot be measured.

---

in some orderly way. They are not free to move about. Diffusion in solids is often so slow that it cannot be measured. Some layers of different kinds of rocks have been in contact with each other in the crust of the earth for millions of years without any appreciable intermingling of the different materials.

An analogy to the order–disorder patterns of molecular behavior in solids, liquids, and gases is the way the members of an audience are arranged during and after a lecture. During a lecture, people, out of courtesy to the speaker, remain in their seats. They may wiggle or scratch or sleep quietly, but they do so in their chosen locations. Occasionally, one or more persons will move a short distance to a vacant seat. Such movement, by analogy, is the way slow diffusion sometimes can occur in solids.

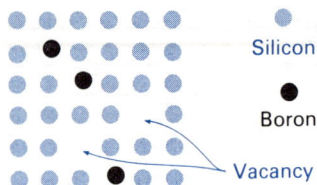

A few boron atoms can occupy vacant locations in a piece of solid silicon. Transistors are made this way.

Units shift locations; each shift opens up a vacancy into which another unit may move. If there are many unoccupied seats, people can gradually work their way through the lecture hall without causing much disruption. In similar fashion, the atoms of one metal are known to accommodate the diffusion of atoms of another metal. The atoms move to occupy vacancies in the orderly arrangement of locations. This is the way the components of transistors or other "solid-state" devices are made. Relatively few atoms of such elements as arsenic or boron are placed in vacant locations within an array of atoms in a solid piece of silicon or germanium.

After a lecture, the members of the audience flow out the door, much as the molecules of a liquid flow from a container. The arrangement of the people is less rigidly established, yet some order is maintained by the close contact of the individuals. After the audience has dispersed all over the surrounding community, they correspond to the gaseous form of matter. The members of the audience have little or no contact; their movements are not restricted by one another.

In this chapter, we will discuss some of the general principles that can be used to explain the properties of solids. Let us start by examining what happens when a solid is changed to a liquid.

MELTING, FREEZING, AND THE HEAT OF FUSION

By this time, it is apparent when we say "oxygen is a gas" or "water is a liquid" or "iron is a solid" that we are describing the form a substance is in *at normal temperatures*. Just as we found that gases can condense to liquids if the temperature is low enough, we also find that liquids can freeze to solids at still lower temperatures. Or if we raise the temperature high enough, we can expect a solid to change to a liquid and, at a still higher temperature, to a gas. Table 12.1 lists the boiling and freezing temperatures for oxygen, water, and iron. If an environment had a temperature of $-200°$ C, oxygen would be thought of as a liquid. In similar fashion, if we were considering an environment of $2000°$ C, we would think of iron as a liquid.

The idea that intermolecular forces of attraction hold molecules together in the liquid form also accounts for molecules being held even more tightly in the solid form. At the lower temperature, when freezing occurs, molecules have less energy of motion, so the attractive forces have even more effect. The effect

Table 12.1   Boiling temperatures and freezing temperatures of oxygen, water, and iron

Substance	Boiling Temperature ($°$ C)	Freezing Temperature ($°$ C)
"gas" oxygen	-183	-219
"liquid" water	100	0
"solid" iron	3000	1530

is so great that the molecules in a solid are held in definite locations; they cannot roll over one another as they do in liquids. In the case of iron, these attractive forces are very much stronger than they are for water or oxygen. The tiny individual particles of iron require the high thermal energy available at 1530° C in order to gain the relative freedom of the liquid form.

When a liquid changes to a solid, we say it *freezes*. When a solid changes to a liquid, we say it *melts*. If you add heat to a solid, its temperature rises until it begins to melt, and then the temperature remains constant until all the solid has changed to liquid. If you cool a liquid, you find that it begins to freeze at the same temperature at which its solid form melts. The temperature at which the solid and liquid forms exist together is called the *melting temperature* or the *freezing temperature*. Just as we found for the liquid ↔ gas transformation, a definite amount of heat is absorbed or given off by the substances when a specific amount of them undergo the liquid ↔ solid transformation. This heat is called the *heat of fusion*

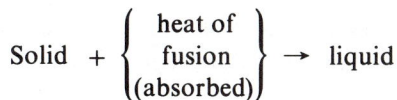

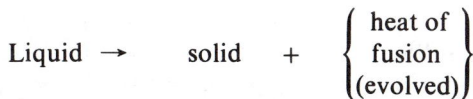

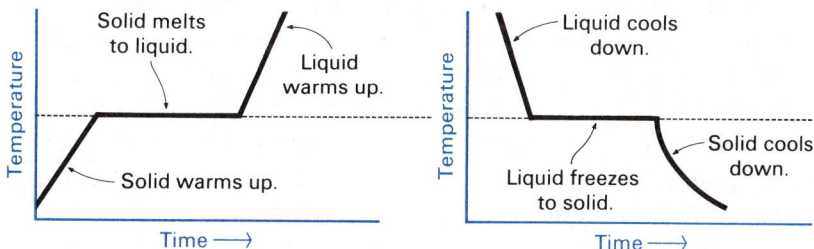

$$\text{Solid} \quad + \quad \left\{ \begin{array}{c} \text{heat of} \\ \text{fusion} \\ \text{(absorbed)} \end{array} \right\} \quad \rightarrow \quad \text{liquid}$$

$$\text{Liquid} \quad \rightarrow \quad \text{solid} \quad + \quad \left\{ \begin{array}{c} \text{heat of} \\ \text{fusion} \\ \text{(evolved)} \end{array} \right\}$$

Heat added to a solid sample.

Heat removed from a liquid sample

When you cool a glass of water by adding ice cubes to it, the ice melts as it absorbs its heat of fusion from the warmer water. The water then absorbs heat from the surroundings. When the ice has all melted, the water continues to absorb heat, and its temperature eventually rises to that of the room. When you make ice cubes in a refrigerator, the exchange of heat goes the other way. The refrigerator has to remove the heat of fusion from the liquid water in the ice-cube tray so that the solid water (ice) can form.

The heat of fusion of water is 80 calorie/gram, or 1440 calorie/mole. You can compare this number with the heat of vaporization of water (540 calorie/gram, or 9772 calorie/mole) to recognize that much less energy is absorbed to accomplish the solid → liquid change than for the liquid → gas change. The reason for this difference is the much greater freedom the molecules attain

The molecules must overcome more of the attractive forces holding them back when they move from liquid to gas.

when going into the gaseous form. The molecules must overcome more of the attractive forces holding them back when they move from liquid to gas. Only a small part of the total attractive force is overcome when the molecules move from solid to liquid.

HOW THE HEAT OF FUSION SAVES FRUIT CROPS

Owners of fruit orchards may make use of the heat of fusion of water to prevent costly frost damage. When a fruit tree is in bloom in the spring, the delicate blossoms, from which fruit later grow, are particularly susceptible to damage by freezing. The moisture within the plant cells expands when it freezes and thus breaks the cell walls. On a night when the air temperature drops to freezing, this damage can be prevented by spraying the trees with a fine mist of water droplets. The liquid water collects on the blossoms. Then the water on the blossoms freezes. Every gram of water releases its 80-calorie heat of fusion when it freezes. This release of heat keeps the blossoms warm enough so that they do not freeze. Moreover, the film of ice insulates the blossoms from further heat loss. This release of heat right where it is needed has been found to be much more effective than the former practice of burning smudge pots in orchards on cool nights. Besides, the air pollution resulting from the smoke is eliminated.

SUBLIMATION OF SOLIDS—
ICE AND DRY ICE

When the molecules of a solid absorb heat, they wiggle or oscillate back and forth around their locations. Sometimes a molecule in a solid, particularly on the surface of the solid, oscillates so fast that it breaks away as a free gas molecule. As more and more molecules do so, the solid changes into a gas without going through the intermediate stages of melting to the liquid form and then evaporating. The direct transformation of solid to gas is called *sublimation*

Dry ice is a familiar substance used to keep ice cream or other frozen foods at a low temperature. Dry ice is the solid form of carbon dioxide, CO_2. It received the name "dry ice" because it does not melt to a liquid. It sublimes directly to CO_2 gas at the low temperature of $-78°$ C at 1 atmosphere of pressure, or 760 torr. We have to specify the pressure under which this change takes place because the external pressure does influence the sublimation of a solid. If solid CO_2 is placed in a closed container so that the pressure of the gaseous CO_2 can build up, it behaves differently. Under a pressure of 5.2 atmospheres, solid CO_2 *melts to liquid CO_2* at a temperature of $-56°$ C. Under higher pressures, the temperature can be even higher. The pressure in some "C-O-2" fire extinguishers is so high that the CO_2 contents are mostly in the liquid form at room temperature.

If you place a piece of dry ice in a glass of water, very vigorous bubbling occurs. The water is not boiling, even though it may appear to be. Nor does the water freeze, as you might expect, even though the dry ice is at so low a temperature. Actually, there is very little direct contact between liquid water and solid CO_2. The CO_2 gas evolves so rapidly that each piece of solid is surrounded by an insulating sheath of gaseous CO_2. The bubbles of CO_2 are formed so rapidly and become so filled with water vapor and small droplets of liquid water that they look like fog or smoke.

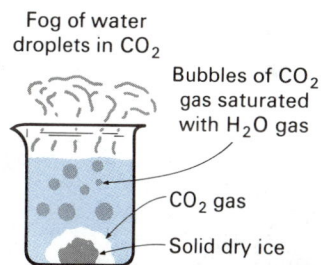

Fog of water droplets in CO_2

Bubbles of CO_2 gas saturated with H_2O gas

CO_2 gas

Solid dry ice

When a flat piece of solid dry ice is placed on a smooth tabletop, you find that in a very short time a small tap or shove sends the solid piece sliding rapidly across the surface. Here again is evidence of the very rapid evolution of CO_2 gas. Very little friction exists between the piece of solid dry ice and the tabletop because a layer of CO_2 gas is always between them.

CO_2 gas also evolves rapidly when your warm skin touches a piece of solid dry ice. The evolving gas gives you some protection, because your flesh and the solid at $-78°$ C do not at first come into direct contact. However, you should not hold a piece of dry ice long with unprotected hands. As the surface of your skin cools, the gas evolves less rapidly. If your skin and the solid do come into close contact, you can be harmed. Flesh freezes solid quickly at temperatures near $-78°$ C. Painful permanent frostbite can result.

Ice, solid H_2O, also tends to sublime to a small extent. If we were to put some ice at a temperature just below 0° C in a closed container, an equilibrium would be established between H_2O molecules leaving as gas molecules and H_2O molecules returning to the solid. The dynamic equilibrium is like that for the vapor pressure of a liquid. The sublimation pressure (the equilibrium pressure of gas over solid) of ice just below 0° C is about 4 torr.

Ice and snow disappear on a windy day, even if the temperature remains below freezing.

Thus ice and snow disappear on a windy day, even if the temperature remains below freezing.

The sublimation of ice is also used in the commercial process of *freeze-drying*. Freeze-dried instant coffee crystals are made by freezing brewed coffee and then removing the H_2O gas molecules that sublime off the frozen coffee solution by pumping with a vacuum pump. The dry solid coffee crystals are left behind. The technique of freeze-drying is especially useful in cases where heating to a high enough temperature to evaporate water might injure the texture, appearance, or flavor of a material.

THE GREAT VARIETY AMONG SOLID FORMS OF MATTER

The variety among solids that we encounter in our daily lives is much greater than the variety among liquids and gases. The solid ground we live on has many forms. So do the structural materials of which buildings are made. Naturally occurring solids, such as rocks and stones, have an infinite variety of shapes and textures. Yet every "rock hound" knows that individual minerals often have a very characteristic shape that can be used to identify a sample. The beautiful appearance of geodes is due to the crystals of quartz formed inside a rock that looks like a stone egg until it is cut open. Such substances as quartz appear as crystals; every quartz crystal, regardless of size, has the same geometric form.

Quartz crystals form inside geodes, often called "thunder eggs."

If you hit a crystal hard enough with a hammer, you break it into smaller chunks. However, if you hit another kind of solid, such as a piece of metal, with a hammer, you will only dent it. Metals can be pounded into different shapes. So although all solids share the property of rigidity (holding one shape as opposed to flowing, as a liquid does), by no means can we say that solids are all alike. What differences in the molecular structure of solids are responsible for such variety in properties?

One important structural difference is that not all solids are made up of molecules. As long as we are talking about gases or liquids, we can always use the word *molecule* to refer to the unit particles. In some cases, such as the noble gases (helium, neon, and so on), the molecule consists of a single atom. Some solids are composed of molecules, for example, solid H_2O and solid CO_2. But molecules are not the only unit particles possible in solids.

Some solids are actually *ionic compounds*. Recall the discussion in Chapter 6 that described such compounds. Salts, of which sodium chloride (table salt) is a familiar example, consist of *ions*. Electrons have been transferred from one kind of atom to another. The resulting solid compound contains no molecules. Instead, a vast array of positively and negatively charged ions make up the solid. No one ion "belongs" to any other ion of opposite charge, as would an atom that forms a covalent bond with another atom in a molecule. We can reasonably expect that the properties of solids made up of ions will differ from the corresponding properties of solids made up of molecules.

We also have encountered examples of solids in which *atoms* are the unit particles. Diamond is an example we described in Chapter 7. Each carbon atom in the solid diamond is bound to four other carbon atoms. The result is an array of atoms located in a definite three-dimensional pattern. Metals, such as iron or copper, also are solids in which atoms, not ions or molecules, are packed together in a regular pattern.

Let us examine some of the forms that solids take and see what general principles they follow. Then we can interpret the properties of solids in terms of these ideas about their structures.

CRYSTALLINE SOLIDS

Crystalline solids are those in which the units that make up the solid, either molecules, atoms, or ions, are arranged in a definite orderly pattern. Every crystal, large or small, of a particular sub-

Crystals

Break!

Grind!

Metal

Dent!

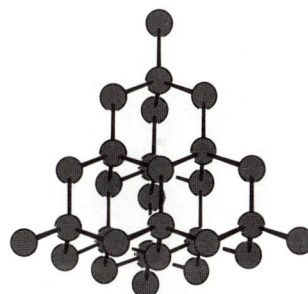

A diamond crystal is made up of carbon atoms.

stance has the same geometric form. For example, if you look at a relatively large crystal of rock salt, sodium chloride, you see that it looks like a cube or a rectangular solid. The faces of the crystal are perpendicular to one another. The angles at the edges and corners are 90°. Even though the rock-salt crystal may not be a perfect cube or rectangular solid, it is obvious that what you see is a piece of one. If you grind up such a large rock-salt crystal, you make smaller pieces. Each of these pieces, too, has the appearance of a rectangular solid. You can grind the crystals further to make a powder and look at the fine particles under a microscope. Again, you find that the tiny pieces have the same rectangular form or shape.

The external shape a crystal adopts is the consequence of the orderly repetition of the same pattern of locations in space of the individual units. A two-dimensional analogy of such a pattern is represented by people in the seats of an auditorium. The people in a block of seats are arranged in rows and tiers. The people in the whole auditorium are arranged in the same pattern of rows and tiers. The point is that, if the whole auditorium (large crystal) were divided into blocks (small crystals), the same pattern of seat arrangement would be observed.

The particular arrangement adopted by the crystal units depends on what they are, that is, on their chemical makeup. For example, in the case of sodium chloride, equal numbers of positive ions and negative ions must be accommodated. Their opposite charges require that each positive ion be surrounded most closely by negative ions, because of the electric forces of attraction. The size of the ions is also a factor. They have to fit into the appropriate places. Figure 12.1 shows how the smaller Na^+ ions and the larger Cl^- ions fit together in crystals of sodium chloride. Notice that the conditions we have mentioned are satisfied by this structure.

One part of Figure 12.1 is a model with spheres of the proper relative size for Na^+ and Cl^-. The other network model in Figure 12.1 may make it easier to recognize why a crystal splits to make smaller crystals of the same characteristic shape. The two portions can be made from one large one by breaking an equal number of attractions between Na^+ and Cl^- ions in the two faces next to each other. Each small crystal has the same number of Na^+ and Cl^- ions, a fact required by the electric neutrality of every piece of sodium chloride.

Given such a structure for solid NaCl, would you expect it to have a high or low melting temperature? Notice how each posi-

Model showing how spheres of
the right size pack together

Cl^-

Na^+

Network model showing
how a crystal splits
along a face of the cube

Cl^-

Na^+

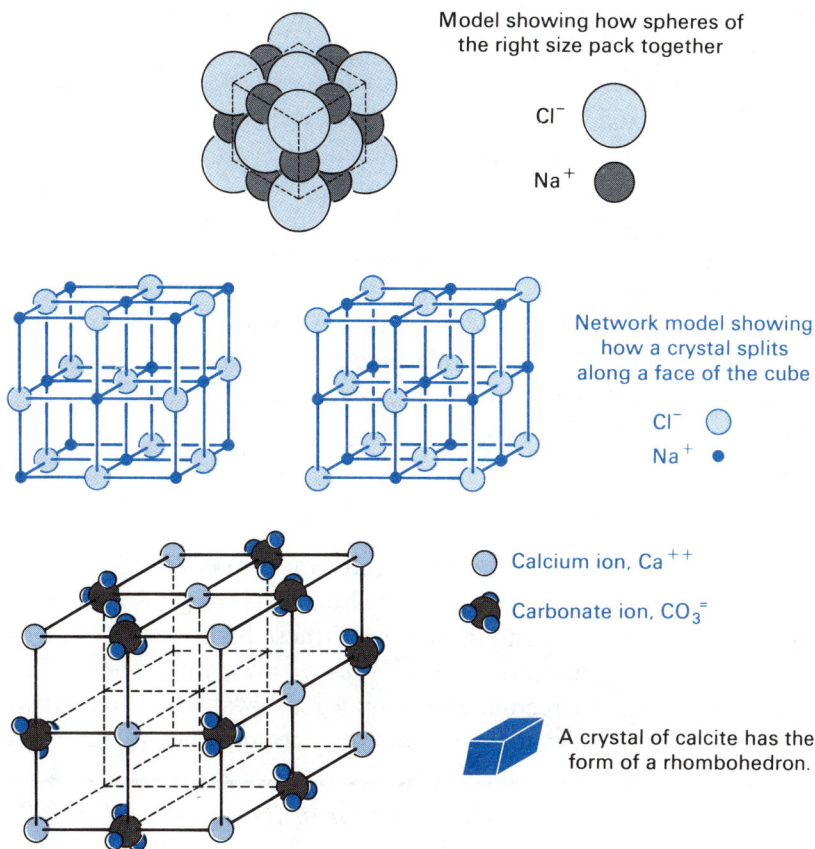

Figure 12.1   The crystal
arrangement of Na^+ ions and
Cl^- ions in solid sodium
chloride.

Calcium ion, Ca^{++}

Carbonate ion, $CO_3^=$

A crystal of calcite has the
form of a rhombohedron.

Figure 12.2   The crystal
structure of calcium car-
bonate, $CaCO_3$.

tive ion is surrounded by negative ions. Remember, too, that ions
of opposite charge exert strong electric forces of attraction on
each other. So we would expect that the ions need large thermal
energies to break loose from their assigned locations. Accordingly, we would expect the melting temperature to be high.
(Sodium chloride melts at 801°C.)

The substance calcium carbonate, $CaCO_3$, occurs in nature in
a great variety of forms. Calcium carbonate is the chief constituent of limestone, marble, and chalk. It too is made up of
ions. Calcium ions, Ca^{++}, which have a double positive charge,
are interspersed with carbonate ions, CO_3^-, which have a double
negative charge. One mineral form of very pure calcium carbonate, called calcite, may be found in the form of large crystals.
Their form is that of a rhombohedron. A rhombohedron looks
like a rectangular solid that has been slightly tilted. The faces are
at an angle that is not 90°. Figure 12.2 suggests how the ions are

arranged in a crystal of calcite. Note again the regular pattern of alternating locations of the two kinds of ions.

Marble is made of many small calcite crystals. When these crystals are mixed with other substances, the marble has a variety of beautiful colors. You know that marble is a very hard substance. It is more difficult to break up a piece of calcite than it is to break up a piece of rock salt. This increased hardness can be explained in terms of the forces holding the ions in their established spots. The fact that each ion has a double electric charge means that the electric forces of attraction among ions of opposite charge is very strong. Consequently, the solid holds together very tightly, more so than rock salt, in which the ions have only single electric charges.

CLAY AND ASBESTOS

The crystal form adopted by other substances may be much more complex than rectangular solids or rhombohedrons. Some crystals are many-sided geometric forms. Others form distinct layers or plates. Still others look like needles or even long fibers. The important point to recognize is that each crystalline substance has a particular orderly pattern for the arrangement of its units

Each crystalline solid has a particular orderly pattern for the arrangement of its units in space.

in space. Every sample of that compound, large or small, looks the same because of that pattern.

The chief elements in the compounds that make up the rocks, sand, clays, and soil of the earth are silicon and oxygen. Other elements, such as aluminum, iron, and calcium, also are combined with the silicon and oxygen. However, the way the silicon and oxygen are arranged in a network often determines the properties of a particular rock or mineral. In hard rocks, the Si and O atoms are arranged in three-dimensional networks like a jungle gym on a children's playground. Moreover, the Si and O atoms are held together by tight covalent bonds. Consequently, rocks and minerals of this type are hard and have very high melting temperatures.

In clays, the silicon and oxygen atoms are arranged in two-dimensional networks. Figure 12.3 suggests this arrangement of atoms. An analogous structure is that of chicken wire or the wire

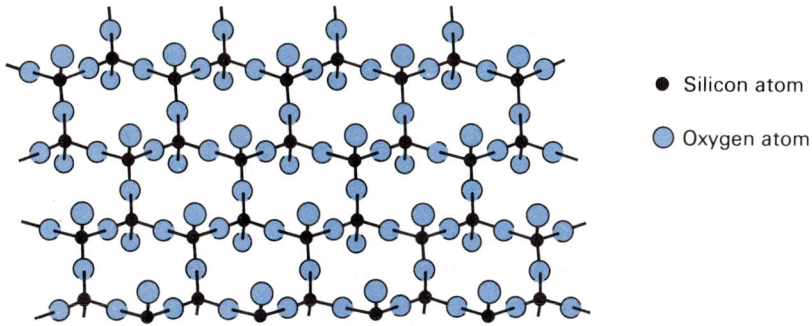

● Silicon atom

○ Oxygen atom

Figure 12.3   The network structure of silicon and oxygen atoms in clay.

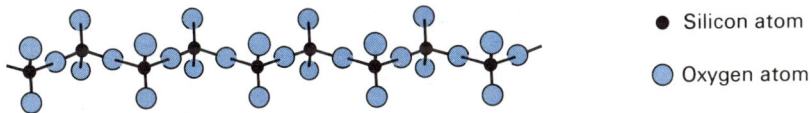

● Silicon atom

○ Oxygen atom

Figure 12.4   The ropelike structure of silicon and oxygen atoms in asbestos.

in a playground fence. Consequently, the crystals in clays are in the form of sheets or plates. Water is held between the layers. This structure is what makes wet clay slippery; the solid sheets or plates slip over one another very readily. When you mold clay into a piece of pottery and then bake it, you drive out the water, and the atoms tend to make bonds from one plate to another. The result is more strength in three dimensions for the pot or vase that has been baked.

Asbestos is a mineral substance in which the silicon and oxygen atoms are linked together in long ropes or chains. Figure 12.4 suggests what this arrangement looks like. The atoms along the chain are tightly held together, but the forces between chains are weak. Consequently, asbestos is a very fibrous material. The long fibers can be pressed together to make sheets, such as those used for building insulating walls in houses. Asbestos combines the two desirable properties of being a good heat insulator and being nonflammable. Asbestos is also used in the linings of automobile or truck brake drums.

Whenever asbestos-containing materials are cut or rubbed in any way, tiny fiber-like crystals of asbestos are set free. If these fibers get into the air humans breathe, they cause a health problem. The fibrous form of the tiny particles makes them difficult to screen out in the upper nasal passages. Consequently, a person breathes them all the way into the lungs, where they lodge and accumulate. This accumulation leads to lung disorders. The awareness of this problem has led in recent years to the adoption of strict safety standards wherever people work with asbestos-containing materials.

Asbestos, a fiber-like mineral in which silicon and oxygen atoms link together in chains.

THE STRUCTURE OF GLASS

Instead of imagining people arranged in the orderly fashion of auditorium seats, consider a tightly packed crowd of people standing at an outdoor rally; this pattern is analogous to that of a *noncrystalline solid.* The people are close together but not in regular tiers and rows. The units in a solid such as glass are arranged in this manner. There may be small regions with some orderly arrangement, but seldom are these regions very large. If you give a piece of glass much of a mechanical shock by dropping it or striking it with a piece of metal, the glass shatters into all sorts of shapes. Remember that a truly crystalline substance, like rock salt, breaks into pieces with the same geometric shape as the original piece. Even though the fine glass of a delicately shaped goblet is referred to as "crystal," the term is not scientifically accurate. Good "crystal" glass actually is made up of many microscopic crystals. These crystals are very small and arranged in no overall order, so that it is not accurate to consider the whole article to be a true crystal.

Glass is made by melting sand (silicon oxide) together with the oxides of other elements, such as sodium, calcium, or lead. As the liquid is allowed to cool, the various atoms are locked into their locations in the solid. If the cooling is uneven or more rapid in one part of the article or sheet than in another, the atoms may be stranded in locations they would not normally occupy if they had been able to roll over one another and adjust their positions more slowly. The result of this forced unevenness in location of the atoms is strains or stresses in the solid glass. When external force is applied, such as a bump by a metal object, the glass breaks along the lines of the stress or strain.

When melted glass is poured out in the form of a sheet, it has a surface tension. Recall from the discussion in Chapter 11 how the molecules on the surface of a liquid are pulled slightly tighter together than are the molecules in the body of the liquid. If the melted glass sheet is suddenly cooled by a blast of cold gas or liquid, the surface tension is frozen into the solid glass. Therefore, the surface of the solid sheet is slightly tougher than the rest of the solid, because the molecules in the surface are more tightly held and are closer together. Furthermore, many short strains or tensions exist all through the surface of the solid. If the surface of the solid glass is scratched, these many strains cause the whole piece to break into many small pieces rather than into big chunks. The glass used in some automobile windows is made in this way. In an accident it is much safer to have the glass shatter into a powder than into larger pieces with knifelike edges. The safety glass for windshields also is made in layers. A middle layer with outer layers of glass bonded to it is made of a tough transparent plastic material. The plastic does not shatter and thus holds the glass together, even if the glass is broken.

METALS ARE FLEXIBLE

Most of the elements in solid form at ordinary temperatures are metals. You are familiar with some of these metals, such as copper or iron in the form of wire or sheets, silver or gold in coins and jewelry, possibly even lead or a mixture of lead and tin in solder. A characteristic property of a piece of heavy wire is its flexibility. You can bend it a little way, and it will spring back to its original straight form. If you push harder, you can make a permanent bend. If you take a piece of wire that has been bent into a U shape, such as a paper clip, and try to straighten it out by grasping the ends and bending it the opposite way, an interesting thing happens. Instead of unbending the U part, two new bends appear in the straight parts of the wire. Metals are flexible, but once they have been deformed, they resist being pushed back into their original form. Thus it is easier to make new bends than to remove old ones. Anyone who has tried to remove a small dent in an automobile body or fender by pounding on the back of the dent has discovered this. What is there about the structure of metals that is responsible for these properties?

A solid piece of an elemental metal, such as a copper wire, consists of atoms packed together in a regular orderly pattern. The spherical atoms are arranged in the same pattern you see displayed by oranges or grapefruit neatly stacked in a bin at a fruit store. Figure 12.5 suggests how such a stack of spheres looks. If the spheres are all the same size, they can be arranged in layers so that each sphere is touching six others in the layer. The next layer is added by putting spheres in the depressions between first-layer spheres. Then another layer is added in a similar regular way. Layers on top of the first two or three repeat the same arrangement, so that every sphere is touching 12 other spheres. The maximum number of contacts any sphere can ever have with other spheres of the same size is 12. This suggests one reason why most metals are so strong; the atoms are arranged in such a way that each atom is surrounded by 12 other atoms with their attractive forces holding it strongly.

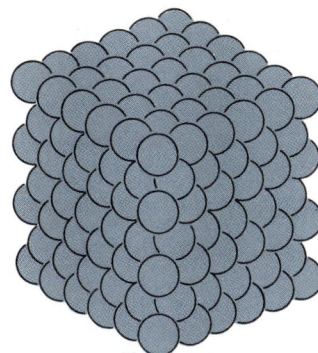

Figure 12.5   The closely packed atoms of a solid metal.

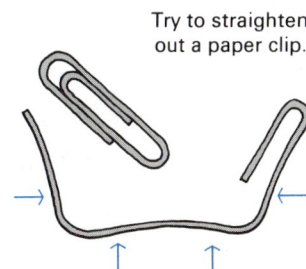

Try to straighten out a paper clip.

New bends can be made more easily than old ones can be removed.

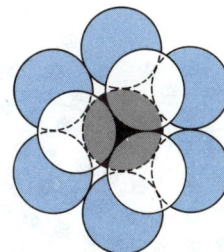

Central sphere ● touches six spheres ● in its layer, three spheres ○ in layer above, and three spheres in layer below (not shown).

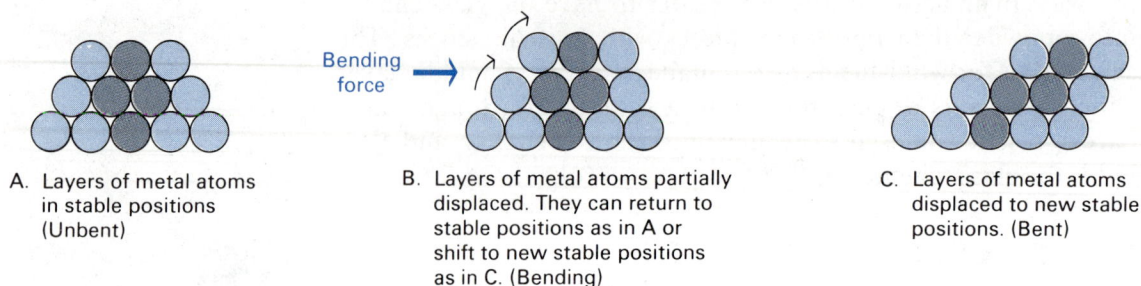

A. Layers of metal atoms in stable positions (Unbent)

Bending force →

B. Layers of metal atoms partially displaced. They can return to stable positions as in A or shift to new stable positions as in C. (Bending)

C. Layers of metal atoms displaced to new stable positions. (Bent)

Figure 12.6   The rearrangement of layers of metal atoms when metal bends.

If you push sideways on a layer of oranges, you can move them a little distance, because they tend to roll up over their neighbors. If you let go, they roll back to where they were. This analogy suggests why a metal wire is flexible. A small force merely tends to roll each layer of atoms a small amount. When the force is relieved, the layers roll back to their original places; the former shape is restored. However, if you push a bit harder on the layer of oranges, you can make it move to a new location. Each sphere rolls over those in the lower layer to occupy new regularly arranged locations. The atoms are pushed into new locations, where they have the same stable pattern of contact with their neighbors. The pattern of contact is the same, but some of the neighbors are different. Figure 12.6 shows how layers of spheres can roll over one another to new stable locations. Note how the atoms marked with X shift to have some new neighbors.

PUSHING CRYSTALLITES INTO DIFFERENT LOCATIONS

Another feature of the structure of metals helps to explain some of their properties. This feature is the fact that a piece of metal is usually made up of very small crystals, called *crystallites,* rather than being made with one large orderly crystal arrangement throughout the whole piece. A piece of solid metal usually is formed by cooling liquid metal that has been poured into the desired form. When the atoms align themselves in positions proper for the solid form, they do so in clusters. Each cluster becomes a crystallite, but its edges may not match exactly with the alignment in the edges of its neighbors. Figure 12.7 suggests, in two dimensions, how these crystallites exist in the solid metal. An analogy is what you see when a marching band is performing on a football field. When they are in the midst of shifting from one formation to another, small groups of players keep their own formation, but the groups go in various directions at various

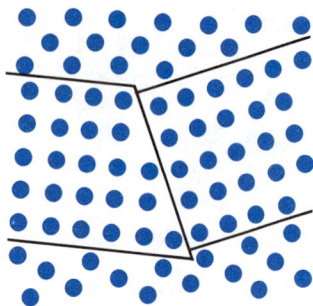

Figure 12.7   Crystallites in metals.

angles. If a whistle blows to stop all the players where they are at that instant, the whole band looks like a jumble of small orderly groups of players.

Another thing happens to keep a piece of metal from being one large orderly crystal. When the solid metal is formed by cooling the liquid metal, there may be places in the crystallites or between them where no atom occupies a site in the otherwise orderly arrangement. The solid metal is left with atomic-sized holes or vacancies, just as some seats may be left unoccupied when a crowd of people quickly fill an auditorium.

When a mechanical stress is put on a piece of metal to bend it, some atoms may be forced into those spots that formerly were vacant. Also the crystallites are moved around. Crystallites become jammed together, interlocked into a tighter arrangement, because some of the vacancies between them are now occupied. More atoms become surrounded by other atoms instead of holes; the result is that all are held more strongly. This interlocking of the crystallites tends to make the piece of metal much less flexible. The crystallites cannot slip or slide over each other. One evidence of this lack of flexibility is the way a bent piece of wire resists attempts to straighten it by bending in the opposite direction. Rather than merely bending back into a straight piece, the wire tends to retain the shape of the bend. The wire acts as though once bent, it is stronger at that place. In fact, trying to straighten the wire usually puts new bends in parts of the wire that formerly were straight.

The same kind of apparent strengthening one particular place in a metal occurs when a sheet of metal is pounded. Anyone who has made jewelry knows that a flexible sheet of metal can be made stiff by pounding. The crystallites are forced into new interlocking arrangements that make the whole sheet more rigid. Just as a bent wire resists attempts to straighten it, we find that once

Once a sheet of metal has been dented, it resists having the dent removed by pounding in the opposite direction.

a sheet of metal has been dented, it resists having the dent removed by pounding in the opposite direction. The realignment of the crystallites by the blow makes the dent less flexible than the rest of the metal.

This same principle of pushing crystallites around is used on a large scale when metal is forged into big articles. Usually the

metal is heated so that the atoms oscillate more in the solid metal. The metal thus becomes softer and the crystallites can be rearranged even more readily. Long before this process was understood in chemical terms, it was the basis of a highly developed practical art. The technique of alternately heating and pounding metal is directly related to much of the course of world history. Before warfare involved guns, superior swords, lances, and battle axes made of hardened steel meant military advantage. In later years, the same was true for armies possessing rifles and cannon made from forged metal.

Our discussion has been based on a simplified picture of metals, but the principles are the same when applied to complicated structures. When two or more metallic elements are mixed together, they form solid *alloys*. Steel is an alloy of iron with other metals, such as nickel, chromium, or carbon. The presence

You can pile grapefruit in a regular pattern and still put a walnut in each hole between the grapefruit.

of other atoms causes changes in the crystal form. An analogy is the way you can pile grapefruit in a regular pattern and still put a walnut in each hole between the grapefruit without disrupting the pattern. But if you tried to pile grapefruit and oranges and lemons together, you would not get very much of a pattern because of the size differences. Different arrangements of the various metal atoms in alloys can be made to give the finished metal a particularly desirable set of properties, such as hardness, inflexibility, or toughness.

THE STRUCTURE OF ICE

Ice floats on the surface of liquid water. Why? When water freezes, it expands; its density decreases. The same mass of water takes up more space in the solid form than it does in the liquid form. If you place a closed bottle full of water or carbonated drink in a freezer, the expansion will break the bottle. If water in the cooling system of an automobile engine is allowed to freeze, the force of the accompanying expansion may be enough to break the engine block or at least rupture the connecting hoses. In the next chapter, we will discuss why substances mixed with the water of a cooling system prevent freezing from occurring.

You may think this property of expanding when freezing is typical of all liquid-to-solid changes. It is not. Water is unique in this property. If you watch a metal or other solid melt in a heated container, you find that, as soon as some liquid is formed, the remaining solid stays on the bottom of the container; the liquid floats on the solid. Our generalized picture of molecules being able to move about in liquids but being restricted to specific locations in solids makes this effect understandable. A crowd of moving people needs more space to move around in than the same crowd needs just standing still, packed close together. Why is water an exception? What unusual feature about water makes it behave differently when it freezes?

The answer to these questions lies in the tendency of the water molecules to establish hydrogen bonds among one another. Coupled with this is the geometrical shape of the H_2O molecule. Figure 12.8 shows how the H_2O molecules are arranged in the solid form, ice. If we look only at the oxygen atoms, we see that they form a network with each oxygen atom arranged tetrahedrally with the others linked to it. The linkages are hydrogen bonds. Every oxygen atom has two hydrogen atoms bonded to it with a covalent bond. But each of these hydrogen atoms is involved in a hydrogen bond to another oxygen. Careful measurement of the angle made by the H—O—H bonds shows it to be 104.5° in the water molecule. This angle is very close to the angle of 109.5° required for the hydrogen bonds to form a regular tetrahedral structure. Very little shift is needed to establish this form.

The open space in the structure in Figure 12.8 has an important implication. The H_2O molecules in ice are *not packed closely together*. They are arranged in an open structure that appears to have six-sided (hexagonal) holes running through it. These holes represent the extra space required in the orderly solid but not

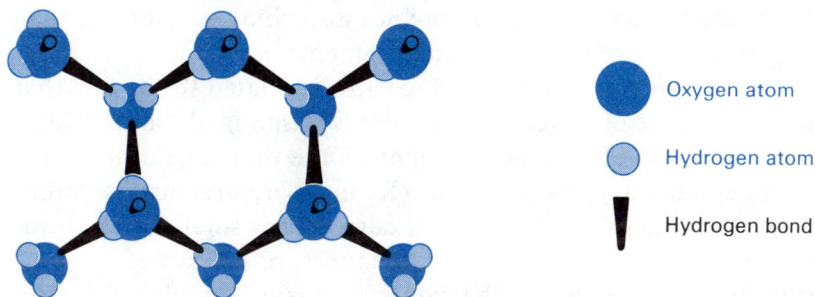

Oxygen atom

Hydrogen atom

Hydrogen bond

Figure 12.8   The structure of ice.

needed for the disorderly arrangement of molecules in the liquid form. Consequently, the same number of molecules takes up more space in the solid form; water expands when it freezes. All ice, from snowflakes to ice cubes to icebergs and glaciers, has regular open spaces between the molecules.

One of the consequences of this structure of ice is the way in which snowflakes or ice crystals, in the form of frost on window-panes, show hexagonal symmetry (the same design repeated in a six-sided figure). Snowflakes show a beautiful variety of dainty designs, but all have a motif with hexagonal symmetry. One of the unsolved mysteries of nature is how each of the six "arms" of these crystals can form just like the others of a particular crystal. Each crystal is different yet symmetrical.

Hexagonal symmetry
in snowflakes

THE FAR-REACHING CONSEQUENCES OF EXPANDING FROZEN WATER

We can hardly imagine what the world would be like if water behaved like other liquids and froze to a more dense solid. As it is, the less dense solid, ice, always forms on the *top* of lakes and rivers when the temperature drops below freezing. On the sur-face, ice forms an insulating layer so that the rest of the lake water does not freeze, and fish or other aquatic life that can adapt to near-freezing temperatures can go on living. When spring comes and the temperature rises, the ice melts. If it sank to the bottom, lakes would freeze from the bottom up, and deep lakes

We can hardly imagine what the world would be like if water behaved like other liquids and froze to a more dense solid.

never would thaw out completely. And the great amount of heat (the heat of fusion) absorbed when ice melts or evolved when liquid water freezes helps greatly to modify the weather. Tem-peratures would change much more drastically if the ice ↔ liquid water transformation were not absorbing or evolving such a large amount of heat in the environment.

The reason you can skate on ice also is related to the fact that the same mass of water takes up less volume in the liquid form than in the solid form. Your weight on the thin edge of the skate blade creates a great pressure. (Remember, pressure is force/ area. The area of the sharp blade edge is very small, so the force of your weight results in a very large pressure.) The effect of pres-sure always is to make the volume of any sample of matter smaller. Because liquid water has a smaller volume than the same mass of solid ice, pressure shifts the equilibrium.

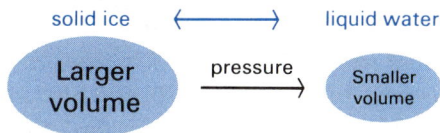

solid ice ⟷ liquid water

Larger volume → pressure → Smaller volume

The thin film of liquid water formed under the skate blade acts as a lubricant, so you can slide over the surface. One theory about the movement of glaciers uses the same reasoning. The tremendous pressure of the weight of glacial ice tends to melt some of it in the lower layers, so the whole mass appears to flow.

The amount of pressure needed to lower the melting temperature of ice (and hence allow the solid → liquid change to occur) is greater at lower temperatures. You can make a snowball out of "wet" snow that is near 0°C, but you cannot make one out of "dry" snow that is very much colder. The snow crystals are in contact at points and edges, so the pressure of your hands exerts great pressures at these points. Snow melts to water that binds the whole ball together when it refreezes as you release the pressure. But if the snow is too cold, the pressure is not great enough to accomplish the necessary lowering of the freezing point. So the snow remains a powder instead of forming into a ball. When you walk on very cold snow, it squeaks—the sound of the crystals slipping over one another rather than melting.

A weighted wire is placed over a block of ice.

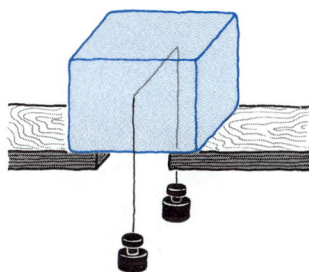

Summary

The properties of matter in the solid form reflect the fact that the unit particles in solids are restricted to some kind of orderly arrangement. These unit particles may be molecules, atoms, or ions with opposite electric charges. The strength of the forces holding these particles in fixed locations is reflected in the solid's heat of fusion. The heat of fusion is the amount of energy that must be absorbed by a given amount of a solid to change it into a liquid.

Some solids are capable of subliming. When a solid sublimes, its molecules go directly into the gaseous form without first melting to a liquid. Dry ice, solid CO_2, is a familiar example of a solid that sublimes at normal atmospheric pressure.

Crystals are solids whose regular geometric shapes come from a particular arrangement of the unit particles in a three-dimensional array. The properties of crystals, such as hardness, are

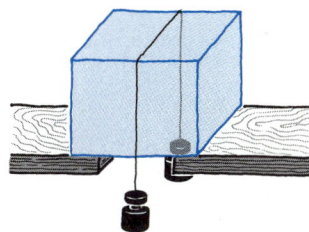

The pressure of the weighted wire on the ice causes it to melt under the wire. As the wire works its way through the block, the ice reforms above it.

influenced by what the units are and how they are arranged. For example, the chemical composition of clay and asbestos is similar. But the arrangement of the atoms leads to sheets or plates in clays and to a fibrous structure in asbestos.

The atoms in metals are arranged in a regular order. Usually, the order extends over only small regions. This leads to the presence of crystallites in metals. When a metal is pounded, adjacent crystallites may be reoriented to produce larger regions of uniformity. This leads to increased strength in the metal piece as a whole.

When liquid water changes to solid ice, the molecules adopt a definite orderly arrangement. This arrangement requires more space to accommodate the molecules than they need in the liquid form. This unique arrangement of the molecules in ice is caused by the hydrogen bonding among H_2O molecules. The expansion of water when freezing, the effect of pressure on the melting process, and the beautiful hexagonal symmetry of snowflakes all can be interpreted as phenomena resulting from this ability of H_2O molecules to form hydrogen bonds. So, too, is the widespread influence exerted on climate and weather by the presence of great quantities of water on the surface and in the atmosphere of the earth.

Glossary

The number in parentheses indicates the text page where you can find the term defined in context.

alloy   a solid mixture of two or more metallic elements (360)

crystalline solid   a solid made up of repeating units (atoms, ions, molecules) arranged in an orderly pattern (351)

crystallite (of a solid)   a very small crystal within a larger sample of a solid (358)

freeze-drying   the removal of water from a substance by freezing of the substance, followed by the sublimation and removal of the water (350)

freezing temperature   the temperature at which the solid and liquid forms of a substance are in dynamic equilibrium with each other (347) *heat of fusion*   the amount of heat given off (or absorbed) when a certain quantity of a substance changes state between liquid and solid (347)

melting temperature   the temperature identical to the freezing temperature of a substance, at which the reverse process takes place (347)

noncrystalline solid   a solid whose subunits are not arranged in a repeating orderly pattern (356)

sublimation   the process by which molecules in a solid enter the gaseous state directly without passing through a liquid phase (349)

Exercises

12.1   At the molecular level, what is the difference between a solid and a liquid?

12.2   What evidence do you have for concluding that the forces holding the molecular-level units together in iron are stronger than those in water?

12.3   Recall the structure of diamond (pure carbon) described in Chapter 7. In view of its great hardness, would you expect diamond to melt at a high or low temperature?

12.4   The graph below is the plot of observed temperatures as a sample of water is heated at a constant rate. The sample at point A is in the form of ice. Tell what form the water has during the time represented by the following line segments. Along some lines water has only one form, along others two forms exist together.

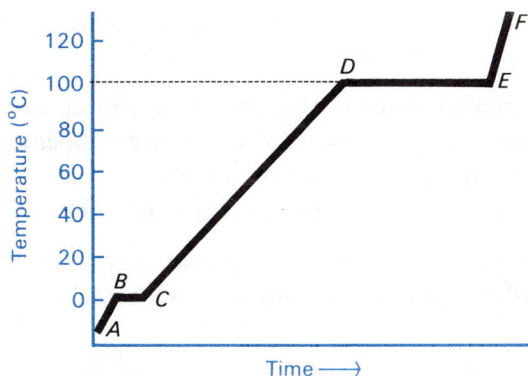

a. A–B   *b.* B–C   *c.* C–D   *d.* D–E   *e.* E–F

12.5   The heat of fusion of water is 80 calorie/gram. The heat of vaporization is 540 calorie/gram. Explain why the heat of vaporization is so much larger than the heat of fusion.

12.6   Before the time of central heating, freezers, and refrigerators, families used to store vegetables such as potatoes in the cellar under a house. A barrel of water was always placed near the vegetables.

> *a.*   What did the water accomplish in the winter time?
>
> *b.*   What did the water accomplish in the summer time?

12.7   *a.*   Why are many plants killed by exposure to temperatures below 0°C?

> *b.*   Why does spraying fruit trees with water on a night when the temperature is near 0°C protect the blossoms from frost damage?

12.8   *a.*   Give a definition for the heat of sublimation.

> *b.*   How do you think the heat of sublimation of a substance compares with its heat of fusion and heat of vaporization?

12.9   If you put some solid dry ice in a balloon and tie it shut, the balloon seems to blow itself up to a larger size just as if you had forced your breath into it. Explain.

12.10   In Alaska in winter, laundry may successfully be hung out to dry, even though it freezes solid by the time it is hung up. Why?

12.11   If ice cubes are placed in a closed freezer, eventually frost will form on the freezer walls. What has happened?

12.12   Explain why ionic crystals generally are found to have high melting temperatures.

12.13   Graphite is a form of carbon that can be used as a lubricant, because it is slippery. Clays also are slippery. From what you learned about the structure of clay in this chapter, what would you expect the structure of graphite to be?

12.14   If a hard rock and a sample of clay both are made up mostly of silicon and oxygen atoms, why are their properties so different?

12.15   It has been found that original glass windowpanes in very old houses have become thicker at the bottom than at the top, though they were made with an approximately uniform thickness. What does this evidence for very slow flowing suggest about the structure of glass?

12.16   *a.*   How is safety glass made?

   b.   What happens to safety glass in automobile accidents?

12.17   Why is the term "crystal" inaccurate when applied to fine glass tableware?

12.18   If you place a cheap aluminum baking pan in a hot oven, often after a few minutes you can hear a slight gong-like sound. When you open the oven, you see that the pan has warped to a bent shape; yet if you remove the pan to cool, it returns to its original shape.

   a.   Describe what has happened at the molecular level.

   b.   Why does the pan not remain bent?

12.19   An attempt to remove dents from a pewter pitcher that has been dropped simply produces more dents instead of removing the original one. Why?

12.20   Explain how the alternate freezing and thawing of water in cracks of porous rocks can act to break them into smaller pieces.

12.21   Several times in this chapter the comment was made that the presence of so much water in the atmosphere and on the surface of the earth has a modifying influence on climate. Put together an organized discussion of this theme.

12.22   Why is the structure of the H_2O molecule responsible for the fact that you have never seen a cubical raindrop or a square snowflake?

13

Solutions— Molecular-Level Mixtures

- ☐ Why are all solutions mixtures, but not all mixtures solutions?
- ☐ How do pressure and temperature affect solubility?
- ☐ How can we predict whether one substance will dissolve in another?
- ☐ Why are low-level applications of DDT a serious environmental hazard?
- ☐ How do soaps and detergents work?
- ☐ How does the presence of solute affect the boiling and freezing of a liquid solvent?
- ☐ Why are the oceans a huge untapped source of raw materials?
- ☐ What is osmosis and why is it important in living systems?

M any of the substances encountered in daily living are solutions. The air we breathe is a solution, a mixture of gas molecules. The water we drink is a solution containing a small amount of dissolved air as well as dissolved mineral substances. Most of the metals we encounter are alloys containing atoms of various metallic elements.

Often we want to make a solution for a particular purpose. For example, we may want to remove a grease spot from an article of clothing by dissolving it. Water will not do the job, but drycleaning fluid will. Water cannot be used to clean a brush that has been used with oil-based paint. But the paint remaining in the brush washes out if we soak the brush in gasoline or paint thinner. We can dissolve a teaspoonful of table salt in a glass of warm water to make a gargle for a sore throat. In contrast, we find that an aspirin tablet does not dissolve in water.

What happens when something dissolves in something else? A solution is a special kind of mixture. The particles mixed together in a solution are as small as they can be, in the form of molecules or ions. We can expect that the ability of one material to dissolve in another is related to the structure of these units, the molecules or ions of the two substances. Our information about the different types of chemical bonding and the structures of gases, liquids, and solids can be used in a more detailed examination of the properties of solutions.

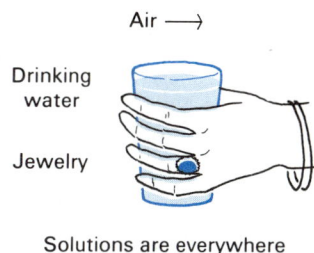

Air ⟶
Drinking water
Jewelry

Solutions are everywhere

SOLUTES DISSOLVE IN SOLVENTS

First we should define some terms that are convenient to use when discussing solutions. The term *solvent* refers to the component of the solution that is in larger proportion. The term *solute* refers to the substance that dissolves when it is added to the solvent. The solute usually is the component of the solution in smaller proportion. The molecules or ions of the solute become separated from one another and are dispersed or scattered throughout the body of the solvent to make the solution. For example, you may use a rag soaked in gasoline (solvent) to dissolve a small smear of tar (solute) from a floor. Or you may add a spoonful of sugar (solute) to a cup of hot coffee. The coffee it-self is a solution of the various substances the hot water (solvent) has dissolved out of the coffee grounds.

The term *concentration* is the measure of the amount of solute in a given amount of solvent. The simplest expression of concentration is percentage by weight. When you buy a 1% solution of iodine at a drug store, you receive a liquid solution the druggist made by mixing 1.0 gram of solid iodine with 99.0 grams of a solution of alcohol and water. Various other ways of expressing concentration are used, some with special meaning. For example, the concentration of ethyl alcohol in liquors is identified as "proof." This usage developed many years ago when a practical test for alcoholic content was to soak gunpowder in the rum or whiskey being tested and then to try to ignite the wet powder. A 50–50 alcohol–water rum or whiskey was the most dilute mixture that would still allow the powder to be burned. So this mixture was called 100 proof. On this scale, then, pure alcohol is 200 proof. Whiskey labeled 90 proof is 45% alcohol.

Most solutions that chemists use when performing reactions for synthesis or analysis are ones in which concentrations are expressed on a *molar* scale. A one molar solution contains one mole of solute in one liter of solution; a 0.10 molar solution contains 0.10 mole of solute in one liter of solution; and so on. Remember that a mole contains Avogadro's number (6×10^{23}) of particles. Therefore, when chemists measure out a volume of a solution with the concentration expressed in molar units, they know how many particles of solute they are working with.

Other terms also are used to describe solutions. *Dilute* and *concentrated* are qualitative terms that can be loosely applied to suggest the concentration of a solution. A dilute solution contains a relatively small amount of solute; a concentrated solution contains a relatively large amount. The more precise terms, *unsaturated* and *saturated,* have more meaning in scientific description. (The use of these terms in describing solutions differs from the way they were used in Chapter 7 to describe hydrocarbons.) Let us illustrate these terms by describing the way a solid dissolves into a liquid.

WHAT HAPPENS WHEN A SOLID DISSOLVES IN A LIQUID?

If you place a large crystal of salt or sugar in a glass of water, it appears to dissolve more slowly than if you grind up the same amount of salt or sugar into a fine powder before you add it to

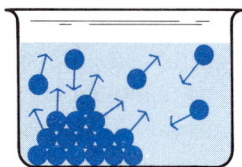

Initially: The rate of particles leaving the solid is faster than the rate of particles returning to the solid (unsaturated).

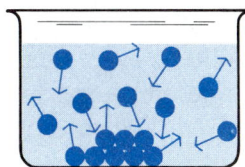

Finally: The rate of particles leaving is balanced equally by the rate of particles returning to the solid (saturated).

Figure 13.1   Dynamic equilibrium of a solid dissolving in a liquid.

the water. Likewise, you find that stirring or shaking the mixture speeds up the dissolving. If you keep adding more solid and continue shaking, more of it dissolves. But eventually you reach a point where additional solid does not appear to dissolve. Figure 13.1 suggests what happens. The particles of the solid break away from the solid and are dispersed in the liquid. They move in random directions. As more and more of them become scattered throughout the solution, the likelihood of some of them hitting the solid and being recaptured by the intermolecular or interionic forces of the solid increases. Eventually a dynamic equilibrium is established; the particles both leave the solid and return to the solid with equal rates.

This picture of dynamic equilibrium in the process of dissolving is quite similar to the one we used to explain the existence of a constant equilibrium vapor pressure over a liquid in a closed container. When this dynamic equilibrium between the solute in solution and the excess remaining undissolved is established, the solution is said to be saturated. A saturated solution holds as much solute as it can, in the presence of excess undissolved solute at that particular temperature. As long as more solute will dissolve, the solution is described as unsaturated.

This model also suggests why a powdered solid dissolves faster than a large crystal: More surface is exposed on the many small pieces, so more particles can escape into the solvent. Mixing also speeds the process because an undisturbed solution soon becomes saturated in the region close to the solid surface. If you mix it or shake it, you move this saturated layer away and bring more fresh solvent near the solid so that more particles can dissolve.

The ability of a solid substance to dissolve in a particular liquid, such as water, is a property characteristic of that substance. The amount that can dissolve to form a saturated solution is called the *solubility* of the substance. At room temperature, a saturated solution of table salt, sodium chloride, contains

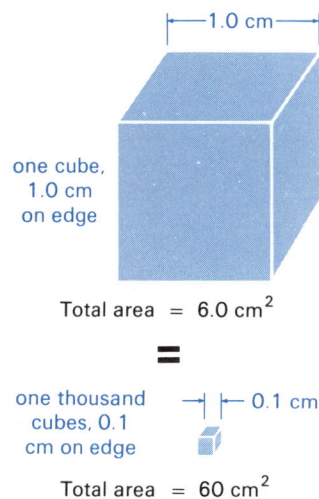

one cube, 1.0 cm on edge

Total area = 6.0 cm^2

=

one thousand cubes, 0.1 cm on edge

Total area = 60 cm^2

380 grams of salt for every kilogram of water. For table sugar, the solubility is even greater, about 2100 grams per kilogram of water. Some substances form saturated solutions when very little has dissolved. For example, a concrete sidewalk dissolves very

A concrete sidewalk dissolves very little in the rain.

little in the rain. It is not quite correct to describe the sidewalk as totally insoluble. Rather, we should say that the solid is only very, very slightly soluble in water. An infinitesimal amount always dissolves, as you can see if you observe a very old sidewalk that has been exposed to weather for many years.

Our use of the terms unsaturated and saturated in connection with solid solutes in liquid solvents is exactly parallel to the way we used these terms when discussing humidity and dew point in Chapter 11. When liquid water evaporates into the atmosphere, water is the solute and the air is the solvent. (The air is already a solution of oxygen, nitrogen, and other gases.) The air becomes saturated with gaseous water molecules when the partial pressure of water in the air is equal to the equilibrium vapor pressure of liquid water. When this equilibrium occurs, no more water can dissolve into the air.

THE EFFECT OF TEMPERATURE ON SOLUBILITY

The model we have used of particles breaking away from solids to become solute particles in liquid solvents should allow us to predict the effect of temperature on solubility. Not only does a hot liquid solvent dissolve a solid solute more rapidly, but also the hot solvent almost always can hold more of the solute in a saturated solution. The solubility of solids in liquids generally increases with increasing temperature. There are a few exceptions to this general rule. These exceptional cases always involve some chemical change in the solute when it leaves the solid form. It becomes a different substance in the solution. We have already described how an increase in temperature means an increase in the thermal energy of the solid particles wiggling and vibrating about their locations in the solid. Consequently, the particles of the solid can overcome the restraining forces and break away to freedom in the solvent more readily. For example, you can always make hot lemonade sweeter than cold lemonade, because the solubility of sugar in hot water is almost double what it is in water at the temperature of ice.

When a substance that is a gas at normal temperature has been dissolved in a liquid, the effect of temperature is different. You know that, when a sip of a cold carbonated beverage warms in your mouth, you feel the bubbles of carbon dioxide gas coming out of solution very rapidly. Another example of the same effect is the much greater frothing and bubbling you observe when you open a warm bottle of a soft drink or beer compared to what occurs when you open a cold one. The gas, carbon dioxide, is less soluble in a warm solution.

The same effect can be observed when water from a faucet is heated in a pan. Both oxygen and nitrogen gases dissolve in water to a small extent. Water supplied to your home has had enough

You can make hot lemonade sweeter than cold lemonade.

contact with air to have formed a saturated solution of the air gases in the water. The saturation level for oxygen in water at 15° C (59° F) is 10 ppm (parts per million). The solubility of nitrogen is slightly lower. If the water temperature rises to 40° C (104° F), the oxygen solubility decreases to about 6 ppm. As the temperature rises, the solubility drops to an even smaller amount. The same is true of the solubility of nitrogen gas in water. This decrease in solubility of the gases of air explains the formation of air bubbles on the bottom of a heated pan before the water temperature has risen to the boiling point.

This same decrease in oxygen gas solubility in water as the temperature rises causes one of the bad effects of thermal pollution. Recall that thermal pollution is the dangerous rise in the temperature of a body of water. The temperature rise is caused by the heat that inevitably must be dumped by a plant that burns fuel to generate electricity. An increase in water temperature means that less oxygen can be held in solution, so that aquatic life is deprived of the oxygen it needs to live. Most fish need a dissolved oxygen level of about 5 ppm.

The decrease in solubility of gases in liquids as the temperature rises relates directly to the fact that increased temperature always increases random molecular motion. As the solute gas molecules move faster, they break away from the solution more readily. Consequently, they tend to go out of solution into the greater freedom of the gaseous form.

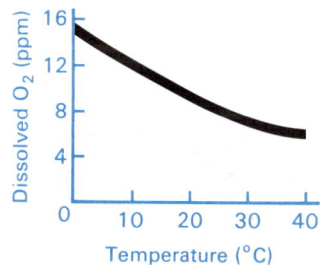

The solubility of O_2 gas in water decreases as temperature increases.

Parts per Million (ppm) and Parts per Billion (ppb)

The term *parts per million,* abbreviated ppm, is an expression of concentration used very often to describe very dilute solutions. The term states how many parts of solute there are in a million parts of the whole solution. Parts per million almost always expresses concentrations on a mass basis. For example, a 10 ppm solution is one in which every million grams of solution contains 10 grams of solute.

The ppm designation is most frequently applied to dilute solutions in water. For example, 1 kilogram (1000 gram) of water contains 1 million milligrams of water; thus

$$1 \text{ kg} = 1 \text{ kg} \times 1000 \, \frac{g}{kg} \times 1000 \, \frac{mg}{g} = 1,000,000 \text{ mg}$$

At normal temperatures, 1 liter of a dilute water solution has a mass of approximately 1 kilogram. So if we have 10 milligrams of solute in 1 liter of solution, we have a concentration of 10 ppm.

$$\frac{10 \text{ mg solute}}{1 \text{ liter solution}} = \frac{10 \text{ mg solute}}{1,000,000 \text{ mg solution}} = 10 \text{ ppm}$$

Thus when we say that the concentration of oxygen in water is 10 ppm, we mean that 1 liter of the solution contains 10 milligrams of dissolved oxygen.

Another example is the measure of fluoride ions to be added to drinking water to make children's teeth more resistant to decay. The appropriate amount is in the range of 0.7 to 1.0 ppm. This means that 1 liter of drinking water should contain only enough soluble fluoride compound to provide 0.7 to 1.0 milligram of fluoride ions.

You can get some idea of how dilute such solutions are by realizing that a 10 ppm solution would mean adding ten drops of vermouth to a 13-gallon martini. A 10 ppm solution also is about the equivalent of ten bites in a lifetime of eating or 50 steps on a hike across the U.S.

In recent years, improvements in analytical techniques have made possible the measurement of solutes at even lower levels than parts per million. In such cases, the unit *parts per billion,* abbreviated ppb, is used. Parts per billion means micrograms of solute per kilogram (or liter) of an aqueous solution. For example, 1 kilogram of water contains 1 billion micrograms of water; thus

$$1 \text{ kg} = 1 \text{ kg} \times 1000 \, \frac{g}{kg} \times 1,000,000 \, \frac{\mu g}{g} = 1,000,000,000 \, \mu g$$

An example of the use of this term is the 50 ppb upper limit of lead concentration allowed in water for domestic purposes by the United States Public Health Service. This figure means

$$\frac{0.000050 \text{ g Pb}}{1 \text{ liter solution}} = \frac{50 \, \mu g \text{ Pb}}{1,000,000,000 \, \mu g \text{ solution}} = 50 \text{ ppb}$$

One ppb is the equivalent of one penny out of 10 million dollars or approximately four persons out of the total population of the world.

PRESSURE AND THE SOLUBILITY OF GASES IN LIQUIDS

The solubility of gases in liquids always increases if the pressure of the gas is increased. Conversely, decreasing the pressure of a gas decreases its solubility in a liquid. This phenomenon also contributes to the bubbling or frothing when you open a bottle of a carbonated soft drink or beer. Carbonated soft drinks are made by forcing carbon dioxide gas to dissolve in the water of the beverage under pressure. In the case of alcoholic beverages, such as beer or champagne, the carbon dioxide gas has been produced by the fermentation that also makes the alcohol in the solution. When a bottle or can of either type of beverage is opened, the outside atmospheric pressure is less than that of the air and carbon dioxide inside the container. Consequently, bubbles of gas always form in the solution and escape. This phenomenon, too, can be explained by imagining the dynamic equilibrium that is established as gas molecules bombard and escape from the surface of the liquid of a saturated solution. Under higher pressure, more gas molecules hit the surface in a given period of time. Consequently, more are captured by the solution. The final equilibrium is established when an equal number leave the solution. If the pressure is lowered, as it is when the bottle is opened, the rate of gas molecules hitting the liquid surface is suddenly decreased, so the rate of molecules leaving exceeds it. Gas leaves the solution.

Pressure equilibrium

CO_2 in solution

CO_2 rapidly comes out of solution

DEEP-SEA DIVING AND THE SOLUBILITY OF GASES

This effect, the increased solubility of gases in liquids as pressure is increased, has a very important physiological consequence for humans who have to breathe air under pressure. Deep-sea divers, or construction workers digging underwater tunnels, or even scuba divers descending to moderate depths, must always allow time to come up to the surface slowly. The weight of the water above a diver at great depths creates a pressure that would collapse the lungs if it were not equalized by the pressure of the air being breathed. (Recall the discussion in Chapter 4.) At this higher pressure, more oxygen and nitrogen from the air dissolve into the blood in the lungs. The same is true of the small amount of argon in the air, but its solubility is so low that the pressure effect is not significant. The blood takes up the additional oxygen through reaction with hemoglobin. But the extra nitrogen gas goes into solution and is carried throughout the body as the blood circulates. If a sudden release of pressure occurs over the whole body, the nitrogen gas comes out of solution from the blood wherever it happens to be. Particularly dangerous is the formation of nitrogen gas bubbles, which may form if the nitrogen comes out of blood in the capillaries. The gas bubbles can cause muscle tensions or even hemorrhaging of blood vessels in the dreaded condition called "the bends." This danger is avoided if the diver ascends slowly. Then the nitrogen-saturated blood has time to recirculate through the lungs, where a gradual shift of the dynamic equilibrium between dissolved nitrogen in the blood and gaseous nitrogen of the air can take place.

For many years, divers who have to descend to great depths have been supplied with synthetic air made up of the necessary oxygen diluted with helium instead of nitrogen. The characteristic solubility of helium in water is much less than that of nitrogen gas. Since less helium than nitrogen dissolves in the blood under comparable high pressure, bubbles are less likely to form in the capillaries.

PREDICTING WHAT WILL DISSOLVE IN WHAT

A broad generalization used as a first basis for predicting what kinds of substances will dissolve in others is "like dissolves like." This rule is not at all unexpected. Consider this analogy: Suppose you go to a party where all the guests have similar ages, interests, and cultural backgrounds. Even though they may not have known one another before, soon they begin to mingle freely. Instead of small, separated, "insoluble" groups, the whole crowd becomes a "solution" of homogeneous interactions. Conversely, a loner may remain "insoluble" all evening because he or she has none of the common interests of the rest of the group. So it is with molecules or ionic substances interacting with solvents to form solutions. If the characteristic structure of a solute is comparable to that of the solvent, a solution will probably be formed. If there is a marked difference between the structures, very little dissolving can take place.

Early in our discussion of the structure of atoms and molecules, we recognized two different ways compounds are formed. In one type, atoms transfer electrons in such a way that ions, electrically charged particles, are formed. The resulting compounds are held together by electric forces of attraction between positive and negative ions. In contrast to this, other compounds are formed of molecules in which atoms share pairs of electrons with other atoms next to them. The covalent molecules also exert some intermolecular attractions on one another. We have learned how the strength of these intermolecular attractions determines whether a substance is a gas, a liquid, or a solid at normal temperatures. Now we can elaborate our model just a bit further to recognize that two kinds of molecules with similar intermolecular forces probably will be soluble or will readily mix with each other. The more similar the two sets of intermolecular attractions, the more compatible and therefore the more soluble the two substances will be in each other.

We would not expect a solid ionic compound to dissolve in a substance made up of covalent molecules. The electric forces of attraction among ions are quite different from the intermolecular attractions holding covalent molecules together. Water, although it is made up of covalent molecules, does dissolve ionic compounds. We will explain this in a later section. Water acts as a solvent in what are called *aqueous solutions*. (*Aqua* is the Latin word for "water.") Before describing how these solutions are formed, let us examine some generalizations about the other broad range of possibilities, the *nonaqueous solutions*

THE ACTION OF DRYCLEANING FLUIDS

A very practical example of the formation of nonaqueous solutions is the use of organic molecules to remove dirt from fabrics. Most stains on fabrics, such as those made by grass, gravy, oil, or grease, involve organic molecules. Dirt is also held onto fabrics chiefly by oil-like or grease-like substances. Consequently, the idea that like dissolves like can be used to choose a solvent to remove the stain or grease. Recall from the discussion in Chapter 7 that oils and greases contain large molecules with formulas such as $C_{20}H_{42}$ or $C_{30}H_{62}$. These are higher members of the same homologous series in which gasoline-like hydrocarbons are found. So we would expect the molecules in oil and grease to be compatible with other hydrocarbon molecules and to dissolve in such solvents.

Ordinary gasoline for automobile engines is not satisfactory as a drycleaning fluid because of the other substances, such as lead tetraethyl, added to improve engine performance. Not only are these hazardous if breathed into the lungs, but they tend to stay in the fabric and leave a ring. However, special mixtures of pure gasoline-like hydrocarbons are used for drycleaning.

Other hydrocarbon molecules modified to contain chlorine, or chlorine and fluorine, are also used. At one time, carbon tetrachloride, CCl_4, was used extensively. However, when this molecule enters a person's lungs, its very ability to dissolve organic materials makes it hazardous. CCl_4 tends to dissolve in the fatlike materials of cell walls. Once in the system, it quickly leads to liver and kidney disorders. Consequently, CCl_4 is used less frequently now in favor of other drycleaning fluids. All similar chlorine-containing substances are toxic, but some are less so. Two typical compounds other than hydrocarbons used for drycleaning fluids have the following structures and names:

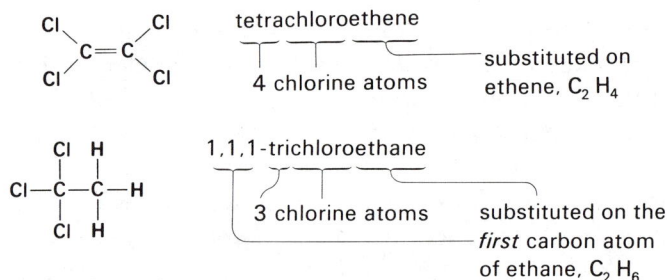

The molecules in drycleaning fluids do not interact with the molecules of the fibers in wool or cotton. Although fabrics soaked in them become wet, the liquid does not adhere as it does when water wets cloth. Hence these "dry" cleaning solvents can readily be recovered to be used again. Some of the newer synthetic fabrics do tend to dissolve slightly in some drycleaning fluids. Special solvents must be chosen to clean such fabrics. It will pay you to heed the warning on the fabric label.

SOLVENTS FOR PAINTS AND ENAMELS

Varnish and oil-based paints used chiefly for covering surfaces that will be exposed to weather are complex mixtures of organic molecules. A typical paint consists of a pigment, or coloring material, mixed with an oil that forms a hard film when the paint dries, and a thinner or solvent that allows the paint to be brushed or spread out. The thinner evaporates as the paint dries. The thinners used in paints are mixtures of hydrocarbon molecules, such as turpentine, $C_{10}H_{16}$, or larger members of the same homologous series from which gasoline is made. Consequently, such paints cannot be washed out of brushes or clothing by using water. Instead, an appropriate solvent that has a molecular structure similar to the thinner must be used. Gasoline, kerosene, and turpentine can act as solvents for paint and varnish that have not yet dried to a hard film.

Nail polish, similarly, is a mixture of organic compounds. The film formed when nail polish dries is chiefly made of many molecules of nitrocellulose strung together. The formula of nitrocellulose can be represented as $(C_6H_7N_3O_{11})_{500-600}$. (Between 500 and 600 of the units represented by the formula within the parentheses are fastened together.) The solvent used as nail-polish remover consists mostly of the compound acetone, a liquid with the structural formula that follows:

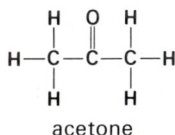

$$
\begin{array}{ccccc}
 & H & O & H & \\
 & | & \| & | & \\
H- & C & -C- & C & -H \\
 & | & & | & \\
 & H & & H &
\end{array}
$$

acetone

Acetone is capable of dissolving many different kinds of organic molecules. Plastics, so much in use to make articles and synthetic fabrics, are all organic molecules. We will discuss some of the structures of the molecules of plastics in a later chapter. Some of these plastics are quite soluble in acetone. This is why an accidental spill of nail-polish remover may damage a fabric or article made of plastic.

THE DDT PROBLEM

The environmental problem of having *pesticides* (poisons that kill insect pests) such as DDT accumulating in the bodies of birds is related to this same solubility. DDT is the abbreviation of the chemical name *d*ichloro*d*iphenyl*t*richloroethane.

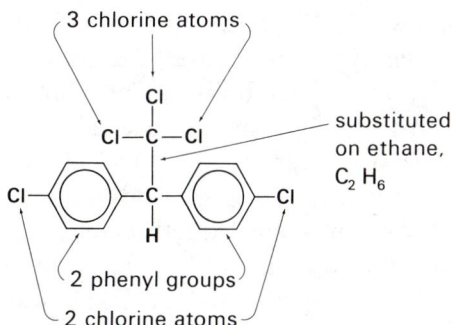

Compare the structure of carbon tetrachloride.

$$
\begin{array}{ccc}
 & Cl & \\
 & | & \\
Cl- & C & -Cl \\
 & | & \\
 & Cl &
\end{array}
$$

We have already mentioned that the molecule carbon tetra-chloride, CCl_4, is soluble in the fatty tissue of living cells. You can see that there is some similarity between the structure of CCl_4 and DDT with all its chlorine atoms. Consequently, we can understand why DDT tends to dissolve in the fatty part of animal tissues. These same structural features are responsible for the fact that very little DDT can dissolve in water, a substance with a very different molecular structure.

The mechanism of DDT's action as a pesticide is not clearly understood. Its widespread use began in the 1950's. Many valuable crops were saved from ravishing insects, and many epidemics of insect-carried diseases were prevented. But what was not realized until a few years ago was the persistence of these molecules in the environment. They are not biodegradable; microorganisms cannot use them for food. Consequently, DDT sprayed on leaves remains, eventually to become part of compost, collected decaying plant material. When worms eat the compost, DDT enters their systems. Birds, in turn, ingest the DDT by eating the worms.

Because of its characteristic solubility, DDT is stored away in the tissues of an organism rather than cast off in aqueous wastes.

Because of its solubility, DDT is stored away in the tissues of an organism rather than cast off in aqueous wastes.

As a consequence, at each stage of the food chain (compost → worms → birds), the amount of DDT retained by the organism increases. For example, a bird may eat hundreds of worms, and the small amount of DDT in each worm is thus multiplied hundreds-fold in the bird.

Large amounts of DDT harm birds and higher animals by changing the action of some enzymes, the catalysts essential for the complex reactions of metabolism. For example, DDT is responsible for a bird's being unable to use calcium compounds to make hard shells on the eggs it lays. The eggshells break prematurely and no offspring are produced. Thus some species of birds, such as brown pelicans, are threatened with extinction. Realization of this harmful pattern has led to very stringent regulations against the widespread use of DDT and other related organic molecules for pesticides. No conclusive evidence has yet linked DDT with harm to humans. However, it is suspected as a substance that can cause some kinds of cancer.

The slight tendency of DDT to dissolve in water was thought at first to be a protection against its spreading to harm forms of life other than insects. However, the concentration of DDT in the fatty tissue of animals and some plants, and the increase in this concentration by a food chain, has exploded this complacent view. Although the water of a lake may contain only 0.01 ppm of DDT, the microscopic animal and plant life, the plankton, may contain as much as 5 or 6 ppm of DDT. Small fish feeding on this plankton are eaten by larger fish to further increase the concentration. The larger fish, then eaten by birds, may have as much as 2500 ppm of DDT in their flesh. No wonder such a diet is lethal to some susceptible species of birds!

Worms eat dead leaves; birds eat worms. DDT concentrates at each step of the food chain.

HOW DOES WATER ACT AS A SOLVENT?

The structure of the water molecule also is responsible for its solvent properties. Figure 13.2 suggests how the dipolar H_2O molecules interact with the ions released into solution when water dissolves table salt, NaCl. Dissolving probably starts with the Na^+ or Cl^- ions on a corner of the solid. These ions are the ones least securely held, because they are least surrounded by ions of opposite charge. An analogy is the way writing wears down a sharp pencil point more rapidly than a dull one. As each Na^+ ion leaves, it becomes surrounded by H_2O molecules. The solvent molecules right next to the ions are not just randomly arranged. Rather, the electric attractions of the positively charged sodium ions cause the negative side of the H_2O dipole to be pointed toward the Na^+ ions. The reverse orientation of the H_2O dipoles occurs around the Cl^- ions. The net result is that each Na^+ ion is somewhat insulated by its covering sphere of H_2O molecules from the attraction of any Cl^- ion that is similarly covered. Therefore, the ions stay separated in the water solution.

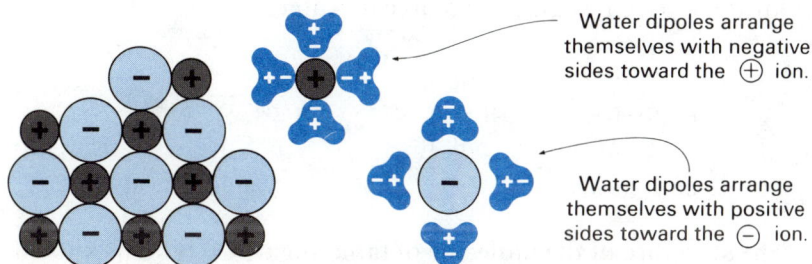

Water dipoles arrange themselves with negative sides toward the ⊕ ion.

Water dipoles arrange themselves with positive sides toward the ⊖ ion.

Figure 13.2 An ionic solid dissolving in water.

This picture of NaCl dissolving is typical of the way various ionic solids tend to dissolve in water. But by no means are all ionic solids highly soluble in water. Other factors—such as the strength of forces within the crystal, the relative size of the ions, the size of the charges on the ions, and the tightness of the sphere of H_2O molecules around them—are involved in determining the extent to which ionic compounds will dissolve into water. For example, calcium carbonate, the other ionic solid whose crystal structure we described in Chapter 12, is almost insoluble in water. The strong electric forces of attraction of the doubly charged ions seem to dominate over other factors. Only 0.014 gram of $CaCO_3$ can dissolve in one liter of water at room temperature. The comparable solubility of NaCl is 380 grams in one liter of water.

Another mechanism by which water acts as a solvent is the formation of hydrogen bonds with some kinds of molecules. Recall the discussion in Chapter 11 of how the H atoms on one H_2O molecule are attracted by the pairs of electrons on the O atom of another H_2O molecule. This same kind of attraction occurs when a potential solute molecule contains an OH group. For example, the molecules of the liquid ethyl alcohol dissolve very readily in water. Ethyl alcohol and water mix in all proportions. "Pure" grain alcohol is a solution of 5% water in 95% alcohol. The structure of ethyl alcohol, C_2H_5OH, is

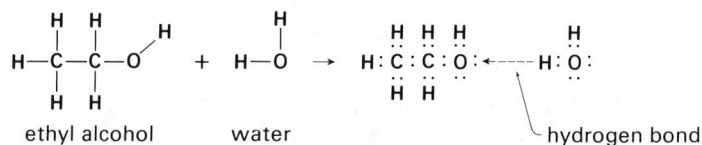

In spite of the fact that the C_2H_5 portion of the ethyl alcohol molecule is typical of an organic compound, and hence not likely to be compatible with the dipolar H_2O structure, the attractive influence of the hydrogen bond has the effect of making the whole molecule soluble in water. A comparable molecule, ethyl chloride, C_2H_5Cl, is quite insoluble in water.

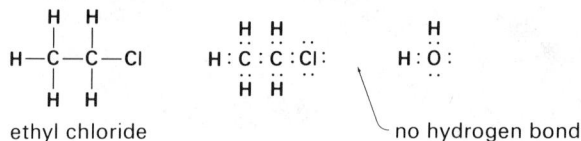

The structure of the molecule of table sugar, $C_{12}H_{22}O_{11}$, known

by the chemical name *sucrose*, suggests that the same mechanism probably is responsible for its solubility in water.

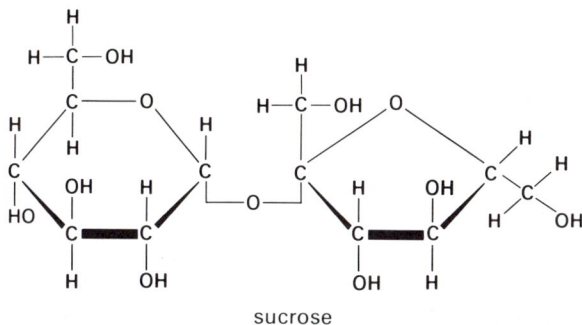

sucrose

In this big molecule there are eight OH groups fastened onto the backbone of strung-together carbon and oxygen atoms. With so many sites for possible hydrogen bonds, sucrose is very soluble in water. In fact, some sugar syrup can be so concentrated that it is more accurate to consider water to be the solute in the sugar solvent.

Some substances dissolve in water by still another mechanism. Water acts as a solvent for some compounds because it reacts chemically with them. When water and these compounds react, the solute is actually changed into new particles, different from those present in the undissolved material. This kind of reaction occurs when the gases carbon dioxide, CO_2, or ammonia, NH_3, dissolve in water. We will examine this process in more detail in the next chapter.

HOW DO SOAPS AND DETERGENTS WORK?

When soaps or detergents are added to water, the ability of the resulting solution to wash away fat or grease is greatly increased. The key to understanding this very useful cleaning method is also the recognition of how the structure of molecules affects their ability to mix with and hence dissolve other molecules. Various kinds of soap have been made and used for centuries. Pioneer families and some who live close to the land today make soap by boiling beef tallow or other kinds of fats with lye. Lye is the compound sodium hydroxide, $NaOH$. $NaOH$, in solid form or water solution, consists of Na^+ ions and OH^- ions. Fats are large organic molecules; $C_{57}H_{110}O_6$ is a typical formula. We will describe the structure of fats in detail in a later chapter. At this point, we need only recognize that the reaction between fats and lye makes soap, a compound that has a structure like the following:

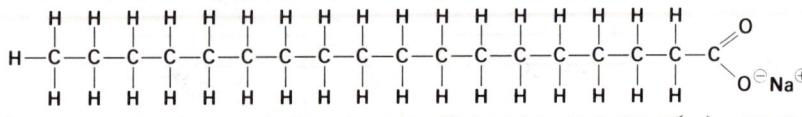

This long covalently bonded part of the molecule dissolves in fat or grease.

This end of the molecule forms ions and dissolves in water.

A great variety of *detergents,* sometimes called "synthetic soaps," are now available. Detergents differ from soaps in that they are not made from fats but rather from materials derived from petroleum. However, both soaps and detergent molecules have the same essential structural features. One end of the molecule is capable of forming ions. The rest of the molecule is a long string of carbon and hydrogen atoms connected with covalent bonds. These two contrasting structures in one molecule allow the molecule to act as a connecting link between fat or grease and water. One end of the molecule dissolves in fat or grease; the other end dissolves in water. The fat or grease is broken up into very small globules that can be imagined to look like the drawing in Figure 13.3. The soap or detergent molecules, because of their dual solubility properties, are arranged around a grease globule like cattle crowded around a feeding trough. The individual globules of grease formed this way may be too small to be seen, but they *are* there. This is an *emulsion.* An emulsion is a dispersion of small particles of one liquid in another. Milk and mayonnaise are emulsions, as is paint before it dries. Although the particles are too small to be seen, they do tend to scatter light, so an emulsion always has a milky or opaque appearance. The individual grease globules are not likely to coalesce, or come together into larger droplets, because each has a shield around it formed by the water-soluble ends of the soap or detergent molecules.

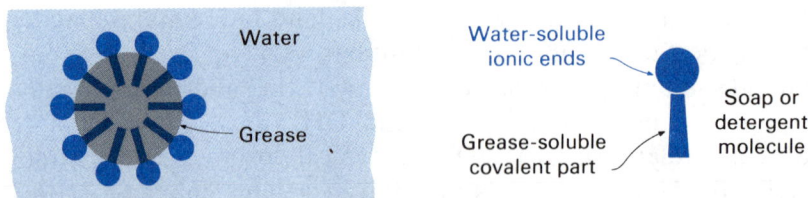

Figure 13.3   Soaps or detergents breaking up grease into small globules in water.

Water

Grease

Water-soluble ionic ends

Grease-soluble covalent part

Soap or detergent molecule

The double solubility characteristic of soap or detergent molecules has another useful effect, especially when they are added to water used to wash fabrics. The soap or detergent tends always to concentrate in the surface of the water. The water-soluble ends of the molecules are in the surface; the less-soluble, covalent ends of the molecules stick out of the surface. This arrangement of the molecules greatly lowers the surface tension of the water. With a lower surface tension, the wash water is able to penetrate much better into the small pores of a fabric. So the wash water wets the fabric more readily. Consequently the dirt, usually held by fat or grease, can be loosened from the fabric with greater ease.

The lower surface tension of soapy water is also responsible for the formation of soap bubbles or suds that can hold trapped air. Pure water does not form such bubbles; the surface tension is so strong that bubbles pull together and collapse as soon as they are formed.

THE ELECTRIC CONDUCTANCE OF WATER SOLUTIONS

Everyone knows the danger in exposed wires that carry electric current. If a person touches exposed wires carrying the usual 110-volt current in a home or building, he or she will get an electric shock and could be killed. The reason is that the moist skin of a person makes as effective a carrier of the electricity as a wire or metal object would if placed to connect the two current-carrying wires. You may have experienced the surprise of causing an electric appliance or automobile battery to short-circuit by accidentally immersing it in or splashing it with water. If a "hot" wire is knocked down by a storm, no one in bare feet should be on the wet ground nearby. The possibility of electric shock is too great a chance to take. Water appears to be a good conductor of electricity. Why? Not all liquids conduct electricity. Some high-voltage transformers are filled with liquid oil because it is an even better *insulator* than air.

If you were to test very pure distilled water in a laboratory experiment to see if it can conduct electricity, you would find that it too is a very poor conductor. *Water conducts electricity well but only when it contains dissolved ionic compounds.* The reason that the moisture on your skin, such as perspiration, conducts electricity is that it always contains dissolved salt and other body substances that form ionic solutions. The salty taste of perspiration comes from these dissolved substances. The same is true of most natural water, such as rain in a puddle. Rainwater contains dissolved substances from the ground. Oil, usually a mixture of covalent carbon compounds, can act as an insulator in a transformer because it cannot dissolve substances made of ions. How then do aqueous solutions of ionic compounds conduct electricity?

Let us examine this process with the aid of Figure 13.4, a diagram of the apparatus used in a laboratory demonstration of *conductance*. A battery is connected by wires to a light bulb and to two pieces of metal. If the two pieces of metal touch each other, the bulb glows, showing that current is being carried through the circuit. The battery serves as an electron pump. Electrons are forced onto one piece of metal (called the negative *electrode*) and removed from the other (called the positive electrode). When the two electrodes are immersed in a container of pure distilled water, but are not allowed to touch each other, the bulb does not glow. No current is being carried. The separation between the electrodes acts the same way as cutting the wire or opening a switch in the circuit. Next a small amount of table salt, NaCl, is added to the water. As soon as the salt dissolves, the bulb glows. The salt solution conducts the current across the gap between the electrodes.

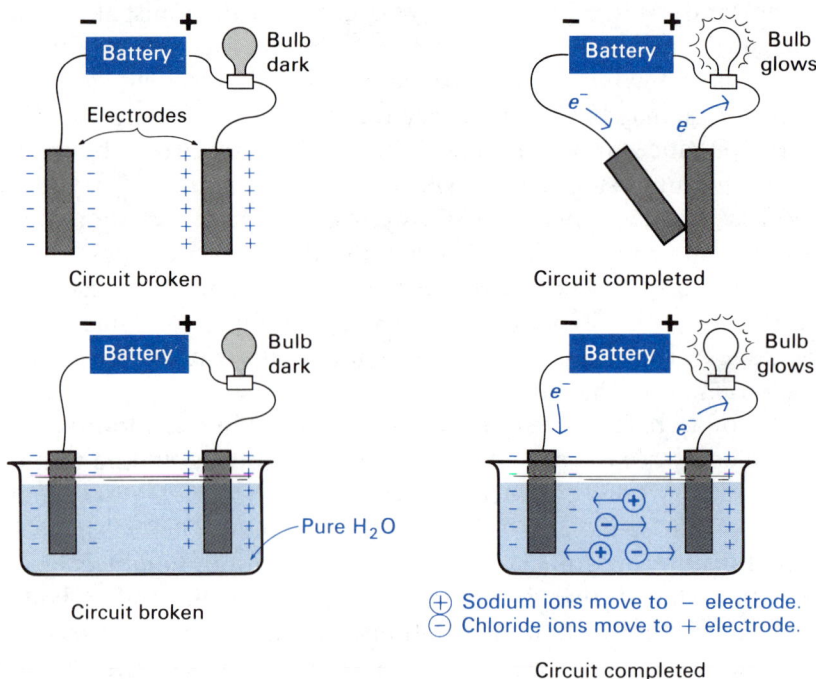

Figure 13.4   Conductance in a solution of an ionic compound in water.

Even though we cannot see the ions of the salt in the solution, we can imagine that the charge on the positive ion means that it will be attracted by the negative electrode. Consequently, it moves through the solution toward that electrode. (Remember, the name *ion* for such an electrically charged particle comes from the Greek word that means "to go.") The negatively charged ions move in the opposite direction through the solution. Thus the electric current is in effect carried through the solution by the moving ions. This kind of conductance is quite different from the flow of electrons through a wire or other piece of metal. But the effect is the same.

The movement or migration of the ions of the solute through the solution only partly explains how the electrical circuit is completed. A simplified picture of what happens at the electrodes is suggested by Figure 13.5. When the positively charged ions reach the surface of the negative electrode, they take on electrons. This addition of electrons changes the positive ions into electrically neutral atoms. These atoms are no longer attracted by the negative charge on the electrode, and so they can leave it to make room for other positive ions to migrate to the electrode surface. While this process is going on, a complementary process is occurring at the positive electrode. There the negatively charged

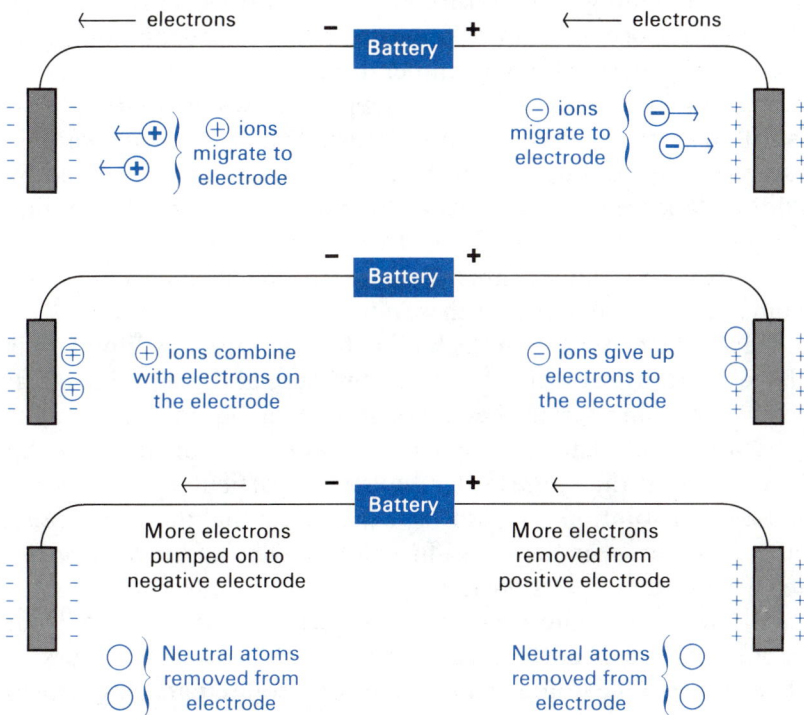

Figure 13.5   The process of electrolysis.

ions come to the electrode surface and give off electrons. This process changes the negative ions into electrically neutral atoms. These atoms then remove themselves from the electrode surface to make room for other migrating negative ions.

You should note that this kind of conductance produces a chemical reaction. In completely general terms, we have the net overall reaction that follows:

At −electrode: A^+ ions + electrons *from* the battery → A atoms
At +electrode: B^- ions → B atoms + electrons *to* the battery
Overall reaction: A^+B^- → A + B

Such a chemical reaction, caused by the action of an electrical current, is called an *electrolysis*. The compound AB has been changed into its separate constituents, A and B.

THE BOILING TEMPERATURES OF SOLUTIONS

Let us return to a further discussion of some generalizations about the properties of liquid solutions. If you have ever boiled a sugar solution or syrup to make candy, you know that the boiling temperature is by no means constant. As the water boils off into steam, the boiling temperature of the syrup rises. This occurs as the solution becomes more concentrated. Instead of boiling at 100° C (212° F), the boiling temperature of the solution may go as high as 145° C (293° F). The freezing point of water solutions also is different from that of pure water. Damage to automobile radiators and cooling systems in cold weather can be avoided by adding "antifreeze" solutions to the radiator water. The cooling system, thus protected, will not freeze to ice even when the temperature drops many degrees below zero. What underlying phenomenon is responsible for these effects?

Recall the definition of the boiling temperature of a liquid. It is the temperature at which the vapor pressure of the liquid is equal to the opposing atmospheric pressure. If a liquid has a vapor pressure lower than atmospheric pressure, it will not boil. So when we find that pure water boils at a particular temperature but a water solution does not, we can suspect that the vapor pressure of water over the solution is lower. *The addition of solute particles always lowers the vapor pressure of the solvent in the solution.* The more solute particles are present, the lower the solvent vapor pressure. Figure 13.6 suggests why this lowering occurs. If the surface of the liquid solution contains solute particles instead

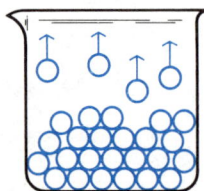

Pure solvent

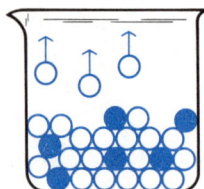

Solution

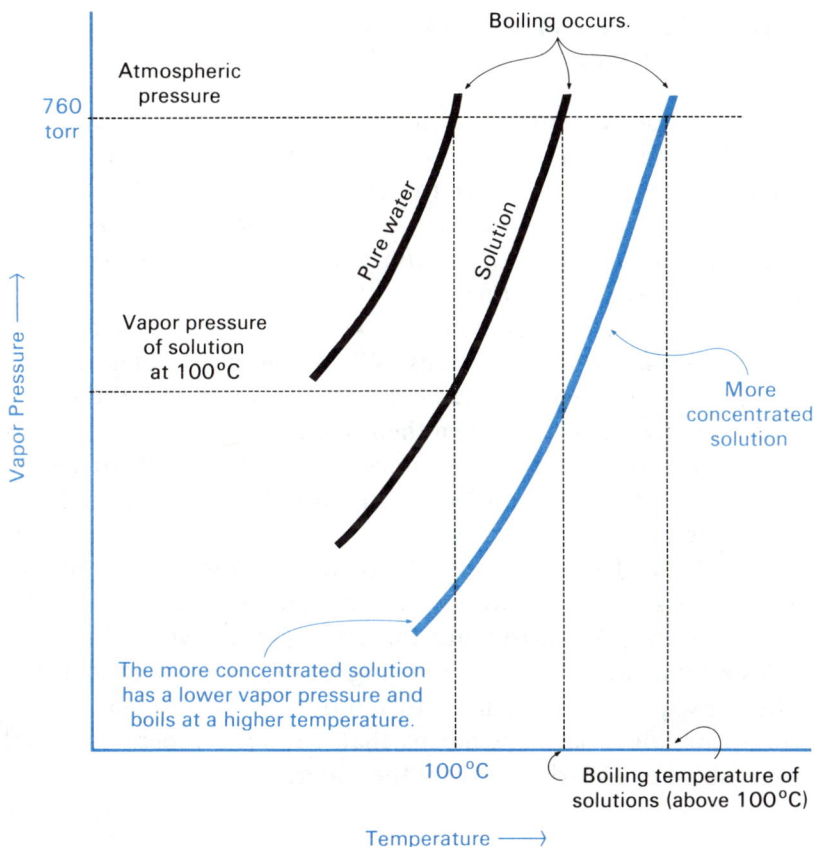

Figure 13.7   The vapor pressure and boiling temperature of pure water and aqueous solutions.

of solvent molecules, fewer of the solvent molecules are in a position to escape from the liquid.

Recall how we interpreted the relationship between boiling and vapor pressure with the aid of Figure 11.4. Figure 13.7 is a comparable diagram suggesting how the vapor pressure of the water (solvent) changes with the temperature of the solution. A solution must be heated to a higher temperature (above 100° C) in order for the vapor pressure of the water in the solution to reach a value equal to the opposing atmospheric pressure. If more solute particles are added, the vapor pressure is lowered still more. Consequently, the solution must be heated to a still higher temperature for boiling to begin. *The addition of solute particles always raises the boiling temperature of a solvent in a solution.* This relationship between vapor pressure lowering and rise in boiling temperature explains why the boiling temperature of a syrup containing dissolved sugar rises as the syrup becomes more concentrated.

THE FREEZING TEMPERATURES OF SOLUTIONS

A different phenomenon is observed when the freezing temperature of a solution is compared with the freezing temperature of the pure solvent. Pure water freezes to ice at 0° C. The freezing temperature of an aqueous solution is always lower than 0° C. *The addition of solute particles always lowers the freezing temperature of a solvent in a solution.*

We can interpret these observations with the aid of our model of molecules and their motions. When a liquid freezes, kinetic energy must be removed from the molecules so that they no longer are free to move. When their average kinetic energy drops below a certain amount, the forces of intermolecular attraction take over and hold the molecules in place. If invader particles of solute are present, as in a solution, these particles also must be slowed down. The solute particles in an aqueous solution do not go into the solid ice. However, they must be slowed down so that their collisions do not disrupt the arrangement the water molecules establish as they are changing from liquid to solid form. The consequence of the need to remove this additional kinetic energy of solute particles means that a lower temperature must be established in order to freeze the water.

CHANGING ICE TO SLUSH

You have encountered one very practical use of adding solutes to lower the freezing point of a solution. When salt, NaCl, or other ionic compounds, such as calcium chloride, $CaCl_2$, are put on ice-covered roads or sidewalks, the ice melts to slush, provided the weather is not *very* cold. When soluble ionic compounds are put on ice, they begin to dissolve in any film of moisture that is present. This dissolving makes a solution in which the water has a very low vapor pressure. Remember that ice, too, has a vapor pressure. This is why ice tends to sublime and disappear on a windy day when the air temperature is slightly below 0° C.

When ice and pure liquid water are in contact at 0° C, a dynamic equilibrium is established. This equilibrium is suggested by part of Figure 13.8. The other part of Figure 13.8 suggests what happens when ice is in contact with a water *solution* at 0° C, as when salt sprinkled on ice mixes with the melted water. The rate of evaporation of water molecules from the liquid solution is less than the rate of sublimation of water molecules from solid

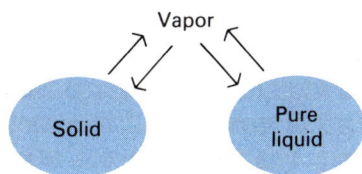

The rate of molecules leaving the liquid and the rate of molecules leaving the solid are both equal to the rate of molecules leaving the vapor. Dynamic equilibrium is established among all three forms.

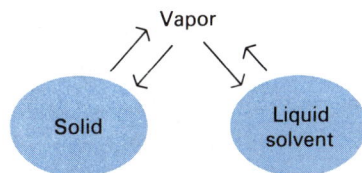

The rate of molecules leaving the solid is equal to the rate of molecules leaving the vapor. However, the rate of molecules leaving the liquid is slower. Consequently, the lack of balance means no dynamic equilibrium is established. The solid gradually changes to liquid by way of the vapor form.

Figure 13.8   Rate of sublimation compared with evaporation rate from pure liquid and a solution.

ice. So no equilibrium among the three forms can exist. The overall result is that water molecules leave the solid and go into the solution. The ice changes to liquid solution by way of the vapor form, just as if it had melted directly. For equilibrium to be reestablished, the temperature of the ice must fall until its sublimation pressure is equal to the vapor pressure of the solution. In effect, the presence of the solution lowers the melting temperature of the ice. Instead of an ice-covered surface, the sidewalk or road becomes covered with slush. If the surrounding temperature is very cold, the slush may refreeze. This happens when the temperature drops below the freezing temperature of the salt solution.

ANTIFREEZE SOLUTIONS

The use of an antifreeze solution instead of pure water in an automobile cooling system is important because, if the water should freeze to ice, the resulting expansion would create pressures that might crack parts of the engine. The liquid substances most frequently used as antifreeze are methyl alcohol, ethylene glycol, or propylene glycol.

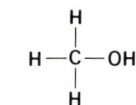

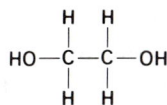

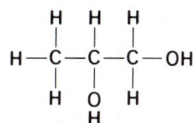

methyl alcohol   ethylene glycol   propylene glycol

All these liquids mix completely with water in all proportions.

You can expect this dissolving in light of the ability of H_2O to form hydrogen bonds and, consequently, to mix with covalent molecules containing OH groups. Methyl alcohol has the disadvantage of boiling at 65° C (153° F), so it tends to distill out of the cooling system when the engine is hot. Ethylene glycol boils at 198° C (389° F). Propylene glycol boils at 189° C (372° F). These temperatures are above the cooling system's temperature, so they do not boil out of the solution. Consequently, solutions containing either ethylene glycol or propylene glycol are called "permanent" antifreeze. Solutions of either compound can be made concentrated enough to protect cooling systems to temperatures as low as −40° C (−40° F). The solutions supplied as commercial antifreeze also contain compounds that help to prevent corrosion and others that can plug up small leaks in the cooling system.

Wrinkled
prunes
in syrup

Prunes swell
up in pure
water.

OSMOSIS

Another property of solutions very important for all living things is the way solutions behave when they are contained by membranes, such as the walls of plant or animal cells. If you soak prunes or raisins in pure water, they quickly lose their wrinkled form and swell up. If, however, they are soaked in a sugar solution or syrup, they remain shriveled up.

The same phenomenon can be observed with red blood cells that have been separated from the plasma (the aqueous portion) of blood. Placed in pure water, red blood cells absorb so much water that they swell up and eventually burst. Placed in a 0.9% sodium chloride solution, red blood cells retain their normal shape. A salt solution of this concentration is referred to as an *isotonic saline* solution. (*Saline* means "salt"; *isotonic* means a solution with the "same", *iso,* effective concentration as body fluids.) Blood plasma is a complex solution of ionic substances such as salt and of modified covalent substances, such as proteins, carbohydrates, fats, and other biochemically important molecules.

Red blood cell in
isotonic saline
solution (0.9% NaCl)

Red blood cell
in concentrated
saline solution

Red blood cell
in pure water

The walls of cells, like the covering of prunes or raisins, are membranes through which molecules of water can pass, but not the solute molecules or ions. Such membranes are called *semipermeable* membranes. (*Semi* means "half"; the solvent "half" of a solution can permeate or pass through, but not the solute "half.") The passage of water or some other solvent across the semipermeable membrane is called *osmosis*. Figure 13.9 suggests what happens. The molecules of solvent pass through the membrane and cause the solution to become more dilute. As more solvent is added to the solution by this process, the volume of the solution increases. If the solution is held in a closed container with a tube extending upward, the liquid rises into the tube. The weight of this height of liquid results in an additional pressure on the body of the solution and on the membrane. This pressure is called the *osmotic pressure*. In the case of the prunes or raisins in water, the osmotic pressure causes them to swell. In the case of the red blood cells, the osmotic pressure created by the dilution of the cell contents may exceed the strength of the cell walls, and hence the cells may burst.

Osmosis is one of the mechanisms by which sap rises in trees. Water enters the roots and passes through the membranes in which the sap is held. The resulting osmotic pressure forces the sap upward in the tree.

In osmosis through semipermeable membranes, solvent molecules always move from a pure solvent or dilute solution into a more concentrated solution. The mechanism of osmosis is not entirely understood. In simple terms, we can think of the membrane as a fence through which solvent molecules can pass, but not solute particles. The kind of solute particles makes no difference. The presence of solute molecules or ions in a solution produces an osmotic pressure that is greater the more solute particles there are in the solution, regardless of kind.

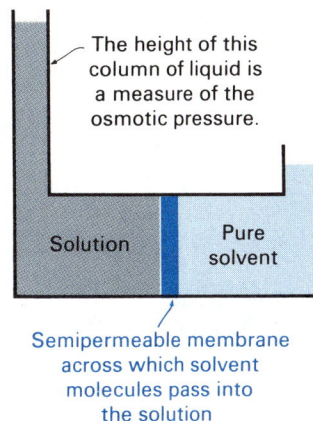

The height of this column of liquid is a measure of the osmotic pressure.

Solution   Pure solvent

Semipermeable membrane across which solvent molecules pass into the solution

Figure 13.9   Diagram illustrating osmosis and osmotic pressure.

OSMOSIS IN PHYSIOLOGICAL PROCESSES

The action of our body organs, such as the kidneys, involves processes regulated by the osmosis of water to and from solutions across membranes. The cells of the body are so sensitive to osmotic processes that any injection into body tissues is always made with a solution that has an osmotic pressure equal to that of blood plasma. (The injected solution must have the same overall concentration of dissolved particles, the same proportion of solute, as normally is present in blood plasma and fluids in the tissues.) Otherwise, the injected solution causes the cells to swell. The cells shrink if the injected solution contains more solute than normally found in the tissue fluids. In shrinking, some water in the cells tends to pass outward through the cell walls to dilute the injected solution. The necessary balance of osmotic pressure is the reason that isotonic saline solutions are used for injections.

Similarly, solutions of glucose, a kind of sugar normally present in blood plasma, can be injected into the bloodstream. Such glucose solutions are made so that the glucose concentration gives the solution the same osmotic pressure as that of blood plasma. Then no damage results to the red blood cells.

THE OCEANS—A SOLUTION OF RESOURCES

Nearly 150 million square miles of the earth's surface is covered by oceans, to an average depth of a little more than 2 miles. Approximately 97% of the water of the earth is in its oceans. This vast quantity of ocean water, more than 300 million cubic miles, is a complex solution of water and mineral substances dissolved over the billions of years since the oceans first appeared on the earth's surface. Ocean water contains approximately 3.5% dissolved solids by weight. As we would expect from the solvent characteristics of water, most of the dissolved material in the ocean consists of ionic compounds. The ionic compound in largest concentration is NaCl, the familiar salt that gives the ocean its taste. In fact, 1 cubic mile of ocean water contains approximately 130 million tons of sodium chloride. The same cubic mile of ocean water contains 15 different elements in amounts greater than 1000 tons. Another 26 elements are found in amounts of 1 ton or more. Figure 13.10 indicates which elements are found in these amounts in ocean water. Probably each of the rest of the naturally occurring elements are present to the extent of at least a few pounds.

Undoubtedly, in the future, as land-based supplies of ores and other minerals become increasingly scarce, we will turn to the oceans as a source. The problem, of course, is the high cost of removing a particularly desired dilute solute from such a large quantity of water solvent. At the present time, salt is the only substance "mined" from the ocean on a large scale. The water is evaporated from the dissolved solids by the free energy of sunshine. The desired NaCl then must be separated from the other less-concentrated ingredients.

Two other elements, bromine and magnesium, also are removed in compounds from the ocean on a small commercial scale. The cost of any process involving recovery of dissolved substance is an energy cost. To recover a dilute substance, a great deal of ocean water must be pumped through a plant, and the desired dissolved substances must be removed by chemical reactions as the water goes through.

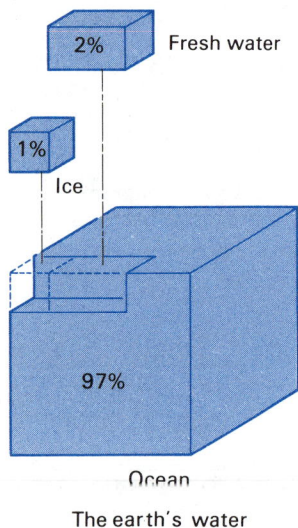

2% Fresh water
1% Ice
97% Ocean

The earth's water

1 kg ocean water

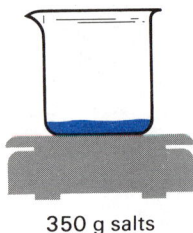

350 g salts

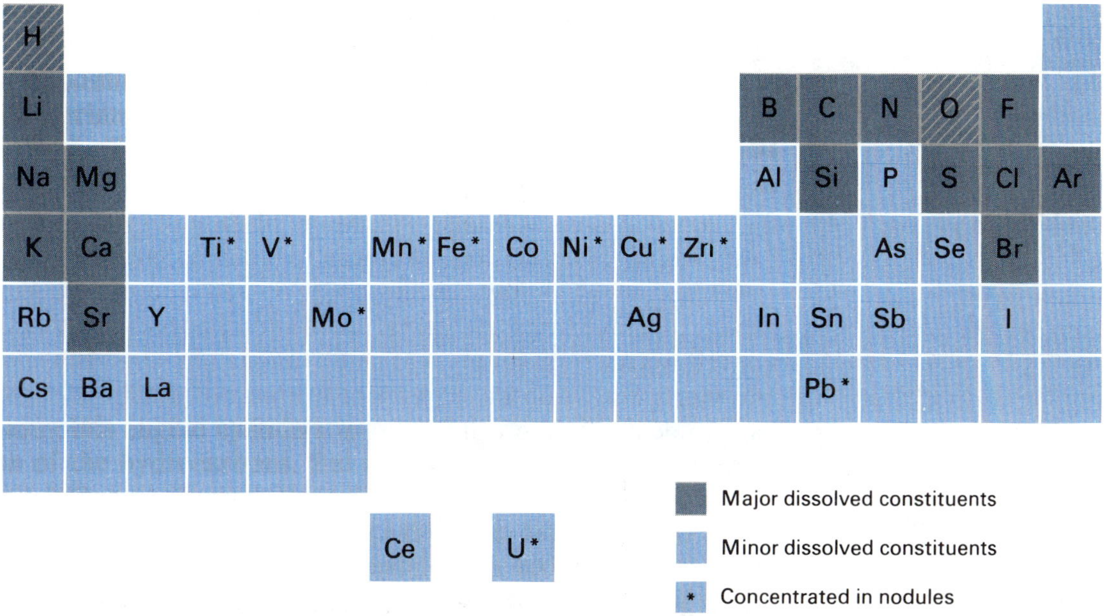

Figure 13.10 The elements found in the ocean.

In recent years, some locations on the ocean floor have been found to contain solid material in which some elements occur in remarkably high amounts. These egg-shaped lumps of solid material are referred to as *nodules*. Nodules are porous solids containing mixtures of the oxides of many valuable metals. Most important are the metals used in the manufacture of steel, such as manganese, iron, titanium, nickel, and molybdenum. Copper and zinc oxides also are found. The various elements found in nodules are indicated in Figure 13.10. Over the eons of geological time, some process apparently has caused these elements to be removed from their dilute water solution and to be concentrated into the solid nodules. When the time comes that the ores in the land surface are used up, we perhaps can meet our needs for the metals found in these nodules by mining the ocean floor. Unfortunately, they are located on the deep ocean floor. Consequently, their recovery will be a complex and costly engineering operation.

Nodules on the ocean floor are a high-grade ore of manganese along with copper, nickel, and cobalt.

USING WATER FROM THE OCEAN

At the present time, and increasingly in the near future, the solution represented by ocean water is being separated more for the value of the solvent—water—than for the value of the dissolved solutes. The demands of the world's human population for water,

Water from the Pacific Ocean becomes potable water in the Point Loma, California desalination plant.

either for direct consumption (*potable* water), for irrigation of agricultural crops, or for industrial uses, often exceeds the supply of fresh water at a particular location. The removal of some or most of the dissolved solutes from ocean water is called the process of *desalination* (removal of saline, salt-like substances). Nearly a thousand plants around the world are accomplishing this, some on a fairly large scale. Many of these start with water from inland sources that has been contaminated with ocean water but is not as high as the ocean in salt content. A great variety of procedures are being used, and research on others is underway. As population and industrialization increase, more and more attacks on the important problem of desalination will have to be made. We will discuss a few of the most widely used methods.

DESALINATION PROCESSES

The oldest desalination processes, and in some respects the simplest in principle, are based on distillation. Solvent water is evaporated from the ocean solution and recondensed to pure water. This process is the one Mother Nature uses on a vast scale to produce rain in the clouds that are moved by winds from ocean to land. A simple distillation device to make use of solar energy needs only a sloping transparent cover over a tray containing sea water. The water evaporates, is condensed on the cover, and drips into collecting ridges around the edge of the tray. However, sunlight is such a diffuse energy source that it seldom can be used to evaporate sea water on a scale large enough to satisfy the needs of large numbers of people in a particular locality. Fuel must be burned to release more concentrated heat. The high heat of vaporization of water means that large amounts of fuel are needed.

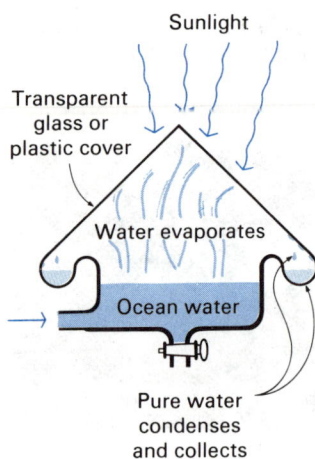

The one-stage evaporation of all the water in a cubic mile of ocean would require burning about half the total coal resources of the United States. Desalination plants using evaporation methods have to be designed to conserve heat. One means of conserving the necessary heat is to use the heat released when the evaporated steam condenses to liquid water to warm up the ocean water in the next batch to be evaporated. Some plants do this in several stages. Each evaporation is done at a successively lower pressure so that the later batches of water can be boiled at a lower temperature.

Another desalination process that has been particularly successful with some brackish waters is *reverse osmosis*. ("Brackish" waters are those more dilute in salt than ocean water, yet still too high in dissolved substances to be useful for irrigation or drinking.) The operation of a reverse osmosis system can be understood by referring back to Figure 13.9. When the salt solution (on the left in Figure 13.9) is placed in a container with an attached piston, pressure can be exerted on the solution by the piston. If the amount of pressure on the solution exceeds the osmotic pressure across the membrane, pure water is forced out of the solution into the reservoir of pure water (at the right of the diagram in Figure 13.9). The energy demands of this system are only for the motors to run the pressure pumps. The chief limitations are in the great strength of the membranes required to withstand large pressures over large areas.

Another method takes advantage of the tendencies of ions to migrate through a solution in response to the electrical attractions of charged electrodes. In this process, called *electrodialysis* the brackish water fills a three-part container. The center section is separated from the side sections by special membranes, as shown in Figure 13.11. The electrodes are located in the side sections. Each of the membranes is different. The one separating the center section from the part holding the negative electrode allows only positively charged ions to pass through. The one separating the center section from the part holding the positive electrode allows only negatively charged ions to pass through. Consequently, the ions in the brackish water of the center section migrate out in one direction or the other, and no additional ions migrate in. This process effectively purifies the water in the center section, which is then discharged.

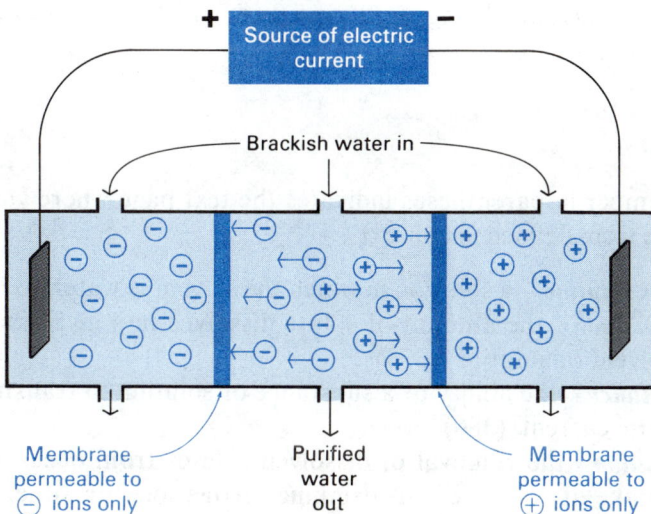

Figure 13.11 Apparatus for purification of brackish water by electrodialysis.

Summary

Solutions contain the particles of solute dispersed throughout the molecules of the solvent. The amount of solute that can be held in a solution is determined by how similar the solute and solvent are; like tends to dissolve like. An increase in temperature tends to make solid or liquid solutes more soluble in liquids but decreases the solubility of gases in liquids. An increase in pressure does increase the solubility of a gas in a liquid.

Water acts as a solvent for ionic substances because of its dipolar character. Water also tends to dissolve molecules with which it can establish hydrogen bonds. Water solutions of ionic compounds display the characteristic property of being able to conduct electric currents.

All solutions in which the solvent is a liquid display a lower vapor pressure, a higher boiling temperature, and a lower freezing temperature than the pure liquid. The degree to which these properties are present depends on how many solute particles of any kind are in the solution. In addition, osmosis is the tendency of a solution to be diluted when it is separated from pure solvent or a more dilute solution by a semipermeable membrane. This passage of solvent molecules, without solute particles, is an important phenomenon in the cells of living things.

The oceans of the earth represent a gigantic reservoir of dissolved mineral substances. In the future, this reservoir may be tapped, but at present the costs in energy are too high. The oceans and natural waters with less salt are being used as a source of water for human, agricultural, and industrial use. Various desalination processes, such as evaporation, reverse osmosis, and electrodialysis, are being used now and will become more important in the future.

Glossary

The number in parentheses indicates the text page where you can find the term defined in context.

aqueous solution   a solution in which the solvent is water (377)

concentration   the amount of solute dissolved in a given amount of solvent or solution (370)

conductance   the ability of a substance or solution to transmit an electric current (386)

desalination   the removal of dissolved solutes from ocean water to make it suitable for drinking, irrigation, or other purposes (396)

detergent a substance whose molecules contain a long-chain nonpolar part and a small polar part that can form ions, used for grease removal because of the differing solubilities of its parts (384)

dilution making a solution less concentrated by addition of solvent (370)

electrode the positively or negatively charged terminal of an electrochemical cell (386)

electrodialysis a method of desalination in which the passage of electric current causes dissolved ions to migrate out of the water being purified (397)

emulsion an opaque mixture (not a solution) in which many tiny droplets of one liquid are suspended in another (384)

isotonicity the state of a solution in which it has the same salt concentration as that of another solution (392)

molarity a measure of concentration, in units of moles of solute per liter of solution (370)

nodule a lump of solid material containing concentrated mixtures of oxides of valuable metals, found in some places on the ocean floor (395)

nonaqueous solution a solution in which the solvent is not water (377)

osmosis the passage of a solvent from a more dilute solution through a membrane into a more concentrated solution (392)

osmotic pressure the pressure built up by a supported solvent column or by a restraining membrane as osmosis occurs (393)

pesticide a poison used to kill insect pests (379)

potability the suitability of a liquid for safe human consumption (396)

reverse osmosis a method of desalination in which pressure greater than the osmotic pressure is exerted on a solution to force pure water across a membrane and out of solution (396)

saturated solution a solution that contains enough solute to be in dynamic equilibrium with excess undissolved solute at a given temperature; no more solute will dissolve (370)

semipermeability the extent to which a membrane permits the passage of solvent but not solute molecules (392)

solubility the amount of a certain solute that can dissolve in a given amount of a certain solvent at a given temperature (371)

solute the component of a solution that is present is the smaller amount (369)

solvent the component of a solution that is present in the greater amount (369)

unsaturated solution a solution in which more solute still can be dissolved (370)

Exercises

13.1   All solutions are mixtures, but not all mixtures are solutions. Explain why this statement is true.

13.2   The directions on a bottle of liquid plant food tell you to make a solution of two droppersful in a gallon of water. You have only a few plants and a quart container to use to make the plant food solution. You find that one dropperful amounts to 14 drops. What do you do to make one quart of solution that has the proper concentration?

13.3   Which sample contains more solute?

a.   One quart of a 5% solution or one pint of a 15% solution

b.   1.5 ounces of a 10% solution or 1.0 ounce of a 5% solution

c.   1.0 liter of a 1.2 molar solution or 1.2 liters of a 1.0 molar solution

13.4   Explain why if you know the molar concentration of a solution, you can figure out how many solute particles are in a certain sample of it.

13.5   A laboratory technician needs to use 0.010 mole of a substance to run a reaction. A 1.0 molar solution of the substance is provided. What volume of the solution does the technician use?

13.6   Explain the following in terms of the behavior of solute molecules and the conditions that affect their motion.

a.   Sugar dissolves faster in coffee or tea if you stir rather than just leave the sugar on the bottom of the cup.

b.   You can dissolve more sugar in hot tea than you can in the same amount of iced tea.

c.   A bottle of carbonated drink starts to bubble as soon as you take off the cap. It bubbles even more in your mouth.

d.   Champagne or other "sparkling" wine should always be served cold.

e.   Honey is a very concentrated solution of sugars and other characteristic compounds in water. The label on a honey jar says that you should not put honey in the refrigerator because it will "crystallize."

f.   The directions on the packet containing a chemical used to develop photographs tell you that you must dissolve the contents in one quart of water at 125° F rather than in cold water.

13.7   Ice cubes made in a refrigerator from water taken out of a faucet always have a cloudy center. The cloudiness is caused by air bubbles frozen into the ice. If water that has been freshly boiled is put in the ice trays, clear ice cubes are formed. Explain the difference.

13.8   How do you explain the fact that water-color paints cannot be mixed with turpentine and the fact that pigments for oil-base paint cannot be mixed with water?

13.9   Which solvent, water or gasoline, would you expect to do the following jobs best?

   a.   Remove marks made by a child's crayon

   b.   Remove the stain made by food coloring on a towel

   c.   Remove tar from an automobile body

   d.   Remove marks made by rubber heels on linoleum

13.10   Which of the following compounds probably will not dissolve in any of the others?

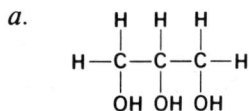

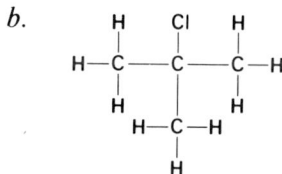

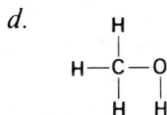

a.

```
        H  H  H
        |  |  |
  H — C — C — C — H
        |  |  |
        OH OH OH
```

b.

```
        H     Cl    H
        |     |     |
  H — C ——— C ——— C — H
        |     |     |
        H   H — C — H   H
              |
              H
```

c.

```
  H — O
      |
      H
```

d.

```
       H
       |
  H — C — O
       |   |
       H   H
```

13.11   Hexachlorophene is a controversial antibacterial agent that no longer may legally be used in cosmetics and other over-the-counter products.

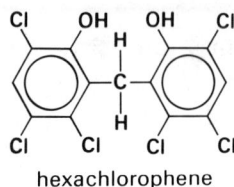

hexachlorophene

On the basis of a comparison of this structure with that of another compound discussed in this chapter, why do you think hexachlorophene may no longer be used in cosmetics?

13.12   Explain why a detergent does a good job of removing grease from a cotton shirt, but has little effect on a rust mark on the same fabric.

13.13   A type of food additive often present in puddings, salad dressings, cheeses, and so on is called a stabilizer or emulsifier. Its function is to keep the water in a food blended with the fats, oils, or other water-insoluble substances that would otherwise separate out into drops or layers. What structural features would you expect to find in a molecule that formed a substance useful as such a stabilizer?

13.14   Some types of packaged foodstuffs can be kept from drying out by including certain types of molecules as additives to help hold moisture. One substance used in this way is called sorbitol. Its formula is:

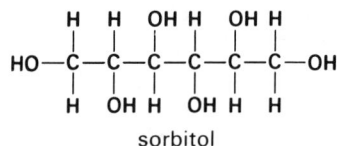

sorbitol

   a.   What structural feature of this molecule would you expect to make it hold on to water?

   b.   What kind of bonds are responsible for its action as an agent for holding water in a food?

13.15   All of the following substances will dissolve in water. Decide, for each solution, whether the aqueous solution so formed will conduct electricity or not. Recall the discussion in Chapter 6 about how the location of elements in the periodic table enables you to predict that compounds between elements are composed of ions.

    a.   table sugar, $C_{12}H_{22}O_{11}$

    b.   potassium chloride, KCl

    c.   methyl alcohol, CH_3OH

    d.   ethylene glycol, $C_2H_6O_2$

    e.   sodium iodide, NaI

    f.   magnesium chloride, $MgCl_2$

    g.   glucose, $C_6H_{12}O_6$

13.16   How does the boiling point of the water in an automobile cooling system containing antifreeze compare with the boiling point of pure water?

13.17   A mixture of ice and salt is used to freeze ice cream in a home churn. Ice cream is made from a water solution of several components. Why is the ice-salt mixture used?

13.18   Why must the temperature in a freezer used for storage of food be kept at 0° F (that is, well below 32° F, the freezing point of water)?

13.19   You may have heard someone say something like: "I learned to love music by osmosis; I grew up in a family where music was being played constantly." Comment on the meaning and accuracy of using the word "osmosis" in this context.

13.20   Explain why dehydrated foods swell up when soaked for a few hours in water.

13.21   The withered leaves and stem of a plant can be restored to firmness by putting water on the roots of the plant. Explain.

13.22   Student nurses were learning the technique of using hypodermic needles to give injections. Several were just about to do so on their own arms when the instructor stopped them and asked what was in the needle. "Oh, it's only sterile distilled water, it can't hurt me," was the general reply. The instructor explained that if they had done so, their arms would have swollen to the size of an egg at the point of injection. Why would this occur?

13.23   Comment on the statement, "A property of water that makes it effective in moderating the weather also is the reason why desalination by distillation of natural waters is so costly." What property is referred to?

14

Acids and Bases— The Give or Take of Protons

☐ What do all acids have in common?

☐ What happens when an acid and a base are combined?

☐ How do we measure acidity and basicity?

☐ What determines the strength of an acid or a base?

☐ What common examples of acid/base reactions exist in the world around us?

☐ Why do oxides of some nonmetals create an environmental hazard?

The term *acid* is familiar to everyone. You may connect it with a sour taste like that of vinegar, lemon juice, or the gastric juices in your own stomach. Or you may react to the term acid as something to be avoided, like the acid in an automobile battery. An acid can eat holes in metal, clothing, and even your own skin if you have the misfortune to spill it on yourself. Acids have been recognized as a type of substance for many centuries. The word *acid* comes from the Latin word for sour. Vinegar and lemon juice, solutions that contain acids, taste sour.

The phrase "the acid test" describes the crucial question or test you apply to find out if something is genuine. This meaning comes down to us from the use of the phrase in ancient times. True gold or silver does not tarnish or become discolored when treated with acid. Metals that may look like them, such as brass, bronze, or shiny iron, change appearance when treated with acid. The Romans tested coins and art objects with acids to reveal fakes or counterfeits.

The term *base* is probably less familiar, at least when it is used to represent a class of substances with a particular chemical property. The origin of the term is more obscure than is the origin of the word acid. The word probably comes from the ancients' attempts to classify substances that reacted with and so destroyed the acid in a solution. The "noble" metals, like gold and silver, do not react with acids, whereas the "base" metals, like lead and iron, do react with and, consequently, use up acids.

You probably have used some bases in your home. Drāno unclogs sink drains because it contains lye, NaOH, sodium hydroxide, a very active base. Another base is household ammonia, the aqueous solution of ammonia gas. It is used to clean floors and windows because it loosens grease and grime. Solutions of both lye and ammonia feel slippery or soapy. They feel this way because they react with and destroy the outer layers of your skin, forming a slippery goo. Consequently, you should protect your hands by wearing rubber gloves if you use such solutions for

Acids in the kitchen

Bases in the kitchen

scrubbing. Antacid tablets, such as Tums and Rolaids, also are bases, although they are much milder than lye and ammonia. Antacid tablets can be eaten to react with excess acid in the stomach and thus to relieve some of the discomfort of indigestion.

The classification of acids and bases is important and useful in chemistry. Up to this point in our discussions, we have based most of our classifications of matter on the structure of the particles. We have related the properties of atoms, molecules, and ions to the way they are built and how their electrons are arranged. Now that we have some understanding of how structural features affect properties, we can take the next step and see how similar reactivities can serve as a scheme for classifying substances. This step means that our definitions (always the place we

Our definitions now can focus on what substances do, *rather than on how they are put together.*

have to start) now can focus more on what substances *do,* rather than on how they are put together.

A definition based on what something or someone does has to be carefully chosen to mean just what it says. That is, it must include every example that it should include, and it must exclude all others. "A basketball player is a person who plays basketball" conveys a more precise meaning than "A basketball player is an athlete." The first definition properly excludes baseball players, weight lifters, and all others included in the broad category athlete. It also allows for the fact that baseball players may, at different times, play basketball. When they do so, they belong in the category of basketball player.

In the same way, when we try to define an acid, we look for something distinctive about what acids do. We want to find a chemical reaction that can identify every acid and every base and exclude everything else.

ACIDS ARE PROTON DONORS

We could start with the definition "Acids have a sour taste." This statement is true for vinegar and lemon juice. But it is not a very useful definition, because it is dangerous and foolish to go around tasting chemicals just to find out if they taste sour. If you tasted a liquid you suspected of being battery acid, and if it really was, you would have a very sore mouth and possibly holes in your teeth.

One test that you can safely make, a simple laboratory experiment, involves seeing what happens when acids and some metals get together. For example, if you put a shiny iron nail in a solution of hydrochloric acid, bubbles of gas form. If you collect some of the gas, you can perform tests to show it is hydrogen gas. You get the same result if you use sulfuric acid instead of hydrochloric. But if you use vinegar or lemon juice, the reaction is so slow that you would have difficulty deciding for sure whether anything had happened. However, this experiment does help direct our search for a definition. If an acid reacts with a metal to produce hydrogen gas, then hydrogen atoms must be part of an acid. It is important to note that we are not saying *all* hydrogen-containing compounds are acids. For example, table sugar and gasoline contain hydrogen atoms. These compounds certainly do not behave as acids.

Another simple experiment yields useful results. You can test solutions of acids for their ability to conduct electric current. The apparatus needed is that sketched in Figure 13.4. When you put a few drops of hydrochloric acid into the water between the electrodes connected to a bulb and battery, the light bulb glows brightly. Remember, this means that the solution contains *ions*. Sulfuric acid in water produces the same result. So do vinegar and lemon juice, although the bulb glows less brightly.

Now, if we put together the ideas that acids contain hydrogen and that their water solutions contain ions, it is reasonable to suspect that some of the ions are hydrogen ions. Indeed, this turns out to be the case. Hence the definition of an acid: *An acid is a substance that gives up or donates hydrogen ions (protons) to some other substance.*

Calling a hydrogen ion (H^+) a proton is not confusing if you recall the makeup of a hydrogen atom. The nucleus of the hydrogen atom (atomic mass = 1, and atomic number = 1) is a single proton. When the H atom loses its electron to become the H^+ ion, its nucleus—what is left—is a proton.

The definition of an acid can be translated into a chemical equation. For example, we expect HCl to break up as

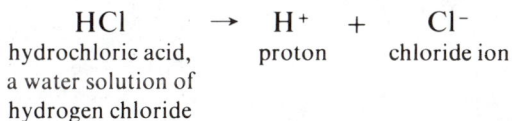

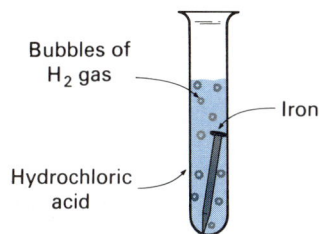

$$HCl \longrightarrow H^+ + Cl^-$$

HCl — hydrochloric acid, a water solution of hydrogen chloride

H^+ — proton

Cl^- — chloride ion

Bubbles of H_2 gas

Iron

Hydrochloric acid

A CLOSER LOOK AT ACIDS
IN WATER SOLUTION

This equation shows correctly that the acid gives up a proton, but it does not give a completely accurate picture of what the proton does. Protons can never exist free and uncombined in a water solution. (Protons can exist free only for a very short time in the gaseous form.) Instead of existing by itself, a proton always fastens onto some other molecule or ion, just as, in a football game, the ball must be in the possession of some player for the game to continue. The quarterback may, like the acid with the proton, either hold onto the ball himself, hand it off, or complete a pass to a teammate for the play to count. Just like a fumble, a free proton is always grabbed by a player in the game.

Where do the protons go when they leave an acid in a water solution? We cannot answer this question by watching the protons. Protons, like molecules and ions, are much too small to see. However, we can do another experiment; its interpretation gives a reasonable picture of what is going on. Let us again use a conductance apparatus, as suggested by Figure 14.1. The first substance we dip the electrodes into is pure hydrogen chloride, HCl. At normal temperatures, HCl is a gas. But if it is cooled below $-35°$ C, it condenses to a liquid. When the electrodes are immersed in liquid HCl, no electricity flows through the circuit. The conclusion is that in pure HCl no ions are present to carry the current.

In Chapter 13, where we first introduced the idea of conductance, we found that water, H_2O, also is a very poor conductor of electric current. The light bulb in the circuit of the apparatus does not glow. This result suggests that no ions are present to carry the current between the electrodes. If we replaced

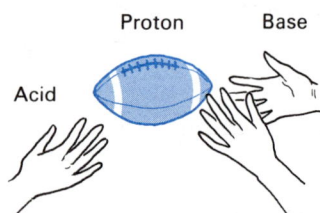

Proton   Base

Acid

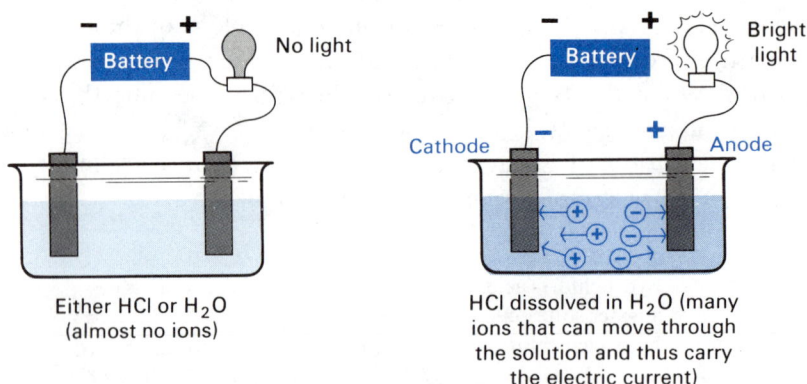

No light

Battery

Either HCl or H_2O
(almost no ions)

Bright light

Battery

Cathode   Anode

Figure 14.1   Nonconductance of HCl or H_2O; good conductance of HCl dissolved in H_2O.

HCl dissolved in H_2O (many ions that can move through the solution and thus carry the electric current)

the light bulb with a very sensitive meter to measure the least little bit of current flow, we would find that there is a very small current passing between the electrodes. Sensitive measurements show that only one out of every 500 million H_2O molecules breaks into H^+ ions and OH^- ions. (These OH^- ions are called hydroxide ions.) For the present, we can neglect this very small number of ions and consider water to consist only of molecules.

However, if HCl and H_2O are mixed at room temperature, the HCl immediately dissolves in H_2O, and the bulb glows brilliantly, indicating the presence of many ions in the solution between the electrodes. We can explain this result with the help of the electron-dot structures we used to show how covalent molecules are formed.

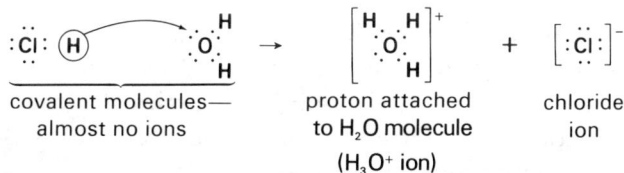

| covalent molecules—almost no ions | proton attached to H_2O molecule (H_3O^+ ion) | chloride ion |

Neither pure HCl nor pure H_2O has an appreciable number of ions in it. But when HCl and H_2O are mixed together, a proton leaves HCl to attach to H_2O. This transfer reaction makes two ions. The proton that shared a pair of electrons with the chlorine atom leaves that pair behind and attaches instead to a pair of electrons on the oxygen atom of a water molecule. All the HCl molecules added to water undergo this reaction. Here is one more way water can act as a solvent. The HCl gas dissolves in and, at the same time, *reacts* with H_2O.

When we described the properties of liquid H_2O, we discussed how the hydrogen bonds among water molecules cause them to form clusters (see Figure 11.9). So if we are going to be completely accurate, we should show that a proton adds to such clusters rather than to a single molecule to form H_3O^+. Instead, a series of ions are formed with formulas such as $H_5O_2^+$ (a cluster of two H_2O molecules plus a proton), or $H_9O_4^+$ (a cluster of four H_2O molecules plus a proton), and so on. This is a much more complicated representation than we need to write every time a proton of an acid is transferred to water in a solution. As long as we recognize that an acid such as HCl dissolves in water by transferring its proton, we can use the simple equation

$$HCl \rightarrow H^+ + Cl^-$$

to mean

$$HCl + water \rightarrow H^+ (attached\ to\ water\ molecules) + Cl^-$$

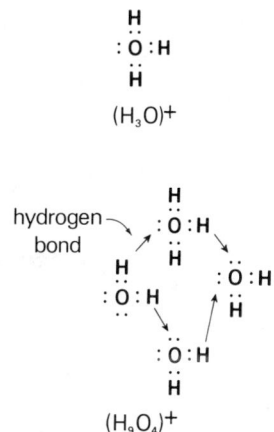

Table 14.1 Common acids and their anions

Acid		Anion	
Name	Formula	Name	Formula
hydrochloric acid	HCl	chloride ion	Cl^-
hydrobromic acid	HBr	bromide ion	Br^-
sulfuric acid	H_2SO_4	hydrogen sulfate ion	HSO_4^-
		sulfate ion	SO_4^-
nitric acid	HNO_3	nitrate ion	NO_3^-
phosphoric acid	H_3PO_4	dihydrogen phosphate ion	$H_2PO_4^-$
		monohydrogen phosphate ion	HPO_4^-
		phosphate ion	PO_4^{3-}
acetic acid	$H(C_2H_3O_2)$, or HOAc	acetate ion	$C_2H_3O_2^-$, or ^-OAc

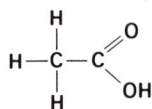

acetic acid
$HC_2H_3O_2$
(HOAc)

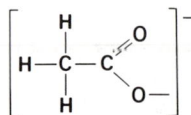

acetate ion
$C_2H_3O_2^-$
(OAc)

So we come back to the definition of an acid that tells best what it is capable of doing: *An acid is a substance that can donate protons.* The word "donate" used here carries the connotation that the proton is transferred to something else, not just set free. A donation is given by a donor to an acceptor.

Table 14.1 lists the formulas of some acids frequently used in chemical laboratories and manufacturing processes. Notice that the conventional way of writing the formulas always puts the hydrogen first. HCl is the formula for hydrochloric acid. Some acids have more than one hydrogen atom that can be given up as a hydrogen ion. Sulfuric acid, H_2SO_4, has two; phosphoric acid, H_3PO_4, has three. On the other hand, perhaps not every hydrogen atom in an acid molecule is released as a hydrogen ion, H^+. This situation is illustrated by the formula for acetic acid, $H(C_2H_3O_2)$. The three hydrogen atoms within the parentheses cannot form hydrogen ions (protons). Only one H in the molecule can become a H^+ ion, so the formula is written to indicate this fact. The abbreviated form, using the symbol OAc^- in place of $(C_2H_3O_2)^-$, can be used to convey this idea in a simpler fashion.

When a proton leaves an acid, a complementary ion, bearing a negative electric charge, also is formed. Such negative ions are called *anions*. In solution, they tend to migrate toward the electrode that has a positive charge. This positively charged electrode is called the *anode*. Ions that have a positive charge are called *cations* because they tend to migrate toward the other electrode, the *cathode*

Table 14.1 lists the formulas and names of the anions that come from some common acids. Anions may be formed from a single atom, as is the chloride ion. Anions also can contain clusters of atoms, as in the case of the nitrate ion, NO_3^-, from nitric acid.

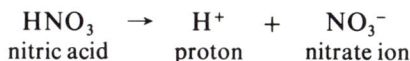

$$HNO_3 \rightarrow H^+ + NO_3^-$$

nitric acid   proton   nitrate ion

Sulfuric acid, H_2SO_4, breaks apart in two steps:

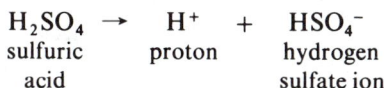

$$H_2SO_4 \rightarrow H^+ + HSO_4^-$$

sulfuric acid   proton   hydrogen sulfate ion

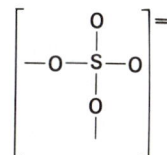

followed by

$$HSO_4^- \rightarrow H^+ + SO_4^=$$

hydrogen sulfate ion   proton   sulfate ion

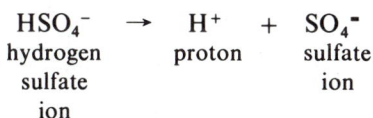

A general equation for the breaking up of an acid is

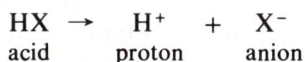

$$HX \rightarrow H^+ + X^-$$

acid   proton   anion

You can see one consistency in the way chemical names are given to the anions listed in Table 14.1. Oxygen-containing acids whose names end in *ic* produce anions whose names end in *ate*. Binary acids (H^+ combined with one other element) whose names end in *ic* produce anions whose names end in *ide*.

BASES ARE PROTON ACCEPTORS

The most appropriate definition of a base refers to what it does with an acid. An acid donates or transfers a proton. *A base is a substance that accepts a proton.* The substance NaOH, sodium hydroxide (lye), is a very effective base. NaOH is a white crystalline solid that readily dissolves in water, just as table salt, NaCl, dissolves. If we immerse the electrodes of a conductance apparatus in an aqueous solution of NaOH, the light bulb glows brilliantly. This proves that the solution contains many ions. These ions are Na^+ cations and OH^- anions. (Note how the name sodium hydroxide identifies the ions in NaOH, just as the name sodium chloride identifies the Na^+ cation and Cl^- anion in NaCl.) When a solution of NaOH is mixed with a solution of hydrochloric acid, HCl, a chemical reaction occurs.

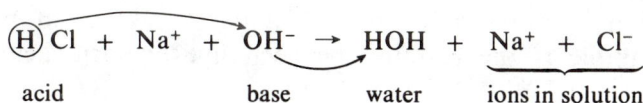

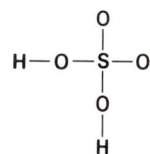

$$(H)Cl + Na^+ + OH^- \rightarrow HOH + Na^+ + Cl^-$$

acid   base   water   ions in solution

sulfuric acid

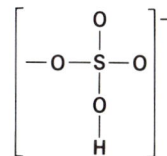

hydrogen sulfate ion

sulfate ion

Sodium hydroxide is called a base because it contains hydroxide ions, OH^-, which accept the protons from the hydrochloric acid solution. The combination of these two ions results in a water molecule. (HOH is just another way of writing H_2O.) The Na^+ ion that came to the party with the OH^- ion and the Cl^- ion that came with the H^+ ion remain in the water solution.

Ammonia, NH_3, is another base. At normal temperatures, NH_3 is a gas that dissolves readily in water. It dissolves by reacting with water, but not in the same way that HCl does. NH_3 takes a proton away from water by the following reaction:

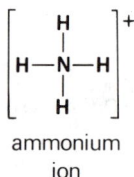

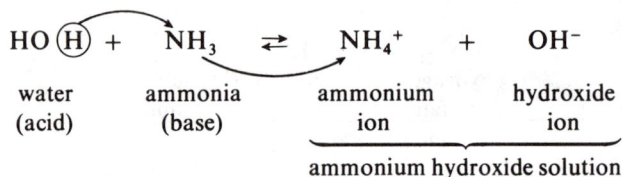

H—N—H
|
H

ammonia

$$HO\,\textcircled{H} + NH_3 \rightleftharpoons NH_4^+ + OH^-$$

water   ammonia   ammonium   hydroxide
(acid)   (base)   ion   ion

ammonium hydroxide solution

$$\begin{bmatrix} & H & \\ & | & \\ H & —N— & H \\ & | & \\ & H & \end{bmatrix}^+$$

ammonium
ion

In this reaction, H_2O has acted in the role of an acid, the proton donor. NH_3 is the base, the proton acceptor. The product of the proton added to the NH_3 molecule is an ion, NH_4^+, called the ammonium ion. We have indicated an arrow pointing in the reverse direction. This arrow suggests that by no means do all NH_3 molecules take on protons from H_2O. Actually, only a small percentage do so. The solution formed by ammonia reacting thus with water is called an ammonium hydroxide solution, which is given the formula NH_4OH. This designation is only a convenient, simplified representation. It does not mean that NH_4OH molecules exist in solution.

NEUTRALIZATION—ACIDS REACT WITH BASES

The reaction of an acid with a base is called a neutralization. This type of chemical reaction is very important. For example, if you allow a solution of Drāno or oven cleaner (both of which contain lye, NaOH) to remain in contact with your skin or fabrics or floor covering, it will cause damage. The first thing to do is wash away, with a lot of water, as much lye as possible. Then the remaining lye can be neutralized by adding vinegar, an acid.

A simple type of neutralization involves the reaction between acids and hydroxides. (Later in this chapter we will discuss neutralization reactions between acids and other kinds of bases.) One example is the reaction between hydrochloric acid and

sodium hydroxide

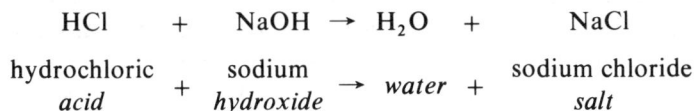

$$HCl \quad + \quad NaOH \rightarrow H_2O \quad + \quad NaCl$$

hydrochloric $+$ sodium $\rightarrow$ water $+$ sodium chloride
acid   *hydroxide*   *salt*

In this equation and others in this section, we write the formulas for the substances in as simple a manner as possible. You now should recognize that writing hydrochloric acid as HCl is a short-cut for "H^+ attached to water molecules $+ Cl^-$ ions." The formula NaCl really means "Na^+ ions $+ Cl^-$ ions in water solution."

The words we have written beneath this example provide a generalization: Acid + hydroxide → water + salt. The term *salt* is broader in meaning than only NaCl. A salt is a compound formed by the anion from an acid (the negative ion combined with H^+) and the cation from the hydroxide (the positive ion combined with OH^-). A completely general equation for this type of neutralization reaction is

$$H(anion) + (cation)OH \rightarrow H_2O + (cation)(anion)$$

For example,

$$HCl + NaOH \rightarrow H_2O + NaCl$$

$$HNO_3 + \quad NaOH \quad \rightarrow H_2O + NaNO_3$$
nitric   sodium   water   sodium
acid   hydroxide   nitrate

$$HCl \quad + \quad NH_4OH \rightarrow H_2O + \quad NH_4Cl$$
hydrochloric   ammonium   water   ammonium
acid   hydroxide   chloride

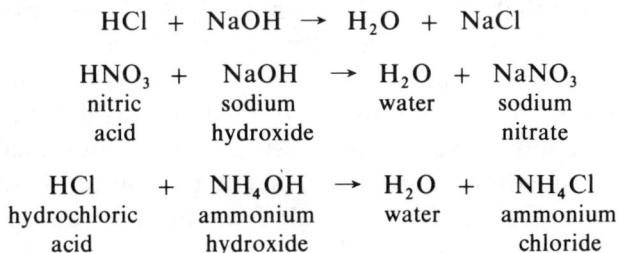

In these examples, note that in nitric acid, HNO_3, the anion is the group of atoms $(NO_3)^-$, called the nitrate anion, with an electric charge of -1. The product of the neutralization reaction with

Table 14.2 Common hydroxides and their cations

Hydroxide		Cation	
Name	Formula	Name	Formula
sodium hydroxide	$NaOH$	sodium ion	Na^+
potassium hydroxide	KOH	potassium ion	K^+
ammonium hydroxide (NH_3 in water)	NH_4OH	ammonium ion	NH_4^+
calcium hydroxide	$Ca(OH)_2$	calcium ion	Ca^{++}
barium hydroxide	$Ba(OH)_2$	barium ion	Ba^{++}

NaOH is the salt $NaNO_3$, the combination of Na^+ cations with NO_3^- anions. The appropriate name for this combination of ions is sodium nitrate. (The convention in naming salts always is to name the cation first.) If you follow the same systematic interpretation for the third example, you will recognize the ammonium ion, $(NH_4)^+$, as a group of atoms that stays together as a cation. That reaction forms the salt ammonium chloride, NH_4Cl.

We are illustrating here another of the ways a generalization greatly simplifies chemistry. Learning one additional feature of

A generalization greatly simplifies chemistry.

this generalization enables you to write correct chemical equations to predict literally thousands of neutralization reactions. We can illustrate this additional idea by writing equations for reactions that involve water solutions of sulfuric acid, H_2SO_4.

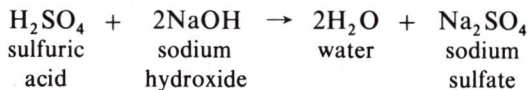

$$H_2SO_4 \; + \; 2NaOH \; \rightarrow \; 2H_2O \; + \; Na_2SO_4$$

sulfuric sodium water sodium
acid   hydroxide   sulfate

The acid H_2SO_4 consists of *two* H^+ ions and *one* sulfate anion, $(SO_4)^-$. Note that, because the formula H_2SO_4 must represent an electrically neutral substance, the charge of the $(SO_4)^-$ anion must be -2. This means that the salt, sodium sulfate, must contain *two* Na^+ in place of the *two* H^+ of the acid. The correct formula for sodium sulfate is Na_2SO_4. The equation then is balanced by using the correct coefficients (2) in front of the formulas for NaOH and H_2O. For every H^+ from the acid, an OH^- from the hydroxide must be added to form a molecule of H_2O. If a neutralization reaction involves phosphoric acid, H_3PO_4, the equation must show that three OH^- are added for every H_3PO_4 used in the equation. The formula for the salt sodium phosphate is Na_3PO_4.

A water solution of lime contains the base calcium hydroxide, $Ca(OH)_2$. Here the Ca^{++} cation combines with *two* OH^- anions. (Calcium, atomic number 20 in the periodic table, is in family 2A. In Chapter 5 you learned that atoms of this family tend to lose *two* outermost electrons to become ions with a charge of $+2$.) The neutralization reaction of $Ca(OH)_2$ with hydrochloric acid, HCl, is written as follows:

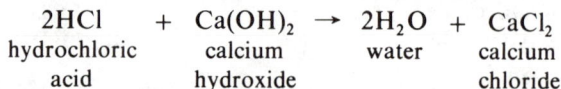

$$2HCl \; + \; Ca(OH)_2 \; \rightarrow \; 2H_2O \; + \; CaCl_2$$

hydrochloric calcium water calcium
acid   hydroxide   chloride

The salt, calcium chloride, has the formula $CaCl_2$; the equation is balanced by using the coefficient 2 in front of HCl and H_2O.

You now have the information and the rules to follow for writing the correct chemical equation for the neutralization of *any* acid by *any* hydroxide. Tables 14.1 and 14.2 list a few of the most common acids and hydroxides encountered in simple chemical studies. You can use the information represented by these formulas to answer Exercises 14.8 and 14.9.

THE ACIDITY AND BASICITY OF WATER SOLUTIONS

The combination of H^+ and OH^- ions to form water is the key reaction in all neutralizations between acids and hydroxides. No matter what anion is in the acid or what cation is in the hydroxide, one product is always H_2O. The other products are the ions of a salt in solution.

Careful measurements show that the combination of H^+ with OH^- ions and the dissociation (the breaking apart) of H_2O molecules is a dynamic equilibrium.

$$H^+ + OH^- \rightleftarrows H_2O$$

In any sample of water, whether pure H_2O or a solution of an acid or hydroxide, some H_2O molecules are constantly breaking apart and some H^+ and OH^- are recombining. This process never stops; at any given moment, the rates of the two opposite reactions are equal.

In pure water, the only source of H^+ is the dissociation of an H_2O molecule, and for every H^+ formed, there also must be one OH^- set free. This balanced situation is what is meant by a *neutral* solution. There is no excess of either H^+ or OH^- in a neutral solution. When the concentration of each of these ions is measured in pure water at 25°C, they are indeed found to be equal and have the very small value 0.0000001 mole/liter. This number can be represented more conveniently, by exponential notation, as 1×10^{-7} mole/liter.

Any process that involves a dynamic equilibrium can be analyzed mathematically to give what is called an equilibrium constant. We need not derive this expression, but we can use it to help answer the question "What happens to the concentration

You now have the rules to follow for writing the chemical equation for the neutralization of any acid by any hydroxide.

of OH⁻ ions in water if extra H⁺ ions are added?" Or vice versa, "What happens to the concentration of H⁺ ions if extra OH⁻ ions are added?" The equilibrium constant for the water–ions back-and-forth shifting is written as follows:

$$K_w = \text{(concentration of H}^+) \times \text{(concentration of OH}^-)$$
$$K_w = [H^+] \times [OH^-]$$

Square brackets around a symbol are used to indicate the concentration of that ion in mole per liter, its molar concentration. So when you see $[H^+]$, you read it as "the concentration of hydrogen ions in mole per liter," and $[OH^-]$ as "the concentration of hydroxide ions in mole per liter."

This relationship for the equilibrium tells us that the product of the two concentrations is always the same. (K_w stands for the *constant* for the formation or dissociation of water and its ions.) If the concentration of H⁺ increases, the concentration of OH⁻ must decrease, and vice versa. You can recall that we encountered this kind of an equation before. The pressure and volume of a gas sample at constant temperature are so related. When the pressure rises, the volume must decrease; an opposite change produces an opposite effect.

Careful measurements of the concentrations of H⁺ and OH⁻ ions in pure water at 25°C give us numerical values we can use to calculate a number for K_w. The numbers are as follows:

$$K_w = [H^+] \times [OH^-]$$
$$K_w = (1 \times 10^{-7}) \times (1 \times 10^{-7})$$
$$K_w = 1 \times 10^{-14}$$

(Remember the rule stated in the box on page 119 for multiplying two exponential numbers. You add the exponents of 10. Thus in this case, $10^{-7} \times 10^{-7} = 10^{-7+(-7)} = 10^{-14}$.)

Table 14.3 gives numerical values for the H⁺ concentration and the corresponding OH⁻ concentration for a wide range of water solutions. The concentrations on each line are ten times the ones above or below that line.

The term *acidity* refers to the $[H^+]$ in a solution; *basicity* or *alkalinity* refers to the $[OH^-]$ in a solution. The word *alkaline* means the same as the word *basic*. The term comes from the word *alkali,* which was used centuries ago to describe minerals that dissolved to produce basic solutions. The acidity of water solutions increases toward the top of the table, as the $[H^+]$ increases. Remember, the number 1×10^{-1} stands for 0.1. This number is

Table 14.3   Acidity and basicity in water solutions

$$K_w = [H^+] \times [OH^-] = 1 \times 10^{-14}$$

	[H⁺]	[OH⁻]	
	1×10^{0}	1×10^{-14}	
	10^{-1}	10^{-13} .... ⟵ 0.1 molar solution	
increasing	10^{-2}	10^{-12} of hydrochloric	
acidity	10^{-3}	10^{-11} acid	
	10^{-4}	10^{-10}	
	10^{-5}	10^{-9}	
	10^{-6}	10^{-8}	
neutral ⟶	10^{-7}	10^{-7} .... ⟵ pure H_2O	
	10^{-8}	10^{-6}	
increasing	10^{-9}	10^{-5}	
basicity (or	10^{-10}	10^{-4}	
alkalinity)	10^{-11}	10^{-3}	
	10^{-12}	10^{-2}	
	10^{-13}	10^{-1} .... ⟵ 0.1 molar solution	
	10^{-14}	10^{-0} of sodium	
			hydroxide

greater than the number 1×10^{-2}, which stands for 0.01, or 1×10^{-7}, which stands for 0.0000001. The basicity of a solution increases toward the bottom of the table, where higher values of [OH⁻] are shown.

You also may note how the numbers in Table 14.3 confirm the relationship $K_w = [H^+] \times [OH^-] = 1 \times 10^{-14}$. As the [H⁺] increases, the [OH⁻] becomes smaller. For example, in a solution of 0.1 molar hydrochloric acid, where the [H⁺] is relatively large, 1×10^{-1}, the [OH⁻] is correspondingly small, 1×10^{-13}. At the bottom of the table, where [OH⁻] has the large value, 1×10^{-1}, in a 0.1 molar solution of NaOH, the [H⁺] is correspondingly small. In pure water, where the solution is neutral, the acidity and basicity are equal (both [H⁺] and [OH⁻] have the value 1×10^{-7}). When the acidity is greater than the basicity, the solution is described as *acidic*. When the basicity exceeds the acidity, the solution is described as *basic* or *alkaline*.

THE pH SCALE OF ACIDITY

Numbers like 1×10^{-3} become monotonous to write and also may seem too complicated to express a simple idea. So a simpler method has been worked out, called the *pH scale* of acidity. Table 14.4 illustrates the relationship between the numerical values of [H⁺] and the pH expression of these values. Notice that

Table 14.4   The pH scale of acidity

		[H$^+$]		pH	
	↑	1 ×	10^0	0	
			10^{-1}	1	← 0.1 molar solution of
	increasing		10^{-2}	2	hydrochloric acid
	acidity		10^{-3}	3	
			10^{-4}	4	
			10^{-5}	5	
			10^{-6}	6	
	neutral ⟶ ··········		10^{-7} ·····	7 ···	← pure H$_2$O
			10^{-8}	8	
	increasing		10^{-9}	9	
	basicity (or		10^{-10}	10	
	alkalinity)		10^{-11}	11	
			10^{-12}	12	
			10^{-13}	13	← 0.1 molar solution of
	↓		10^{-14}	14	sodium hydroxide

the pH is the number of the negative power of 10 in the concentration value. When [H$^+$] = 1 × 10^{-3}, the pH is 3; when [H$^+$] = 1 × 10^{-9}, the pH is 9; and so on. The important thing to remember is: *acidic solutions have pH values less than 7; basic solutions have pH values greater than 7.* A low pH means the solution is acidic; a high pH means it is basic.

One other feature of this scale is important to remember. Each unit of change in pH means a tenfold increase or decrease in the concentration of H$^+$. If you were to add one drop of a dilute hydrochloric acid solution to pure water, you would change the pH from 7 to 6. However, you would have to add *ten* drops of hydrochloric acid to that solution to change the pH to 5.

THE pH OF FAMILIAR SOLUTIONS

The pH scale is widely used to express the acidity or basicity of solutions you may encounter in your activities. For example, most plants thrive best in soil containing moisture that is slightly alkaline, or basic. Such soils have a pH between 7 and 8. If the pH is 9 or more, the soil is too alkaline for the plants to grow well. However, some common houseplants, such as African violets, azaleas, and ferns, do better in soil that has a pH of 5 or 6. Such soils are slightly acidic.

Figure 14.2 gives the pH of a variety of familiar water solutions, such as common beverages, some drugs, and household solutions. The values given are approximate, but they give some idea of the range of acidity or basicity to expect.

Figure 14.2 (pH scale, Acidic / pH scale / Basic):

- 0
- 1 — Battery acid
- — Gastric juice in stomach
- 2
- — Vinegar
- — Aspirin
- 3 — Coke
- — Orange juice
- — Grape juice
- — 7-Up
- 4 — Beer
- — Sauerkraut juice
- 5 — Shaving lotion
- — Coffee
- 6 — Alka Seltzer
- — Hair tonic
- — Mouthwash
- 7 — Pure water
- — Rubbing alcohol
- — Swimming pool water
- 8
- 9
- — Clorox
- — Rug cleaner
- 10 — Tums
- — Milk of magnesia
- 11 — Ammonia
- 12
- 13 — Hair remover
- 14 — Oven cleaner

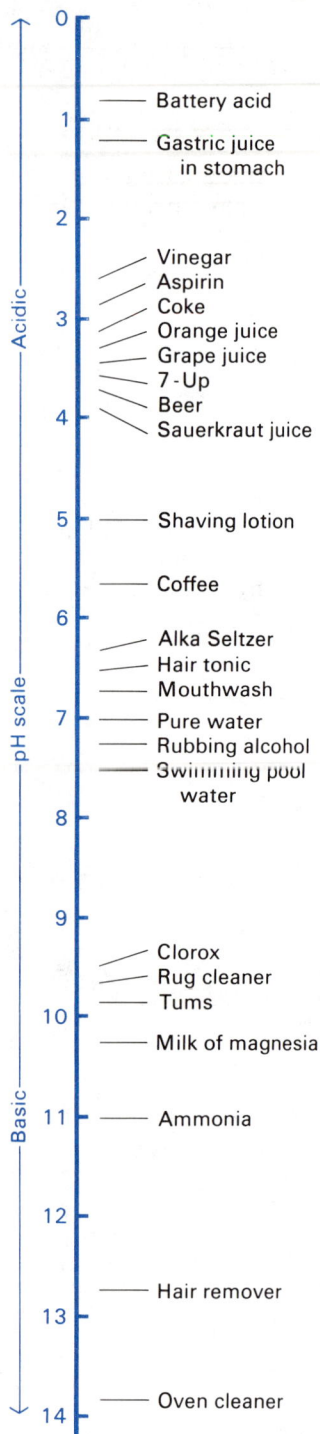

Figure 14.2   pH values of familiar water solutions.

Cacti thrive in alkaline
soil that has pH = 8.

African violets thrive in
acidic soil that has pH = 5.

INDICATORS—THE MEASUREMENT OF pH

Some substances in water solutions change color depending on the pH of the solution. Many such substances are dyes found in growing things. The coloring material in plants often behaves this way. The dye has one color in an acidic solution and another in a basic solution. You may have noticed how the color of some jellies, such as grape or blackberry, changes when a dish smeared with jelly is washed in soap or detergent. The jelly itself is slightly acidic. Dishwater is usually slightly basic, made so by the added soap or detergent. Consequently, washing the dish neutralizes the acid in the jelly. In the neutralization process, the coloring matter in the jelly changes hue from purple to dark blue. The dye is a large complex organic molecule that alters its structure slightly in response to changes in the $[H^+]$ of the solution in which it is placed.

Natural dyes with this property of color change have been used for centuries to determine the acidity or basicity of solutions. They are called *acid–base indicators*. One of the substances of this type is litmus, a powder obtained from lichens. (The name *litmus* probably derives from old Norse words for "dye moss.") A piece of paper containing litmus is pink in acidic solutions, blue in basic solutions. The color changes at a pH of about 7. Other indicators change color at other pH values, usually over a range of one or two pH units. Another indicator often used is phenolphthalein. (This is the compound in Ex-Lax that acts as a laxative.) Phenolphthalein is colorless at pH values lower than 8 and changes to a red color at the higher pH values of more basic solutions.

Various indicators can be combined in solution to produce a range of colors in response to a range of pH changes. Paper that has been soaked in such a solution and then dried can be used as a "universal indicator." A drop of a solution to be tested is put

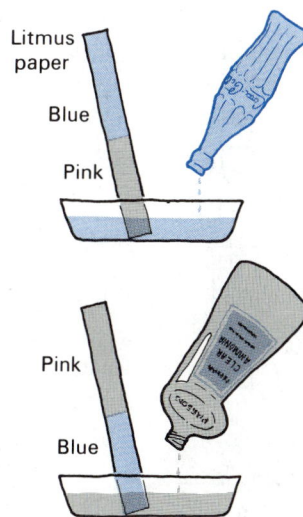

Litmus
paper

Blue

Pink

Pink

Blue

on this prepared paper. Its color, compared to those on a standard chart, allows pH to be estimated to within one unit over the range from pH 0 to 14.

The pH of a solution can be measured by methods other than by the use of indicators. The device most frequently used in a laboratory is called a "pH meter." It is an instrument that measures the hydrogen ion concentration by means of balanced electrical circuits. With a pH meter, more precise measurements are possible than with the use of indicators. Using a pH meter also allows you to measure the pH of a colored solution that would obscure the color of an added indicator.

Indicators also can be used to tell when a neutralization reaction has been completed. For example, suppose you have a solution that you know contains acid, but you do not know how much. You can find out how much acid is present by measuring the amount of a base required to neutralize it. You do not want to add more base than is necessary to do the job. First you add a few drops of indicator solution to the acid sample. This addition gives the whole sample the color characteristic of that indicator in acid solution. Then you gradually add a solution of the base, stirring thoroughly after each addition. When just enough base has been added to cause the first change in the color of the solution, you know that the correct amount has been added to make the solution neutral. Such a procedure, which is performed with containers that allow you to measure the volumes of the solutions, is called a *titration*. Titration is a very important technique in quantitative chemical analysis.

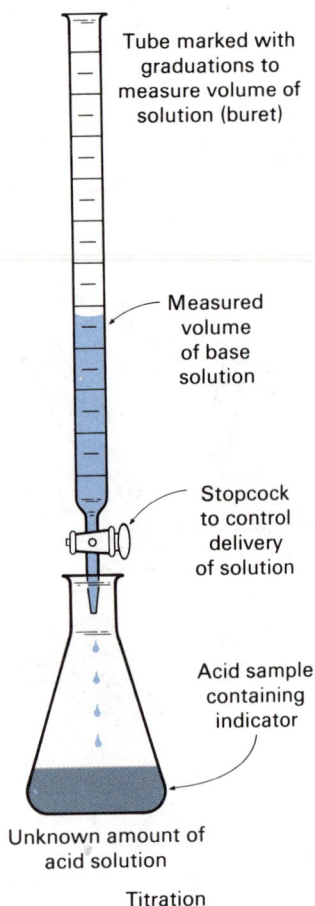

STRONG AND WEAK ACIDS

The definition of an acid as a proton donor can be refined to make it more useful (just as the definition of a basketball player as one who plays basketball can be refined by calling the person a center, a guard, or a forward). Some acids have a stronger tendency to give off protons than others do. The solution of sulfuric acid in your automobile battery causes painful burns on your skin. Yet you can drink orange juice, a solution containing citric acid, or eat vitamin C, which is the compound ascorbic acid, without harm. The pH of battery acid is about 0.5, which means that the $[H^+]$ is very high. The pH of orange juice is about 3.7, whereas that of a solution of vitamin C is about 6.5. The higher pH values indicate correspondingly lower values of the $[H^+]$ in these solutions. An acid that produces a high $[H^+]$ in water solu-

Tube marked with graduations to measure volume of solution (buret)

Measured volume of base solution

Stopcock to control delivery of solution

Acid sample containing indicator

Unknown amount of acid solution

Titration

tion is called a *strong* acid; one that produces a lower $[H^+]$ is called a *weak* acid.

We can visualize what is happening in water solutions of these acids if we use a chemical equation, as we did to represent how acids dissolve in water. When we put hydrogen chloride gas into water, every one of the HCl molecules reacts to make a H^+ ion (attached to clusters of H_2O molecules) and a Cl^- ion. We say that the HCl becomes 100% ionized.

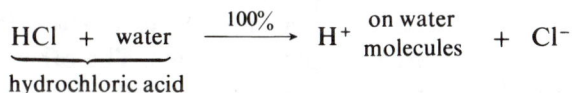

$$\underbrace{HCl + water}_{\text{hydrochloric acid}} \xrightarrow{100\%} H^+ \begin{smallmatrix}\text{on water}\\\text{molecules}\end{smallmatrix} + Cl^-$$

Hydrochloric acid (HCl in water), sulfuric acid (H_2SO_4 in water), nitric acid (HNO_3 in water), and phosphoric acid (H_3PO_4 in water) are all strong acids. These strong acids are found more often in chemical laboratories than in stores and homes. Because they are strong acids, they often react violently with other substances when used in concentrated solutions. However, they are by no means laboratory curiosities. The United States' chemical industry produces over 30 million tons of H_2SO_4 each year; the others are made in smaller amounts but in comparable magnitude. These acids are used in one way or another by just about every manufacturing industry—from steel, fertilizer, and plastic manufacture to textile, food, and drug production.

EQUILIBRIUM IN SOLUTIONS OF WEAK ACIDS

When we write the equation to show what happens when water reacts with acetic acid (the acid in vinegar) or citric acid (the acid in orange juice), we have to indicate a difference. Only about 2% of these molecules react to form ions in water.

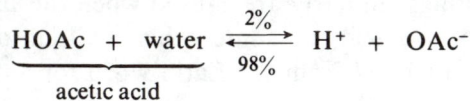

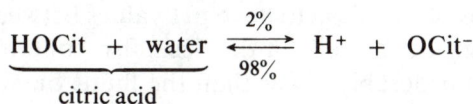

$$\underbrace{HOAc + water}_{\text{acetic acid}} \underset{98\%}{\overset{2\%}{\rightleftharpoons}} H^+ + OAc^-$$

$$\underbrace{HOCit + water}_{\text{citric acid}} \underset{98\%}{\overset{2\%}{\rightleftharpoons}} H^+ + OCit^-$$

The formula for citric acid is $C_6H_8O_7$; it is an organic molecule with a complicated structure. The essential idea of its ionization can be represented by using the symbol $OCit^-$ for the anion it

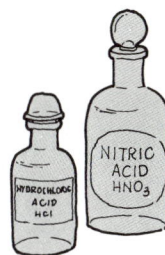

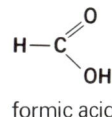

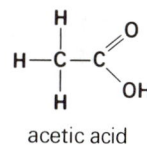

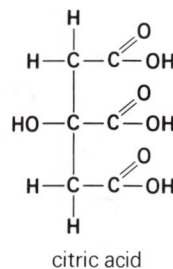

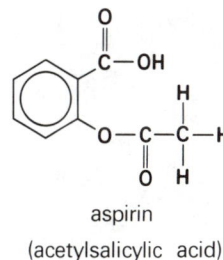

formic acid

acetic acid

citric acid

aspirin
(acetylsalicylic acid)

Structural formulas of some weak acids

forms, just as we can use OAc⁻ as a convenient abbreviation for the anion formed by acetic acid. We also can simplify the equation by writing H^+ instead of "H^+ attached to water molecules."

In these equations, arrows are written pointing forward and backward. What we have here is another illustration of dynamic equilibrium. Here we have molecules changing into ions and vice versa. Molecules and ions both are in the solution. Acetic acid and citric acid molecules are constantly breaking apart into ions, and the ions are recombining into molecules at an equal rate. Equilibrium is established when only 2% (2 out of 100) of the acid molecules have broken apart into ions. The other 98 out of 100 remain in the solution as molecules, so that the solution holds fewer H^+ ions than it would if a strong acid were present. Consequently, $[H^+]$ is less and the pH is higher than for a hydrochloric acid solution, where no equilibrium exists to pull the H^+ ions back out of solution.

The many acids you encounter among foods and drinks are weak acids. (Note the variety of examples between about pH 3 and 6 in Figure 14.2.) Almost all naturally occurring acids involved in the complicated chemical reactions of metabolism are weak acids. The equilibria between the molecules and ions of these acids are essential to the balance of life processes.

PAIN AND pH

The sensation of pain in humans and animals comes from the response of nerve endings to a change in the pH of the solutions around them. The pH of blood and intercellular fluids in tissues is 7.4. This pH is just about the value of "neutral" pure water. If you spill vinegar or lemon juice on a cut finger, you experience the sting that means your nerve endings are encountering a solution of much lower pH (more acidic) than normally surrounds them. Insect stings similarly are caused when the insect forces a solution of formic acid, or some other acid, into the tissues. Formic acid (so named from the Latin word for ants, the source from which the compound was originally extracted) is a weak acid. Solutions of formic acid have pH values between 2 and 3.

The fluids inside the cells in the flesh also contain weak acids. Their pH is considerably lower than the fluids outside such cells. Consequently, when anything happens to break the walls of cells in the flesh, such as cutting or burning the flesh or some toxic process within the cells, the acidic fluid normally held inside the cells escapes. When this low pH fluid comes in contact with nerve endings, it causes a feeling of pain. Studies have shown that even as small a drop in pH as that from 7.4 to 6.2 (a very slightly acidic solution) can be detected by the irritation it causes. If the solution in the tissues that is causing the irritation or pain can be neutralized, pain is relieved. Salves or creams used to treat such skin troubles as poison ivy or sunburn use slightly basic solutions to accomplish this pain-relieving neutralization reaction.

CARBON DIOXIDE IN WATER

When we were discussing the role of CO_2 as a constituent of the atmosphere (for example, its role in the greenhouse effect), we mentioned the fact that the atmosphere contains only 0.03% CO_2. One reason why this percentage is so small is that CO_2 is more soluble in water than O_2 and N_2 are. At normal temperatures, when the pressures of the gases are equal, CO_2 is nearly 30 times more soluble in water than is O_2. This greater solubility is caused by the ability of CO_2 to react chemically with H_2O, whereas O_2 molecules merely mix with H_2O molecules. CO_2 dissolves from the atmosphere into the water that falls as rain. CO_2 also reacts with compounds in the soil to form minerals such as limestone, $CaCO_3$, calcium carbonate.

A clue to the kind of solution formed when CO_2 dissolves in water is the fact that pure rain water has a pH of about 5.7. This suggests that rain is a solution of a weak acid. The acid is H_2CO_3, carbonic acid, formed by the reaction

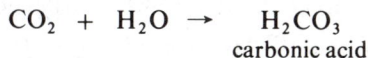

$$CO_2 + H_2O \rightarrow H_2CO_3$$
$$\text{carbonic acid}$$

Carbonic acid is a very unstable compound; it readily decomposes into CO_2 gas and H_2O. It has never been isolated as a pure compound. Carbonic acid is a weak acid and dissociates into ions, H^+ and HCO_3^-. HCO_3^- is called the *bicarbonate ion*. This ion also is called the *hydrogen carbonate ion,* but the simpler name bicarbonate is more frequently used. The term *bi*carbonate, as you can recognize, implies that the ion is *half*way toward being a carbonate ion. The bicarbonate ion, in turn, can dissociate into a second H^+ and CO_3^-; CO_3^- is called the carbonate ion. All these species are related to one another by the series of equilibria represented by the equation

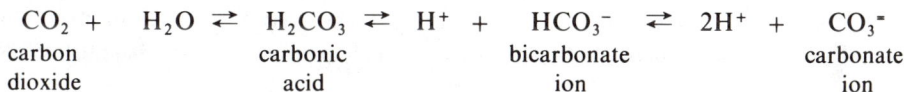

$$CO_2 + H_2O \rightleftarrows H_2CO_3 \rightleftarrows H^+ + HCO_3^- \rightleftarrows 2H^+ + CO_3^-$$

| carbon dioxide | | carbonic acid | | | bicarbonate ion | | | carbonate ion |

The equilibria shown by this equation are important in explaining a great variety of experiences, from the burp that accompanies relief from indigestion with baking soda to the hard, scaly substance that clogs a steam iron if you use tap water in it. Let us explore some of the chemistry that is involved.

USES OF BAKING SODA

Baking soda is a familiar white powder with many uses in the home. Baking soda is the compound sodium bicarbonate, $NaHCO_3$. Sometimes it is used directly in recipes for baked goods along with some weakly acidic solution such as buttermilk. More often it is used as a constituent in baking powder. Baking powder also contains some substance that acts as an acid when dissolved in the moisture of the batter that is to be baked. (One baking powder on the market contains tartaric acid, the weak acid in grapes.) As the food bakes, the HCO_3^- ions from the baking soda act as a base toward the H^+ ions from the acids. The equation for the reaction is part of the equilibria shown above.

$$HCO_3^- \; + \; H^+ \; \rightleftharpoons \; H_2CO_3 \; \rightleftharpoons \; H_2O \; + \; CO_2$$

bicarbonate   acid   gas bubbles
(base)

Added heat forces the reaction $\rightarrow$

We have written the arrows with uneven lengths to suggest that the equilibrium is forced to the right by the heat of the baking process. The left-to-right reaction occurs more than the right to left reaction. The CO_2 gas forms in bubbles throughout the baking batter and causes it to rise. Both the CO_2 gas and the H_2O, in the form of steam, are driven off. An analogous equilibrium shift is what may occur when two children of equal weight are playing on a seesaw. While both remain seated, the seesaw is in balance, in equilibrium. But when one child jumps off, the other hits the ground with a thump. The equilibrium is destroyed, and the reaction goes in one direction only. The same effect is accomplished when the bubbles of CO_2 gas leave the solution. Then the equilibrium is pulled out of balance (toward the right in the above equation).

Bubbles of CO_2 form in a biscuit when it bakes—therefore it rises.

$$HCO_3^- + H^+ \longrightarrow CO_2 + H_2O$$

Baking soda can be used to neutralize any acids that may be accidentally spilled. Thus damage to flesh, cloth, or other articles can be avoided. Baking soda also is useful in absorbing some odors from the air in a refrigerator. The molecules responsible

for the odors are weak acids and are neutralized by the bicarbonate ions.

Indigestion usually is a symptom of a stomach disorder called hyperacidity (too much acid). Something has caused the stomach to secrete excess acid into the gastric juices. Baking soda can be used as an effective *antacid* (*anti* means "against"). The reaction that occurs involves the neutralization of this excess acid (H^+ ions) by the HCO_3^- ions in baking soda, as shown in the above equation (going toward the right). CO_2 gas is the product of the reaction. This CO_2 gas is responsible for the burp that necessarily follows using baking soda as an antacid.

ORIGINAL
Alka-Seltzer
12 TABLETS   For UPSET STOMACH with HEAD-
IN 12 FOIL PACKS ACHE or BODY ACHES and PAINS

Each tablet contains
1904 mg Sodium
Bicarbonate
1000 mg Citric Acid

HARD WATER AND LIMESTONE

Municipal water supplies in many parts of the country are taken from wells, which are also the usual source of water in rural areas. Water in such wells comes from underground, where it has been in contact with minerals. Dissolved minerals in water are what make such water "hard." A principal reaction responsible for this hardness is the dissolving of such minerals as limestone, $CaCO_3$, in water that also contains dissolved CO_2. The reaction is

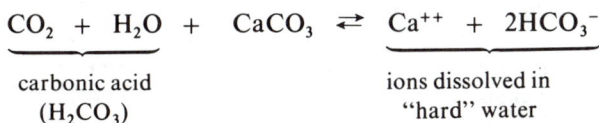

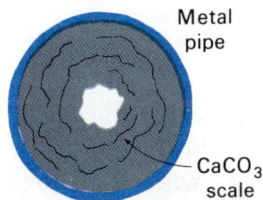

$$HCO_3^- + H^+ \longrightarrow CO_2 + H_2O$$

$$\underbrace{CO_2 + H_2O + CaCO_3}_{\substack{\text{carbonic acid} \\ (H_2CO_3)}} \rightleftarrows \underbrace{Ca^{++} + 2HCO_3^-}_{\substack{\text{ions dissolved in} \\ \text{"hard" water}}}$$

Metal
pipe

CaCO₃
scale

Pipes that carry untreated hard water almost fill with scale after a few years.

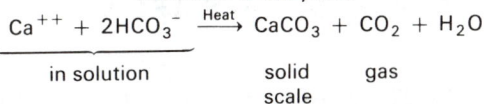

$$\underbrace{Ca^{++} + 2HCO_3^-}_{\text{in solution}} \xrightarrow{\text{Heat}} \underset{\substack{\text{solid} \\ \text{scale}}}{CaCO_3} + \underset{\text{gas}}{CO_2} + H_2O$$

We have indicated arrows going both ways because the reverse reaction is what happens when hard water is heated. The heat drives off the CO_2 gas from the solution, and this forces the equilibrium to the left in the equation. Solid, hard $CaCO_3$ forms as a result. If the reaction takes place in a steam iron or tea kettle, the solid $CaCO_3$ forms as a scaly substance covering the metal surface, which may eventually plug up holes.

A steam iron or tea kettle can be cleaned by putting vinegar in it and heating. Remember, vinegar is a solution of acetic acid.

The heating speeds up the reaction. The reaction that occurs can be written

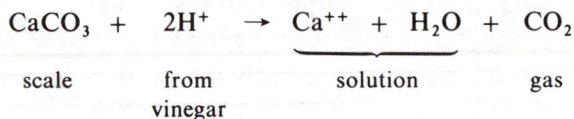

$$CaCO_3 \;+\; 2H^+ \;\rightarrow\; \underbrace{Ca^{++} \;+\; H_2O} \;+\; CO_2$$

    scale   from   solution   gas
            vinegar

The CO_2 gas bubbles off. The calcium ions and unreacted vinegar in solution, along with pieces of the dirty scale knocked loose by the reaction, should be thoroughly rinsed out with water before the iron or kettle is used again.

Limestone, $CaCO_3$, is the substance most frequently used by gardeners to help regulate the acidity of soil. The soil in a forest—filled with humus made by decaying leaves, wood, and bark—usually is quite acidic. The pH of such soil may be as low as 4.5. The nutrient organic matter in humus is good for plants, but few can thrive in a medium with such a low pH. The addition of limestone effectively neutralizes some of this acidity. The H^+ ions are removed by combining them into HCO_3^- ions. Consequently, the pH of soil so treated is raised to the range of 6 to 7. Soil with this pH is more suited for the growth of most plants. The equation for the reaction is written

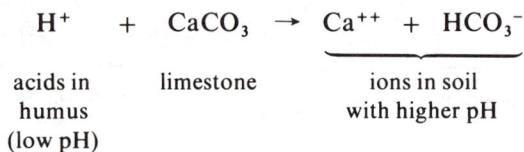

$$H^+ \;+\; CaCO_3 \;\rightarrow\; \underbrace{Ca^{++} \;+\; HCO_3^-}$$

acids in   limestone   ions in soil
humus   with higher pH
(low pH)

The features of a marble statue show the effects of the reaction $CaCO_3 + 2H^+ \rightarrow Ca^{++} + H_2O + CO_2$ caused by rain with a pH of 4–5.

OXIDES OF METALS ARE BASIC

The oxides of metals are generally basic in character rather than acidic. The most easily recognized examples of this property are metal oxides that dissolve in water by reacting with it to form hydroxides. For example, the oxides of sodium, potassium, and barium react with water in the following way:

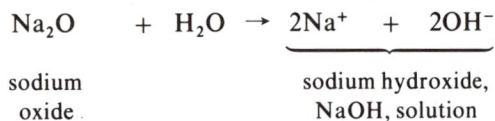

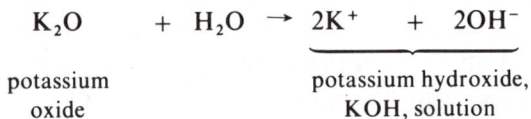

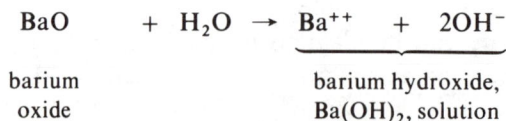

$$Na_2O \;+\; H_2O \;\rightarrow\; \underbrace{2Na^+ \;+\; 2OH^-}$$

sodium   sodium hydroxide,
oxide   NaOH, solution

$$K_2O \;+\; H_2O \;\rightarrow\; \underbrace{2K^+ \;+\; 2OH^-}$$

potassium   potassium hydroxide,
oxide   KOH, solution

$$BaO \;+\; H_2O \;\rightarrow\; \underbrace{Ba^{++} \;+\; 2OH^-}$$

barium   barium hydroxide,
oxide $Ba(OH)_2$, solution

The solutions formed by NaOH, KOH, and $Ba(OH)_2$ are typical of what are called *strong* bases. These solutions have very high concentrations of OH^- ions (pH in the range of 12 to 14). With acids, they perform neutralization reactions like those represented by the equations on page 413.

Other metals form oxides that are only partially soluble in water. For example, magnesium oxide can be mixed with water to form the milky-looking suspension called *milk of magnesia*. Magnesia is an old common name for the compound MgO. A suspension is a mixture of visible small solid particles in water; the particles settle out slowly. Some of the MgO reacts with water as follows:

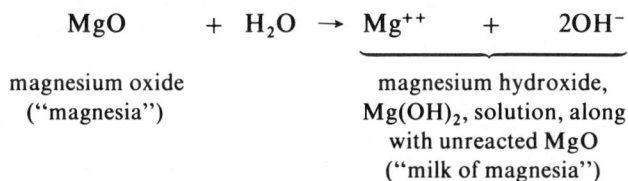

$$MgO \quad + \quad H_2O \quad \rightarrow \quad Mg^{++} \quad + \quad 2OH^-$$

magnesium oxide ("magnesia") | magnesium hydroxide, $Mg(OH)_2$, solution, along with unreacted MgO ("milk of magnesia")

Aluminum oxide, Al_2O_3, reacts even less with water, but it too can be ground into fine particles that can be mixed with water to form a suspension. A small amount of Al_2O_3 reacts as follows:

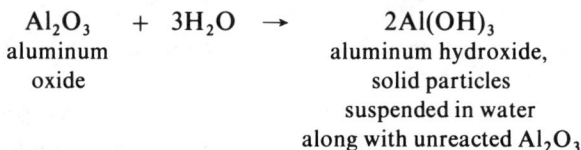

$$Al_2O_3 \quad + \quad 3H_2O \quad \rightarrow \quad 2Al(OH)_3$$

aluminum oxide | aluminum hydroxide, solid particles suspended in water along with unreacted Al_2O_3

The suspensions formed by the partially reacted oxides MgO and Al_2O_3 do not contain a high concentration of OH^- ions. The pH of milk of magnesia is about 10; that of aluminum oxide suspension is lower yet, about 8. Consequently, these are called weak bases. (Recall how the terms strong and weak were used to distinguish acids that produced many or few H^+ in solution.) However, both of these substances also can act to neutralize an acid. The equations for these reactions can be written

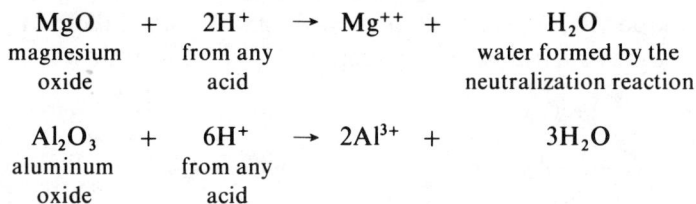

$$MgO \quad + \quad 2H^+ \quad \rightarrow \quad Mg^{++} \quad + \quad H_2O$$

magnesium oxide | from any acid | water formed by the neutralization reaction

$$Al_2O_3 \quad + \quad 6H^+ \quad \rightarrow \quad 2Al^{3+} \quad + \quad 3H_2O$$

aluminum oxide | from any acid

These equations represent the way either of these substances acts as an antacid. They neutralize excess stomach acid. It would be very dangerous to use a strong base such as NaOH as a stomach antacid. The OH⁻ ions would neutralize the acid, but the high concentration of OH⁻ ions in the solution would hurt your mouth and throat. And any excess beyond that involved in neutralization would severely damage the tissues. Magnesium oxide and aluminum oxide are the active agents in most of the antacids available at drug counters in stores. These compounds are safe antacids because at no time do their solutions have a very high concentration of OH⁻ ions. They are weak bases.

The oxides of most metals do not react with water to form solutions or suspensions of hydroxides. However, they do react with acids by a neutralization reaction. For example, if you place an old, dark-colored penny in a solution of hydrochloric acid, you will soon have a shiny, new-appearing penny. The black film covering an old penny is copper oxide, CuO. The equation for its removal is

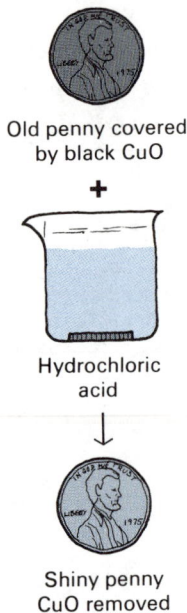

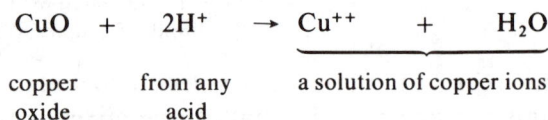

$$CuO \; + \; 2H^+ \; \rightarrow \; \underline{Cu^{++} \; + \; H_2O}$$

copper   from any   a solution of copper ions
oxide   acid

A similar reaction occurs when you use a flux as an aid in soldering to make a firm connection. The term flux means "to flow." A flux is a compound used to help the melted solder flow. Solder is a mixture of tin and lead that melts at a low temperature. If the article being soldered has a surface covered by an oxide film, the melted solder cannot come in contact with the metal to make a tight seal. Solder flux contains acidic compounds, such as rosin. The acid in the flux reacts with the basic oxide coating on the metal and removes it, so the melted solder can flow onto the metal more smoothly.

The steel industry and manufacturing operations involving steel use tremendous quantities of acids, such as sulfuric acid, H_2SO_4. Acid is used to clean steel before it is painted or otherwise coated. The acid is held in large tanks into which a steel sheet or finished article is dipped to dissolve the coating of rust (iron oxide, Fe_2O_3), as well as grease or dirt that adheres to the oxide-coated surface. Iron rust reacts with acid as follows:

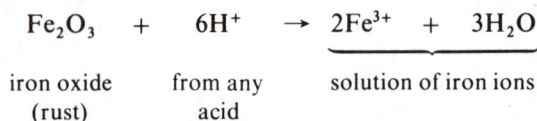

$$Fe_2O_3 \; + \; 6H^+ \; \rightarrow \; \underline{2Fe^{3+} \; + \; 3H_2O}$$

iron oxide   from any   solution of iron ions
(rust)   acid

Inevitably, the discarded waste water from such a plant contains

Old penny covered
by black CuO

+

Hydrochloric
acid

↓

Shiny penny
CuO removed

some of this acid. Before the water is allowed to go into a lake or sewer system, the extra acid in it must be neutralized, or else the waste water will be harmful to the environment into which it is dumped. The acid must be neutralized by reaction with a base. Lime, $Ca(OH)_2$, is often used, because lime is cheaper than other bases.

OXIDES OF NONMETALS ARE ACIDIC

In contrast to the basic character of metal oxides, the oxides of nonmetals generally display acidic characteristics. One example we have already discussed is the way CO_2, the oxide of the nonmetal carbon, reacts with water to form a solution of carbonic acid.

In Chapter 3, where we first introduced the term "oxides of nonmetals," and again in Chapter 8, where we discussed coal as a fuel, we referred to the problems caused by sulfur oxides in the atmosphere. You can now understand what is meant by a report stating that the pH may be as low as 3 or 4 in the rain falling on highly industrialized areas such as western Europe or the eastern United States. Such high acidity in rain is caused by the sulfur oxides, SO_2 and SO_3, that dissolve in the atmospheric moisture.

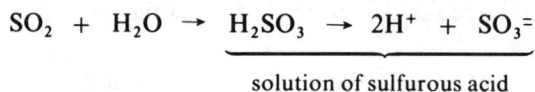

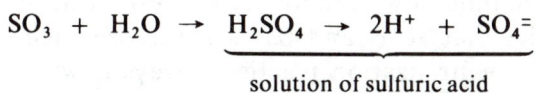

$$SO_2 + H_2O \rightarrow \underbrace{H_2SO_3 \rightarrow 2H^+ + SO_3^=}_{\text{solution of sulfurous acid}}$$

$$SO_3 + H_2O \rightarrow \underbrace{H_2SO_4 \rightarrow 2H^+ + SO_4^=}_{\text{solution of sulfuric acid}}$$

We can also relate another part of the discussion in Chapter 8 to the ideas of this chapter. The treatment used to remove SO_2 from the gas and smoke of a coal burner is essentially a neutralization reaction. The acidic SO_2 in the gases that otherwise would go out into the air is reacted with basic compounds, such as $CaCO_3$ or Na_2CO_3. The reactions are

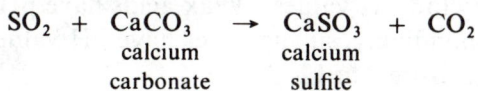

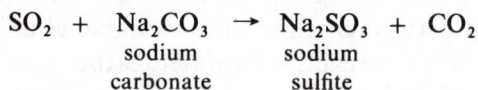

$$SO_2 + \underset{\substack{\text{calcium} \\ \text{carbonate}}}{CaCO_3} \rightarrow \underset{\substack{\text{calcium} \\ \text{sulfite}}}{CaSO_3} + CO_2$$

$$SO_2 + \underset{\substack{\text{sodium} \\ \text{carbonate}}}{Na_2CO_3} \rightarrow \underset{\substack{\text{sodium} \\ \text{sulfite}}}{Na_2SO_3} + CO_2$$

The sulfur-containing compounds, $CaSO_3$ and Na_2SO_3, are

solids that can be collected out of the gas stream. The CO_2 goes up the stack.

Abandoned coal mines also represent a source of environmental danger, because of the same chemical reactions that produce a low pH in rain. When water seeps into an old mine, it dissolves the SO_2 that has slowly been formed by the action of oxygen from the air on the sulfur in the coal. The drainage from such locations has a high acidity. As the acid-containing water moves through streams and over the ground, the acid reacts further with minerals. When the water finally reaches a lake or river, it may be badly polluted. The solution to this combination of natural events is to seal the mines tightly or to keep the water from entering in the first place.

Summary

Our discussion of acids and bases illustrates the use of a broad general principle to interpret many specific observations. Not only do we thus "find out what is going on," but we can also use these ideas to predict and perform many beneficial reactions. For example, knowing that sulfur oxides are acidic helps us predict that they can be reacted with basic substances before they pollute the atmosphere.

Acids are proton donors; bases are proton acceptors. Consequently, even though we cannot see the protons transferring, we can recognize that acids and bases neutralize each other. One product of a neutralization reaction always is water. The combination of H^+ ions with OH^- ions produces H_2O molecules.

Water solutions in which the $[H^+]$ exceeds the $[OH^-]$ are called acidic solutions. The size of the $[H^+]$ can be expressed on a scale called the pH scale. Acidic solutions have low pH values, from 0 to 7. Solutions in which the $[OH^-]$ exceeds the $[H^+]$ are called basic, or alkaline, solutions. Such solutions have high pH values, from 7 to 14. Neutral water has a pH of 7. Solutions of strong acids have very low pH values. Weak acids have pH values from 2 to 7. Correspondingly, strong bases have pH values from 12 to 14; weak bases, from 7 to 12.

Oxides of metals have basic properties; oxides of nonmetals have acidic properties. One important nonmetal oxide is carbon dioxide, CO_2. When it reacts with water, carbonic acid is formed, which can break up into bicarbonate and carbonate ions. Many reactions involve the equilibria that exist among CO_2 and the ions it forms in water solution. Other important nonmetallic oxides are SO_2 and SO_3, because of the potential environmental damage their water solutions may cause. Avoiding this hazard by removing sulfur oxides from the gases emerging from coal-burning plants is an application of the principles of acid–base chemistry.

Glossary

The number in parentheses indicates the text page where you can find the term defined in context.

acid   a substance that donates hydrogen ions to some other substance (407)

acid-base indicator   a large molecule whose color changes at a given pH because of slight changes in its structure (419)

acidity   the concentration of hydrogen ions in a solution (416)

anion   a negative ion, so named because it is attracted to the anode (410)

anode   the positively charged electrode of a battery or voltage source (410)

antacid   a substance swallowed to reduce the acidity of the gastric juices (425)

base   a substance that accepts hydrogen ions from an acid (411)

basicity   the concentration of hydroxide (OH^-) ions in a solution (416)

cathode   the negatively charged electrode of a battery or voltage source (410)

cation   a positive ion, so named because it is attracted to the cathode (410)

equilibrium constant for water   a number that represents the balance between the concentrations of H^+ and OH^- in a solution (416)

neutrality   the condition of a solution in which $[H^+] = [OH^-] = 10^{-7}$ mole/liter, equal to a pH of 7.0 (415)

pH scale   a shorthand notation of concentration of H^+ in solutions. pH values above 7.0 are basic; those below 7.0 are acidic. (417)

strength (of acid or base)   the extent to which an acid (base) produces H^+ (OH^-) in a solution (420)

titration   the determination of the acidity of a substance by the addition of a measured quantity of a base (or vice versa, a base tested with the addition of an acid) until neutrality of the solution is achieved (420)

Exercises

14.1   In a speech accepting the Democratic nomination for President in July 1952, Adlai Stevenson said: "More important than winning the election is governing the nation. That is the test of a political party: the acid, final test." What is meant by the phrase "acid test" in this context?

14.2   President Harry Truman once said, "If there is one basic element in our Constitution, it is civilian control of the military." Contrast the meaning of the term "basic element" used in this context with the way the term has been used in this chapter.

14.3   Substance X receives a proton from substance Y. Which is the acid and which is the base?

14.4   Explain why the salt sodium sulfate, Na_2SO_4, contains two sodium ions, whereas the salt sodium hydrogen sulfate, $NaHSO_4$, contains only one sodium ion.

14.5   Why would it be incorrect for a recipe to tell you to add one teaspoonful of *a* salt to food you are cooking?

14.6   A water solution containing 0.10 mole of hydrochloric acid is added to a water solution containing 0.10 mole of sodium hydroxide.

   a.   What molecules are formed by the neutralization reaction that occurs?
   b.   Why do no crystals of the salt sodium chloride appear on the bottom of the vessel?
   c.   What could you do to obtain crystals of solid sodium chloride?

14.7   Classify each of the following as an acid, a base, or a salt.

   a. $NaNO_3$ *b.* H_2SO_4 *c.* Na_3PO_4
   d. KOH *e.* HBr

14.8   Complete the following table with the formula of the salt that is formed when the hydroxide at the top of the column reacts with the acid at the left. (The filled-in examples can be used as helpful hints.)

	NaOH	Ca(OH)$_2$	NH$_4$OH	Ba(OH)$_2$
HCl				
H$_2$SO$_4$		CaSO$_4$	(NH$_4$)$_2$SO$_4$	
HNO$_3$				Ba(NO$_3$)$_2$
H$_3$PO$_4$		Ca$_3$(PO$_4$)$_2$		

14.9   Write complete balanced chemical equations for each of the reactions represented by the salts whose formulas you supplied in 14.8. The following equations are given for the filled-in examples:

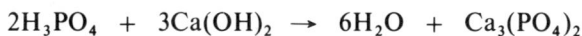

$$H_2SO_4 \ + \ Ca(OH)_2 \ \rightarrow \ 2H_2O \ + \ CaSO_4$$

$$H_2SO_4 \ + \ 2NH_4OH \ \rightarrow \ 2H_2O \ + \ (NH_4)_2SO_4$$

$$2HNO_3 \ + \ Ba(OH)_2 \ \rightarrow \ 2H_2O \ + \ Ba(NO_3)_2$$

$$2H_3PO_4 \ + \ 3Ca(OH)_2 \ \rightarrow \ 6H_2O \ + \ Ca_3(PO_4)_2$$

14.10   A certain solution contains equal numbers of hydrogen ions and hydroxide ions.

   a.   What term can be used to describe the solution?

   b.   What is the pH of the solution?

14.11   How would you read the symbol $[Cl^-]$ in words?

14.12   The pH of several solutions are given. In each case, tell whether the solution is acidic, basic, or neutral.

   a.   pH 2   *b.*   pH 10   *c.*   pH 6.5

   d.   pH 7.0   *e.*   pH 7.3

14.13   If the $[OH^-]$ increases in an aqueous solution, does the $[H^+]$ increase or decrease? Is this relationship direct or inverse?

14.14   When you develop a print of a photograph, you first place the print in developer solution (a basic solution). Then you move the print into a solution called a "stop bath" (a solution of acetic acid and other substances).

   a.   What type of chemical reaction occurs in the stop bath solution?

   b.   Some stop bath solutions turn blue when they are exhausted, so that you are warned to make up fresh solution. How is this color change accomplished?

14.15   Fresh cow's milk has a pH of 6.4. Milk that has gone sour has a pH in the range of 3 to 4. What kind of chemical substances have been produced in the milk by the action of the bacteria that caused it to go sour?

14.16   Tomato juice has a pH of about 4. How does this information tie together with the taste of tomato juice?

14.17   The pH of a cosmetic is in the range of 4.5 to 5.5. The manufacturer states that the product is "pH balanced" to be compatible with the natural pH of human skin. Is skin very acidic, slightly acidic, slightly basic, or very basic?

14.18   If you have an accident involving the spilling of a strong acid, the first thing you should do is flood the area with a lot of water. How does this procedure help? (Consider what you are doing to the concentration of hydrogen ions.)

14.19   If you have spilled strong acid on your skin, you should first wash it off with a lot of water. Then you should cover the affected skin area with a paste of baking soda, $NaHCO_3$, in water.

> *a.*   Describe the chemical reaction between the baking soda and the acid.
>
> *b.*   Write a balanced chemical equation for the reaction.

14.20   If you burn your finger on a hot stove and have no burn ointment in the house, you can ease the pain of the blister by immediately covering it with a paste of baking soda, $NaHCO_3$. In light of the text discussion about pH and pain, explain why baking soda brings relief to a burned finger.

14.21   The antacid marketed under the trade name Rolaids has as its effective ingredient a combination of sodium carbonate, Na_2CO_3, and aluminum hydroxide, $Al(OH)_3$. The formula for this compound can be represented as $NaAl(OH)_2CO_3$. Complete and balance the equation for the neutralization reaction it performs:

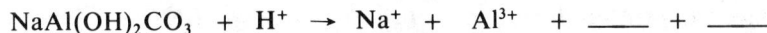

$$NaAl(OH)_2CO_3 \;+\; H^+ \;\rightarrow\; Na^+ \;+\; Al^{3+} \;+\; \underline{\quad} \;+\; \underline{\quad}$$

14.22   If you repeatedly water a potted plant with hard water, you may raise the pH of the soil in the pot and hinder the growth of the plant. What chemical reaction is responsible for this increase in the alkalinity (basicity) of the soil? (Remember what ions are dissolved in hard water.)

14.23   Grass will not grow under a large pine tree where the needles fall, accumulate, and decompose.

> *a.*   Why will grass not grow in this soil even after the accumulated dead pine needles are cleared away?
>
> *b.*   What should be added to the soil so that grass can grow on it?

14.24   The compound sodium carbonate, Na_2CO_3, is a white, water-soluble solid sold in stores as "washing soda." A water solution of this compound feels slippery on your fingers.

> *a.*   What does this sensation suggest about the concentration of OH^- ions in the solution?

b. If you measure the pH of the solution, you find it is about 11.5. Complete the following chemical equation and use it to explain why the pH is so high.

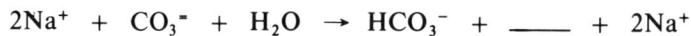

$$2Na^+ + CO_3^- + H_2O \rightarrow HCO_3^- + \underline{\quad\quad} + 2Na^+$$

14.25 One method of brightening the bottom of copper pots is to rub them with a cloth soaked in vinegar or with the inside of the rind left from squeezing a lemon. Why does this method work?

14.26 When the acid drainage from abandoned mines reaches groundwater and enters streams, it can make the pH too low for organisms to live there. One proposal to counteract this problem is to dump large amounts of limestone in these mines. Why would this procedure correct the trouble?

15

Oxidation and Reduction—The Give or Take of Electrons

☐ What happens when something is oxidized?

☐ What is the connection between oxidation and reduction?

☐ What is the connection between the reactivity of metals and the advance of human technology?

☐ How does a battery function as an "electron pump"?

☐ How can we store an electric current in a battery?

☐ How can electricity be used to release metals from their ores?

I n Chapter 14, we discussed how the idea of acids and bases, one of the broad principles of chemistry, ties together a wide variety of observations and information. In this chapter, we will explore a similar broad theme—oxidation and reduction. Instead of a model involving the transfer of invisible protons from one substance to another, we will develop a comparable model that involves the transfer of equally invisible electrons.

You may already expect that a general principle must connect at least some chemical reactions (atoms) with phenomena involving electricity (electrons). In Chapter 6, we described how atoms of some elements combine by a transfer of electrons. In Chapter 8, we described a fuel cell, a device in which the transfer of electrons caused by a chemical reaction can be harnessed to perform electrical work. The discussion in Chapter 13 of conductance in aqueous solutions included a description of the reverse process: A flow of electrons can be harnessed to perform a chemical reaction. We will build on this information and develop additional ideas. For example, what happens when iron rusts? And assuming an understanding of the rusting process, what can be done to stop, or at least control, this spontaneous chemical reaction that costs industrialized societies billions of dollars a year? How is iron ore, a compound of iron similar to rust, converted into iron metal at the rate of more than half a ton per person per year in the United States? How do batteries work? And how does heating tarnished silverware in a solution of baking soda in an aluminum pan remove the tarnish? Answers to such varied questions as these all illustrate the principles of oxidation and reduction.

WHAT IS OXIDATION?

The word *oxidation* refers to a kind of chemical reaction. Unlike the term "acid," which refers to a substance that behaves in a certain way, oxidation refers to a *process* by which a substance is changed into something else. The original meaning of the term derives from the reaction that occurs when something burns. In the simplest sense, oxidation is the combining of something with oxygen. For example, if a piece of copper metal is heated by a gas flame so that air can reach the metal, the surface of the copper turns black. A coating of black solid copper oxide, CuO, forms. The atoms of copper undergo an oxidation reaction.

Copper metal reacts in a similar way with sulfur to form CuS, copper sulfide, or with chlorine to form copper chloride, $CuCl_2$. The green color you see on copper roofs or bronze statues exposed to the weather is a mixture of CuO, CuS, and $CuCl_2$, along with copper carbonate, $CuCO_3$. The surface of the copper metal is covered with a film of these compounds formed by reaction with air, smoke, and sometimes the salt in moist air blown from an ocean. An examination of the chemical equations for these reactions shows that the common feature is what happens to the copper.

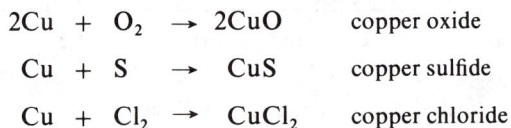

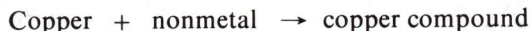

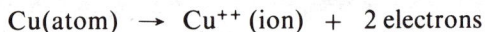

$$2Cu + O_2 \rightarrow 2CuO \qquad \text{copper oxide}$$

$$Cu + S \rightarrow CuS \qquad \text{copper sulfide}$$

$$Cu + Cl_2 \rightarrow CuCl_2 \qquad \text{copper chloride}$$

$$\text{Copper} + \text{nonmetal} \rightarrow \text{copper compound}$$

$$Cu(\text{atom}) \rightarrow Cu^{++}(\text{ion}) + 2 \text{ electrons}$$

In each of these reactions, the copper forms a compound by losing two electrons to one or more atoms of a nonmetal. This common feature is the basis for classifying all these reactions as oxidation reactions, even though oxygen is not necessarily involved.

An oxidation is a reaction in which a substance loses electrons.

All metals form compounds by oxidation reactions; the characteristic chemical property of metals is to lose electrons and thus to become positively charged cations. Therefore, we can anticipate that a discussion of oxidation will include such important reactions as the rusting of iron. Metallic iron forms the compound iron oxide, Fe_2O_3, in the presence of oxygen and moisture. The corrosion of *any* metal is an oxidation reaction; metal atoms lose electrons to form compounds.

WHAT IS REDUCTION?

A *reduction* is a reaction that effectively is the reverse of an oxidation. The original meaning of the word was the removal of oxygen from a compound. Practical application of reduction reactions is almost as old as history itself. Few metals are found uncombined in the free state in the natural world. The ores from which metals are produced usually are oxides, sulfides, or carbonates of the metallic elements. Centuries ago people discovered that some of these compounds could be roasted or heated with charcoal to yield metals.

Few metals are found uncombined in the natural world.

Modern technology involves refined and special methods, but the essential chemistry is the same. For example, the reaction in a blast furnace takes place between the ore—iron oxide—and the gas—carbon monoxide, CO. CO is formed by the partial combustion of the carbon in coke with a stream of hot air. The coke has previously been made by heating coal in the absence of air to drive off moisture, tar, and other volatile constituents in the coal. The equation for the reaction in a blast furnace can be written

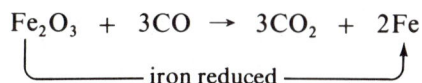

$$Fe_2O_3 + 3CO \rightarrow 3CO_2 + 2Fe$$

iron reduced

The removal of oxygen from Fe_2O_3 to make Fe metal involves the change

$$2Fe^{3+} + 6\,electrons \rightarrow 2Fe$$

This equation illustrates the general definition of a reduction:

A reduction is a reaction in which a substance gains electrons.

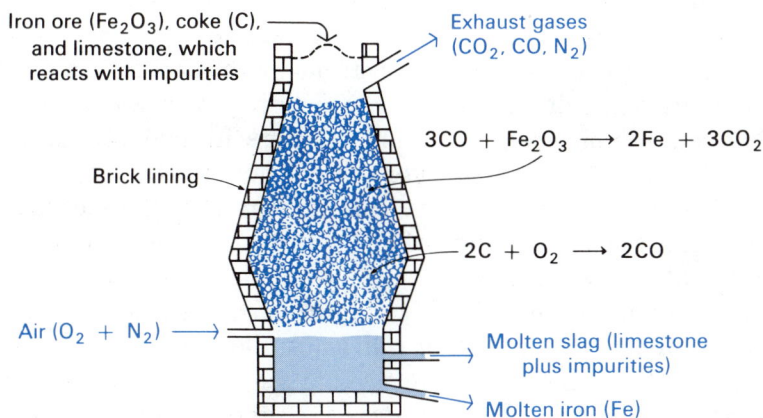

Iron ore (Fe_2O_3), coke (C), and limestone, which reacts with impurities

Exhaust gases (CO_2, CO, N_2)

$$3CO + Fe_2O_3 \rightarrow 2Fe + 3CO_2$$

Brick lining

$$2C + O_2 \rightarrow 2CO$$

Air ($O_2 + N_2$) →

Molten slag (limestone plus impurities)

Molten iron (Fe)

A blast furnace used to reduce iron ore

OXIDATION AND REDUCTION
MUST OCCUR TOGETHER

The definitions of oxidation as the giving up of electrons and reduction as the taking on of electrons imply an important relationship between these two reactions. They always must occur together. Just as in an exchange of a gift between friends, giving and receiving must occur at the same time. Therefore, whenever you see evidence of oxidation, such as rusting or tarnishing, you know that something else must have been reduced at the same time. Electrons lost by one substance are gained by the other. You can see how oxidation and reduction fit together by reexamining the equation that describes the simple combination of copper with sulfur.

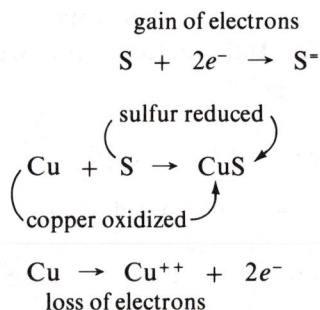

$$\text{gain of electrons}$$
$$S + 2e^- \rightarrow S^=$$

$$\text{sulfur reduced}$$
$$Cu + S \rightarrow CuS$$
$$\text{copper oxidized}$$

$$Cu \rightarrow Cu^{++} + 2e^-$$
$$\text{loss of electrons}$$

The balanced equation accounts for the transfer of two electrons from a copper atom to a sulfur atom.

In the case of the reactions in a blast furnace, the reduction of the iron from iron oxide to metallic iron is accompanied by the oxidation of carbon from CO to CO_2. We have shown the equation for the gain of six electrons by two Fe^{3+} to make two Fe atoms. However, the loss of electrons by the carbon atoms may be more difficult to recognize. Carbon atoms are covalently bonded with one oxygen atom in CO and with two oxygen atoms in CO_2. Pairs of electrons are shared, not transferred, between the carbon and oxygen atoms. It is not correct to think of the CO molecule as a combination of C^{++} with $O^=$ ions. Nor can the CO_2 molecule be written as C^{4+} ions combined with two $O^=$ ions. However, it is correct to think that the carbon atom loses at least some influence over its four outermost electrons when it shares any or all of them with one or two oxygen atoms. In the CO molecule, the C atom can be considered to share *one* pair of electrons ($2e^-$) with an O atom. In the CO_2 molecule, the C atom can be considered to share *two* pairs of electrons ($4e^-$) with two O atoms. Therefore, the consequence is the same: The carbon atoms effectively lose electrons in the reaction.

Each of the 3C atoms shares 2 more electrons,
$3 \times 2e^- = 6e^-$ effectively lost.

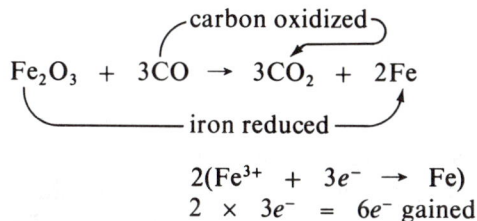

carbon oxidized

$$Fe_2O_3 + 3CO \rightarrow 3CO_2 + 2Fe$$

iron reduced

$$2(Fe^{3+} + 3e^- \rightarrow Fe)$$
$$2 \times 3e^- = 6e^- \text{ gained}$$

Many oxidation–reduction reactions are very complicated, yet all involve the same essential process: an exchange of electrons. For example, the stimulation and response of nerve cells in your body involve oxidation–reduction reactions among complex molecules. So do the reactions of metabolism, by which glucose, $C_6H_{12}O_6$, reacts with oxygen to produce CO_2 and H_2O. Even though the mechanism may be very complex, the reaction can be represented in a simplified manner.

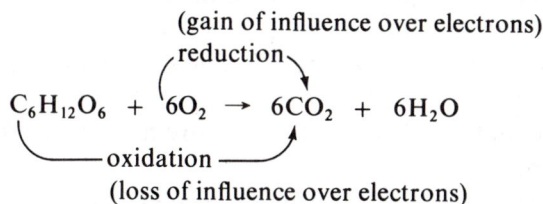

(gain of influence over electrons)
reduction

$$C_6H_{12}O_6 + 6O_2 \rightarrow 6CO_2 + 6H_2O$$

oxidation
(loss of influence over electrons)

In the oxidation part of this reaction, the carbon atoms are forced to share more pairs of electrons in CO_2 molecules than they do in the $C_6H_{12}O_6$ molecule. So we can consider the carbon atoms to have effectively lost electrons. The oxygen atoms of the O_2 molecules effectively have gained an influence over more electrons by forming the covalent bonds in the molecule CO_2. So the oxygen is thereby reduced.

GERMICIDES AND BLEACHING AGENTS

Killing germs by adding chlorine to water is another example of a complex oxidation–reduction reaction. The atoms in the Cl_2 molecule have a strong attraction for electrons. Consequently, they gain them by disrupting complex molecules involved in the metabolism reactions of the living microorganisms, the germs.

A chemical that kills germs is called a *germicide*

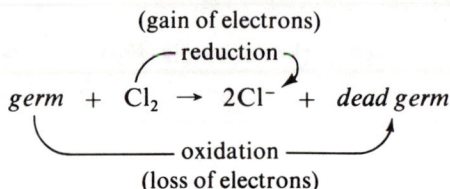

$$\text{germ} + Cl_2 \rightarrow 2Cl^- + \text{dead germ}$$

(gain of electrons)
reduction

oxidation
(loss of electrons)

Chlorine is a very powerful producer of such oxidation reactions. This is why Cl_2 is a very dangerous chemical. Many large organic molecules are susceptible to disruption by chlorine. Among these molecules are some kinds of dye molecules. Therefore, solutions of chlorine are effective as bleaching agents:

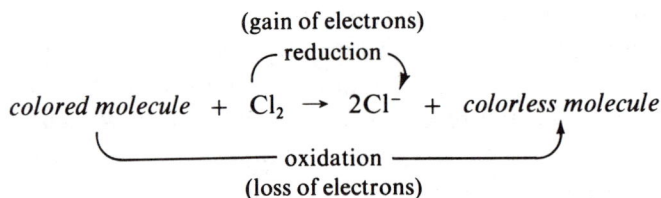

$$\text{colored molecule} + Cl_2 \rightarrow 2Cl^- + \text{colorless molecule}$$

(gain of electrons)
reduction

oxidation
(loss of electrons)

These simplified equations should make clear to you why the addition of sodium chloride, salt, NaCl, even though it contains the element chlorine, cannot be effective as a germicide or bleach. Adding chloride ions, Cl^-, to a solution cannot accomplish the necessary oxidation. The Cl^- ion already has its electron; it can take on no more.

The danger of having chlorine molecules oxidizing everything they encounter can be lessened by using dilute solutions made by reacting Cl_2 gas with sodium hydroxide, NaOH. Most laundry bleaches, such as Clorox, contain NaOH. The chlorine is combined into a compound called sodium hypochlorite, NaOCl, in the solution. This substance decomposes so that delivery of the powerful chlorine can be controlled as needed. Another similar compound that can be used as a germicide is calcium hypochlorite, $Ca(OCl)_2$. The common name for this chemical, chloride of lime, suggests the way it is made. Chlorine, Cl_2, is combined with $Ca(OH)_2$. ($Ca(OH)_2$ is the compound formed when lime, CaO, is put in water.)

Laundry
Bleach
(NaOCl)

H.T.H.
High Test
Hypochlorite
$Ca(OCl)_2$

Bleaches and germicides release chlorine in solutions.

METALS IN ANCIENT HISTORY

The large majority of the elements are metals. With a few exceptions—such as gold and, in a few places, silver and copper—metals are found in the natural world as ores, compounds of metals. In spite of this lack of metals found free in nature, even the humans of primitive cultures used metals for tools, weapons, and other articles. The extracting of metals from their ores existed as an art for centuries before modern times. Some of the workmen King Solomon summoned to build his temple in about 950 BC knew about gold, silver, copper, mercury, lead, tin, iron, and zinc. Mixtures of metals, *alloys,* such as bronze (copper and tin) or brass (copper and zinc) also were known.

At the height of their civilization, Romans used lead to make water pipes and cooking pots. An interesting speculation, based on modern knowledge, is that the harmful effects of lead poisoning on the health of the upper-class Romans may have played a

---

Harmful effects of lead poisoning may have played a role in the fall of Roman civilization.

---

role in the fall of Roman civilization. Lead, in harmful amounts, could have come into their diets from the lead water pipes and cooking utensils. The general level of the health, the ability to bear children, and the capacities of the leaders of the nation declined. The poorer Romans had no water pipes in their houses, and they cooked in clay pots. The same was true for the invading barbarians. Consequently, the weakened leaders were overcome and the nation along with them. Archeologists have found Roman coins that contain varying amounts of lead in the gold and silver. Emperors in financial difficulty apparently debased the currency of the empire by mixing the cheaper lead into the alloys of the more expensive metals.

Producing metals from ores remained an art until the nineteenth century. The ancients knew *what* to do but little of *why* to do it or how to improve these processes. Metals were produced by following recipes that changed little during thousands of years. Improvements came about through trial and error. Seldom were specific chemical reactions recognized as such, much less understood. The *science* of metals began when chemistry began, at about the beginning of the nineteenth century. Changes in substances began to be interpreted in terms of how atoms underwent chemical reactions. Gradually observations grew into laws, and theories developed to interpret the laws. Now answers to questions of "Why?" (theories) guide decisions on "What to do?" (experiments) to provide the metals so indispensable to modern living.

Table 15.1   Ores, reduction processes, and uses of typical metals

	Metal	Ore	Reduction of Ore	Description and Uses
Most reactive	Sodium Na	NaCl salt, brine	Electrolysis: $2NaCl \rightarrow 2Na + Cl_2$	Soft white metal used as coolant in nuclear reactors
	Magnesium Mg	$MgCl_2$ seawater, brines	Electrolysis: $MgCl_2 \rightarrow Mg + Cl_2$	Light metal used in alloys for structures such as aircraft
	Aluminum Al	Al_2O_3 bauxite	Electrolysis: $2Al_2O_3 \rightarrow 4Al + 3O_2$	Covered by corrosion-resistant film; used in packaging, alloys for buildings, aircraft
	Zinc Zn	ZnS zinc blende	Heat + air + carbon: $ZnS + 2O_2 + C$ $\rightarrow Zn + SO_2 + CO_2$	Alloys for diecasting, brass solders; used to "galvanize" steel
	Iron Fe	Fe_2O_3 hematite	Heat + CO gas: $Fe_2O_3 + 3CO$ $\rightarrow 2Fe + 3CO_2$	Alloys with other metals to make steel structures, vehicles, machines
	Tin Sn	SnO_2 cassiterite	Heat + carbon: $SnO_2 + C \rightarrow Sn + CO_2$	Coating steel for "tin cans"; alloys, bronze, solder
	Lead Pb	PbS galena	Heat + air + carbon: $PbS + 2O_2 + C$ $\rightarrow Pb + SO_2 + CO_2$	Heavy metal used in storage batteries, bronze, solder
	Copper Cu	Cu_2S chalcosite	Heat + air: $Cu_2S + O_2 \rightarrow 2Cu + SO_2$	Used in electrical conductors, bronze, brass
	Mercury HgS	HgS cinnabar	Heat + air: $HgS + O_2 \rightarrow Hg + SO_2$	Dense liquid metal used in electrical devices, barometers, thermometers
Least reactive	Gold Au	Au gold nuggets	Metal found in natural state	Highly reflective and inert metal used in jewelry, coinage

THE REACTIVITY SERIES OF THE METALS

Table 15.1 summarizes some essential information about ten common metals. The chief ore of each is listed along with an indication of the chemical reaction used to reduce the metal from its ore. Some of the chief uses of each metal also are listed. The order of the metals, going from top to bottom, is a significant feature of this table. The metals are arranged in order of decreasing reactivity. The most reactive of the ten metals is sodium, Na, at the top of the list. The least reactive of the ten is gold, Au, placed at the bottom.

When a metal reacts to form a compound, it is oxidized. The reverse reaction, a reduction, occurs when a metal is produced from one of its compounds. A general equation representing these two reactions is

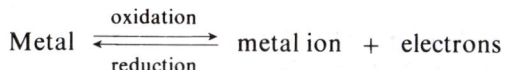

$$\text{Metal} \underset{\text{reduction}}{\overset{\text{oxidation}}{\rightleftharpoons}} \text{metal ion} + \text{electrons}$$

The greater the tendency of a metal to be oxidized, the less its tendency to be reduced from its compounds. This relationship also means that the compounds of a reactive metal are correspondingly very stable. Sodium, at the top of Table 15.1, is so reactive that it has to be stored in oil to keep air away from it. Otherwise, it spontaneously forms compounds with the oxygen and moisture in air. Sodium compounds are very stable; it is difficult to break them down. In contrast, gold, at the bottom of the table, is quite unreactive. Gold is found free in nature because it does not combine readily with other elements. The compounds

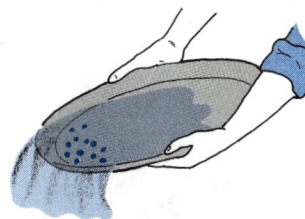

"Panning" for gold—metal nuggets separate from sand because gold metal is more dense than sand.

The greater the tendency of a metal to be oxidized, the less its tendency to be reduced from its compounds.

it does form are relatively unstable. Consequently, if such compounds had been formed at some early time in the earth's history, they long since would have decomposed to yield free gold.

Table 15.2 suggests a summary of the general information implied by arranging the metals in a reactivity series. The metals

Table 15.2   The reactivity series of the metals

Stability of compounds (ores)	Tendency of atom to lose electrons	Tendency of atom to be oxidized	React-ivity		Tendency of com-pounds (ores) to be reduced	Tendency of ion to take on electrons
---------- (Greatest) ---------- ↑	↑	↑	↑	Na (Least) Mg Al Zn Fe Sn Pb	↓	↓
↑	↑	↑	↑	Cu Hg	↓	↓
---------- (Least) ----------				Au (Greatest)		

that form the most stable compounds are at the top. These metals are the most reactive; their atoms have the greatest tendency to lose electrons and thus to be oxidized. The metals at the bottom are the least reactive; their compounds arc thc least stable; their ions have the greatest tendency to take on electrons and thus be reduced.

HUMAN CIVILIZATION AND THE REACTIVITY SERIES

There is a connection between the reactivity series of the metals and the order of the major epochs of human history. This connection is suggested by Table 15.3.

During the Stone Age, the earliest period of human history, people were nomads or wanderers. They used few things that could not be found "ready-made" on the earth. No doubt they came upon pieces of gold among the stones they found that could be fashioned into crude tools. Silver and copper also can occasionally be found in natural metallic form. These metals have been known from the earliest times. Their small tendency to react and the instability of their compounds mean that nature provides them free. People did not have to devise chemical processes using energy to produce these metals.

After thousands of years, some people began to wander less. Early agriculture and other uses of natural resources led to the establishment of the first settled communities. The invention of the wheel during this period of history marks the beginning of

Table 15.3　Human civilization's climb up the reactivity series

Metal		Epoch	Technology
Sodium	Na		
Magnesium	Mg	Modern Age	Light metals produced by use of electricity
Aluminum	Al		
Zinc	Zn		
Iron	Fe	Iron Age	Metal ore heated with charcoal
Tin	Sn		
Lead	Pb		
Copper	Cu	Bronze Age	Metals produced from ores by simple heating
Mercury	Hg		
Gold	Au	Stone Age	Metal found as a free, uncombined metal in nature

human civilization. During this period of history, some fortunate or perhaps particularly inquisitive person discovered that some kinds of rocks could be heated to yield metals. A naturally occurring compound of copper, Cu_2S, only needs to be heated in air to produce copper metal.

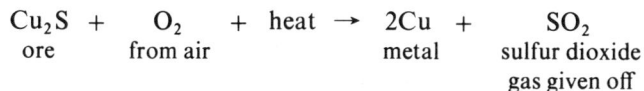

$$Cu_2S \; + \quad O_2 \quad + \; heat \; \rightarrow \; 2Cu \; + \qquad SO_2$$
$$\text{ore} \qquad \text{from air} \qquad\qquad\qquad \text{metal} \qquad \text{sulfur dioxide}$$
$$\text{gas given off}$$

Bronze was formed from ores of copper and tin by this uncomplicated process. The slightly greater reactivity of the metals that make bronze meant people had to exercise some ingenuity and acquire some knowledge to win them from nature. To make bronze, people had to do more than just pick up free samples, as could be done with gold, a less reactive metal. This epoch in history is known as the Bronze Age.

At a still later date, the slightly more difficult reaction required for winning iron from its ores was discovered. This discovery involved knowing not only the proper kind of rocks to use, but also that charcoal is needed. One kind of iron ore is the compound iron carbonate, $FeCO_3$. When $FeCO_3$ is heated with charcoal, iron is produced.

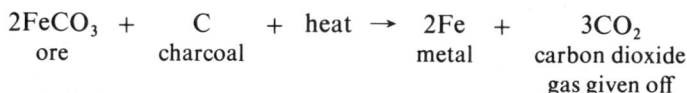

$$2FeCO_3 \; + \quad C \quad + \; heat \; \rightarrow \; 2Fe \; + \qquad 3CO_2$$
$$\text{ore} \qquad \text{charcoal} \qquad\qquad\qquad \text{metal} \qquad \text{carbon dioxide}$$
$$\text{gas given off}$$

Another kind of iron ore is Fe_2O_3. This ore reacts in the same way.

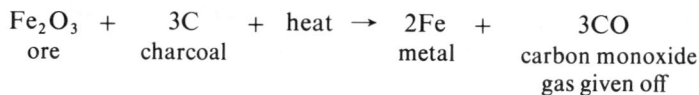

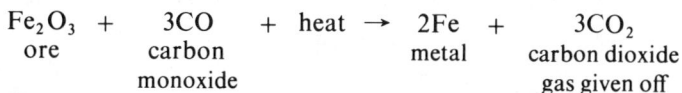

$$Fe_2O_3 \; + \quad 3C \quad + \; heat \; \rightarrow \; 2Fe \; + \qquad 3CO$$
$$\text{ore} \qquad \text{charcoal} \qquad\qquad\qquad \text{metal} \qquad \text{carbon monoxide}$$
$$\text{gas given off}$$

$$Fe_2O_3 \; + \quad 3CO \quad + \; heat \; \rightarrow \; 2Fe \; + \qquad 3CO_2$$
$$\text{ore} \qquad \text{carbon} \qquad\qquad\qquad \text{metal} \qquad \text{carbon dioxide}$$
$$\text{monoxide} \qquad\qquad\qquad\qquad \text{gas given off}$$

The metallurgy of iron was exploited to provide the many various tools and weapons for the civilizations of the Iron Age. The great period of Greek history, about 1000 BC, was part of the early Iron Age.

Human civilization remained on this rung of the reactivity series ladder until late in the industrial revolution. Not until the modern era did the more difficult task of reducing ores of the light metals, aluminum and magnesium, become a large-scale possibility. The much greater reactivity of these metals means their compounds are very stable. Solving the more difficult problem of extracting these metals from their ores required more knowledge and technical resources than people had before the nineteenth century.

ELECTRICAL ENERGY USED TO PRODUCE ALUMINUM METAL

The use of electricity was the key. In 1886 Charles M. Hall, an American chemist, developed a successful process for producing aluminum from its very stable compound, aluminum oxide, by an electrolysis process. A diagram of the apparatus used is shown in Figure 15.1. In this process, a mineral called cryolite is heated to its melting point. Purified bauxite, an aluminum ore composed of aluminum oxide, Al_2O_3, is added to make a solution in this hot liquid. Electricity is passed into the solution. Electrodes made of carbon are dipped into the cell; the other side of the electric circuit is connected to the body of the cell. The electricity causes the Al_2O_3 to break down into metallic aluminum, which collects on the bottom of the hot cell as a liquid. The oxygen collects at the carbon electrodes, where it combines with the carbon to make carbon monoxide gas.

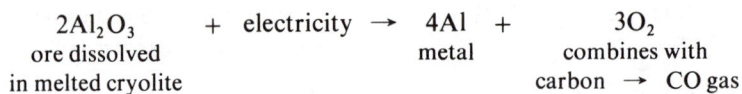

$$2Al_2O_3 \quad + \text{ electricity } \rightarrow \quad 4Al \quad + \quad 3O_2$$
ore dissolved   metal   combines with
in melted cryolite   carbon $\rightarrow$ CO gas

Aluminum and magnesium, another metal made from its ore by using electricity, are the metals typical of the modern era. Alloys of aluminum and magnesium are lightweight but very strong. Their use in aircraft has made modern air transportation possible. In spite of the high reactivity of aluminum, it can be

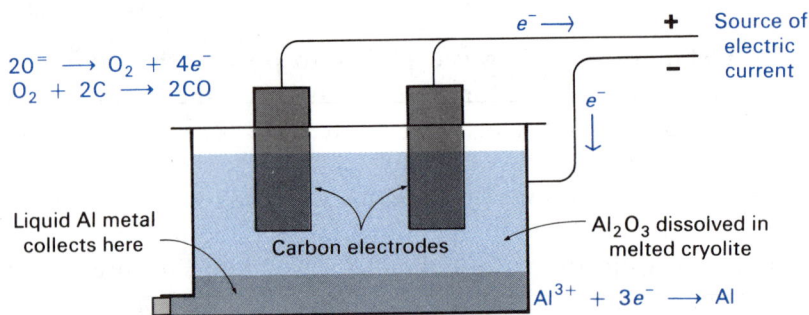

Figure 15.1 The reduction of aluminum from its ore in an electrolysis cell.

used in the metal form because the metal surface is always protected by a tough coating of aluminum oxide.

The great stability of aluminum oxide means that producing aluminum metal from its ore requires a great expenditure of energy. The greater chemical stability of aluminum ores over iron ores means modern civilizations must use five times as much energy to produce a ton of aluminum as to produce a ton of iron. The great cost in energy is why the recycling of aluminum from used cans and containers is such an important conservation measure. Used aluminum metal can be melted down and reformed into new articles. Aluminum melts at the high temperature of 660°C (1220°F). Heat energy must be supplied to reach this melting temperature. However, the amount of energy required for this step in recycling is far less than that needed to reduce new aluminum metal from its ore.

HYDROGEN'S PLACE IN THE REACTIVITY SERIES

In our discussion in Chapter 14, leading up to the definition of an acid as a proton donor, we mentioned the reaction that occurs between acids and metals. If you place a piece of iron in a hydrochloric acid solution, hydrogen gas is produced. If you go one step further and evaporate the water from the remaining solution, the solid compound, iron chloride, $FeCl_2$, separates out. The equation for the reaction is

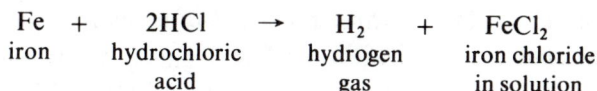

$$Fe + 2HCl \rightarrow H_2 + FeCl_2$$

| iron | hydrochloric acid | hydrogen gas | iron chloride in solution |

Sulfuric acid works the same way, except that iron sulfate, $FeSO_4$, is formed.

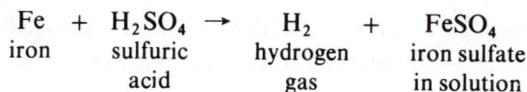

$$Fe + H_2SO_4 \rightarrow H_2 + FeSO_4$$

| iron | sulfuric acid | hydrogen gas | iron sulfate in solution |

A similar reaction occurs when the metal zinc, Zn, is placed in either hydrochloric or sulfuric acid. There is one notable difference. The reaction is more vigorous; bubbles of H_2 gas are formed faster.

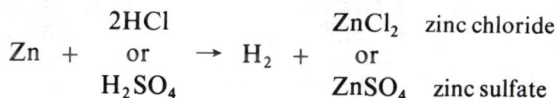

$$Zn + \begin{matrix} 2HCl \\ or \\ H_2SO_4 \end{matrix} \rightarrow H_2 + \begin{matrix} ZnCl_2 \\ or \\ ZnSO_4 \end{matrix}$$

zinc chloride

zinc sulfate

When you put a piece of magnesium metal, Mg, into either acid, again a similar reaction occurs. This time the reaction is even more vigorous.

$$Mg \ + \ \begin{matrix} 2HCl \\ or \\ H_2SO_4 \end{matrix} \ \rightarrow \ H_2 \ + \ \begin{matrix} MgCl_2 \quad \text{magnesium chloride} \\ or \\ MgSO_4 \quad \text{magnesium sulfate} \end{matrix}$$

Figure 15.2 compares the reactions of an acid with Mg, Zn, and Fe. A general representation of what happens can be written as follows:

$$\overbrace{Metal \quad + \quad 2H^+ \quad \rightarrow \quad H_2 \quad + \quad M^{++}}$$

(gain of electrons)
reduction

| atom | hydrogen ions from acid | hydrogen molecules | metal ion |

oxidation
(loss of electrons)

In each case, the metal atoms have been oxidized. They lose electrons to form metal ions. The hydrogen ions take on the electrons to form two hydrogen atoms that combine into a hydrogen molecule. The total overall reaction is a transfer of electrons, an oxidation and a reduction.

This interpretation of what happens also ties together the vigor of reaction with the position of the metals in the reactivity series (see Table 15.2). Of the three metals, Mg is nearest the top. It is the most reactive of the three. The H_2 gas is produced most vigorously by the reaction of magnesium with an acid.

Another idea is suggested by these experiments. *Hydrogen behaves chemically as the metals do.* Hydrogen ions are reduced when the metal atoms are oxidized. Likewise, a comparison of the formulas—for example, a comparison of H_2SO_4 with the product $MgSO_4$—suggests that hydrogen and a metal can change

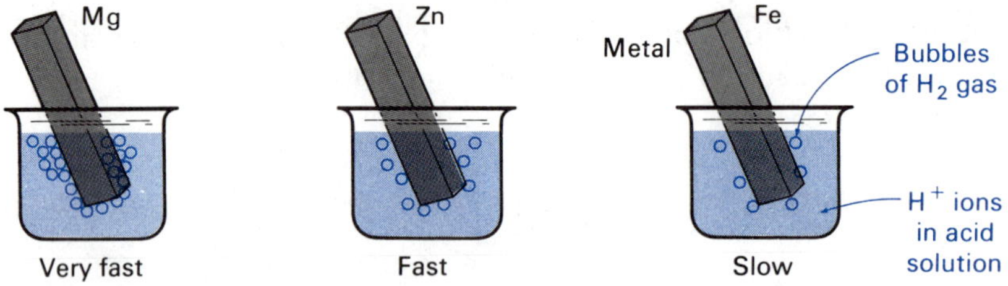

Figure 15.2   A comparison of the rates of reaction between metals and acid.

Mg — Very fast

Zn — Fast

Metal — Fe — Slow

Bubbles of H_2 gas

H^+ ions in acid solution

places in compounds. So if hydrogen behaves as a metal, the next question to ask is "Where does hydrogen fit in the reactivity series?"

To answer this question, we need to develop one more idea about the series. What we are dealing with is the balance between the tendency of the atoms to be oxidized and the tendency of the ions to undergo the reverse reaction and be reduced. Magnesium atoms' tendency to be oxidized must be the greatest of the three metals we placed in the acids. The reaction in which magnesium was oxidized was the most vigorous. The tendency for zinc to be oxidized was smaller and that for iron, even smaller.

If we follow this line of reasoning, the next experiment would be to try to react tin with an acid. Tin is located in the series just below iron. Tin placed in acid reacts to give hydrogen gas, but with a less vigorous reaction than occurs between iron and an acid. Lead also reacts, but very slowly. The reaction is slow, even if we speed up the reaction by heating the mixture. (Remember, all reactions are speeded up by increasing the temperature.)

However, when we try to react copper metal with acid, no hydrogen gas is produced. Figure 15.3 suggests the way these reactions compare with those of the other metals with acids (as in Figure 15.2). Even heating the acid or using a very concentrated solution of acid with copper does not produce the same result as the other metals above copper in the series produce. So our tentative conclusion is that hydrogen's reaction tendencies place it in the series between lead and copper.

We can test this idea and confirm the place we have assigned hydrogen by doing one more experiment. If copper does not react with hydrogen ions, then what about the reverse reaction? Hydrogen atoms or molecules may be able to react with copper ions. In an experiment to test this idea, hydrogen gas, H_2, is passed over copper oxide, CuO, in a heated tube. Figure 15.4 shows the kind of apparatus that can be used. Black copper oxide, CuO, is placed in a glass tube. Hydrogen gas is introduced from a tank so that it passes over the pieces of CuO. When the

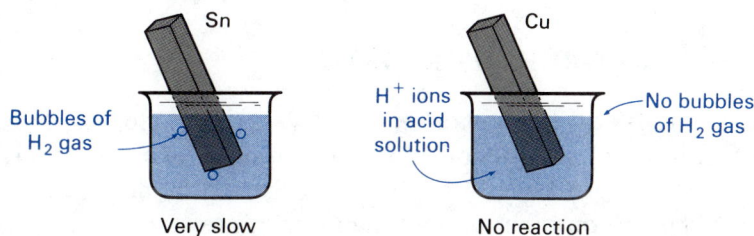

Figure 15.3 Further comparison of the rates of reaction between metals and acid.

Figure 15.4   The reduction of copper oxide to copper metal by reaction with H_2.

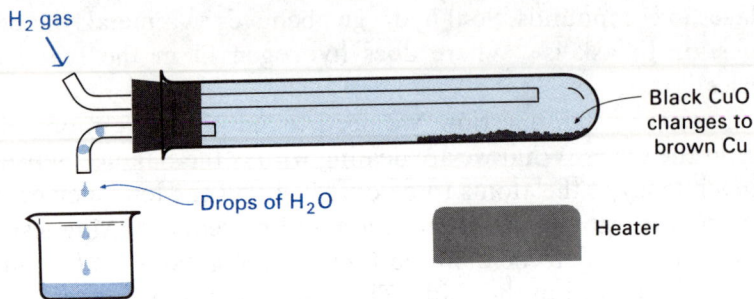

tube is heated, two changes occur. The black CuO changes to brown-colored pieces of copper. At the same time, drops of liquid H_2O form in the cooler part of the apparatus and drip out. The reaction that occurs is represented as follows:

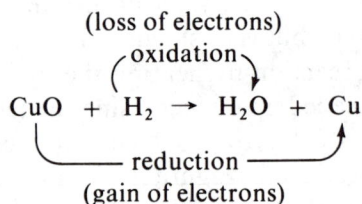

$$\text{CuO} + \text{H}_2 \rightarrow \text{H}_2\text{O} + \text{Cu}$$

(loss of electrons)
oxidation (over H_2)

reduction (under CuO → Cu)
(gain of electrons)

Table 15.4

Opposing tendencies for metals and hydrogen to gain or lose electrons

$Na^+ + e^-$	Na
$Mg^{++} + 2e^-$	Mg
$Al^{3+} + 3e^-$	Al
$Zn^{++} + 2e^-$	Zn
$Fe^{++} + 2e^-$	Fe
$Sn^{++} + 2e^-$	Sn
$Pb^{++} + 2e^-$	Pb
$2H^+ + 2e^-$	H_2
$Cu^{++} + 2e^-$	Cu
$Hg^{++} + 2e^-$	Hg
$Au^{3+} + 3e^-$	Au

These results demonstrate that hydrogen atoms (in the H_2 molecule) can give off electrons to the ions of metals below hydrogen in the reactivity series. When hydrogen ions (from an acid solution) are put in contact with atoms of metals higher in the series, the spontaneous reaction is in the other direction. The hydrogen ions accept electrons from the more reactive metal atoms.

Table 15.4 represents the reactivity series, including hydrogen, in a manner that suggests the degree to which one tendency wins out over the other. If we use arrows of equal length for ($2H^+ + 2e^- \rightleftarrows H_2$), then those metals above hydrogen have longer arrows pointing to the left. The metals above hydrogen tend to form ions more readily than hydrogen does. Below hydrogen, the longer arrows point to the right. The ions of these metals—copper, mercury, and gold—tend to accept electrons to form atoms more readily than hydrogen ions do.

A PECKING ORDER FOR METALS

Observations of the behavior of the individuals in a flock of poultry reveal the social phenomenon known as a "pecking order." Each bird fits into a hierarchy. Each member of the flock submits to pecking and domination by the bigger, more aggressive members. This submission carries with it the right to peck the weaker members of the flock. This idea of a hierarchy, or ordering of individuals in terms of their comparative behavior, is often used as a figure of speech. For example, some people speak of a pecking order of the nations in the United Nations. We can extend this analogy to suggest a parallel with the properties the metals and their ions display. Metals at the top of the reactivity series are at the top of the pecking order. Each metal is capable of

Magnesium

Each metal is capable of displacing a metal below it in the reactivity series from its compounds.

---

displacing a metal below it from its compounds. The reactions we have just discussed illustrate this general principle.

Atoms of zinc spontaneously react with compounds of hydrogen (acids) to displace hydrogen atoms. Atoms of hydrogen in turn are capable of reacting with compounds of copper to displace copper atoms.

Hydrogen compound

$$Zn + 2H^+ \rightarrow Zn^{++} + H_2$$

$$H_2 + Cu^{++} \rightarrow 2H^+ + Cu$$

A general representation of this "pecking order" is

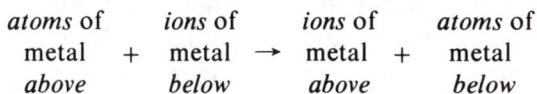

atoms of metal above	+	*ions* of metal below	→	*ions* of metal above	+	*atoms* of metal below

Magnesium compound

If you use this rule for predicting spontaneous reactions, you can tell which of the following reactions can occur:

$$Zn + CuSO_4 \rightarrow ?$$

$$Cu + ZnSO_4 \rightarrow ?$$

The location of the metals in the reactivity series shows zinc above copper. So you can correctly predict the following reactions:

Hydrogen driven off

$$Zn + CuSO_4 \rightarrow Cu + ZnSO_4$$

$$Cu + ZnSO_4 \nrightarrow \text{ no reaction}$$

REMOVING TARNISH FROM SILVERWARE

An illustration of how this pecking order can be used to practical advantage involves the household task of removing the tarnish

from articles made of silver. Tarnish consists of dark-colored compounds of silver. Silver sulfide, formed by reaction of the metal with sulfur-containing foods, such as eggs, is a chief constituent of the tarnish. Tarnish can be removed by putting the

Tarnish can be removed by putting the silver articles in contact with some aluminum foil in a pan containing boiling water and baking soda, $NaHCO_3$.

silver articles in contact with some aluminum foil in a pan containing a boiling water solution of baking soda, $NaHCO_3$. The silver becomes bright and shiny; the aluminum foil darkens and partially disintegrates. The essential chemical reaction is between aluminum atoms (higher in the series) with the ions of silver (lower in the series) that are in the compounds of the tarnish.

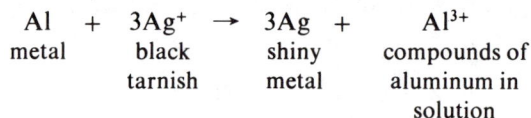

$$Al + 3Ag^+ \rightarrow 3Ag + Al^{3+}$$

| metal | black tarnish | shiny metal | compounds of aluminum in solution |

The surface of any piece of aluminum is covered by a film of aluminum oxide. Normally, this oxide film protects the metal from chemical reactions with other substances. But the oxide film partially dissolves in a basic solution (high pH). The baking soda is added to make the solution basic (pH greater than 7). Consequently, the aluminum metal and silver tarnish can come into contact, and the spontaneous reaction occurs. The aluminum ions that are formed end up either in solution or combined into a dark-colored mixture of solid aluminum compounds.

CORROSION AND RUSTING

Whenever any piece of metal loses its shiny appearance and becomes covered by a dull coating of some compound of the metal, we say that the metal has *corroded*. Sometimes the corrosion is severe; the metal appears to be eaten away. The rusting of iron is the most familiar example of corrosion. Another example of corrosion is shown by the appearance of the metal surfaces of worn-out flashlight batteries. If old batteries are left in the flashlight casing, the metal covering of the battery may develop holes. If the contents of the battery, along with moist air, come in contact with a flashlight casing made of iron metal, the casing also rusts badly.

Tarnished

Baking soda

Al foil

Clean and shiny

$$4Fe + 3O_2 \longrightarrow 2Fe_2O_3$$

In many cases, such as the rusting of an iron nail or an automobile body, the chemical reactions involved in the corrosion process are complicated and not completely understood. But the essential process can always be recognized as an oxidation of the metal. Somehow, *the metal atoms lose electrons and become positively charged metal ions.* Something else must take on these electrons to become negatively charged ions. Seldom does a completely dry piece of metal become corroded. Moisture usually has to be present for corrosion to occur. The water in a film of moisture always contains something dissolved in it that *takes on the electrons released when the atoms of the metal become metal ions.*

The simplest example of a corrosion reaction is the way acids (water solutions containing H^+ ions) react with metals. This is why rain that falls through an atmosphere polluted with sulfur oxides causes damage to metals. Sulfuric acid in rain reacts with metals just as sulfuric acid in a laboratory solution reacts with a metal.

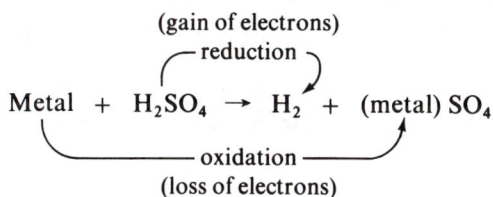

$$\text{Metal} + H_2SO_4 \rightarrow H_2 + \text{(metal) } SO_4$$

(gain of electrons)
reduction

oxidation
(loss of electrons)

The same essential reaction occurs when acid-containing foods, such as tomato juice or citrus fruit juices, come in contact with metal. The metal surface inside the cans such foods are sold in usually is covered by a protective lining of lacquer. Such food should never be stored in open metal cans. If the protective lining of the can is scratched or broken in any place, the acid-containing solution comes in contact with the metal. And oxygen

The corrosion reaction that most affects our daily lives is the rusting of iron.

from the air dissolves in the liquid, further increasing the likelihood of corrosion at exposed metal surfaces. The metal ions released by the corrosion are then absorbed into the food, making it poisonous to humans.

The corrosion reaction that most affects our daily lives is the rusting of iron. The economic impact of this one chemical reaction is tremendous. Probably at least one-fourth of the annual output of the iron and steel industry is needed to replace articles worn out by rusting. Understanding the causes and prevention of rusting is the object of much research. We need not explore in detail the complex reactions that occur when iron rusts. But we can recognize how the essential process is an important illustration of oxidation–reduction reactions.

Rust forms only when iron is in contact with *both* water and oxygen. Oxygen from the air can dissolve in the water to a slight extent. This solution reacts with iron in a series of steps. The first step involves the loss of two electrons by iron atoms to form Fe^{++} ions. Then later, another electron is lost.

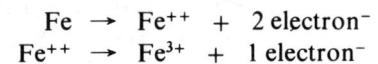

$$Fe \rightarrow Fe^{++} + 2 \text{ electron}^-$$
$$\underline{Fe^{++} \rightarrow Fe^{3+} + 1 \text{ electron}^-}$$

Overall reaction: $Fe \rightarrow Fe^{3+} + 3 \text{ electron}^-$

The final product is iron hydroxide, $Fe(OH)_3$, which is one form of rust. When $Fe(OH)_3$ dries out, it forms iron oxide, Fe_2O_3, the compound we have previously referred to as iron rust. A single equation for the overall process of forming $Fe(OH)_3$ is

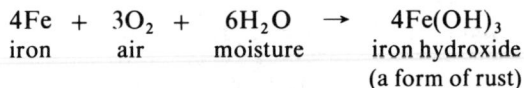

$$\underset{\text{iron}}{4Fe} + \underset{\text{air}}{3O_2} + \underset{\text{moisture}}{6H_2O} \rightarrow \underset{\substack{\text{iron hydroxide} \\ \text{(a form of rust)}}}{4Fe(OH)_3}$$

If we take this equation apart into two equations, each representing half of what is going on, we have

Oxidation: $4Fe \rightarrow 4Fe^{3+} + 12 \text{ electrons}^-$

Reduction: $3O_2 + 6H_2O + 12 \text{ electrons}^- \rightarrow 12(OH)^-$

The second of these two equations explains why both oxygen and water must be present for rusting to occur. Oxygen molecules dissolved in water are very good electron acceptors. They accept electrons in the presence of H_2O molecules to form hydroxide ions, OH^-, as shown in the above reduction equation. Consequently, this reaction causes the metal atoms to give up electrons. Then the metal ions so formed combine with the hydroxide ions in solution.

Splitting the above equation into two halves helps to explain another peculiar thing about rusting. Rust spots always seem to "grow." Painting over a rust spot does no good; rust must always be removed before painting if the metal is to be protected. Metal may appear to be eaten away at a place other than where the rust forms—for example, on automobile bodies or chassis

$$O_2 + 2H_2O + 4e^- \longrightarrow 4OH^-$$
(reduction)

O$_2$ dissolves in moisture

Rust spot

Iron eaten away here

Paint on metal surface

$Fe \longrightarrow Fe^{3+} + 3e^-$
(oxidation)

(Electrons flow through metal)

Metal

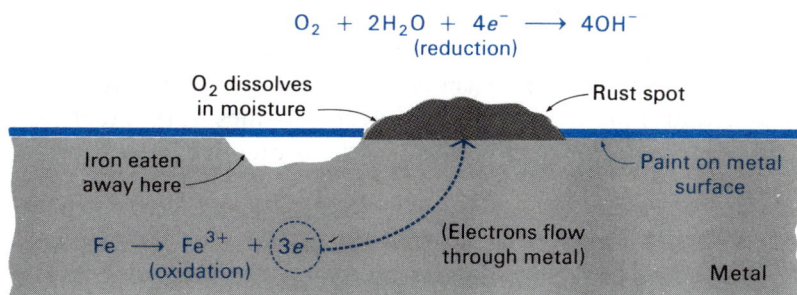

Figure 15.5   The corrosion of iron under the paint near a rust spot.

where the paint or undercoating may have been broken or cracked, as on a dented fender. Moisture and oxygen can reach the iron at these breaks. But the electrons the O$_2$ molecules take up may have been released by iron atoms changing to ions at some other location. The electrons so released can flow through the metal. Consequently, pits or holes can form in the iron surface under the paint. The moisture that seeps in also allows the metal ions to migrate through the solution they form in water. The rust spot grows at the place where the paint is broken, but the damage occurs under the painted surface. This situation is illustrated by Figure 15.5.

RUST PREVENTION

Rusting is such a serious problem that, wherever possible, steps must be taken to prevent it or at least to reduce the extent of rusting. Nothing can be done to alter the tendency for iron atoms to lose electrons. That is a fundamental property of the atoms. However, one thing that can be done is to cover the surface to keep oxygen and moisture away so that there is nothing to accept electrons.

One obvious way of sealing out oxygen and moisture is to paint the surface or to cover it with grease or oil. The inside of cans to be used for food or beverages are often covered with a lacquer or plastic film.

Another method is to cover the surface with a thin layer of chromium metal. Such surfaces are referred to as chromium-plated surfaces. The chromium is applied as a thin film, usually by an electrolysis process. Chromium forms a very tough and unreactive coating of chromium oxide on its surface. This film is so thin that the high luster of the underlying metal can be seen. However, the film is thick enough to keep oxygen and moisture from getting through. Consequently, no electron-accepting process can occur, and therefore the iron metal does not oxidize. Alloys of iron and chromium have this same property. The most frequently used stainless steel is an alloy of 18% chromium, 8% nickel, and 74% iron. Stainless steel corrodes very much less than other types of steel because of the protective layer of oxide formed by the chromium in it.

TIN CANS

Another metal used to coat iron and steel is tin. The familiar "tin can" is made of steel covered with a very thin layer of tin. Tin is a less reactive metal than iron, so its surface does not oxidize so readily. Moreover, the tin oxide coating that does form with oxygen and moisture is tough, like the coating of chromium oxide. The tin oxide coating on the thin layer of tin thus protects the iron of the can.

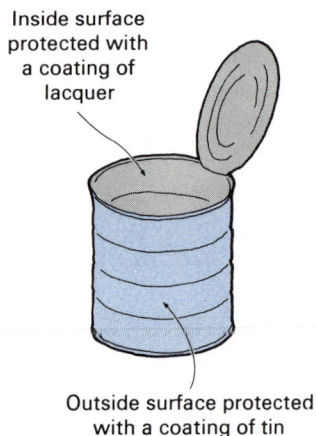

However, if the tin coating is scratched or scraped so that the underlying iron is exposed, the iron rusts even faster than if the tin were not there. Anyone who has ever seen a pile of discarded tin cans knows how rusty they look. The reason for this greater amount of rust is related to the locations of the two metals in the reactivity series, the "pecking-order" principle. Iron is above tin in the series, just as iron is above hydrogen. We have already found that a spontaneous reaction occurs between iron atoms and hydrogen ions. (Recall that acid on iron produces iron ions and molecules of hydrogen gas.) Iron also reacts the same way with tin ions. A small amount of the tin compounds in the coating dissolves in any moisture on the can. When this solution comes in contact with the iron metal, the spontaneous reaction is

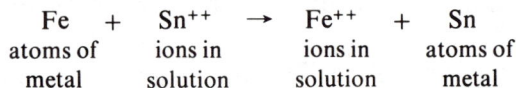

$$\underset{\substack{\text{atoms of}\\\text{metal}}}{\text{Fe}} + \underset{\substack{\text{ions in}\\\text{solution}}}{\text{Sn}^{++}} \rightarrow \underset{\substack{\text{ions in}\\\text{solution}}}{\text{Fe}^{++}} + \underset{\substack{\text{atoms of}\\\text{metal}}}{\text{Sn}}$$

The Fe^{++} ions react readily with oxygen and water to form rust. So we see that the tin, instead of protecting the iron, makes it rust faster once the layer of tin is broken.

GALVANIZED IRON

Another method of protecting iron is the use of a thin coating of the more reactive metal zinc. Sheets of iron or steel so covered are called *galvanized iron* and are used extensively for roofing and to make water pipes and such containers as troughs and buckets.

Inside surface protected with a coating of lacquer

Outside surface protected with a coating of tin

Zinc forms a protective coating of zinc oxide and other zinc compounds on its surface. But it serves another purpose, too. Zinc is a more reactive metal than iron. Note that zinc is above iron in the reactivity series (see Table 15.2). Therefore, if the zinc coating on iron is punctured, the effect is just the opposite of that produced by tin. Iron ions that might be formed by any oxidation of the iron metal are changed back into iron atoms by the zinc. The spontaneous reaction is

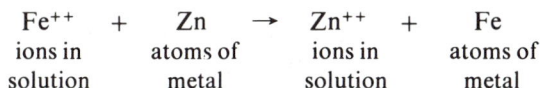

$$\underset{\substack{\text{ions in} \\ \text{solution}}}{Fe^{++}} + \underset{\substack{\text{atoms of} \\ \text{metal}}}{Zn} \rightarrow \underset{\substack{\text{ions in} \\ \text{solution}}}{Zn^{++}} + \underset{\substack{\text{atoms of} \\ \text{metal}}}{Fe}$$

So as long as any coating of zinc metal remains, it is preferentially oxidized ahead of the iron. Here is another illustration of the "pecking-order" principle. When no more zinc atoms remain to react with iron ions, rust begins to form.

Another procedure for protecting such structures as underground tanks from corrosion also makes use of a more reactive metal than iron (above it in the reactivity series). A piece of magnesium is buried in the ground and connected to the tank by a good electrical conductor, as shown in Figure 15.6. Any dissolved oxygen in the ground water causes the magnesium to react instead of corroding the iron. The reduction reaction, forming OH^- ions from O_2 and H_2O, may occur at the surface of the tank. However, the electrons to accomplish this reduction come from the more reactive magnesium rather than the less reactive iron. Gradually, the piece of magnesium is eaten away by the oxidation reaction. When the magnesium is all gone, a fresh piece is put in its place and connected to the tank.

PUSHING ELECTRONS AROUND

Much of our discussion in this chapter has dealt with the tendency of metal atoms to lose electrons and the tendency of metal

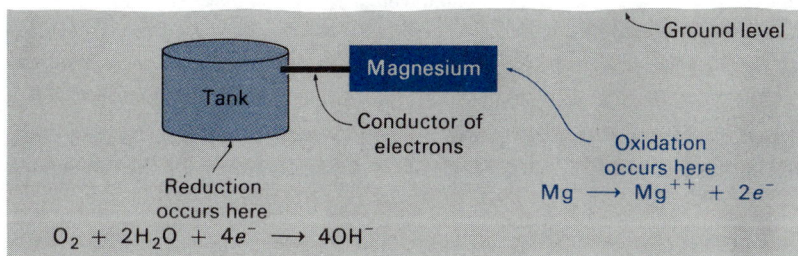

Figure 15.6   The corrosion of magnesium instead of iron.

ions to accept electrons. Differences in these tendencies of different metals are responsible for what we have called the "pecking order" of the metals in the reactivity series. For example, zinc is above copper in the series. Consequently, you can predict that zinc atoms will react spontaneously with copper compounds to displace copper atoms. Figure 15.7 shows what happens if a piece of shiny gray zinc metal is placed in a solution of copper sulfate. The color of copper sulfate solution is a deep blue. Soon after the zinc is placed in the solution, two things begin to happen: The blue color of the solution gets lighter, and the zinc metal beneath the surface of the solution becomes covered with a dark brown color. After some time has passed, if the zinc metal is shaken, a pile of dark brown powder falls to the bottom. Moreover, the zinc metal is eaten away, and the solution no longer has a blue color. The reaction that occurs is represented by the following equation:

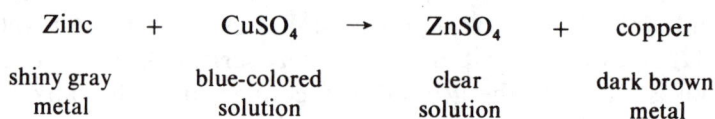

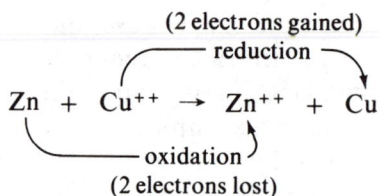

$$\underset{\substack{\text{shiny gray} \\ \text{metal}}}{\text{Zinc}} \quad + \quad \underset{\substack{\text{blue-colored} \\ \text{solution}}}{\text{CuSO}_4} \quad \rightarrow \quad \underset{\substack{\text{clear} \\ \text{solution}}}{\text{ZnSO}_4} \quad + \quad \underset{\substack{\text{dark brown} \\ \text{metal}}}{\text{copper}}$$

$$\text{Zn} + \text{Cu}^{++} \rightarrow \text{Zn}^{++} + \text{Cu}$$

(2 electrons gained) — reduction

oxidation (2 electrons lost)

Zinc, the more reactive metal, delivers electrons to copper ions. This delivery of electrons occurs on the surface of the zinc metal, where the metal touches the solution.

Now we can ask: Can the electrons be allowed to flow through a wire to some other location? We have seen how electrons can

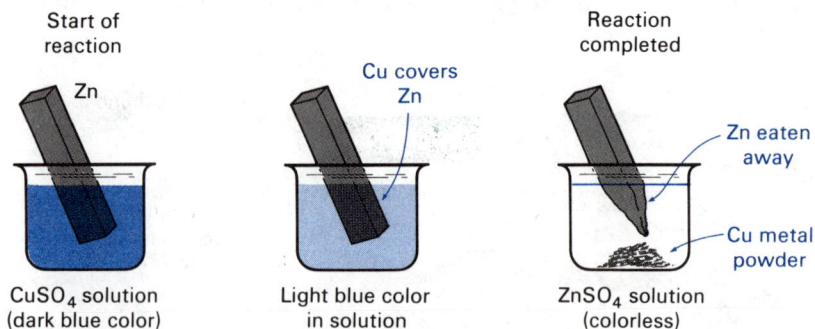

Figure 15.7   The reaction between zinc metal and copper sulfate solution.

Start of reaction

Zn

CuSO$_4$ solution (dark blue color)

Cu covers Zn

Light blue color in solution

Reaction completed

Zn eaten away

Cu metal powder

ZnSO$_4$ solution (colorless)

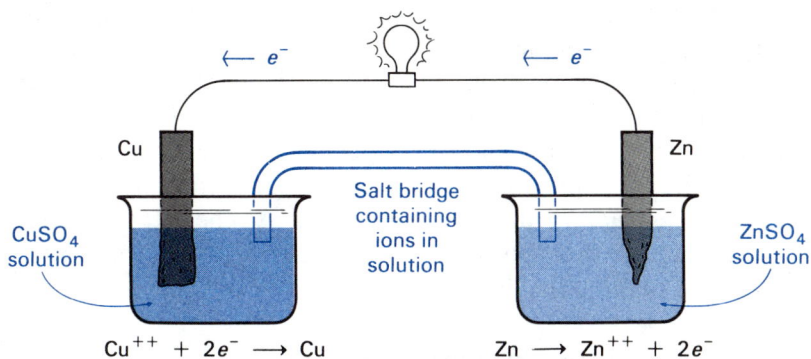

Figure 15.8   A simple battery in which zinc atoms are oxidized and copper ions are reduced.

flow from a magnesium bar to an iron tank so that the tank is protected from corrosion. The oxidation of the magnesium occurred at one place, and the reduction of oxygen with water, at another. If we could arrange a setup in which the zinc is oxidized

If we could arrange a setup in which the zinc is oxidized at one place and the copper ions reduced at another, perhaps the flow of electrons could be used to do useful work.

at one place and the copper ions reduced at another place, perhaps the flow of electrons could be used to do electrical work. The arrangement of apparatus shown in Figure 15.8 accomplishes this process. Zinc metal dips into a solution containing zinc ions in one vessel. Copper metal dips into a solution containing copper ions in another vessel. A wire runs from the zinc to a light bulb and on to the copper. The solutions in the two vessels are connected by a glass tube, called a salt bridge. The salt bridge contains a solution of some salt, which provides ions able to migrate through the bridge thus formed between the two vessels. This salt bridge and its solution are necessary to complete the electrical circuit. (Because ions have electric charges, they migrate through a solution, thus carrying the electrical current.) When all the connections are made, the light bulb glows. As the reaction continues, the zinc metal is gradually used up, and the piece of copper metal becomes larger. And the copper sulfate solution becomes lighter in color, just as it did in the direct reaction shown in Figure 15.6.

What we have constructed in this experiment is called an *electrochemical cell*. We can think of an electrochemical cell as a simple battery. (The term *battery* is more accurately used to describe a combination of several electrochemical cells connected together.) An electrochemical cell harnesses a chemical reaction to produce a flow of electrons (electricity). When zinc atoms become zinc ions, the electrons released by the process are pushed through the light bulb. Electrons go on from the bulb to the copper, where they are taken on by copper ions. This experiment illustrates the principle on which any electrical battery operates. *A battery is essentially an electron pump.* A chemical reaction pushes electrons through the wires of a circuit to do the electrical work of making the light bulb glow.

DRY CELLS AND MERCURY CELLS

The kind of apparatus we used to harness the spontaneous tendencies of zinc and copper ions to transfer electrons is not very practical to use outside a laboratory. You need a simpler device to use in your flashlight or transistor radio. You want a battery that can be carried around easily without spilling solutions. We will describe how two types of common batteries are constructed. One is the so-called *dry cell,* and the other is the *mercury cell.* These cells are commonly referred to as "batteries," even though they are often made up as single cells. Although such devices as flashlight batteries are simple, the chemical reactions that go on inside them are complicated. However, the essential principle on which they operate is the same as we have described for the zinc–copper cell. An oxidation takes place at one location in the cell, and a reduction, at another location.

Figure 15.9 shows how a dry cell is made. The outside can is made of zinc metal. This can is the negative terminal of the cell, from which electrons are pushed through a circuit. The positive terminal, to which electrons flow back from a circuit, is a carbon

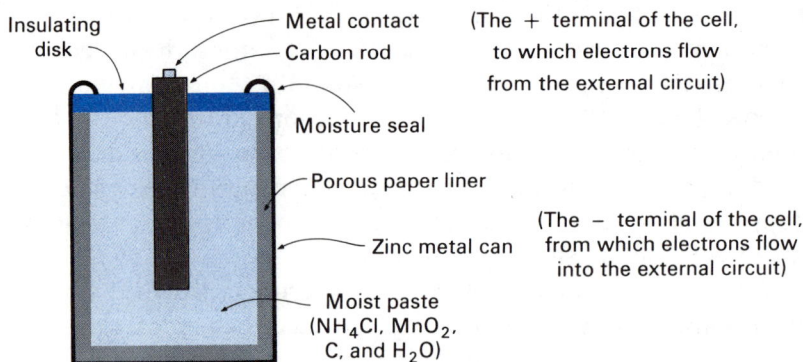

Insulating disk
Metal contact
Carbon rod
Moisture seal
Porous paper liner
Zinc metal can
Moist paste (NH_4Cl, MnO_2, C, and H_2O)

(The + terminal of the cell, to which electrons flow from the external circuit)

(The − terminal of the cell, from which electrons flow into the external circuit)

Figure 15.9   The construction of a dry cell.

rod. The body of the cell is packed with a moist paste. (If the cell were truly "dry," it would not work. The moisture inside, H_2O, makes it possible for ions to move and complete the circuit, just like the salt bridge in Figure 15.8.) The paste contains two black solids, powdered carbon and the essential ingredient, manganese dioxide, MnO_2. Ammonium chloride, NH_4Cl, is also present, dissolved in the moisture. When the cell is operating, the following two reactions occur:

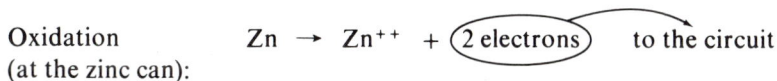

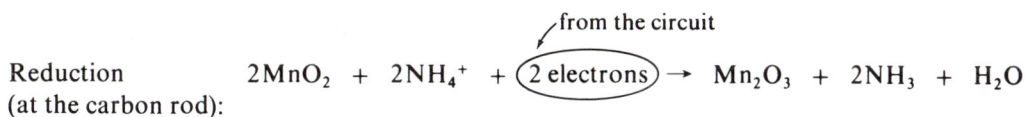

Oxidation (at the zinc can):

$$Zn \rightarrow Zn^{++} + \boxed{2\ electrons} \quad \text{to the circuit}$$

Reduction (at the carbon rod):

from the circuit

$$2MnO_2 + 2NH_4^+ + \boxed{2\ electrons} \rightarrow Mn_2O_3 + 2NH_3 + H_2O$$

Overall reaction:

$$Zn + 2MnO_2 + 2NH_4^+ \rightarrow Zn^{++} + Mn_2O_3 + 2NH_3 + H_2O$$

The voltage of a cell or battery is a measure of what can be thought of as the electron pressure. The voltage is the push given electrons by the cell or battery. A flashlight dry cell has a voltage of about 1.5 volt. If four such cells are connected together with the positive terminal of one wired to the negative terminal of the next, a battery with a voltage of 6 volts is produced (4 × 1.5 volt = 6.0 volt).

The mercury cell is a type particularly well-suited for use in electronic equipment. It produces a current of 1.3 volt that stays very constant as long as the cell is working. Figure 15.10 shows the essential features of this type of cell. Zinc, again, is used as the more reactive metal; its atoms are oxidized to zinc ions. The zinc is mixed with mercury metal to form a layer in the top of the cell. (Any such mixture of another metal with mercury is called

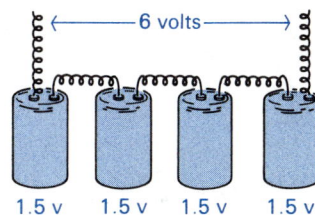

6 volts

1.5 v   1.5 v   1.5 v   1.5 v

Four dry cells wired together deliver a 6-volt current.

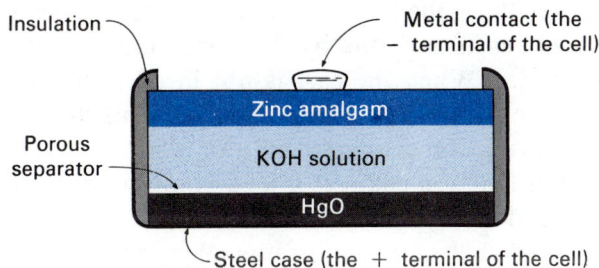

Insulation

Metal contact (the − terminal of the cell)

Zinc amalgam

Porous separator

KOH solution

HgO

Steel case (the + terminal of the cell)

Figure 15.10 The construction of a mercury cell.

an *amalgam.*) Zinc in this form has the advantage of not becoming coated with a covering of zinc oxide. In the presence of potassium hydroxide, KOH, the zinc ions form a mixture of complex substances in solution. The other essential material in the cell is mercury oxide, HgO. The Hg^{++} ions are reduced to mercury atoms, Hg. The overall chemical reaction that occurs when the cell delivers current can be written as follows:

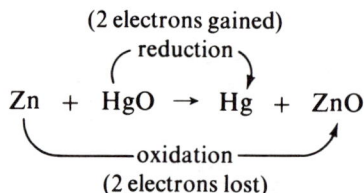

$$
\begin{array}{c}
\text{(2 electrons gained)} \\
\overset{\displaystyle\frown}{\text{reduction}} \\
\text{Zn} \;+\; \text{HgO} \;\rightarrow\; \text{Hg} \;+\; \text{ZnO} \\
\underset{\displaystyle\smile}{\text{oxidation}} \\
\text{(2 electrons lost)}
\end{array}
$$

THE LEAD STORAGE BATTERY

A dry cell or a mercury cell wears out after it has been used for a while. The reacting substances are used up by the chemical reactions that push the electrons through a circuit. A worn-out cell has to be discarded and replaced by a new one. This situation does not hold for the kind of battery called a *storage battery*. A storage battery is designed completely differently, as the name "storage" implies. When such a battery stops pushing electrons, it can be recharged by connecting it to a source of electricity in such a way that electrons are forced back into the battery. This forcing of electrons can be accomplished easily by using a charging current with a higher voltage than the voltage the battery would produce. Recharging reverses the chemical reactions of the battery. The original chemical substances are reformed by this process, and the battery is ready for work again. In this way the battery "stores" the electricity forced into it.

The type of storage battery you are most familiar with is the lead storage battery, the kind used to start automobiles and other vehicles. Golf carts are run by electrical motors powered by storage batteries. When the cart is not in use, the batteries are recharged to be ready for the next use. In a lead storage battery, the oxidation reaction changes lead atoms into lead ions, Pb^{++}. The reduction reaction changes a lead compound, lead dioxide, PbO_2, into lead ions, Pb^{++}. The battery also contains a solution of sulfuric acid. Let us examine how this lead storage battery works.

Terminal ■ ■ Terminal

e^- ↑ ↓ e^-

Oxidation:
$$Pb + SO_4^= \longrightarrow$$
$$PbSO_4 + 2e^-$$

Reduction:
$$PbO_2 + SO_4^= + 4H^+ + 2e^- \longrightarrow$$
$$PbSO_4 + 2H_2O$$

PbO₂

Pb

H₂SO₄ solution

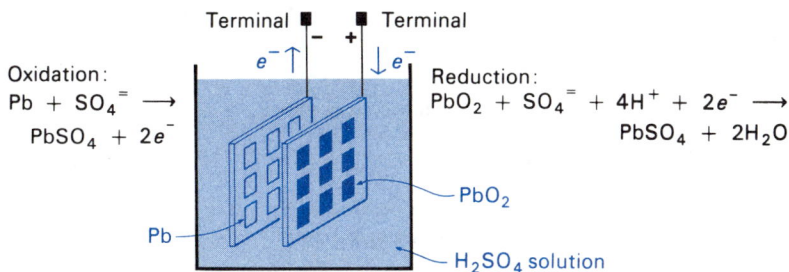

Figure 15.11 Two plates of a lead storage battery and the chemical reactions that occur when the battery delivers electric current.

A single cell of a lead storage battery is made by dipping a series of plates into a solution of sulfuric acid. The plates are made in the form of grids, like the plates of a waffle iron. Pieces of lead dioxide, PbO_2, are fastened in the holes of one plate. A spongy form of lead metal is put in the holes of the other plate. The plates are separated from one another by sheets of porous material, such as woven glass fibers (Fiberglas). Figure 15.11 suggests an enlarged view of one pair of plates. Many such pairs are stacked together, along with the Fiberglas separators, in a container made of hard plastic material. Each such cell produces a current of 2 volts. Six cells connected together make a 12-volt battery, the type usually used in automobiles.

The oxidation reaction shown in Figure 15.11 changes Pb atoms into Pb^{++} ions. However, H_2SO_4 solution contains SO_4^- ions, which combine with the Pb^{++} ions to form a solid compound, lead sulfate, $PbSO_4$. Consequently, as the reaction proceeds, the spongy lead in the waffle-like grid changes right in that spot to solid $PbSO_4$, which stays in the grid. At the surface of the other plate, the compound PbO_2 takes on electrons and likewise changes to solid $PbSO_4$ that stays where it is formed. An equation to represent the overall combined reaction is as follows:

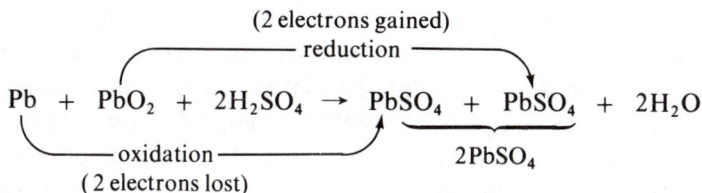

12 volts

A lead storage battery is made from six cells, each of which delivers 2 volts. The battery delivers a 12-volt current.

(2 electrons gained)
reduction

$$Pb + PbO_2 + 2H_2SO_4 \rightarrow PbSO_4 + PbSO_4 + 2H_2O$$

oxidation
(2 electrons lost)

2PbSO₄

The advantage of having solid $PbSO_4$ produced in place on each plate is shown by considering how the above reaction is reversed when the battery is charged. At one plate, electrons are forced back onto the lead ions in lead sulfate to make lead atoms. On the other plate, PbO_2 is reformed from the $PbSO_4$ on that

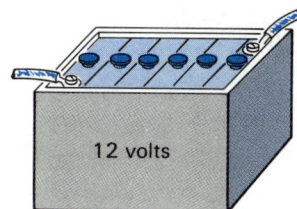

plate. The equation for this overall reaction is the reverse of the one we just wrote.

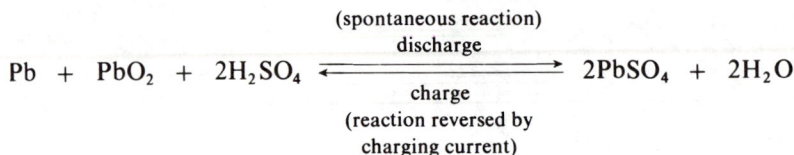

$$\underset{\substack{\text{charge} \\ \text{(reaction reversed by} \\ \text{charging current)}}}{\overset{\substack{\text{(spontaneous reaction)} \\ \text{discharge}}}{Pb + PbO_2 + 2H_2SO_4 \rightleftharpoons 2PbSO_4 + 2H_2O}}$$

This equation shows another change that takes place when the battery discharges: H_2SO_4 is used up, and H_2O is produced. Sulfuric acid, the water solution of H_2SO_4, is more dense than water. The lower the concentration of H_2SO_4, the lower the density of the solution. In a fully charged battery, the density of the acid is about 1.3 gram/milliliter (the density of pure water = 1.0 gram/milliliter). As the battery discharges, the acid becomes more dilute. The density of the acid solution decreases. Usually when the density of the solution reaches 1.2 gram/milliliter, the battery needs to be recharged. Recharging reverses the process. A simple way of telling when the battery is ready for service

A simple way of telling when the battery is ready for service again is to test the density of its acid solution.

is to test the density of its acid solution. When enough H_2SO_4 has been remade to raise the density of its solution to 1.3 gram/milliliter, the operator knows that enough Pb and PbO_2 have been reformed on the plates for the battery to work properly again.

Users of lead storage batteries are warned never to let a battery dry out. If the battery heats up, as it will in hot weather, water evaporates out of the acid solution. The evaporation removes water, so that the amount of liquid in the battery decreases. The $PbSO_4$ formed on the plates may dry out and fall off the plates. If this happens, the $PbSO_4$ is gone from where it must be for successful recharging. Then the battery cannot possibly be recharged and must be discarded. You can avoid this problem by adding extra water from time to time to replace the water evaporated from the battery. Distilled water should be used rather than water from a tap. Tap water may contain dissolved substances, such as those present in "hard" water, that will harm the battery. Undesirable reactions may occur between the dissolved ions in the hard water and the lead or sulfate ions in the battery.

Summary

Oxidation and reduction are terms used to describe chemical reactions in which electrons are transferred. Oxidation and reduction always must occur together. A spontaneous reaction takes place when an atom or ion with a high tendency to lose electrons encounters another atom or ion with a high tendency to gain electrons.

Natural resources of metals (ores) usually occur as compounds. Therefore the process of obtaining metals from their ores involves various kinds of reduction reactions. The spontaneous tendency of metals to revert back to compounds results in rusting and corrosion. Various methods of combating corrosion all involve the essential process of stopping or slowing down the reactions by which the metal atoms lose electrons.

Atoms of metals all tend to lose electrons, but with varying degrees of intensity. Consequently, the metals can be arranged in order of their reactivity. A great many practical uses can be made of the comparable tendencies of metal atoms to be oxidized or their ions to be reduced. Electrochemical cells can be built to harness the drive of two substances to exchange electrons. By separating the oxidation reaction from the reduction reaction, electrons can be pushed through an electrical circuit connected to the two terminals of a battery.

The discussions in this chapter, like those in the previous chapter dealing with acids and bases, illustrate once more how useful broad principles are in helping to organize scientific knowledge. Information in the form of facts standing alone is seldom useful or even interesting. Like unassembled bricks, facts can be built into useful structures by using the mortar of ideas. One such very important idea is that oxidation and reduction involve the give and take of electrons. This idea is vital to our technology, our standard of living, and our conservation of resources.

Glossary

The number in parentheses indicates the text page where you can find the term defined in context.

amalgam   a solution (alloy) of mercury with another metal (464)

battery   a series of electrochemical cells connected to produce a larger voltage (461)

corrosion   the oxidation of a metal surface, which yields metal ionic compounds (454)

dry cell   a type of electrochemical cell based on the oxidation-reduction reaction of zinc and manganese dioxide (462)

electrochemical cell   a device arranged to make use of the electrons that flow in an oxidation-reduction reaction to do electrical work (461)

galvanizing   coating sheet iron or steel with the more reactive metal, zinc, which is preferentially oxidized, preventing rust (458)

germicide   a substance used to kill germs (442)

mercury cell   an electrochemical cell based on the oxidation-reduction reaction of zinc with mercury oxide, HgO (462)

oxidation   a reaction in which a substance loses electrons (437)

reduction   a reaction in which a substance gains electrons (438)

storage battery   a battery that can be recharged by reversal of its chemical reactions with an outside electric current (464)

Exercises

15.1   What is the connection between the chemical symbol for the element lead and the origin of such words as plumbing and plumb line?

15.2   Fill in the blanks with the word "oxidized" or "reduced."

a.   A sulfur atom gains two electrons. The sulfur atom is _____ in this process.

b.   Iron atoms give up electrons to oxygen atoms. The iron atoms are _____ and the oxygen atoms are _____.

c. Cl_2 represents chlorine in a(n) _____ state, while Cl^- represents chlorine in a(n) _____ state.

d.   When an atom loses electrons, it is_____.

e.   When an atom gains electrons, it is_____.

f.   When a metal is produced from its ore, the metal is _____.

g.   When a bronze statue is exposed to the weather, it becomes covered with a green coating. The metals in the bronze have been _____.

h.   When an automobile body rusts, the iron in the steel is _____.

15.3   Identify what is oxidized and what is reduced in each of the following reactions.

a. $Fe_2O_3 + 3CO \rightarrow 2Fe + 3CO_2$

b. $3Fe + 2O_2 \rightarrow Fe_3O_4$

c. $Zn + CuSO_4 \rightarrow ZnSO_4 + Cu$

d. $2Mg + CO_2 \rightarrow 2MgO + C$

e. $2HgO + heat \rightarrow 2Hg + O_2$

f. $Mg + H_2SO_4 \rightarrow MgSO_4 + H_2$

15.4   Why must the water in a swimming pool be treated with chlorine or sodium hypochlorite rather than with sodium chloride?

15.5   Why must special "colorfast" dyes be used on fabrics made into suits to be worn in swimming pools?

15.6   A camping handbook says that if you are unsure of the safety of your water supply, you should add a few drops of laundry bleach to a gallon of water to make it suitable for drinking. What is accomplished by adding the bleach?

15.7   Why will chlorine bleach not remove the mark left on a white shirt caused by drying it on an iron wire clothesline?

15.8   A good carbon steel knife must be dried as soon as it is washed, but a stainless steel knife does not require such care. Explain.

15.9   Metallic sodium reacts vigorously with cold water to produce hydrogen gas.

$$2Na + 2HOH \rightarrow H_2 + 2NaOH$$

Magnesium metal does not react with cold water in the same way. But hot magnesium does react with steam to produce hydrogen gas. How does this information fit together with the location of the two metals in the reactivity series (see Table 15.2) and with the fact that water can act as a very weak acid?

15.10   Calcium metal reacts with cold water in the same way sodium metal does; bubbles of hydrogen gas are produced. Where, approximately, does calcium belong in the reactivity series of the metals?

15.11   Hydrogen is a very flammable gas. If a fire should break out in a building where magnesium alloys are being fabricated, water should not be used to put out the fire. (Instead, the fire should be smothered with foam or sand.) What is the reason for this precaution?

15.12   Pewter is a metal alloy made chiefly of lead. Pewter mugs should never be used as containers for acidic drinks, such as orange or tomato juice. In fact, it is unwise to use pewter articles as containers for any food. The danger of lead poisoning is too great. Explain why this precaution should be taken.

15.13   Use the idea of the "pecking-order" principle among metals to explain why a coating of zinc helps protect iron against rusting.

15.14   Why does a tin can rust rapidly, once the coating of tin is broken so that both tin and the underlying iron are exposed to oxygen and moisture?

15.15   Old houses were built with galvanized iron pipes in the plumbing system. After years of use, some of the pipes invariably get holes in them from rust. If a piece of rusted pipe is replaced with a piece of modern copper pipe, the remaining iron piping rusts even faster. Explain why this happens. (Remember where iron and copper are relative to each other in the reactivity series.)

15.16   Explain the statement: "A battery is an electron pump."

15.17   The essential features of the operation of a hydrogen–oxygen fuel cell were discussed in Chapter 8. Describe the similarities and differences between fuel cells and batteries.

15.18   A magazine article about the Indianapolis 500 auto race included the statement: "Silver-zinc batteries were used to start the motors in the racing cars." Do you think such batteries are made with zinc oxide and silver or with silver oxide and zinc? (Locate the two metals in the reactivity series.)

15.19 *a.* What happens to the H_2SO_4 in a lead storage battery when the battery discharges?

b. What happens when the battery is recharged?

c. The freezing temperature of a solution depends on how much solute is in it. Why may an old discharged lead storage battery freeze in very cold weather, while a new one does not?

16

An Introduction to the Chemistry of Carbon Compounds

☐ How does the idea of functional groups help to organize the information of organic chemistry?

☐ How may alcohols become a key to reducing demands on petroleum resources?

☐ What class of organic compounds gives flowers their fragrances?

☐ How does the chemistry of fats and oils demonstrate the consistency which exists in the natural world?

☐ How do organic acids and organic halides react differently from their inorganic counterparts?

☐ How do sulfa drugs work?

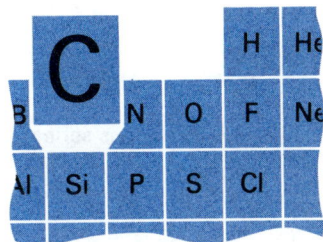

uch of the discussion in Chapter 7, which dealt with the structure of molecules, involved compounds of carbon. In this chapter, we will take a further look at some of the kinds of compounds carbon forms. Recall why carbon is such a special element: Carbon atoms, more than those of any other element, are able to form stable, covalent bonds among themselves. Carbon atoms share pairs of electrons with other carbon atoms to form chains and rings in great variety. Some chains are short; some are very long. Some are continuous; others are branched in many different ways. Carbon is often referred to as the ubiquitous element, meaning that it appears to be everywhere. Recall also that carbon is the element central to life. This is why the early chemists, who found carbon in so many living things around them and in compounds made by living things, used the term *organic* chemistry to describe the chemistry of carbon compounds.

The discussion in this chapter builds on that of Chapter 7. You may benefit from reviewing the information presented there, especially the main ideas and some of the convenient special vocabulary. For example, we dealt there with hydrocarbons, compounds containing only carbon and hydrogen. The stringing together of carbon atoms represents the skeleton or backbone of the hydrocarbon molecules. When the same kind of atoms are arranged differently, the molecules so formed are called isomers, meaning that they contain equal parts. A series of compounds that all have the same type of structure but that differ by a —CH₂—unit is called a homologous series, meaning that the compounds have a similar structure or function.

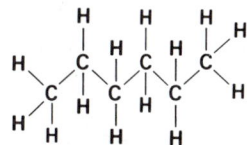

normal hexane (chain)
C_6H_{14}

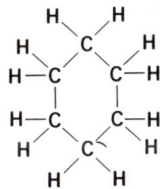

cyclohexane (ring)
C_6H_{12}

THE IDEA OF FUNCTIONAL GROUPS

One of the useful features of the homologous series idea is that it provides an efficient way to classify substances. This classification is based on the fact that observation of the chemical properties of one member of a homologous series allows us to predict that the other members of the series will have very similar properties. Similarity of properties occurs because of similarity of

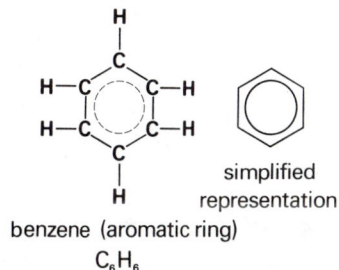

simplified representation

benzene (aromatic ring)
C_6H_6

473

CH_4	methane
C_2H_6	ethane
C_3H_8	propane
C_4H_{10}	butane
C_5H_{12}	pentane
C_6H_{14}	hexane

C_nH_{2n+2}

Alkane homologous series

structures. All the members of the alkane, or paraffin, series (molecules with the generalized formula C_nH_{2n+2}) behave the same way. Although under ordinary conditions most of them burn readily in air, they do not readily react with other substances.

A hydrocarbon molecule that contains two carbon atoms joined by a double bond can react in different ways from those molecules in which all the carbon atoms are joined by single bonds. Two carbon atoms joined by a double bond are sharing

---

Carbon atoms, more than those of any other element, are able to form stable, covalent bonds among themselves.

---

four electrons (two pairs) between them. Thus we find that C_3H_8, propane, does not react with Br_2, bromine, at room temperature, whereas C_3H_6, propene, does react by an addition reaction.

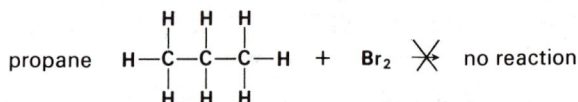

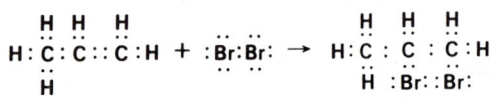

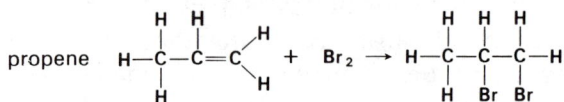

propane and Br_2 → no reaction

propene + Br_2 →

The bromine atoms add to the propene molecule at the double bond. Whenever *any* molecule contains a C═C double bond, the molecule reacts in similar ways. The presence of a C═C double bond in a molecule makes it *function* or behave in a particular way. Hence, the C═C is referred to as an example of a *functional group.*

An organic molecule, because it usually is made up of many atoms, may undergo many different kinds of reactions. The variety of possibilities increases as we go from the relatively simple hydrocarbons to molecules containing other elements, such as oxygen or nitrogen, in addition to carbon and hydrogen. The idea of functional groups helps us classify and predict the kinds of reactions molecules can undergo. *A functional group is a combination of atoms in a molecule that makes the molecule react (function) in a particular way.* The other parts of the molecule generally have only a slight influence in modifying this function. For example, we can predict that butene, C_4H_8, because it contains the functional group, the C=C double bond, will react with Br_2 in a similar way, but at a slightly different rate. Indeed it does react, as shown by the following equation:

Isomers

Normal butane, C_4H_{10}

isobutane, C_4H_{10}

$$H-\underset{\underset{H}{|}}{\overset{\overset{H}{|}}{C}}-\underset{\underset{H}{|}}{\overset{\overset{H}{|}}{C}}-C=C\overset{H}{\underset{H}{\diagdown}} \quad + \quad Br_2 \quad \rightarrow \quad H-\underset{\underset{H}{|}}{\overset{\overset{H}{|}}{C}}-\underset{\underset{H}{|}}{\overset{\overset{H}{|}}{C}}-\underset{\underset{Br}{|}}{\overset{\overset{H}{|}}{C}}-\underset{\underset{Br}{|}}{\overset{\overset{H}{|}}{C}}-H$$

butene

We will concern ourselves with only a few examples to illustrate the idea of functional groups. The functional groups we will discuss in this chapter are important for understanding the structures of the very big molecules described in the next two chapters. In Chapter 17 we will discuss such big molecules as Nylon and rubber; in Chapter 18 we will examine the structure of the big molecules essential to life, such as carbohydrates, proteins, and nucleic acids.

A typical example of a functional group is the OH group. Whenever a molecule contains the OH group attached to a carbon atom, it behaves in a characteristic way. This function (way of reacting) is quite different from the way we found the OH^- ion, called the hydroxide ion, behaving in an inorganic substance, such as sodium hydroxide, NaOH. (Recall, for example, that OH^- ions are good bases, and therefore, they neutralize acids.) Molecules with the OH group attached to a carbon atom do not behave this way. The OH group attached to a carbon atom is called a *hydroxyl group.* Molecules containing this functional group are called *alcohols.*

Table 16.1 shows the structures of a few of the functional groups we will discuss in this and the next two chapters. In the column headed "General Formula," we use the symbol R to stand for the rest of the organic molecules. This is a convention chemists use to emphasize the idea that a functional group establishes the chemical characteristics of all molecules containing it. For example, C_2H_5OH, C_3H_7OH, and C_4H_9OH are all alcohols. The symbol R can stand for C_2H_5, C_3H_7, or C_4H_9 in the general formula ROH, which applies to all three. Examples are given in Table 16.1, along with the names of the compounds. The few names you will need to know, along with a few rules to follow in naming compounds, will be introduced as we go along.

Ethyl alcohol
C_2H_5OH

Alcohol functional group

Methyl alcohol
CH_3OH

Table 16.1   Typical examples of functional groups encountered in organic chemistry

Name	General Formula	Structural Formula	Example	
Alcohol	R—OH	R—O—H	$H-\overset{\overset{\displaystyle H}{\mid}}{C}-\overset{\overset{\displaystyle H}{\mid}}{C}-OH$	ethyl alcohol
Amine	R—NH$_2$	R—N⟨^H_H	$H-\overset{\overset{\displaystyle H}{\mid}}{\underset{\underset{\displaystyle H}{\mid}}{C}}-\overset{\overset{\displaystyle H}{\mid}}{\underset{\underset{\displaystyle H}{\mid}}{C}}-N\langle^H_H$	ethyl amine
Acid	R—COOH	$R-C\langle^O_{OH}$	$H-\overset{\overset{\displaystyle H}{\mid}}{\underset{\underset{\displaystyle H}{\mid}}{C}}-C\langle^O_{OH}$	acetic acid
Halide	R—X (X = F, Cl, Br, I)	R—X	$H-\overset{\overset{\displaystyle H}{\mid}}{\underset{\underset{\displaystyle H}{\mid}}{C}}-\overset{\overset{\displaystyle H}{\mid}}{\underset{\underset{\displaystyle H}{\mid}}{C}}-Br$	ethyl bromide

NAMING ALKYL GROUPS

The fact that carbon forms such a multitude and variety of compounds means that chemists have had to devise elaborate rules for naming compounds. Writing out an entire structural formula every time a particular carbon-containing compound is mentioned is a very inefficient way for chemists to communicate with one another. We will illustrate the way the system works with simple examples. This procedure also will allow us to identify our examples by name.

Recall that the names given the members of the paraffin or saturated hydrocarbon series all ended in *ane.* The series is called the *alkane* series. Molecules derived from this series by having a functional group in place of a hydrogen atom are called *alkyl* compounds. You can see how the system works by imagining on paper that one H atom is removed to leave the rest of the molecule as an alkyl group. For example:

methane   methy*l* group   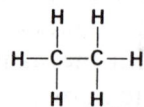 eth*ane*   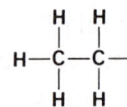 ethy*l* group

Notice that the name given the *alkyl group* uses the prefix of the hydrocarbon name and changes the suffix *ane* to *yl*. Thus whenever the CH_3 group appears in the structural formula of a compound, it is called a *methyl* group; C_2H_5 is an *ethyl* group.

Table 16.2 lists the simplest alkyl groups, along with their parent hydrocarbons. The point at which some functional group can be attached to the alkyl group is indicated in this table by an arrow. This representation is used only in this table to help you see the differences in structure. The arrow need not be shown when you write the formulas of molecules containing alkyl groups.

You notice that the possibility of isomers (the same number and kind of atoms arranged differently) comes into the picture sooner than it did with the hydrocarbon chains. Although there are two isomers of C_4H_{10}, the butanes, there is only one possible molecule with the formula C_3H_8, propane. But there are two possible forms of C_3H_7—the normal propyl and the isopropyl groups. A functional group may attach to an *end* carbon atom or to the *middle* one of a C_3H_7 group. For example, two *different* propyl alcohols exist, both with the simple formula C_3H_7OH.

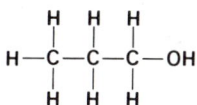

Normal propyl alcohol

Alcohol functional group

normal propyl alcohol
(boiling temperature 97° C)

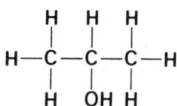

isopropyl alcohol
(boiling temperature 82° C)

Isopropyl alcohol

Remember that having to write three-dimensional structures on a two-dimensional paper surface may lead to the mistake of thinking that more than one possible isomer of normal propyl alcohol exists. Thus

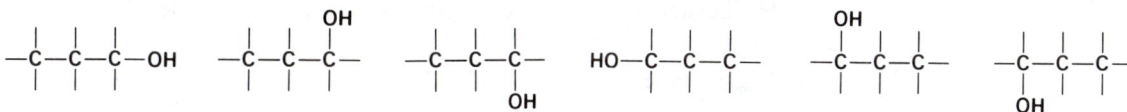

normal propyl alcohol

are all ways to represent the *same* compound. (We have omitted the H atoms attached to the C atoms for convenience.) Only *one* compound exists with the OH fastened to an *end* carbon atom.

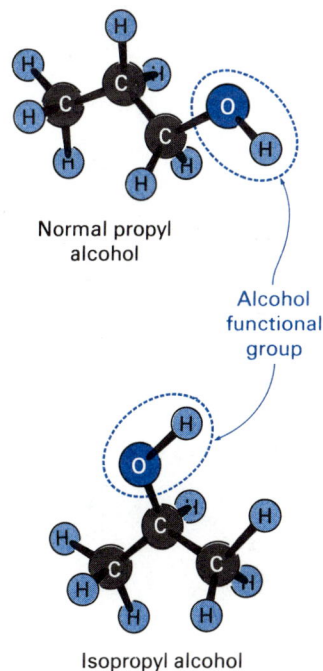

Table 16.2   The simplest alkyl groups

Parent Alkane	Structural Formula	Common Name
CH_4 methane	H—C→ (with H above, H below)	methyl
C_2H_6 ethane	H—C—C→	ethyl
C_3H_8 propane	H—C—C—C→	propyl (normal propyl)
	H—C—C—C—H	isopropyl
C_4H_{10} normal butane	H—C—C—C—C→	butyl (normal butyl)
	H—C—C—C—C—H	secondary butyl
C_4H_{10} isobutane	H—C—C—C→ (with H—C—H above center)	isobutyl
	H—C—C—C—H (with H—C—H above center)	tertiary butyl
C_5H_{12} pentane	*C_5H_{11}→	pentyl
C_6H_{14} hexane	*C_6H_{13}→	hexyl
C_7H_{16} heptane	*C_7H_{15}→	heptyl
C_8H_{18} octane	*C_8H_{17}→	octyl

*These formulas are not structural formulas. There is such an increasing variety of isomers possible that it would require needless space to show all the structural formulas. The simplified formulas here are given only to illustrate how the names of the alkyl groups are related to the names of the parent hydrocarbons from which they are derived.

As you move down to the butyl groups, derived from two different butanes, C_4H_{10}, you notice more possible variety of structure. The details of this increasing variety as you move on to even larger alkyl groups need not be a concern. The variety of possibilities also complicates the process of naming the compounds. Therefore, chemists have established a more comprehensive official system that, for them, is preferable to the common names we are using here. However, these common names will be satisfactory for our discussions without going into greater detail.

ETHYL ALCOHOL

It is appropriate to start a brief survey of organic chemicals with an example of an alcohol, because ethyl alcohol, C_2H_5OH, is probably the first organic compound to have been deliberately manufactured by humans. Certainly, the first cave dweller who discovered that drinking some fermented fruit juice made the cave feel less uncomfortable did not know that yeast had turned sugar into alcohol. But the art of making wine and beer developed as people found how to arrange the best conditions for this chemical reaction to take place. This process has been used and improved ever since those prehistoric times. Ethyl alcohol is known as grain alcohol, because it can be distilled from a fermenting mash made from corn, wheat, barley, or any other source of carbohydrates. Potatoes are often used in Europe, and rice, in Asian countries. Growing yeast cells produce enzymes that act as catalysts for the chemical reactions involved in the fermentation process. Starch, a carbohydrate in the grain, first breaks down to form glucose, a simple kind of sugar molecule. The glucose then is changed into ethyl alcohol and carbon dioxide.

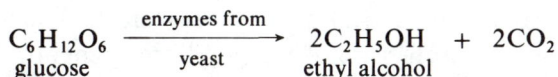

$$C_6H_{12}O_6 \xrightarrow[\text{yeast}]{\text{enzymes from}} 2C_2H_5OH + 2CO_2$$
$$\text{glucose} \qquad\qquad \text{ethyl alcohol}$$

The reaction slows down as the alcohol begins to accumulate. When the alcohol concentration has risen to about 15%, the fermentation stops, because the yeast cells cannot live in a solution with a higher concentration of alcohol. Therefore, wine produced by simple fermentation of fruit juices cannot contain more than a maximum of about 15% alcohol. (More alcohol can be added to make so-called fortified wines, such as sherry.) Alcohol can be removed from the fermenting mixture by distillation. It then can be used for industrial purposes or blended with various flavorings in proper proportion to make whiskey or other liquors.

Ethyl alcohol is a colorless liquid that boils at 78° C (173° F). It mixes in all proportions with water. This solubility is to be expected, because the structure of ethyl alcohol, C_2H_5OH, and water, HOH, are sufficiently similar to allow a great deal of hydrogen bonding to occur. (Recall the discussion in Chapter 13 of ethyl alcohol as an example of how one substance dissolves in another.) Water and ethyl alcohol have such a great affinity that removing all the water from an ethyl alcohol solution is very difficult. So-called "pure" ethyl alcohol is actually a solution with about 5% water in it.

Ethyl alcohol can properly be classed as a drug, because when you drink it, your central nervous system is affected. When you drink alcoholic beverages, the ethyl alcohol quickly finds its way into your blood, where it causes dilation of blood vessels and a lowering of blood pressure. Consuming three bottles of beer or three 1.5-ounce portions of 90-proof whiskey results in an alcohol concentration of about 0.10% in the blood of a 160-pound person. Muscular coordination and mental judgment begin to be impaired by this amount of alcohol, so 0.10% alcohol in the blood of a motor vehicle driver has been chosen as the legal definition of intoxication in most states for arrest and prosecution by traffic officers. Higher levels of alcohol in the blood cause increasing loss of judgment and control of muscles. A blood level of 0.5% ethyl alcohol may cause death. The liver has the job of removing poisonous substances, such as alcohol, from the blood. Later in this chapter, we will examine the chemical reaction by which the liver accomplishes this process. Prolonged, excessive intake of ethyl alcohol overloads the liver, and causes severe permanent damage.

Ethyl alcohol has industrial uses as a solvent and as a starting material for the manufacture of many different chemicals. Prior to the 1940's, most of the ethyl alcohol for industrial use was produced by fermentation of substances produced by growing plants. Such raw materials as corn, byproducts from the conversion of wood into paper, and molasses residues from sugar refining were used. But since that time, increasing amounts of ethyl alcohol have been made more cheaply from the hydrocarbon ethene, C_2H_4. Ethene is a cheap, abundant raw material produced as a byproduct of petroleum refining. In the 1960's and early 1970's, approximately 85% of industrial ethyl alcohol came

"Pure" alcohol
95% ethyl alcohol,
5% water

90-proof whiskey
45% ethyl alcohol,
55% water

Denatured alcohol
95% ethyl alcohol,
5% water, plus traces
of very poisonous
substances

Beer
3%–7% ethyl alcohol,
97%–93% water

from petroleum. The alcohol molecule is made by an addition re-action in which the H and OH of water, HOH, add to the $C=C$ bond. This addition reaction can be summarized by the following equation:

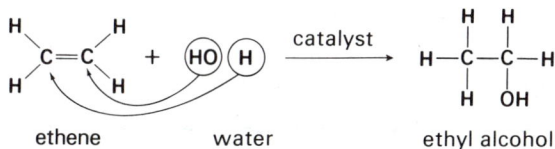

The taxation of beverages containing ethyl alcohol has always been a source of revenue for governments. Consequently, ethyl alcohol sold for other than beverage purposes is deliberately con-taminated with small amounts of poisonous substances so that it cannot be consumed. Ethyl alcohol so treated is called *denatured* alcohol and is much cheaper than beverage alcohol.

ETHENE FROM ETHYL ALCOHOL— SUNLIGHT INTO PETROCHEMICALS

The reverse of the above reaction, the production of ethene, C_2H_4, from ethyl alcohol, C_2H_5OH, also can be accomplished. The key to this reversal is to use a different catalyst and different conditions of temperature and pressure. Some who predict future trends see this reaction assuming increasing importance for our industrial economy in future years. Ethene is a very versatile starting material for the wide range of materials known as *petro-chemicals,* chemicals derived from petroleum. In Chapter 17 we will describe the use of ethene to manufacture various plastics. Drugs, textiles, and paints are other products that come from petrochemicals.

As the price of crude oil rises, and as the available supplies de-crease, the conversion of ethene into ethyl alcohol may no longer be either economically attractive or the best use of our natural resources. On the other hand, the reverse reaction, ethyl alcohol

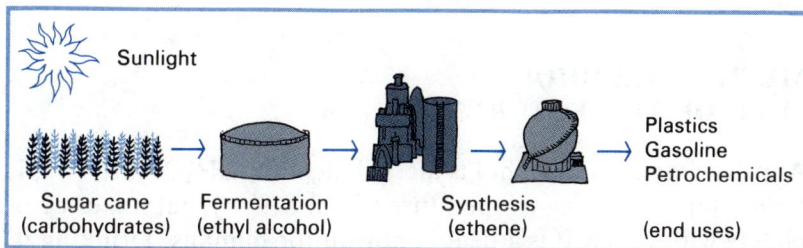

into ethene, may help to save some of our diminishing supplies of petroleum, an important fossil fuel.

The production of ethene from ethyl alcohol is one means of using sunlight as a source of energy. Ethyl alcohol can be made

---

The production of ethene from ethyl alcohol is one means of using sunlight as a source of energy.

---

by fermenting substances produced by growing plants. And growing plants use the energy of sunlight to accomplish the reactions of photosynthesis. Some varieties of plants, such as sugar cane, are relatively efficient at storing sunlight as potential energy in the carbohydrates they manufacture from CO_2 and H_2O. So growing sugar cane, fermenting it to ethyl alcohol, and converting the ethyl alcohol into ethene is an effective and direct method of using sunlight to make the desired product, ethene, C_2H_4, from the abundant raw materials CO_2 and H_2O.

Photosynthesis: $6CO_2 \ + \ 6H_2O \ + \ \text{sunlight} \ \rightarrow \ C_6H_{12}O_6 \ + \ 6O_2$

Fermentation: $C_6H_{12}O_6 \ \rightarrow \ 2C_2H_5OH \ + \ 2\,CO_2$

Ethene synthesis: $C_2H_5OH \ \rightarrow \ C_2H_4 \ + \ H_2O$

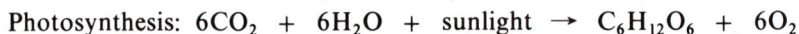

This overall conversion is "direct" only in the sense that it avoids using up the fossil fuel, petroleum, produced over eons of geological time by some similar sequence of reactions.

The large-scale use of this reaction sequence will involve much planning and costly facilities. But with the outlook for the future suggesting depleted supplies of petroleum and increasing demands for ethene, it may become economically feasible in a few years. It also has the obvious advantage of conserving resources. Some predictions suggest that by the 1980's, this procedure may become one of the first large-scale uses of solar energy to supply a raw material to meet the demands of our technological industry.

METHYL ALCOHOL—A FUEL OF THE FUTURE?

Another important alcohol is methyl alcohol, CH_3OH. It, too, is a clear liquid (boiling temperature 65° C, 149° F) that mixes completely with water. It is a deadly poison for humans. Drinking it, or even breathing its vapors, damages the nervous system; blindness occurs first. This compound is known as wood alcohol, because for many years its chief source was a process in which hardwoods, such as oak, were heated in the absence of air. Thus treated, the wood decomposes and the methyl alcohol boils off. For the past 50 years, however, most methyl alcohol has been made by combining carbon monoxide and hydrogen gases, in the presence of a catalyst, under pressure at high temperatures. For example,

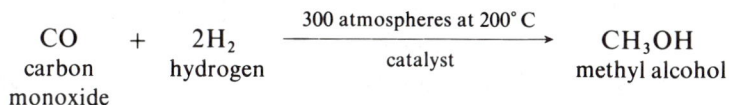

$$\underset{\substack{\text{carbon} \\ \text{monoxide}}}{CO} + \underset{\text{hydrogen}}{2H_2} \xrightarrow[\text{catalyst}]{\text{300 atmospheres at 200° C}} \underset{\text{methyl alcohol}}{CH_3OH}$$

Some people believe this reaction will assume increasing importance as a source of fuel. Methyl alcohol burns to release almost half as much energy as an equal amount of gasoline. Moreover, it is a liquid that dissolves in gasoline. Therefore, methyl alcohol can be used, mixed with gasoline, in automobile engines without making expensive changes in engine design. Consequently, using a mixture of methyl alcohol and gasoline may be a means of conserving supplies of petroleum, because less gasoline would be needed. (Look back at Exercise 8.20.)

The raw material for manufacturing methyl alcohol is anything that can be burned to produce carbon monoxide, CO. Perhaps this process is one way to make better use of waste paper, cardboard, wood, and plastics that are now burned merely to get rid of them. There also is a long-term possibility that coal can be converted to carbon monoxide and then on to methyl alcohol for automobile fuel more easily than coal can be converted to petroleum.

OTHER EXAMPLES OF ALCOHOLS

Other alcohols you may encounter are isopropyl alcohol and normal butyl alcohol. Isopropyl alcohol is used as rubbing alcohol because it is a good germicide (bacteria killer). Normal butyl alcohol is widely used as a solvent for lacquers.

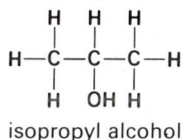

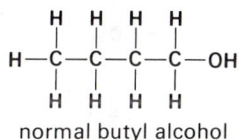

Isopropyl alcohol is made from the hydrocarbon propene, C_3H_6, in the same way that ethyl alcohol is made from ethene, C_2H_4.

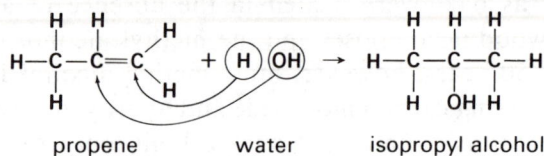

$$H-\overset{\overset{\displaystyle H}{|}}{\underset{\underset{\displaystyle H}{|}}{C}}-\overset{\overset{\displaystyle H}{|}}{C}=C\overset{\displaystyle H}{\underset{\displaystyle H}{}} + \boxed{H}\,\boxed{OH} \rightarrow H-\overset{\overset{\displaystyle H}{|}}{\underset{\underset{\displaystyle H}{|}}{C}}-\overset{\overset{\displaystyle H}{|}}{\underset{\underset{\displaystyle OH}{|}}{C}}-\overset{\overset{\displaystyle H}{|}}{\underset{\underset{\displaystyle H}{|}}{C}}-H$$

propene   water   isopropyl alcohol

You should notice one difference between the equation for this reaction and the one for the formation of ethyl alcohol from ethene. It makes no difference which of the carbon atoms in ethene takes on the OH group (HOC_2H_5 is the same as C_2H_5OH). But in the case of the reaction starting with propene, there is a difference. The OH goes to the *middle* carbon atom, which has only one H atom attached to it. The general rule for this type of reaction is that the OH goes to the carbon holding the fewest H atoms. The H of the water goes to the end carbon atom that already has two H atoms attached to it. Normal propyl alcohol, $CH_3CH_2CH_2OH$, with the OH attached to the *end* carbon atom, is *not* formed by this reaction. Normal propyl alcohol must be made by a completely different and more complicated process, starting with a different substance. This kind of difference in reactivity is the type organic chemists use to advantage when they want to synthesize one particular molecule selectively, and reduce or prevent the synthesis of others.

CARBOXYLIC ACIDS

The functional group characteristic of organic acids is called the *carboxyl group* $-C\overset{\displaystyle O}{\underset{\displaystyle OH}{}}$. Accordingly, organic acids are called *carboxylic acids* in contrast to the mineral acids such as hydrochloric acid, HCl, sulfuric acid, H_2SO_4, or nitric acid, HNO_3. Carboxylic acids can be formed by the reaction of oxygen with alcohols. The following general equation shows an example of how acids are made from alcohols:

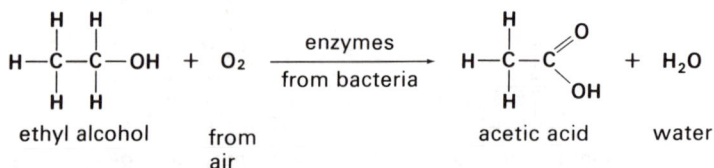

$$H-\overset{\overset{\displaystyle H}{|}}{\underset{\underset{\displaystyle H}{|}}{C}}-\overset{\overset{\displaystyle H}{|}}{\underset{\underset{\displaystyle H}{|}}{C}}-OH + O_2 \xrightarrow[\text{from bacteria}]{\text{enzymes}} H-\overset{\overset{\displaystyle H}{|}}{\underset{\underset{\displaystyle H}{|}}{C}}-C\overset{\displaystyle O}{\underset{\displaystyle OH}{}} + H_2O$$

ethyl alcohol   from   acetic acid   water
                     air

Here is another reaction that has been known since the beginning of human history. Undoubtedly, those primitive cave dwellers who discovered that fruit juices could ferment to alcohol also experienced the frustration of having their wine (containing ethyl alcohol) turn to vinegar (containing acetic acid). The fermentation produced by yeast is an anaerobic process (a reaction that takes place in the absence of air). The CO_2 gas bubbling from the fermenting mixture helps keep the air away. When fermentation stops, air is no longer kept away. Various bacteria inevitably present in the mixture then cause the oxygen from the air to convert the alcohol into the acid. Vinegar is a solution containing about 4% to 5% acetic acid in water. Other flavoring and coloring compounds may also be in the vinegar solution. These compounds were already present in the wine from which the vinegar was made.

Acetic acid is typical of many carboxylic acids in being a weak acid. Recall that a weak acid is one that does not readily give up protons, H^+ ions. Water solutions of carboxylic acids generally have pH values in the range of 3 to 6. By contrast, the strong inorganic acids, at the same concentration, give water solutions pH values in the range of 0 to 3.

Like all acids, carboxylic acids can be neutralized by bases to form *salts*. For example, acetic acid reacts with sodium hydroxide to form the salt sodium acetate. (Note that the naming of the salt follows the rule that *ic* acids make *ate* salts.)

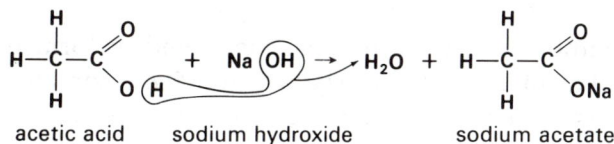

acetic acid   sodium hydroxide   sodium acetate

Another carboxylic acid you may encounter under unpleasant circumstances is formic acid. This compound is part of the irritating solution injected under the skin when bees and other insects sting. As we noted in Chapter 14, the feeling of pain results when nerve endings encounter a solution with a low pH. Formic acid is the simplest carboxylic acid.

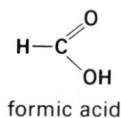

formic acid

A comparison of the formulas for formic acid and acetic acid

Acetic acid
CH_3COOH

Carboxyl group

Formic acid
$HCOOH$

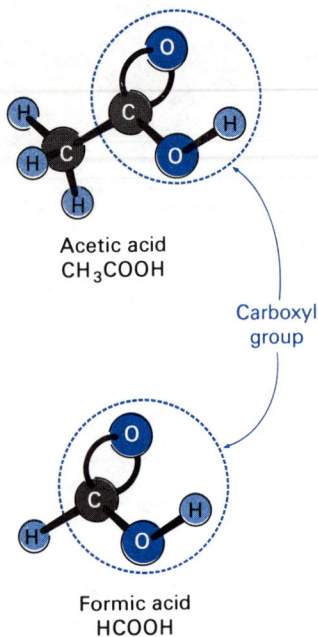

shows that in formic acid a hydrogen atom is attached to the carbon of the carboxyl group. In acetic acid, a CH_3 group is attached to the carbon of the carboxyl group.

You can expect to find a carboxyl group in the formula of any organic compound that gives a water solution with a low pH. (There are a few exceptions to this rule that need not concern us.) This is the case even if the organic molecule is only slightly soluble. Aspirin is an example. (You find aspirin at a pH of about 3 in Figure 14.2 of Chapter 14.) Citric acid, found in many fruits, is another example. The molecule of citric acid contains three carboxyl groups. Even though citric acid contains three groups that can give up a proton, it is still a weak acid, because the groups do not tend to give up protons readily.

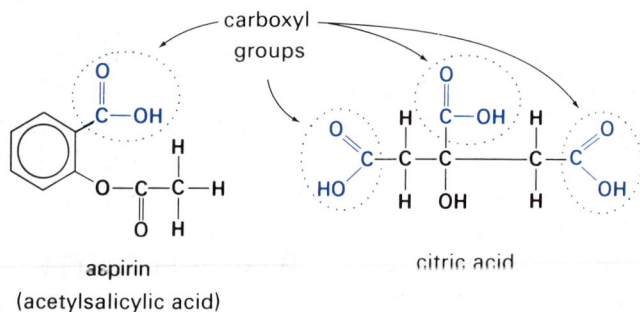

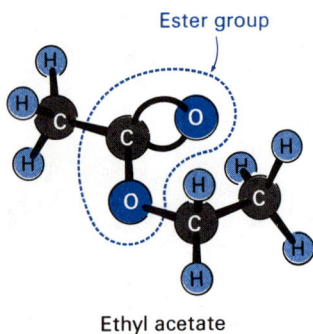

carboxyl groups

aspirin
(acetylsalicylic acid)

citric acid

Ester group

Ethyl acetate

ESTERS

Another important class of organic compounds is formed by the reaction of alcohols with carboxylic acids. These compounds are called *esters*. Water, H_2O, is another product of the reaction. We can illustrate the formation of an ester by showing the equation for the reaction between ethyl alcohol and acetic acid to form ethyl acetate.

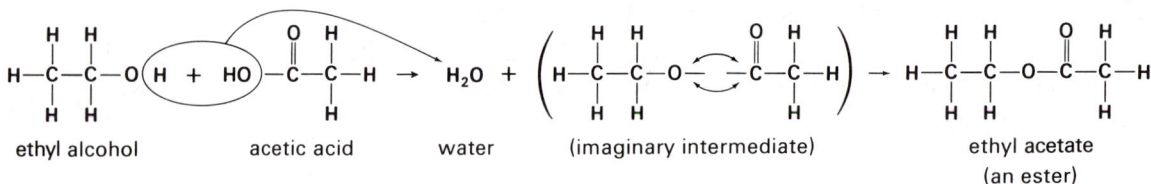

ethyl alcohol   acetic acid   water   (imaginary intermediate)   ethyl acetate (an ester)

Esters are named in the same manner as salts of acids are named, even though there is no similarity in the properties of esters and salts. The alkyl group from the alcohol is named first (ethyl), followed by the name of the acid, with the suffix *ic* changed to *ate* (acetate).

We have used the device of showing a "lasso" around combining atoms and intermediate structures in parentheses to suggest how the structures are changed. This is another example of "on-paper" chemistry rather than real "test-tube" chemistry. The actual mechanism is known to be more complicated. One fact that suggests this is the need to add a strong inorganic acid, such as H_2SO_4, to make the reaction work. However, all you need be concerned with is recognizing how the structure of the ester is built from the structures of the alcohol and the acid. The OH of the carboxyl group combines to form water with the H that was bonded to the O of the alcohol group. (Notice that none of the H atoms attached directly to C atoms of the alcohol react.) Then the remaining fragments of the original acid and alcohol combine to form the ester.

You may well ask why water is not formed from the OH of the alcohol combining with the H of the acid. There is strong evidence to suggest that the reaction does not go this way. For example, if the reaction is run with an alcohol containing the isotope of oxygen with an atomic mass of 18, none of this extra-heavy oxygen is found in the water produced by the reaction. The extra-heavy oxygen-18 is found in the ester. Hence, we conclude that the correct way to write the reaction is the way shown.

Another example of the formation of an ester is

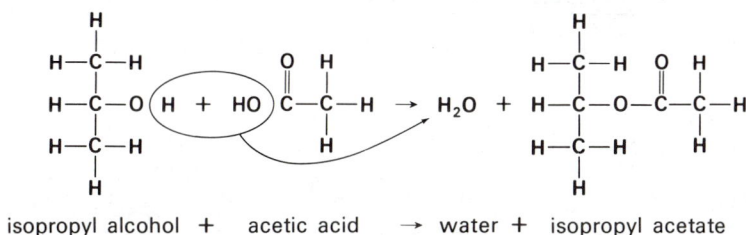

Ester group

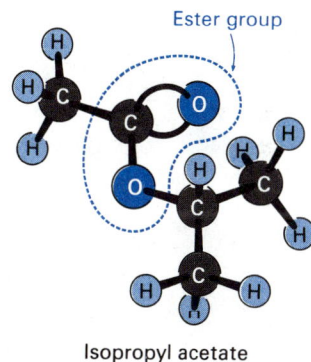

isopropyl alcohol + acetic acid → water + isopropyl acetate

Isopropyl acetate

Notice again how you can be sure the structural formula of the ester is correct. The alcohol reacts through its OH functional group. In this case, the OH is on the *middle* carbon atom of the isopropyl alcohol. Consequently, it is here in the molecule that the combination is made into the ester.

Esters formed from relatively small alcohols and acids are clear liquids. (By small, we mean a total of about 20 carbon atoms.) For example, ethyl acetate boils at 77° C (171° F). Ethyl acetate is used as one component of nail-polish remover because it is a good solvent for lacquers of that type. Esters generally have a pleasant odor. Many of the aromas and flavors of fruits and flowers come from the presence of volatile esters (volatile means easily vaporized). Table 16.3 lists the formulas of a few esters that are familiar because of their fragrance. In all of these you can find the ester functional group:

$$R-O-\overset{\displaystyle O}{\overset{\displaystyle \|}{C}}-R'$$

In this generalized formula, R and R' stand for separate alkyl or other groups of carbon and hydrogen atoms.

Esters of long-chain alcohols with long-chain acids are waxy solids that melt at a low temperature. Beeswax is a mixture of such esters as

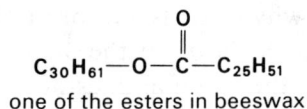

$$C_{30}H_{61}-O-\overset{\displaystyle O}{\overset{\displaystyle \|}{C}}-C_{25}H_{51}$$

one of the esters in beeswax

Other naturally occurring waxes—for example, carnauba wax, used in protective coatings for automobile bodies—are also mixtures of long-chain esters.

Table 16.3   Formulas, names, and aromas of some esters

Ester	Formula	Aroma
ethyl butyrate	$C_2H_5-O-\overset{O}{\overset{\|}{C}}-C_3H_7$	apricot
octyl acetate	$C_8H_{17}-O-\overset{O}{\overset{\|}{C}}-CH_3$	orange
isoamyl acetate	$H-\overset{CH_3}{\underset{CH_3}{\overset{\|}{\underset{\|}{C}}}}-\overset{H}{\underset{H}{\overset{\|}{\underset{\|}{C}}}}-\overset{H}{\underset{H}{\overset{\|}{\underset{\|}{C}}}}-O-\overset{O}{\overset{\|}{C}}-CH_3$	banana
butyl butyrate	$C_4H_9-O-\overset{O}{\overset{\|}{C}}-C_3H_7$	pineapple
methyl salicylate	$CH_3-O-\overset{O}{\overset{\|}{C}}$ (benzene ring, HO—)	wintergreen

FATS AND OILS

Fats are naturally occurring substances that are an important class of foods. Butter and meat fat are examples. They are usually light in color, have a greasy feel, and generally melt at a fairly low temperature to become an oily liquid that floats on water. Fats do not dissolve appreciably in water. Fats usually are of animal origin, found in various kinds of meat and milk. You can guess that vegetable *oils* are liquid at room temperature. However, because of their similar chemical structure, their properties are similar to those of fats. Corn oil, obtained by pressing kernels of corn, or soybean oil, obtained from soybeans, are examples.

Early in the nineteenth century, chemists discovered that, if fats were heated with an inorganic acid such as hydrochloric acid, they broke down into other substances. One type of product obtained by this reaction always behaved as a weak acid. Hence, these new acids were called *fatty acids,* because they came from fats. It is now known that fats are esters of a particular alcohol called *glycerol.* Glycerol is a *tri* functional alcohol; its formula contains *three* alcohol, OH, groups. Glycerol is also known by the common name glycerine. It is often used in hand lotions to soften rough skin.

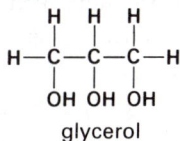

Esters of glycerol

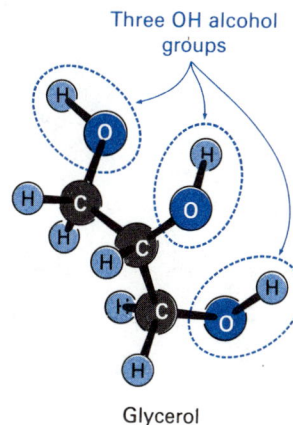

Three OH alcohol groups

Glycerol

$$\begin{array}{c} \text{H} \quad \text{H} \quad \text{H} \\ | \qquad | \qquad | \\ \text{H}-\text{C}-\text{C}-\text{C}-\text{H} \\ | \qquad | \qquad | \\ \text{OH} \quad \text{OH} \quad \text{OH} \end{array}$$

glycerol

The reaction that occurs when a fat is heated with water in the presence of a dilute inorganic acid as a catalyst can be represented by the following equation:

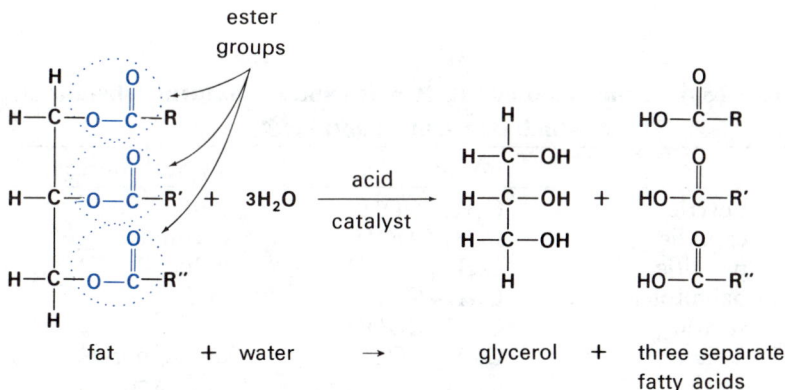

ester groups

| fat | + | water | → | glycerol | + | three separate fatty acids |

In this equation, the symbols R, R′, and R″ are used to represent

three separate alkyl groups. Fats may be mixed esters of glycerol rather than being formed from only one kind of fatty acid. Note that there are three ester functional groups in the fat molecule. Each of the fatty acids has formed an ester with one of the OH groups of the glycerol molecule.

If you compare this equation for the reaction of fats to make glycerol and fatty acids with the one on page 486, which describes the formation of an ester, you see that here we have the reverse of the reaction shown there. A reaction such as this, in which water adds instead of being split off, is called a *hydrolysis* reaction. This particular hydrolysis reaction is important, because it is the first step your body takes in the digestion of the fats you eat. The products of the reaction, glycerol and fatty acids, may be broken down further and eventually oxidized to CO_2 and H_2O by the process of metabolism. Or your body may put glycerol and fatty acids back together again and store the remade molecules of fat in your tissues. We will discuss the use and storage of fats as holders of potential energy for humans and other animals in a later chapter. At this point, we will simply stress that fats are important examples of esters that occur in nature.

Only about 20 different fatty acids are found to be common to the fats living things manufacture. Table 16.4 lists a few of the common fatty acids. Palmitic acid is the most abundant fatty acid in nature; oleic acid is the most widely distributed. This narrow range of variety is one of the remarkable consistencies about the chemistry of life. Of all the many possibilities, nature synthesizes only these relatively few building blocks in plant and animal tissues.

The consistency extends beyond the simple number of fatty acids. All the R groups in fats are alkyl groups; that is, they are

Table 16.4   Some common fatty acids and the natural substances from which they can be obtained

Name	Formula	Typical source
butyric	C_3H_7COOH	butter
caprylic	$C_7H_{15}COOH$	coconuts
palmitic	$C_{15}H_{31}COOH$	palm oil
palmitoleic	$C_{15}H_{29}COOH$	milk
stearic	$C_{17}H_{35}COOH$	beef
oleic	$C_{17}H_{33}COOH$	olives, pork
linoleic	$C_{17}H_{31}COOH$	safflowers
linolenic	$C_{17}H_{29}COOH$	soybeans

chains of carbon atoms, not rings. Most are long, continuous chains, although some branching occurs. All the R groups have an odd number of carbon atoms, so that the fatty acids themselves (RCOOH) all contain *even* numbers of carbon atoms. This fact ties in with the fact that the two-carbon C—C unit is the molecular fragment you and all other animals are able to use in the chemical reactions of metabolism.

The R groups in some fatty acids contain C=C bonds. Recall that hydrocarbons containing C=C bonds are called unsaturated molecules. Fatty acids containing from one to four C=C bonds have been found. This unsaturation in the R group of the fatty acid influences the properties of the ester formed by combining the fatty acid with glycerol. The greater the amount of unsaturation, the lower the melting point of the ester. This structural difference is responsible for one substance being a solid or semisolid fat at an ordinary temperature and another being a liquid oil. Generally, vegetables make oils, animals manufacture fats.

It is possible, by using high pressures and suitable catalysts, to cause hydrogen gas (H_2) to add to the C=C bonds in liquid oil molecules to change them into solid fat molecules. Such a reaction is called *hydrogenation*. We first mentioned this reaction in Chapter 7 when we introduced the term unsaturated hydrocarbons and described what can happen at the C=C bond. Solid fats derived from vegetables, such as the product sold under the brand name Crisco, are made this way. Likewise, oleomargarine is made chiefly from mixtures of purified vegetable oils, such as soybean or corn oils, that have been partially hydrogenated. (By partially, we mean that some but not all the C=C bonds have taken on H atoms.) Other substances are added to oleomargarine so that moisture can be held in tiny droplets spread throughout the body of the material. Such a dispersion of fine droplets of one liquid (in this case, water) in another liquid or solid (the oil or fat) is called an *emulsion*. Flavoring and coloring agents also are added.

In recent years, some evidence has been found to suggest that the human body is less able to manufacture cholesterol from unsaturated fatty acids than it can from the saturated kind. Cholesterol is a big organic molecule (formula $C_{27}H_{45}OH$) found in the blood and tissues of animals. When there are excessive cholesterol deposits in the tissues of blood vessels, the walls of the blood vessels stiffen and become hard. In fact, the name cholesterol means "bile solid," because it was first isolated from gall-

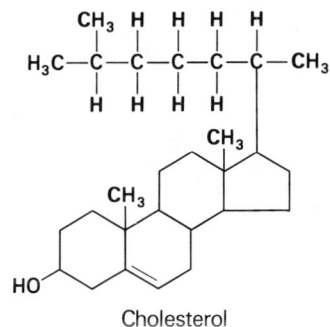

Cholesterol

stones. Thus if a person tends to synthesize an excessive amount of cholesterol, he or she may lessen the possible danger from hardening of the arteries by avoiding saturated fats in the diet. Instead, that person should get the fatty acids that his or her body needs for normal metabolism from unsaturated vegetable oils.

AMINES AND AMIDES

Amines are a class of compounds in which nitrogen atoms are incorporated into organic molecules. Amines are derived from ammonia, NH_3, by replacing one or more of the hydrogen atoms with organic groups. Compare the following amine structures with that of ammonia:

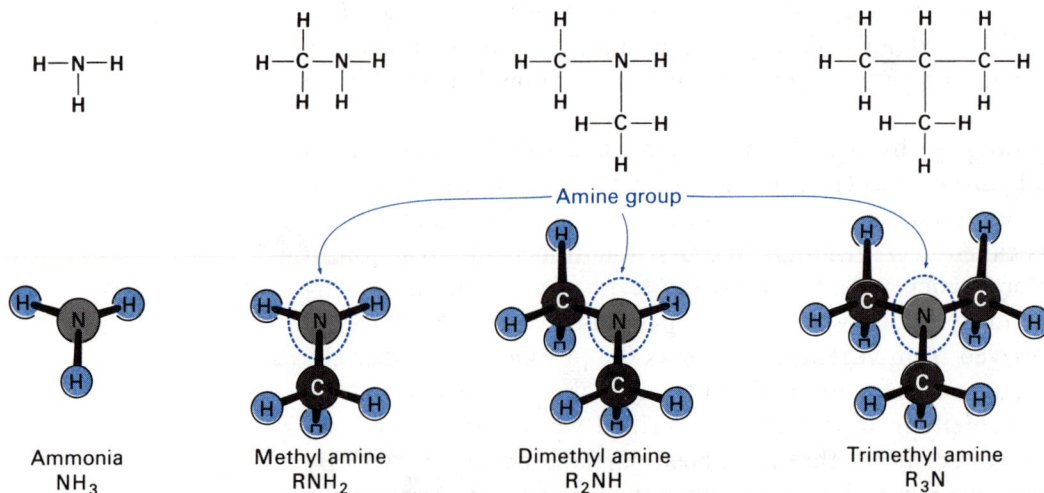

Ammonia NH_3 — Methyl amine RNH_2 — Dimethyl amine R_2NH — Trimethyl amine R_3N

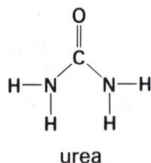

urea

Like ammonia, the amines are soluble in water and produce basic solutions (pH greater than 7). Because one, two, or three of the H atoms on ammonia can be replaced, the prefix *di* (for two) or *tri* (for three) is used in the name to designate the structure. Amines generally have an unpleasant odor, usually suggesting decaying fish or other animal matter. Animal protein contains nitrogen. Amines result from the decomposition of protein or its metabolism by microorganisms involved in the decay of animal matter. Urea, a waste product found in urine, is a molecule containing two amine groups.

Amines react with acids to form *amides,* in much the same way that alcohols react with acids to form esters. We will find in later chapters how important this reaction is in the production of some polymers (big molecules formed by the joining together of many small ones). Nylon is an example of a material formed from an amine and an acid. So are proteins, the big molecules so important to living things.

The general equation for the formation of an amide can be written as follows:

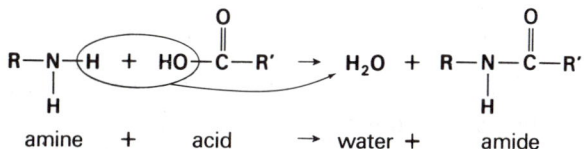

Amide group

Acetamide

$$CH_3\overset{O}{\overset{\|}{C}}-NH_2$$

Acetamide

$$R-N-H \; + \; HO-\overset{O}{\overset{\|}{C}}-R' \; \rightarrow \; H_2O \; + \; R-N-\overset{O}{\overset{\|}{C}}-R'$$

amine + acid → water + amide

You can see the similarity between this equation and that for the formation of an ester.

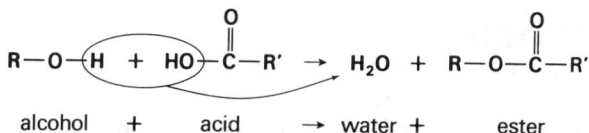

$$R-O-H \; + \; HO-\overset{O}{\overset{\|}{C}}-R' \; \rightarrow \; H_2O \; + \; R-O-\overset{O}{\overset{\|}{C}}-R'$$

alcohol + acid → water + ester

The most famous, or infamous, amide in recent years is *ly*sergic acid *d*iethylamide, commonly known in LSD. The formula can be written as follows:

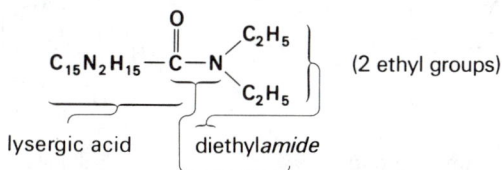

$$C_{15}N_2H_{15}-\overset{O}{\overset{\|}{C}}-N\overset{C_2H_5}{\underset{C_2H_5}{}}$$ (2 ethyl groups)

lysergic acid   diethyl*amide*

LSD is a hallucinogen, a drug that produces hallucinations. The mechanism of the action of LSD on the brain is the object of extensive research, but no theory has yet been proposed that completely explains its action. Apparently, LSD acts to block some reaction involved in transmitting impulses from one brain cell to another. The structure of the amine part may or may not have anything to do with this action. But it has been found that, if the molecule is modified so that only one ethyl group (C_2H_5) and a hydrogen atom (H) are attached to the nitrogen, a molecule is formed that has none of the physiological effects that LSD does.

The amine functional group also can attach to rings of carbon atoms as well as to alkyl groups. Organic compounds containing carbon atoms hooked together in rings are referred to as *aromatic* compounds. Benzene, C_6H_6, is the simplest aromatic hydrocarbon. We discussed the special features of benzene's structure in Chapter 7. The molecule that results when the amine

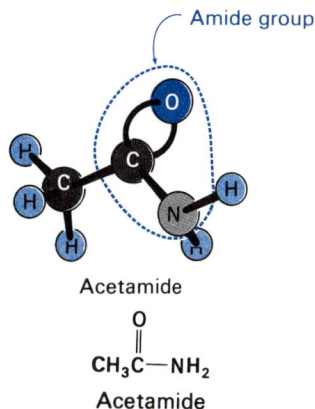

functional group, NH_2, replaces an H atom on the benzene ring is called aniline.

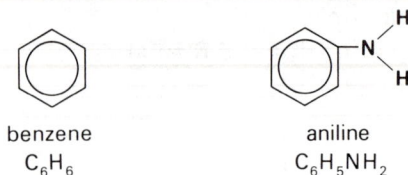

benzene
C_6H_6

aniline
$C_6H_5NH_2$

Aniline is a liquid that boils at 184°C (363°F), is only slightly soluble in water, and is highly poisonous to humans. It is a very important starting material for the manufacture of dyes, compounds used in rubber manufacture, drugs, and photographic developers.

SULFANILAMIDE AND SULFA ANTIBIOTICS

Sulfanilamide is a derivative of aniline. The following diagram shows the structure of the molecule. You can see how the name sulfanilamide tells what the parts of the molecule are.

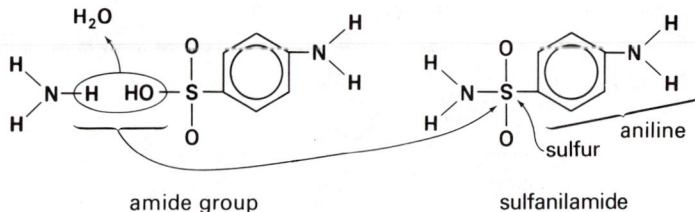

amide group   sulfanilamide

Sulfanilamide was the first of the many modern sulfa antibiotics found to be effective in killing bacteria. The effectiveness of sulfanilamide is related to its structure. Its structure is quite similar to a compound called *para*-aminobenzoic acid, used by bacteria to make another molecule called folic acid, which is essential to their life process.

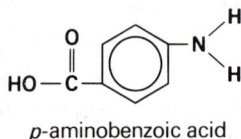

p-aminobenzoic acid

The sulfanilamide takes the place of *para*-aminobenzoic acid in the sequence of reactions catalyzed by the enzymes in the bacteria. Thus an important chemical reaction of the bacteria's metabolism is completely changed. The result is a disruption of the normal sequence of reactions, and the bacteria die because they are unable to manufacture folic acid.

Humans also need folic acid for their metabolism. However, because humans do not have the enzymes bacteria possess to make folic acid from other substances, folic acid must be included in their diets. Folic acid is one of the B vitamins present in a balanced diet. Yeast, wheat germ, soybeans, and eggs all contain folic acid. Probably, because of this lack of enzymes to make use of either sulfanilamide or *p*-aminobenzoic acid, sulfanilamide is not lethal to humans, unlike its effect on bacteria.

Since the discovery in the 1930's of the effectiveness of sulfanilamide as a killer of disease germs, thousands of molecules, with different but similar structures, have been synthesized and tested. Some have proved useless, but others have turned out to be even more effective or more specific in their ability to kill various kinds of bacteria. Most of these retain the *sulfa* part of the name but change the *amide* part to indicate the change in structure. Sulfaguanidine or sulfathiazole are two examples of such variations.

This research illustrates the way chemical knowledge of molecular structure can lead to very beneficial technology. The very

The effectiveness of sulfanilamide is related to its structure.

poisonous compound aniline is changed into other compounds by modifying its structure. Modification of the structure in turn leads to modification of the physiological properties of the resulting compounds.

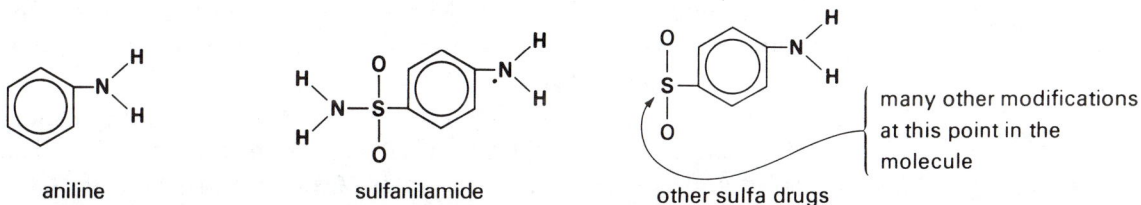

aniline   sulfanilamide   other sulfa drugs

many other modifications at this point in the molecule

Aniline is poisonous to humans and bacteria alike. But modifying its structure produces molecules that are much less poisonous to humans than they are to bacteria. We must say "less poisonous" rather than "non-poisonous," because most antibiotics produce some undesirable side effects in nearly everyone.

Even though the details of just how an antibiotic molecule accomplishes its purpose are very seldom known completely, antibiotics can be used to destroy disease bacteria and save human lives. The original discovery of the effectiveness of sulfanilamide resulted from the accidental observation that a certain dye used in the preparation of specimens for microscope slides killed bacteria. Knowledge of the structure of the dye molecule and its reactions led to the finding that sulfanilamide could be used to kill bacteria. Then chemists followed the lead suggested by the structure of the sulfanilamide molecule to synthesize other molecules useful as antibiotics. More knowledge gradually accumulates about how the antibiotic stops the life process of the bacteria without badly affecting the life of the human, the host to the bacteria. In the future, we may have enough knowledge of the complicated processes going on in living things to design a molecule that we can expect to have a limited, specific action as an antibiotic without adversely affecting the human metabolism. Just as the construction of a house starts on an architect's drawing board, so a drug to save life may start on an organic chemist's drawing board.

ORGANIC HALIDES

The elements of family 7A of the periodic table are known as the *halogens*, which means "salt formers." Sodium chloride, NaCl, is the most familiar salt. Accordingly, the compounds of these elements are called *halides*. In the halides that are salts, typical in organic compounds, the elements fluorine, F, chlorine, Cl, bromine, Br, and iodine, I, are combined as ions. By contrast, when these elements combine with carbon in organic molecules, the bonds they form are covalent bonds; the atoms share one pair of electrons. However, the term halide is still used as a name for the compounds of the halogens with groups of atoms containing carbon and hydrogen. A more descriptive name for these compounds is *organic halides*.

The properties of the organic halides are correspondingly quite different from the properties of salts. Sodium chloride, NaCl, is a solid and is very soluble in water. Carbon tetrachloride, CCl_4 (the prefix *tetra* means "four"), is a liquid made up of molecules with none of the dipole characteristics of water molecules. Consequently, CCl_4 and all other covalently bonded organic halides are not soluble in water.

Ions of sodium chloride, formed by the transfer of one electron (ionic bonds)

Molecule of carbon tetrachloride, formed by sharing pairs of electrons (covalent bonds)

Organic halides, as a class of compounds, are an almost limitless variety of molecules. Organic chemists very frequently use them as starting materials or make them as intermediate steps for the synthesis of other compounds. We have already mentioned various organic halides. For example, in discussing the general topic of how substances dissolve (see Chapter 13), organic chlorine compounds were mentioned as drycleaning agents.

Alkyl halides, formed by the attachment of a halogen to an alkyl group, generally are toxic (poisonous). Some of them, in low concentration, can act as anesthetics. For example, chloroform, $CHCl_3$, was one of the first general anesthetics to be administered by having a patient breath in its vapor. However, prolonged breathing of chloroform or its closely related chemical cousin, carbon tetrachloride, causes death.

An interesting exception to the toxicity of organic halides are the alkyl compounds that contain fluorine along with other halogens. Such compounds are known by the trade name Freon. Apparently, fluorine atoms form such strong covalent bonds with carbon atoms that any organic molecule containing fluorine is very stable and does not decompose or react with other substances. One of the reasons that Freon compounds are used as fluids in refrigerator or air-conditioner compressors and are put under pressure as the propellent gas in aerosol cans is that they are *not* poisonous to humans. However, the unusual stability of these fluorine-containing molecules makes them last a long time in the upper atmosphere. You may recall, from the discussion in Chapter 10, how this stability can lead to trouble. The absorption of ultraviolet light can cause the chlorine atoms in Freon to break off. These chlorine atoms can then go on to destroy the ozone, O_3, in the upper atmosphere.

The formulas of some typical Freon molecules are

$$\text{Cl}-\underset{\underset{\text{F}}{|}}{\overset{\overset{\text{Cl}}{|}}{\text{C}}}-\text{F} \qquad \text{Cl}-\underset{\underset{\text{Cl}}{|}}{\overset{\overset{\text{Cl}}{|}}{\text{C}}}-\text{F} \qquad \text{Br}-\underset{\underset{\text{F}}{|}}{\overset{\overset{\text{F}}{|}}{\text{C}}}-\text{F}$$

dichlorodifluoromethane   trichlorofluoromethane   bromotrifluoromethane

The last of these examples, $CBrF_3$, is used in fire extinguishers. It is kept as a gas, under pressure, to be released from a cylinder pointed toward a fire. The gas blankets a fire and thus keeps oxygen away. Because it is not poisonous to humans and is so effective in small amounts, it is used in fire extinguishers carried on passenger aircraft.

The compound ethyl chloride, C_2H_5Cl, is sometimes used as a local anesthetic. For example, it can be sprayed on a dislocated finger to relieve the pain of resetting. Ethyl chloride boils under normal pressure at about 12°C (54°F). Consequently, when the liquid is sprayed on the skin, it evaporates so rapidly that the quick cooling also has an anesthetic effect.

ALDEHYDES

The functional group for the *aldehyde* class of organic compounds is written as follows:

$$-C\!\!\begin{array}{c}\nearrow O\\\searrow H\end{array}$$

The two simplest aldehydes are acetaldehyde and formaldehyde.

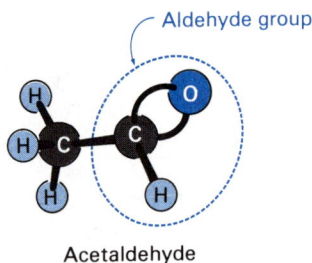

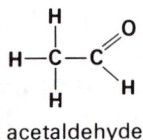

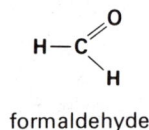

acetaldehyde   formaldehyde

Aldehyde group

Acetaldehyde

Compare these formulas of aldehydes with the following formulas of acids:

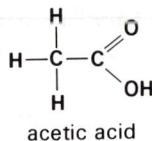

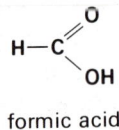

acetic acid   formic acid

The name formaldehyde suggests the relationship of its structure to that of the formic acid molecule. Similarly, acetaldehyde is related to acetic acid.

The relationship is more than one of appearance. Recall that acetic acid can be made by the reaction of oxygen with ethyl alcohol. In this process, acetaldehyde is an intermediate product, as follows:

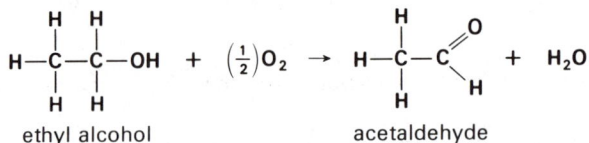

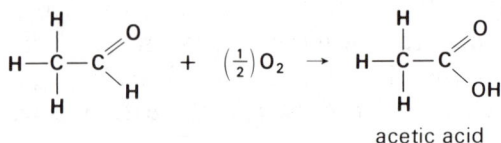

ethyl alcohol   acetaldehyde

acetic acid

When a person drinks ethyl alcohol, as we mentioned before, the liver is the organ called upon to rid the body of this undesirable substance. The liver produces enzymes capable of acting as catalysts to accomplish the oxidation of the alcohol, first to acetaldehyde and then to acetic acid. Eventually, the acetic acid is oxidized even further to CO_2 and H_2O. However, the reactions are relatively slow, and if a large amount of alcohol has been consumed, the liver's mechanism becomes overloaded. Although all the details of what happens are not clearly understood, there is considerable evidence to suggest that many of the actions, and sometimes nausea, of an intoxicated person are caused by the build-up of amounts of acetaldehyde produced by the first step of the oxidation process. A similar line of reasoning suggests that the very poisonous property of methyl alcohol, CH_3OH, is related to the fact that, in the body, methyl alcohol is first converted to formaldehyde, CH_2O. Formaldehyde then interferes with the normal chemistry of metabolism very quickly.

Formaldehyde is used to preserve biological specimens. It has a penetrating odor that most people consider unpleasant. It is a very reactive chemical and is used as a starting material for various syntheses. Large amounts of formaldehyde are used in the production of plastics. We will discuss some of these products and the reactions by which they are produced in the next chapter.

KETONES

Ketones are a class of compounds closely related to aldehydes.

The functional group is represented thus:

$$\begin{array}{c} O \\ \parallel \\ -C- \end{array}$$

Whereas the aldehyde group contains one H atom attached to the carbon atom, in ketones the carbon atom is attached to two alkyl groups. Acetone is the simplest ketone; methyl ethyl ketone is another example.

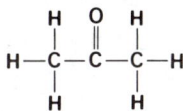

Ketone group

Acetone

acetone

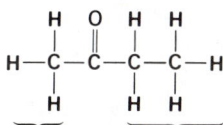

methyl ethyl ketone

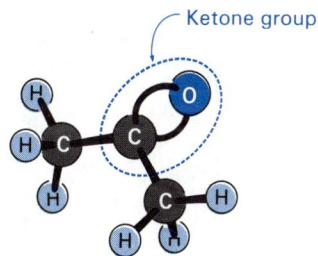

Acetone is widely used as a solvent, particularly for some types of lacquers. Acetone boils under 1 atmosphere pressure at 56°C (133°F). Acetone vapor ignites in air very easily if any flame is present. Consequently, the use of acetone as a solvent can be dangerous unless precautions are taken to avoid igniting the vapor. There are arguments in favor of using methyl ethyl ketone in place of acetone in such products as nail-polish remover. Although methyl ethyl ketone also is highly flammable, it boils at 80°C (176°F) and consequently does not tend to evaporate as readily at ordinary temperatures. It is almost as good a solvent for lacquers, but it does not dissolve oils from the skin quite as extensively as acetone does.

Table 16.5　A summary of functional groups introduced in this chapter

Name	Structural Formula	Example	
acid	$R-C\!\!\begin{smallmatrix}\nearrow O\\ \searrow OH\end{smallmatrix}$	$H-\overset{\overset{H}{|}}{\underset{\underset{H}{|}}{C}}-C\!\!\begin{smallmatrix}\nearrow O\\ \searrow OH\end{smallmatrix}$	acetic acid
alcohol	$R-OH$	$H-\overset{\overset{H}{|}}{\underset{\underset{H}{|}}{C}}-\overset{\overset{H}{|}}{\underset{\underset{H}{|}}{C}}-OH$	ethyl alcohol
aldehyde	$R-C\!\!\begin{smallmatrix}\nearrow O\\ \searrow H\end{smallmatrix}$	$H-\overset{\overset{H}{|}}{\underset{\underset{H}{|}}{C}}-C\!\!\begin{smallmatrix}\nearrow O\\ \searrow H\end{smallmatrix}$	acetaldehyde
alkene	$\;\;C\!\!=\!\!C\;\;$	$\overset{H}{\diagdown}C\!\!=\!\!C\overset{H}{\diagup}$ (H below both)	ethene (ethylene)
amide	$R-\overset{\overset{O}{\|}}{C}-N\!\!<$	$H-\overset{\overset{H}{|}}{\underset{\underset{H}{|}}{C}}-\overset{\overset{O}{\|}}{C}-N\!\!\begin{smallmatrix}\nearrow H\\ \searrow H\end{smallmatrix}$	acetamide
amine	$R-N\!\!<$	$H-\overset{\overset{H}{|}}{\underset{\underset{H}{|}}{C}}-\overset{\overset{H}{|}}{\underset{\underset{H}{|}}{C}}-N\!\!\begin{smallmatrix}\nearrow H\\ \searrow H\end{smallmatrix}$	ethyl amine
ester	$R-C\!\!\begin{smallmatrix}\nearrow O\\ \searrow OR'\end{smallmatrix}$	$H-\overset{\overset{H}{|}}{\underset{\underset{H}{|}}{C}}-\overset{\overset{H}{|}}{\underset{\underset{H}{|}}{C}}-O-\overset{\overset{O}{\|}}{C}-\overset{\overset{H}{|}}{\underset{\underset{H}{|}}{C}}-H$	ethyl acetate
halide	$R-X$	$H-\overset{\overset{H}{|}}{\underset{\underset{H}{|}}{C}}-\overset{\overset{H}{|}}{\underset{\underset{H}{|}}{C}}-Br$	ethyl bromide
ketone	$R-C\!\!\begin{smallmatrix}\nearrow O\\ \searrow R'\end{smallmatrix}$	$H-\overset{\overset{H}{|}}{\underset{\underset{H}{|}}{C}}-\overset{\overset{O}{\|}}{C}-\overset{\overset{H}{|}}{\underset{\underset{H}{|}}{C}}-H$	acetone

Summary

Organic chemistry, the chemistry of carbon compounds, can be organized in a helpful way by using the idea of functional groups. A functional group can be recognized by its particular structure. Table 16.5 summarizes the formulas of the various functional groups introduced in this chapter, along with examples and their names.

Glossary

The number in parentheses indicates the text page where you can find the term defined in context.

alcohol an organic molecule containing the hydroxyl group, OH, attached to a carbon atom (475)

aldehyde an organic compound that contains the functional group $-C{\overset{\displaystyle O}{\underset{\displaystyle H}{}}}$ (498)

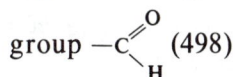

alkyl group a group of atoms derived from an alkane molecule by the removal of one H atom (476)

amide a compound having the general formula $R-\underset{\underset{\displaystyle H}{|}}{N}-\overset{\overset{\displaystyle O}{\|}}{C}-R'$, resulting from the reaction of a carboxylic acid with ammonia or an amine (492)

amine an organic compound having the general formula $R-N{\overset{\displaystyle R''}{\underset{\displaystyle R'}{}}}$, derived from ammonia by replacement of one or more of its hydrogens (492)

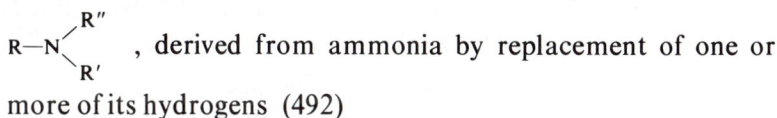

carboxyl group an acid functional group containing carbon, oxygen, and hydrogen having the general formula $-C{\overset{\displaystyle O}{\underset{\displaystyle OH}{}}}$ (484)

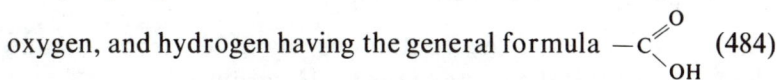

carboxylic acid an organic compound that contains the carboxyl group; most carboxylic acids are weak acids (484)

ester a compound having the general formula $R-C{\overset{\displaystyle O}{\underset{\displaystyle O-R'}{}}}$, resulting from the reaction of a carboxylic acid with an alcohol (486)

fat a naturally occurring water insoluble substance that is a triester of saturated long-chain carboxylic acids with glycerol (489)

fatty acid　a long-chain carboxylic acid obtained by hydrolysis of the ester linkages in fats or oils (489)

functional group　a combination (group) of atoms in a molecule that makes the molecule react (function) in a particular way (474)

halogen　any element of family 7A of the periodic table: usually F, Cl, Br, and I (496)

hydroxyl group　the functional group OH attached to a carbon atom (475)

hydrolysis　the reaction of a molecule of water with another molecule to produce two smaller molecules (490)

ketone　an organic compound having the general formula

$$R—C—R' \quad (499)$$
$$\overset{\|}{\underset{O}{}}$$

oil　a naturally occurring low-melting insoluble substance that is a triester of fatty acids with glycerol, its long-chain fatty acids usually are unsaturated (489)

organic halide　an organic compound that contains one of the halogens: fluorine, chlorine, bromine, or iodine (496)

petrochemical　a chemical substance derived from petroleum (481)

salt　the product (other than water) of a neutralization reaction between an acid and a base (485)

Exercises

16.1　*a.*　What is the difference between an alkane molecule and an alkyl group?

b.　Illustrate the difference by writing the structural formulas for methane and the methyl group.

c.　Illustrate the difference by drawing the structural formulas for the propane molecule and for the normal propyl and isopropyl groups.

16.2　The name of the continuous-chain 20-carbon alkane is eicosane. What is the name of the $C_{20}H_{41}$ group?

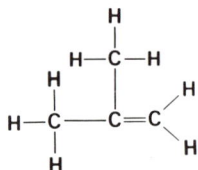

16.3　The molecule isobutene has the structural formula shown in the margin. Draw the structural formula of the product formed by its reaction with each of the following molecules.

　　a.　Br_2　　*b.*　Cl_2　　*c.*　H_2

16.4　Draw the structural formula of the product formed when each of the following molecules adds to the C=C bond of isobutene. Use the text example of the reaction forming isopropyl alcohol from propene as a guide to decide where the H atom goes in each of the reactions.

    a. HOH   *b.* HCl

16.5 Explain what is meant by the statement "A solution of methyl alcohol in ethyl alcohol is denatured alcohol."

16.6 What is meant by saying that a fermentation reaction takes place under anaerobic conditions?

16.7 If pure cider (with no preservative added) is kept for a long time in a corked jug, the cork may be blown off. What has happened?

16.8 A cookbook suggests that the remains of an opened bottle of wine can be left at room temperature for a few days and then used as vinegar in salad dressing. Explain why this can be done.

16.9 Draw the structural formula of the ester formed when acetic acid reacts with each of the following alcohols:

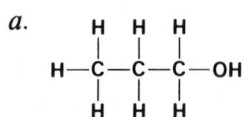

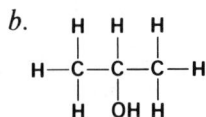

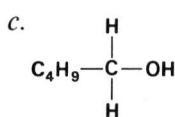

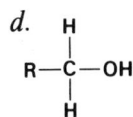

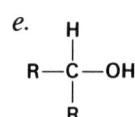

a.
$$H-\overset{\overset{\displaystyle H}{|}}{\underset{\underset{\displaystyle H}{|}}{C}}-\overset{\overset{\displaystyle H}{|}}{\underset{\underset{\displaystyle H}{|}}{C}}-\overset{\overset{\displaystyle H}{|}}{\underset{\underset{\displaystyle H}{|}}{C}}-OH$$

b.
$$H-\overset{\overset{\displaystyle H}{|}}{\underset{\underset{\displaystyle H}{|}}{C}}-\overset{\overset{\displaystyle H}{|}}{\underset{\underset{\displaystyle OH}{|}}{C}}-\overset{\overset{\displaystyle H}{|}}{\underset{\underset{\displaystyle H}{|}}{C}}-H$$

c.
$$C_4H_9-\overset{\overset{\displaystyle H}{|}}{\underset{\underset{\displaystyle H}{|}}{C}}-OH$$

d.
$$R-\overset{\overset{\displaystyle H}{|}}{\underset{\underset{\displaystyle H}{|}}{C}}-OH$$

e.
$$R-\overset{\overset{\displaystyle H}{|}}{\underset{\underset{\displaystyle R}{|}}{C}}-OH$$

16.10 Consider the following structural formula of an ester. Draw a circle around the part that came from the acid when it was synthesized.

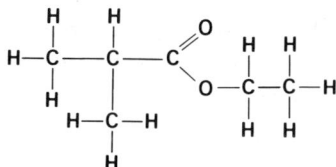

16.11 The ester that gives rum its flavor and aroma has the structural formula:

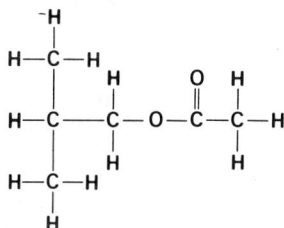

Draw the structural formulas of the acid and the alcohol from which this ester is formed.

16.12   *a.*   Why is glycerol called a trifunctional alcohol?

b.   What kind of an alcohol is ethylene glycol, used as an antifreeze in automobile radiators? Its formula is:

$$
\begin{array}{cc}
H & H \\
| & | \\
H-C-C-H \\
| & | \\
OH & OH
\end{array}
$$

16.13   Explain in terms of chemical structure the difference between animal fats and vegetable oils.

16.14   Explain what is meant by saying that the first step in the digestion of fats is a hydrolysis reaction.

16.15   How does the chemical composition and structure of fats and oils offer evidence for continuity and evolution in all forms of living things?

16.16   Examine the formulas of the fatty acids listed in Table 16.4. Which of these fatty acids contain unsaturated chains of carbon atoms?

16.17   Paraffin wax is obtained from petroleum and consists of large molecules of the paraffin homologous series. Beeswax has properties that are different, as anyone who has made candles knows. What is the explanation for this difference in terms of the chemical structure of the molecules?

16.18   Soap can be made by boiling beef fat with a water solution of lye, NaOH. The equation for the reaction is similar to the one for the hydrolysis of a fat, except that NaOH reacts instead of HOH (H_2O). Beef fat, called *stearin,* may be represented as the ester of only one fatty acid, stearic acid.

a.   Complete the following equation.

$$
\begin{array}{l}
H \\
| \\
H-C-O-C-C_{17}H_{35} \\
\quad\quad\quad \parallel \\
\quad\quad\quad O \\
| \\
H-C-O-C-C_{17}H_{35} \quad + \quad 3NaOH \rightarrow \quad\quad\quad + \\
\quad\quad\quad \parallel \\
\quad\quad\quad O \\
| \\
H-C-O-C-C_{17}H_{35} \\
| \\
H
\end{array}
$$

fat   lye   glycerol   soap

b. In Chapter 13, where we described the way the structure of a soap molecule is responsible for its effectiveness in removing grease from fabrics, a soap was called a salt of a fatty acid. Justify this statement.

16.19 The names of many fatty acids are related to their sources in nature. The fat of what animal contains the acids caproic, caprylic, and capric? (Hint: The animal is also a symbol of the zodiac.)

16.20 Identify the functional group in each of the following molecules by name.

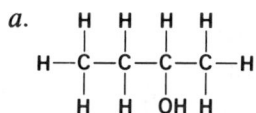

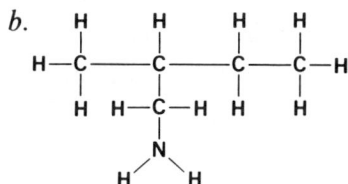

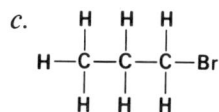

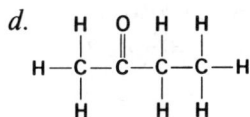

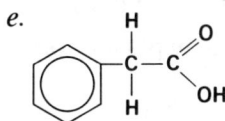

a.

H—C—C—C—C—H
(H H H H / H H OH H)

b.

H—C — C — C — C—H
(H H H H / H H—C—H H H / N H H)

c.

H—C—C—C—Br
(H H H / H H H)

d.

H—C—C—C—C—H
(H O H H / H H H)

e.

(benzene ring)—C—C
(H / H / O OH)

16.21 The disease ketosis is diagnosed by finding molecules of a certain type of organic compound in the urine of affected persons. Draw the structural formula for an example of the kind of compound involved.

16.22 A substance used as an ingredient in special solutions for cleaning contact lenses has the following structural formula.

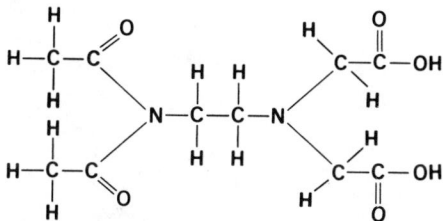

The chemical name for this substance is ethylene diamine tetraacetic acid, abbreviated EDTA. Show by a diagram how the name accounts for the way the parts of the molecule are put together.

17

Giant Molecules

☐ Why could we call our times the Age of Plastics?

☐ Is Mother Nature a polymer chemist?

☐ How does the molecular structure of the monomers affect the properties of the resultant polymers?

☐ How were polymer chemists able to create synthetic rubber?

☐ Why does Nylon dry more quickly than cotton?

☐ Why do waste plastics present a unique disposal problem?

In an earlier chapter, we suggested that the Modern Age is the age of light reactive metals (aluminum and magnesium), in contrast to the Stone, Bronze, and Iron Ages, earlier periods in the history of civilization. Of course, this description of the Modern Age is too narrow and oversimplified to be completely accurate. But it does emphasize the idea that the materials we live with are possible because people are using scientific knowledge never before available. Calling the present period in history the Space Age or Jet Age carries the same implication. Still another name might be used to describe this period: the Age of Plastics. Here we have a label that focuses attention on the subject of this chapter. Plastics are made from giant molecules. Plastics, like light metals and jet transportation, have been made possible by the use and application of accumulated scientific knowledge. The use of plastics, perhaps more than any other new material, marks the technology of the Modern Age.

The word *plastic* means "capable of being shaped or formed." The word also has come to mean a specific type of synthetic organic (carbon-containing) material. Plastics come in fibers, sheets, and blocks and can be molded into nearly any desired shape. You have already learned that carbon atoms can form stable bonds among themselves better than the atoms of any other element can. So you will not find it surprising that most of the giant molecules in plastics have skeletons or backbones of carbon atoms bonded to one another. Some articles manufactured out of plastic, such as a case for a radio, can be considered one giant molecule. The object gets its strength from the covalent bonds that tie all its atoms together. Plastics are made by chemists from simpler naturally occurring substances, such as coal, petroleum, air, water, and salt. But giant molecules are by no means only man-made. Many natural substances also are giant molecules. Wood, rubber, cotton, starch, and the proteins that constitute animal flesh are all examples of nature's ability to produce giant molecules.

In this chapter, we will discuss the structure and properties of some giant molecules. In particular, we will explore some of the general principles relating the properties of materials to the architecture of the molecules. This kind of knowledge has enabled chemists to synthesize materials with specific properties. They often function better for particular uses than natural materials do. For example, rubber gloves, gasoline hoses, and shoe soles made of Neoprene, a synthetic rubber, last longer than ones made from natural rubber. Plastic garden hose is lighter in weight, yet lasts longer in sun and weather than hose made from rubber does. Polyester fabrics wrinkle less than cotton or wool ones do. Plastic dishes are lighter in weight and break less easily than china ones. Many synthetic giant molecules have become so indispensable to us that we take them for granted.

GIANT MOLECULES—POLYMERS

The term *polymer* accurately describes the kind of giant molecules from which plastics are made. *Polymer* comes from the Greek words that mean "made up of many" (*poly*) "parts" (*mer*). Compare the term *isomer*, which means "made from the same" (*iso*) "parts" (*mer*). Polymers are made by stringing together many identical small molecular units called *monomers* (*mono* means "one" or "single").

You may imagine one kind of polymer molecule to be like a freight train. Many cars (the monomers) are coupled together to form the train (the polymer). Long freight trains of the same kind contain approximately the same number of cars, although some variation in train length is possible. So it is with the length of polymer chains. Another kind of variety is possible. A long train may be made up of only one kind of car, depending on its purpose (coal cars, automobile carriers, or refrigerator cars). Or various kinds of cars may be linked together. Some polymers are made from only one kind of monomer. An example is polyethylene, the material used to make most of the plastic bags with which you are familiar. Just one kind of monomer, ethene (also called ethylene), is strung together. Other plastics are made from two or more kinds of monomers, just as a freight train can be made up of various kinds of cars. For example, the tough, hard, metallike material called ABS is made from substances that have the chemical names *a*crylonitrile, *b*utadiene, and *s*tyrene. ABS is used to make a variety of articles, from football helmets to furniture.

A polymer molecule may contain thousands of atoms. Polymers are bigger molecules than any we have discussed previously. The molecular mass of most of the molecules we have discussed

A freight train is an analogy of a polymer.

is small. For example, the molecular mass of ethene, C_2H_4, is 28 (2 × 12 for each C atom, plus 4 × 1 for each H atom). One mole of ethene, approximately $6 × 10^{23}$ molecules, weighs 28 grams. Even big molecules, such as those in beef fat, have molecular masses of only about 900. So a mole of beef fat weighs 900 grams (0.9 kilogram, or about 2 pounds). However, a molecule of rubber, a naturally occurring polymer, may have a molecular mass of almost 1 million. Therefore, a mole of rubber weighs 1000 kilograms, or about 1 ton.

With molecules as big as this, you can expect that many of the properties of a substance composed of polymers depend on how the chains are made and how they are arranged. The variety of possible arrangements makes the analogy between a polymer and a freight train much too simple. You can apply what you know about chains of carbon atoms to predict that some polymer chains can be branched. (This is impossible for a train of cars.) You also can conceive of branched chains hooking together to form cross-linkages among chains. This also occurs in some polymers with structures like giant nets or tangles in two or three dimensions. So before we discuss the specific chemical makeup of various polymers, let us take a look at some of the general features we can expect to find in the behavior of such big molecules.

THE ARRANGEMENT OF POLYMER CHAINS

It is not difficult to imagine several possible ways long-chain molecules can be arranged so that materials made from the same monomer units will have different properties. Figure 17.1 suggests, diagramatically, what some of these possibilities are.

The simplest possible form is a completely random coiling of the chains (part A in the figure). This arrangement may remind you of a bowl of cooked spaghetti or a can of worms. The chains can slip and slide over one another easily. There is not much form to the pile; it is quite flexible, yet it sticks together. This is the kind of arrangement the polymer chains have in a plastic used to make bags and wrappings. They are flexible but can be torn rather easily.

Part B suggests an arrangement that may occur at some locations in a pile of randomly coiled polymer chains. At some places, several chains may be lined up parallel to each other for a short distance. In these regions, there is an orderly arrangement

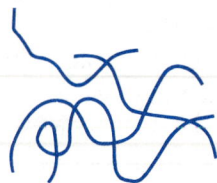

A. Flexible chains, randomly coiled

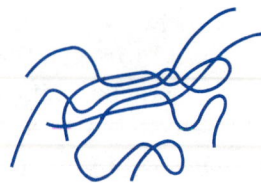

B. Chains lined up parallel in a small region to form a crystallite

C. Cross-linked chains forming a network

D. Stiff chains, cannot slip over each other

Figure 17.1   Possible arrangements of long-chain polymer molecules.

of atoms. Recall that orderly arrangement is typical of the way atoms are arranged in a crystal. Such a small region of order in an otherwise disorderly pile of polymer chains is called a *crystallite* (a little crystal). In Chapter 12, we used the term crystallite to describe similar small regions in metals.

Within a crystallite, the atoms all are in an orderly, regular array. Consequently, in a crystallite the atoms are, on the average, closer together than they would be in a random pile. Because they are closer together, the attractive forces among them are stronger, and the crystallite as a whole is stronger. The orderly arrangement of atoms always means that the molecular structure resists a force trying to break it apart. Consequently, the presence of crystallites in a mass of polymer gives the entire structure more strength than it would have if the chains were in a completely random arrangement. You can observe a similar effect in a pile of spaghetti that has begun to dry. The strands are sticky, and where they are lined up together, they stick together. This kind of arrangement that lines chains up in parallel fashion is accomplished in the polymers used to make automobile steering wheels.

Part *C* of Figure 17.1 suggests what you may recognize as the difference between a piece of macramé and a tangled pile of string. In the macramé, the strands are tied together and cross-linked in a definite way. Consequently, the whole piece has considerable strength, as a rope does, although it may remain flexible. The more cross-linking in three dimensions, the stiffer you expect the piece to be. This is the kind of molecular arrangement

in the polymers used for telephones, radio and television cabinets, and so on. When a polymer with the random arrangement of part *A* in Figure 17.1 is treated to create some, but not a lot of, cross-linkages, it becomes tougher yet retains some flexibility. This is true of the polymers used to make squeeze bottles.

Another possible modification is suggested by part *D*. If the chains are stiff and rigid instead of flexible, the whole structure made from the polymer is stronger. You would expect a tangle of

---

A tangle of stiff wire holds its shape better than a tangle of flexible string.

---

stiff wire to hold its shape better than a tangle of flexible string. Polymers with this type of molecular arrangement also are able to hold their shape when the temperature rises. Remember, an increased temperature always causes the units in solids to wiggle about more. The less rigid the chain, the more this increased atomic or molecular wiggling can lead to flexibility, softening the polymer as the chains slip over one another, or even melting it. Polymers with very rigid chains do not soften at ordinary temperatures. They are used for electrical insulation and building materials.

With such a variety of possible forms, chemists can combine these features of arrangement to varying degrees to produce a product with a particular set of desired properties. Let us move on to consider some of the ways in which the molecular structure of the monomers, the links that make the chains, influence the arrangement the chains will have.

The chemical reactions by which polymers are formed can be classified into two general types: *addition polymerization* reactions and *condensation polymerization* reactions. First we will consider the types of polymers formed by addition reactions. Later in the chapter, we will describe the other type.

POLYETHYLENE: AN EXAMPLE OF AN ADDITION POLYMER

Addition polymers are those made from monomers containing a $C=C$ bond. Recall that molecules containing the $C=C$ functional group can add reactants at the carbons of the double bond. In the 1930's, chemists discovered that when the gas ethene, C_2H_4 (the simplest molecule containing a $C=C$ bond), is heated under pressure, a polymer forms. The C=C bond of one molecule is disrupted in such a way that another molecule can add to it. Appropriately, this type of polymer is called an *addition polymer*. The sequence of reactions by which they are made is called *addition polymerization*

We can represent the process by which ethene forms a polymer in the following simplified diagram, which suggests, with structures drawn in only two dimensions, how a fragment of a long polymer chain is formed.

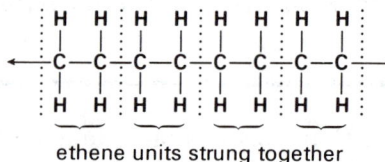

ethene units strung together

The arrows extending from each end of the diagram suggest that the chain continues with many such fragments linked together.

Addition polymerization is a chain reaction that occurs in a series of steps. Recall that a chain reaction is one in which each step sets off the next one. This kind of reaction is sometimes called the "domino effect," because it acts like a line of dominoes set on end. Each falling domino knocks down the next one in line. (Another example of a chain reaction is nuclear fission.)

Addition polymerization can be initiated (started) by a type of molecular fragment called a *free radical*. A free radical is a piece of a molecule with an unpaired electron on the end. We can represent it by the formula R· (the dot stands for a single electron). Free radicals can be formed from ethene itself by partially breaking the double bond with heat, ultraviolet light, X rays, or gamma rays. Or some other molecule that readily breaks apart into free radicals can be added. Free radicals are very reactive fragments. This is not surprising. Remember that electrons tend to establish themselves in pairs, if at all possible. Consequently, a free radical readily reacts with something so that the single electron can become part of a pair in a covalent bond. A free radical is attracted to the pileup of electrons in the C=C bond and attaches to it. The result is the formation of a new free radical.

A chain reaction proceeds as does the domino effect.

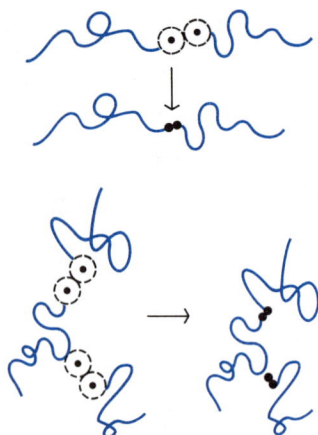

Free radicals combine to form longer, sometimes branched, chains.

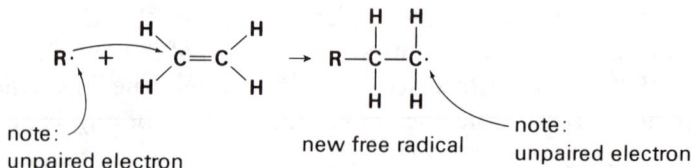

note: unpaired electron

new free radical

note: unpaired electron

This new free radical then reacts with another ethene molecule, as suggested by the following equation:

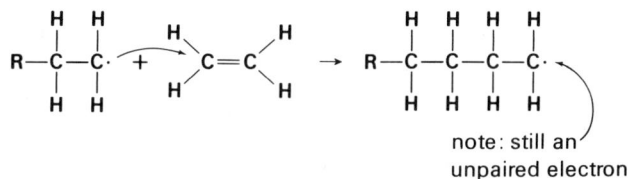

```
    H  H            H           H   H  H  H  H
    |  |             \         /    |  |  |  |
 R—C—C· +            C=C     →  R—C—C—C—C·
    |  |            /         \     |  |  |  |
    H  H           H           H    H  H  H  H
                                    note: still an
                                    unpaired electron
```

This process repeats, causing the chain to grow very rapidly. Each addition produces a new free radical with one more ethene unit. Hundreds to thousands of ethene molecules add to the growing chain. The chain reaction stops when two growing free-radical chain fragments encounter each other and form a bond, as follows:

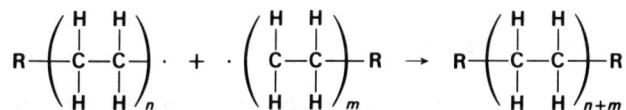

```
     / H  H \              / H  H \            / H  H \
    |  |  |  |            |  |  |  |          |  |  |  |
 R—|—C—C—|— ·  + ·  |—C—C—|—R  →  R—|—C—C—|—R
    |  |  |  |            |  |  |  |          |  |  |  |
     \ H  H /n             \ H  H /m          \ H  H /n+m
```

In this representation, n and m stand for very large numbers. The

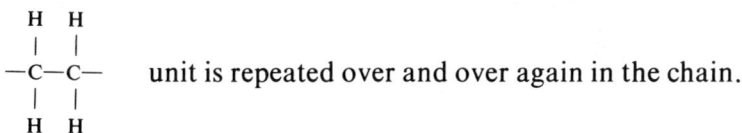

```
  H  H
  |  |
—C—C—
  |  |
  H  H
```

unit is repeated over and over again in the chain.

If conditions are such that many free radicals are formed at the same time, the likelihood is increased that the following reaction can occur at various places along the chain:

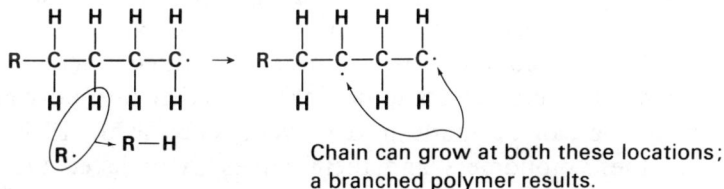

```
    H  H  H  H              H  H  H  H
    |  |  |  |              |  |  |  |
 R—C—C—C—C·     →       R—C—C—C—C·
    |  |  |  |              |  |  |  |
    H (H) H  H             H    H  H
       |                        Chain can grow at both these locations;
   (R·) — R—H                   a branched polymer results.
```

This process leads to chain branching. When branched-chain free radicals combine, cross-linking results. In this way, it is possible to change flexible, pliable polyethylene into a stiff, hard material by exposing it to high-energy radiation. The radiation produces many free radicals which in turn induce abundant cross-linking of the polymer. Cross-linked polyethylene is a good material for unbreakable dishes.

Polyethylene is far ahead of all other plastics in terms of production. Many billions of pounds are produced each year in the

Astroturf

United States and other countries all over the world. One reason so much is produced is its versatility; so many different kinds of polyethylene polymers, with different properties, can be made. Another is the fact that ethene is now cheap and abundant. But it is a byproduct of petroleum refining. We may be forced to adjust our demands for petroleum and its byproducts downward in response to conservation, economics, and international politics. When this happens, we may be forced to get along with fewer of the familiar convenience articles made from polyethylene, such as plastic bags, squeeze bottles, Frisbees, and Astroturf. The smaller amount of ethene available from decreased petroleum refining, even including the more costly amounts that may come from other sources such as conversion from ethyl alcohol, may have to be used for more important purposes.

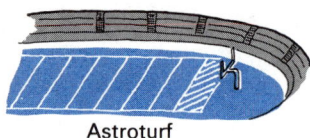

VINYL POLYMERS: OTHER EXAMPLES OF ADDITION POLYMERS

A great variety of molecules can be made by substituting some other atom or group of atoms for one of the hydrogen atoms on the ethene molecule.

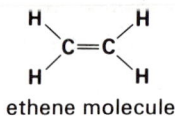

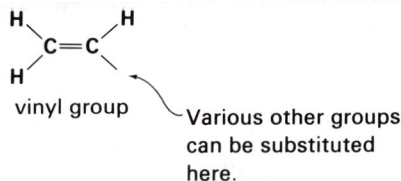

$$
\begin{array}{cc}
\underset{H}{\overset{H}{}}C=C\underset{H}{\overset{H}{}} & \underset{H}{\overset{H}{}}C=C\underset{}{\overset{H}{}}
\end{array}
$$

ethene molecule   vinyl group

Various other groups can be substituted here.

The $H_2C{=}CH$ group is called the *vinyl group*. Compounds with some other group of atoms attached to the vinyl group are generally called vinyl compounds. All these substituted ethene derivatives can act as monomers to be linked together into polymers, just as ethene can be made into polyethylene. Table 17.1 lists some of these monomers and their names. Also listed are the trade names and typical uses of some familiar polymer materials made from them. The polymer formula is written to show the structure of the repeating unit in the polymer chain. (The arrows going through the parentheses around each structure suggest at what point each unit is joined to the next. The subscript n used in these formulas stands for some very large number, like 100,000.)

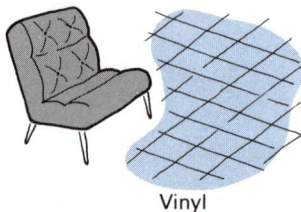

The first example in Table 17.1 is the polymer polyvinyl chloride, PVC. So many uses have been found for this material that its name is often shortened to just vinyl. One example is vinyl

Vinyl

Table 17.1 Formulas of some vinyl-type plastic materials

Monomer	Polymer Formula	Trade Names and Typical Uses
vinyl chloride		Polyvinyl chloride (PVC), Geon (film, insulation, rainwear, floor tiles)
styrene		Polystyrene, Styrofoam (foam, molded articles)
vinyl acetate		Polyvinyl acetate (PVA) (adhesives, latex paint, chewing gum)
acrylonitrile		Acrilan, Orlon (fibers, rugs, clothing)
propene		Polypropylene, Polyolefin (fibers, molded articles, outdoor carpeting)
tetrafluoroethene		Teflon (nonstick coatings, bearings, gaskets)
methyl methacrylate		Lucite, Plexiglas (transparent molded articles, windows, lamp globes)

upholstery covering that can be made to look like leather and wear as well yet is much cheaper. Vinyl chloride is a gas at ordinary temperatures. (It boils under 1 atmosphere pressure at $-14°C$.) The polymerization is accomplished under high pressure. Precautions must be taken to be sure the apparatus does not leak so that no vinyl chloride gas escapes into the air breathed by humans. Recent evidence indicates that the monomer vinyl chloride can cause cancer in animals and humans. Once formed into the polymer, the vinyl chloride molecules are so tightly held that articles made from polyvinyl chloride are completely safe to use. However, excessive heating or burning of articles made from PVC may be hazardous because of the decomposition or partial depolymerization that may occur.

The properties of the polymer polystyrene, or Styrofoam, reflect the structure of its monomer, styrene, and illustrate what we meant in a preceding section by "stiff" chains. You are familiar with many articles made from rigid Styrofoam, such as ice buckets, heat-insulating picnic boxes, packing material for fragile articles, and so on. The bulky benzene ring structure attached to the polymer backbone makes the chain lumpy. Consequently, chains set next to one another cannot easily slide over one another.

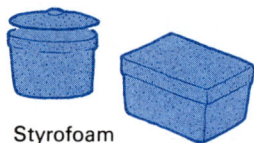

Styrofoam

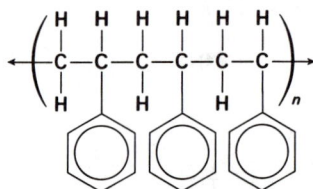

Styrofoam. Note the bulky benzene ring attached to alternate carbon atoms along the chain.

During the polymerization process, as the monomers join together to become polymers, the material changes from a liquid to a solid form. While it is in the liquid form, other chemicals are added that are capable of decomposing into gases. Many gas bubbles form, so the liquid becomes a foam. The liquid foam turns to a solid foam as the polymerization reaction is completed. The result is a very light but rigid material. Styrofoam softens when heated slightly and melts at higher temperatures. Thus small chunks can be molded together into a desired shape.

Also included in Table 17.1 are two examples of monomers derived from ethene, C_2H_4, by replacing more than one of the hydrogen atoms by another atom or group. One example is tetrafluoroethene, which polymerizes to make Teflon. Teflon is an extremely tough plastic, very little affected by other chemicals or by high temperatures. Its most familiar use is for coating utensils so that food will not stick when cooking. Teflon is another illustration of the way the presence of fluorine atoms makes the resulting molecule very unreactive. (Recall the earlier discussion of the lack of reactivity except in the presence of ultraviolet light of

The fluorine-carbon bond is very stable and gives Teflon its desirable properties.

Teflon

Freon molecules, such as CCl_2F_2.) The carbon-atom backbone of the Teflon polymer molecule is completely covered by fluorine atoms. The fluorine–carbon bond is very stable; it does not break easily to allow the polymer to react with other chemicals. This stability gives Teflon its desirable properties.

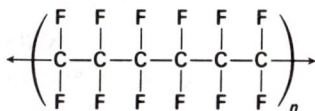

Teflon. Note how the chain is completely covered by fluorine atoms attached to carbon atoms.

The final example in Table 17.1 involves another monomer derived from ethene by substituting more than one H atom with other groups of atoms. This monomer, called methylmethacrylate, polymerizes to form the familiar clear plastic as transparent as glass but so tough that it is practically unbreakable. It is sold under the trade names Lucite and Plexiglas. This material illustrates a type of plastic called a *thermoplastic*. A thermoplastic material can be softened (made more plastic) by heat, yet it becomes hard again when cooled.

Lucite

The effect of bulky side groups attached to the polymer backbone gives Lucite or Plexiglas an interesting property. Articles made from this material can be warmed slightly and forced into some other shape. (A strip can be bent or twisted, for example.) Cooling the article freezes it into its new shape. But if it is reheated slightly, the article gradually returns to its original form; it acts as if it remembered its original shape and tends to relieve the stresses and strains that were introduced by bending it around.

This effect is less of a mystery if we recognize the features of the molecular structure of the polymer. Note the polymer formula in the last entry in Table 17.1. Two side groups, one quite bulky, are attached to the chain at regular intervals, at alternate carbon atoms along the chain.

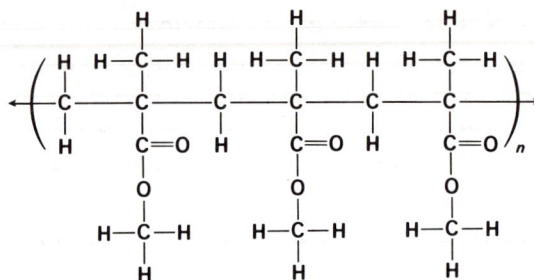

Lucite or Plexiglas. Note the two side groups attached to alternate carbon atoms along the chain.

When a sheet or rod of Lucite or Plexiglas is first formed, the polymer chains fit next to one another so that the bulky side groups can be accommodated. When the article is heated, the chains are able to wiggle with some freedom, so that the article will bend when forced. The chains slip and slide to new locations, but strains and stresses are produced when the bulky side groups of adjacent chains bump into one another. Cooling freezes the chains into this strained arrangement. When the article is reheated without any bending force applied, the chains relieve the strain by slipping back to their original locations, where everything fit together in the first place.

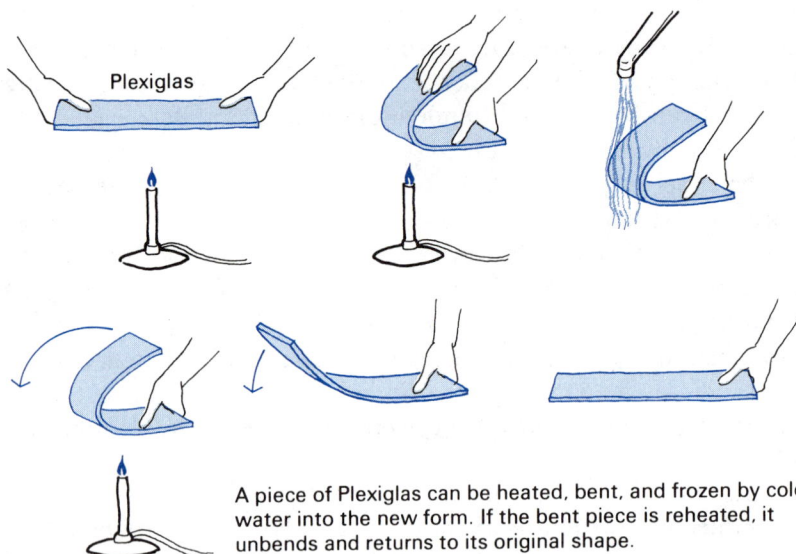

A piece of Plexiglas can be heated, bent, and frozen by cold water into the new form. If the bent piece is reheated, it unbends and returns to its original shape.

COPOLYMERS ARE FORMED FROM MIXTURES OF MONOMERS

All the various vinyl-type monomers undergo the same kind of addition polymerization reaction. There are some differences in reaction tendency and response to the presence of catalysts. But the essential way the molecules add to one another is the same. Consequently, it is possible to combine the molecules of two or three different monomers into a polymer, just as we can find a freight train made up of several different kinds of cars, all coupled together by the same kind of fastenings. This kind of a polymer is called a *copolymer* (a cooperative building of a long chain).

The properties of a copolymer depend on the proportions in which the two or more monomers are mixed. All the factors that influence the properties of a polymer—such as chain length and degree of cross-linking—can be changed by changing the proportions of monomers in a copolymer. Consequently, polymer chemists have one more means of varying composition to achieve a particular set of desirable properties in the product.

An example of copolymers that you may encounter is Dynel, which can be formed into fibers resembling human hair for wigs. Dynel is a copolymer of vinyl chloride and acrylonitrile in the proportions of nine to one. Another copolymer is Vinylite, used for fabrics. It is made from an eight-to-one mixture of vinyl chloride and vinyl acetate.

NATURAL RUBBER

Rubber, one example of a polymer made by nature, has a long and fascinating history. Rubber balls were among the curiosities from the New World (Central and South America) taken back to England and the European continent by sixteenth-century explorers. The natives of Central America made rubber from the sticky fluid they found oozing from broken places in the bark of some kinds of trees. This fluid, called *latex,* is an emulsion of rubber particles in water. The Indians caused it to coagulate by working the latex into a ball over a smoky fire. The same thing happens when children rub the milky fluid from milkweed, goldenrod, or dandelions between their fingers. The name *rubber* began to be used for this substance in the late eighteenth century, when people found that it could be used to rub out pencil marks.

As early as 1826, Michael Faraday found that, if he heated rubber in the absence of air, a hydrocarbon molecule with the

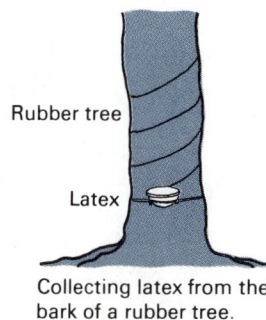

Rubber tree

Latex

Collecting latex from the bark of a rubber tree.

formula C_5H_8 distilled out of the rubber. This product, at ordinary temperatures, is a liquid that reverts to a rubber-like solid on standing. In about the year 1860, the structure of this C_5H_8 molecule, called isoprene, was determined. A few years later, it was identified as the monomer from which nature makes the polymer rubber.

The polymerization of isoprene to rubber can be represented as follows:

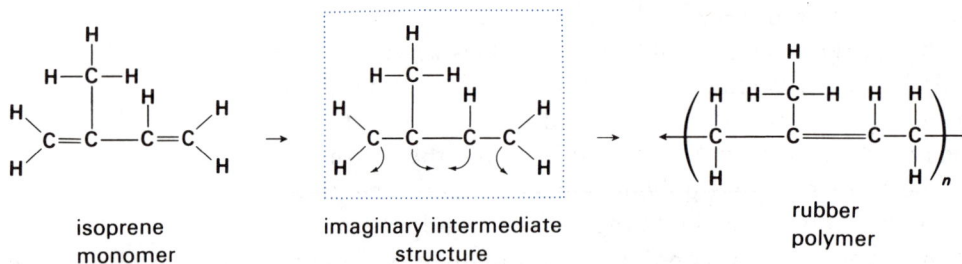

isoprene
monomer

imaginary intermediate
structure

rubber
polymer

The imaginary intermediate structure is drawn to suggest how the two C=C bonds in isoprene change position to form the polymer. The curved arrows under the carbon atoms of the imaginary intermediate structure suggest the way one bond of each double bond breaks and then reforms with a different carbon atom. Note that one of the C=C bonds of isoprene is used to string the polymer together. The other double bond remains in each C_5H_8 unit of the polymer. (In the case of ethene and the vinyl compounds, the monomers have only one C=C bond, so the polymer units contain none.)

THE VULCANIZATION OF RUBBER

The presence of a C=C bond at repeated intervals in the rubber polymer has one very important consequence. The C=C bond is the spot where chemical action occurs in the process called *vulcanization.* Raw rubber tends to be sticky and to pull apart without being very elastic. It has none of the strength, resilience, or elasticity that we recognize as typical of rubber bands and automobile tires. In 1839 Charles Goodyear discovered that raw rubber could be heated with sulfur to make a much stronger material. He named his process after the Roman god Vulcan, the god of fire. (Forging a metal article in a fire made it stronger, so the gain in strength was attributed to Vulcan.)

The chemical reaction between sulfur and rubber can be represented as a reaction that cross-links one rubber polymer chain to another. The principle involved in the reaction can be represented as follows:

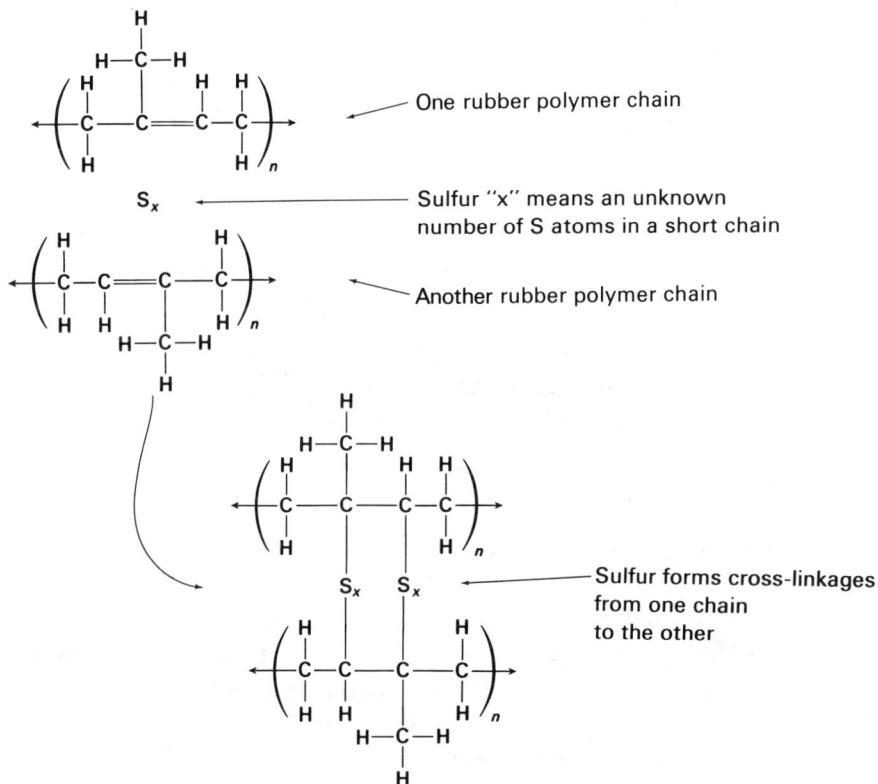

One rubber polymer chain

Sulfur "x" means an unknown number of S atoms in a short chain

Another rubber polymer chain

Sulfur forms cross-linkages from one chain to the other

The presence of these cross-linkages from one chain to another gives vulcanized rubber its elasticity and strength. By no means is every C=C bond in rubber involved in cross-linking. The polymer chains in raw rubber are too randomly coiled to make possible the lining up of every C=C group opposite another one in another chain. However, the cross-linkages that do form hold the chains together when rubber is stretched. Figure 17.2 suggests how the cross-linkages act. At various points, the random coils in unstretched rubber are cross-linked. When rubber is stretched, the polymer chains are pulled into straighter alignment. They are kept from being pulled past one another—and apart—by the cross-linkages. This alignment makes the rubber strong, and the tendency of the chains to return to their random coiled form makes the rubber elastic.

An old rubber article often appears to have cracks in it. When stretched, it may fall apart. What has happened is that oxygen or other chemicals have reacted with some of the remaining C=C

Randomly coiled
polymer chains

Sulfur cross-linkages
between chains

A. Unvulcanized rubber

B. Vulcanized unstretched rubber

Elongated chains kept from
slipping further by the
sulfur cross-linkages

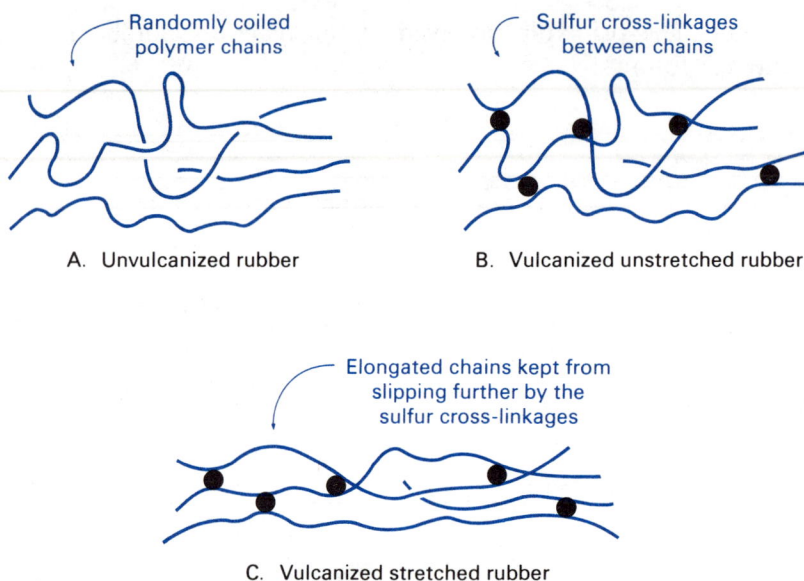

Figure 17.2 Schematic
representation of how
vulcanizing imparts strength
and elasticity to rubber.

C. Vulcanized stretched rubber

bonds in rubber. As a result, the long chains break into smaller segments that are no longer able to return to their original coiled form after they have been stretched. Ozone is particularly hard on rubber in this way. The ultraviolet part of sunlight also causes the large polymer molecules to break into smaller pieces. Therefore, tires are made from rubber to which special compounds called antioxidants have been added. These react with oxygen or ozone more readily than the $C{=}C$ bond does, so the weakening effect on the rubber is lessened. Rubber in tires also has small particles of carbon (like soot) mixed into it to make the tires wear longer.

CIS–TRANS ISOMERS

Although chemists knew that rubber is a polymer of isoprene for almost 90 years, all attempts to make the rubber polymer from the isoprene monomer in a laboratory failed. Not until 1954 were catalysts discovered that allowed humans to duplicate nature's accomplishment. The difficulty was related to a subtle but essential detail in the structure of the polymer chain around the $C{=}C$ bond. We can see what this detail is by examining the possible structures of molecules formed by substituting other atoms for two of the hydrogen atoms in ethene, the simplest compound containing a $C{=}C$ bond.

In Chapter 7, where we first discussed the structure of molecules, ethene, C_2H_4, was introduced as an example of a flat, planar molecule. All the atoms are in one plane. When we draw the structural formula of ethene, we are drawing what amounts to a top view of a model of ethene.

ethene

The model of ethene shows another important feature of its structure. The C=C bond is stiff; the model cannot be twisted about the C=C bond, as can a model of ethane, C_2H_6, which contains a C—C bond. The C_2H_4 molecule is rigid. Therefore, if one of the hydrogen atoms on each carbon atom of C_2H_4 is replaced by another atom, *two different* molecules can be formed. These two molecules are called *cis–trans isomers*. The term *cis* comes from the Latin word meaning "next to"; *trans* is the Latin word meaning "across." One such pair of isomers both have the formula $C_2H_2Cl_2$.

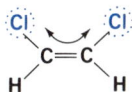

Top view of a model of ethene

cis isomer boils at 60° C
(The two Cl atoms are on the *same* side of the molecule.)

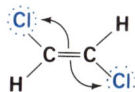

trans isomer boils at 47° C
(The two Cl atoms are on *opposite* sides of the molecule.)

Side view of a model of ethene

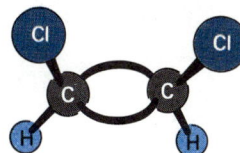

You should note that this difference in structure is responsible for considerable differences in the properties of the two isomer molecules (boiling temperatures at 60°C for the *cis* isomer and 47°C for the *trans* isomer).

In the polymer of natural rubber, all the isoprene units are built in the *cis* form around the C=C bond. The geometry of this form is suggested by the structure shown in Figure 17.3. Only two isoprene units are shown; the polymer extends in either direction. The long chain can have a twisted or a coiled structure, because it can twist or turn around the single bond formed where two isoprene units join together. Before the right catalysts were discovered, the polymers chemists made from isoprene were a mixture of the *cis* and *trans* forms. Consequently, they did not behave as natural rubber does.

cis—$C_2H_2Cl_2$

Nature's apparent preference for one arrangement rather than another may seem to be a very small detail of molecular structure. However, it has a very significant effect on the properties of the resulting polymer materials. Nature does make an all-*trans* isoprene polymer. This substance, called gutta-percha, is hard and tough and has none of the elasticity or resiliency of rubber. The first phonograph records were made from gutta-percha. Now it is used for the covers of golf balls, some kinds of electrical insulations, and a root-canal filling in teeth.

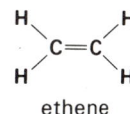

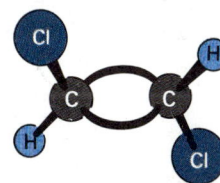

trans—$C_2H_2Cl_2$

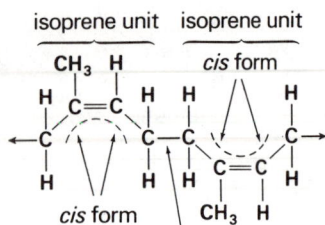

isoprene unit   isoprene unit

cis form

cis form

Rotation can occur around this bond, so the chain can twist and coil.

Figure 17.3   Natural rubber —a polymer of *cis*-isoprene.

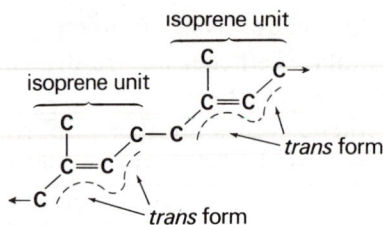

isoprene unit

isoprene unit

trans form

trans form

Gutta-percha is a polymer of *trans*-isoprene.

There are many examples of nature's preference for one form or the other of the two possible isomers. Often the point of difference is deep in the structure of a complicated molecule. For example, the molecule *retinal* is involved in the initial chemical reactions of the intricate processes by which our eyes see things. When light reaches the eye, the light energy changes the *cis* form of retinal into the *trans* form. We can represent the structures in outline form as follows:

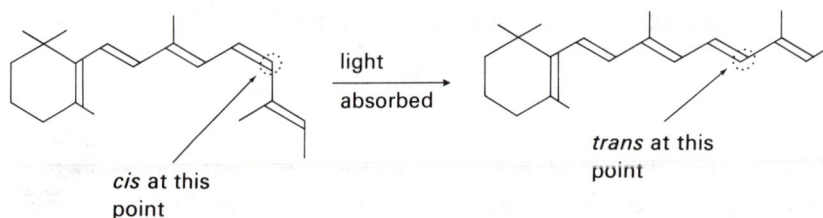

light absorbed

cis at this point

trans at this point

SYNTHETIC RUBBER

For many years, the chief source of raw material for rubber was the latex produced by rubber trees. During the nineteenth century, British botanists, working in the botanical gardens in England with plants from the jungles of Brazil, developed superior strains of rubber trees. These trees became the stock for the huge plantations in what was then the British Empire in the Far East. For many years, the world's needs for rubber were met from these and other tropical sources.

During the present century, many of our technological advances have depended on rubber. Automobiles and trucks roll on rubber tires. Flexible rubber conveyor belts are used to move coal, scrap metal, freight, and sometimes people. Each time a world-wide war occurred to disrupt world trade, the rubber industry was threatened with being shut off from supplies of rubber latex. The source of this raw material was far across the world from where the rubber was needed. Consequently, the rubber industry began a long and vigorous research program whose goal was the development of synthetic rubber. German and Russian chemists tried, with very limited success, during World War I. Just before World War II, the threat of enemy control of the Far Eastern rubber supply spurred intense research activity in the laboratories of the United States and its allies. The success of the

---

A logical starting material for synthetic rubber is petroleum.

---

industry in developing and producing a satisfactory synthetic rubber was essential to military victory. Since that time, the quality and amount of synthetic rubber produced has continued to grow. As is so often true, developments forced by military necessity greatly enrich our lives in times of peace. All the tires and other rubber articles we use are better and cheaper because they contain large amounts of synthetic rubber.

Because rubber is made from giant hydrocarbon molecules, a logical starting material for synthetic rubber is petroleum. Because chemists were unable at that time to make the monomer of natural rubber, isoprene, polymerize totally in the *cis* form, they turned to other monomers. They found that the one that works best is a close relative of isoprene called *butadiene*.

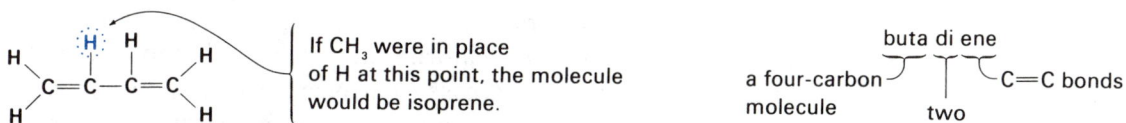

If CH_3 were in place of H at this point, the molecule would be isoprene.

buta di ene

a four-carbon molecule   two $C=C$ bonds

The synthetic rubber made in the greatest amounts is a *copolymer* of butadiene with styrene, called SBR (*styrene-butadiene-rubber*). The polymerization can be represented as follows:

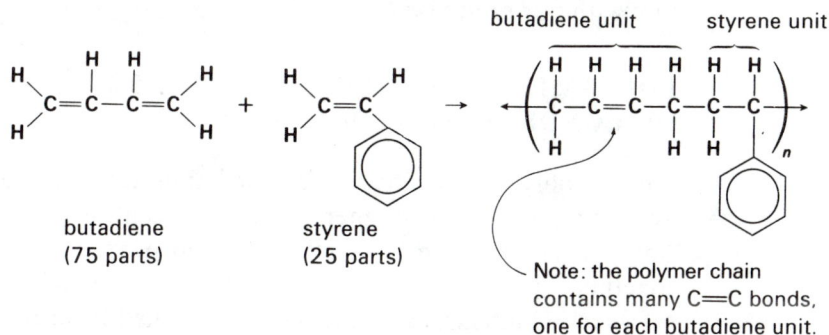

butadiene
(75 parts)

styrene
(25 parts)

butadiene unit   styrene unit

Note: the polymer chain contains many $C=C$ bonds, one for each butadiene unit.

Various proportions of the two monomers can be used to make the copolymer. The most frequently used proportions are 75% butadiene and 25% styrene. You can see that the polymer molecule contains many C=C double bonds (one for each butadiene unit), like those in natural rubber. This feature means that this synthetic rubber can be cross-linked by vulcanization, like natural rubber.

As chemists learned more and more of the general principles behind the reactions that turn monomers into polymers, many varieties of synthetic rubber were developed. Some of these special kinds of rubbers have properties that make them superior for specific uses. One example is Neoprene, made from a monomer called chloroprene.

If CH$_3$ were in place of Cl at this point, the molecule would be isoprene.

chloroprene → Neoprene rubber

The presence of the chlorine atom in the molecule makes Neoprene less soluble than natural rubber in oil or grease. Remember, from Chapter 13, the generalization that "like tends to be soluble in like." Natural rubber, made only of carbon and hydrogen atoms, without any chlorine atoms in it, is more like the hydrocarbons of oil or grease than Neoprene is. Consequently, oil or grease tends to soak into an article made of natural rubber. This causes the article to swell up and become less resistant to scratching or tearing apart. Articles made from Neoprene do this less, so it is a better material for shoe soles and gasoline hoses. The same is true of rubber gloves that may come into contact with fats or grease.

POLYAMIDE POLYMERS: EXAMPLES OF CONDENSATION POLYMERS

All the polymers we have discussed thus far are examples of addition polymers (polymers made by additions at the C=C bond of the monomer). Another method of stringing molecules together is possible. This method is the formation of *condensation polymers*. Condensation polymers are formed from two different monomers by a reaction that joins them at the same time that water, H_2O, is formed. One example makes use of the reaction between the amine functional group and the acid functional group to form an amide. You can refer back to the discussion in Chapter 16 to recall how this works. The formation of an amide is called a condensation because the two molecules "condense" together and split out water, H_2O, as a second product.

If a molecule contains *two* amine groups, one at either end, it is capable of joining with an acid at both places. Such an amine is called a *di*functional amine. If the acid molecule likewise contains *two* acid groups (a *di*functional acid), it also can react in two directions. This feature is the basis for forming long chains. The principle is as simple as having a group of men and women link arms to form a human chain. Each man is *di*functional (two arms). So is each woman. Each person links on either side with a person of the opposite sex. The human chain can thus grow longer until it involves every person of the group. A polymer is formed from two kinds of difunctional monomers.

Nylon is made in this way. One of the monomers is a small, six-carbon straight-chain molecule with an amine group on each end. The other monomer is a small, six-carbon straight-chain molecule with an acid group on each end. The polymerization reaction involves forming an amide linkage between the monomer units. The resulting polymer is appropriately called a *polyamide* The reaction is represented by the diagram in Figure 17.4. (The symbol *n* again is used to mean that the structure is repeated over and over many times.)

One of the properties of Nylon that makes it very useful as a fiber is its great tensile strength (ability to hold a weight without breaking). A Nylon rope 1 inch in diameter can be used to suspend a load of $5\frac{1}{2}$ tons. A steel rope weighing the same as the

Figure 17.4 The formation of Nylon, a polyamide condensation polymer.

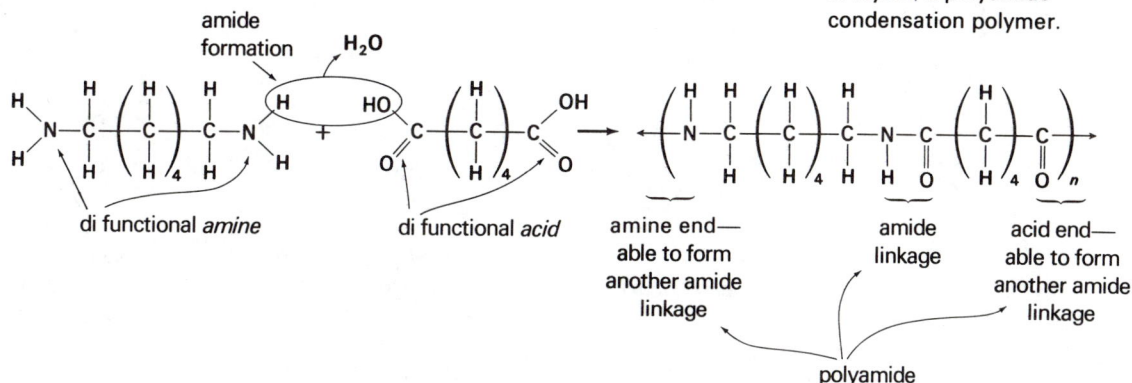

1-inch Nylon rope

Steel rope weighing the same as the Nylon rope

5½ tons

1½ tons

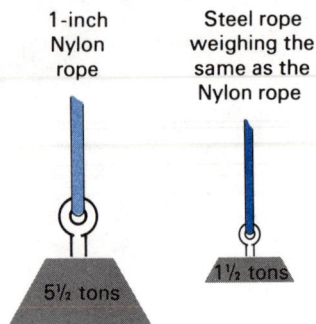

Nylon one can hold only 1½ tons. The reason for this remarkable strength in Nylon fibers is another illustration of the effect of hydrogen bonding. (Recall that hydrogen bonding is the attraction of a hydrogen atom in one molecule for the electrons of an oxygen atom in another molecule. Hydrogen bonding is responsible for the solubility of alcohols in water, for example.) Figure 17.5 suggests, in diagram form, how hydrogen bonds are formed between one polymer chain and another. The result of this bonding between the chains is a great resistance to forces pulling the chains past one another. The polymer chains act as if they were twisted together like many threads in a strong rope.

POLYESTER POLYMERS

Another example of condensation polymers are those made from difunctional alcohols and difunctional acids. The principle is quite like that by which polyamides are made, except that in this case the polymer is a polyester. (Recall that alcohol + acid →

Figure 17.5   How hydrogen bonds hold one Nylon polymer chain to another.

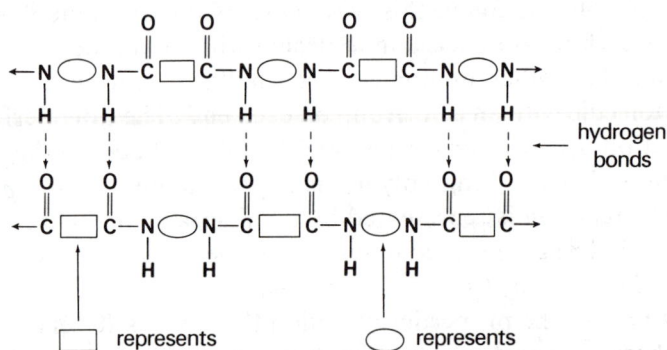

Figure 17.6   The formation of Dacron, a polyester condensation polymer.

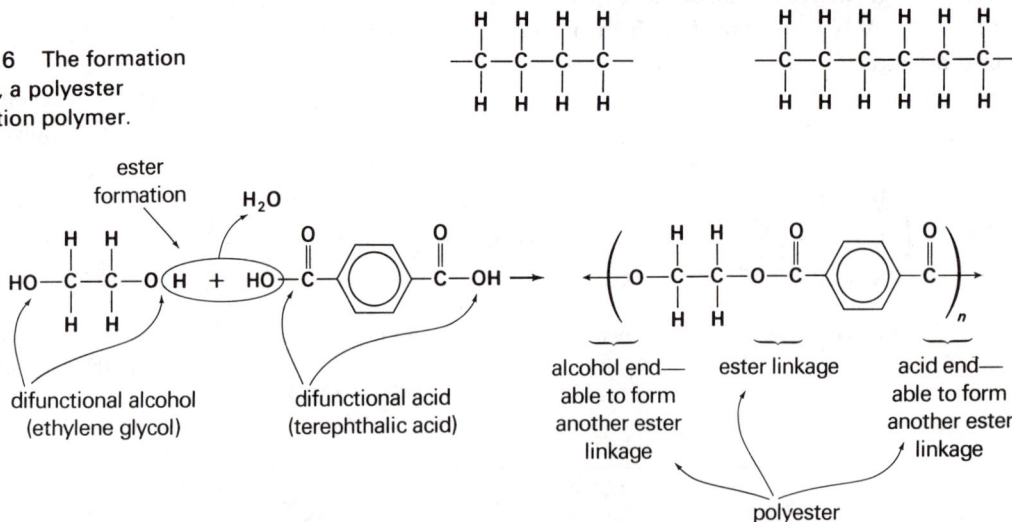

ester formation H_2O

difunctional alcohol (ethylene glycol)

difunctional acid (terephthalic acid)

alcohol end— able to form another ester linkage

ester linkage

acid end— able to form another ester linkage

polyester

ester + water.) One familiar example of a polyester-type polymer is the material Dacron, used in textiles. A chemically similar material, Mylar, is used in the form of thin sheets or tapes, such as magnetic tapes for recorders and computers. Figure 17.6 suggests how the Dacron polyester polymer is formed from the difunctional alcohol, ethylene glycol, and the difunctional acid, terephthalic acid. Again, as in the polyamide formation, a molecule of water splits off as each ester linkage forms. Each end of the growing chain is able to react further, so the polymer grows to a great length.

One of the properties of fabrics made from either polyamides (for example, Nylon) or polyesters (for example, Dacron) is the ease with which they dry after they have been wet. This property is related to their structure. These polymer molecules contain no hydrogen atoms capable of hydrogen bonding with the oxygen atoms in water. Consequently, there is no way for water to be held to these molecules by chemical forces. All the hydrogen atoms along the polymer chains are tightly bonded to carbon atoms, so they are not capable of forming hydrogen bonds with other molecules. There are no OH groups in either polymer molecule. Hydrogen atoms of OH and NH groups are the ones capable of forming hydrogen bonds. In Nylon, almost all the hydrogen atoms on the NH groups are already involved in hydrogen bonds to other polymer chains.

The same ease of drying is a property of fabrics made of Acrilan or Orlon. Table 17.1 lists these as the trade names of the vinyl-type addition polymers made from acrylonitrile monomers. You can see from the structure of the polymer shown in Table 17.1 that, again, no hydrogen atoms in the polymer are capable of forming hydrogen bonds with water. Consequently, water does not adhere strongly to fabrics made from these polymers.

Articles made of cotton dry much more slowly than those made from these synthetic fibers. Cotton contains a naturally occurring polymer, cellulose. We will describe the structure of the cellulose molecule in the next chapter. At this point, we can anticipate one feature of the molecular structure of cellulose. Many OH groups are located on the polymer chain of cellulose. Consequently, hydrogen bonds can be formed and the water can adhere more strongly to cotton fabrics.

POLYMERS THAT FORM NETWORKS

Most of the discussion thus far has dealt with how polymer chains are formed. We have seen various examples of the way the properties of a polymer are related to the structures of the monomer molecules from which it is made. For example, it seems reasonable to find that a polymer like Nylon, suitable for making fibers, is formed by fastening together two kinds of short straightchain molecules. The monomer molecules hook end-to-end, like cars of a freight train, because that is where their functional groups, the amine and the carboxylic acid groups, are located. Now let us look at examples of polymers that build in the form of networks. The networks they form are three-dimensional, more like a sponge than a two-dimensional fishnet or a onedimensional chain or rope.

One familiar type of plastic material made from network polymers is the *alkyd resins*. The term *alkyd* refers to the kind of polyesters we have already described. The term *resin* refers to a hard, tough, solid material, such as the tar or gum from some trees that for centuries has been used to make varnish. Alkyd resins are used for automobile enamels, paints, adhesives, and material for false teeth. Most Fiberglas used to make boats or truck bodies contains alkyd resins. The alkyd resin makes bonds through and over the glass fiber or glass cloth, thus is a reinforcing material.

Alkyd resins are polyesters. The alcohol used is the trifunctional alcohol glycerol. Recall that fats, discussed in the previous chapter, are esters of glycerol with fatty acids. Alkyd resins are esters formed by the reaction of glycerol with *di*functional acids. The reaction between the OH of the glycerol and the COOH of the acid links the molecules together into a polymer, just as with polyesters like Dacron. However, the presence of a *third* OH on the glycerol allows the polymer to grow in *three* directions instead of just two. This feature of the structure of the glycerol monomer leads directly to a highly cross-linked polymer. Moreover, the other monomer containing the acid groups is a big, bulky molecule. Consequently, the only way a third one can fit around the glycerol is to join at an angle to the plane of the other two ester linkages. So the polymer network grows in three dimensions, as well as three directions. Figure 17.7 suggests how this happens.

Alkyd resins are tough and strong because of their network structure. The coating on an automobile body shows this toughness. Even though the metal may be bent or dented, the coating gives with the metal; it sticks to the bend or dent without cracking.

Another type of network polymer is the kind known by the trade name Bakelite. This material was the first synthetic poly-

The breeze fills sails made of Dacron mounted on a hull made of plastic-impregnated Fiberglas.

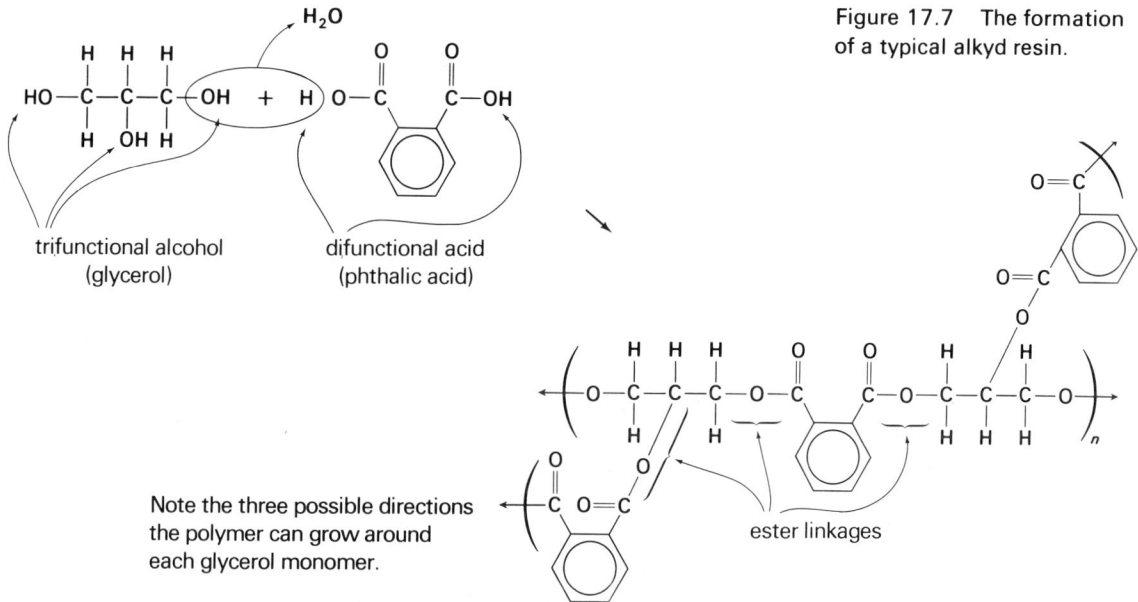

Figure 17.7 The formation of a typical alkyd resin.

trifunctional alcohol
(glycerol)

difunctional acid
(phthalic acid)

Note the three possible directions the polymer can grow around each glycerol monomer.

ester linkages

mer to be made. (Note that we say first *synthetic* polymer, because naturally occurring polymers have been on earth since life began.) In 1907, the Belgian chemist Leo Baekeland discovered how to control the reaction between phenol and formaldehyde to make a plastic material. The structures of the monomers are as follows:

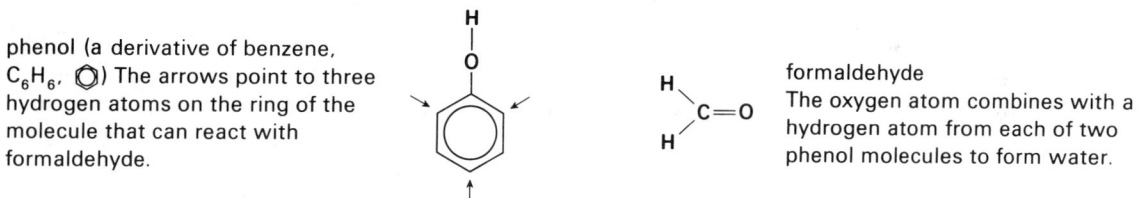

phenol (a derivative of benzene, C_6H_6, ◯) The arrows point to three hydrogen atoms on the ring of the molecule that can react with formaldehyde.

formaldehyde
The oxygen atom combines with a hydrogen atom from each of two phenol molecules to form water.

Figure 17.8 suggests how the polymer grows. Note again that a *tri*functional molecule (phenol) can react by losing hydrogen atoms from three places on the ring. This circumstance makes growth in three dimensions possible.

One of the advantages of Bakelite is that the polymerization reaction can be stopped when it is only partly finished. This process makes the material in the form of a powder that can be put into a metal mold made in the shape of some desired article (such as a radio cabinet). Then the mold can be heated to complete the polymerization. Covalent bonds form among the particles of powder. The whole article is thus one giant polymer molecule. This type of polymer is called a *thermosetting plastic,* because the

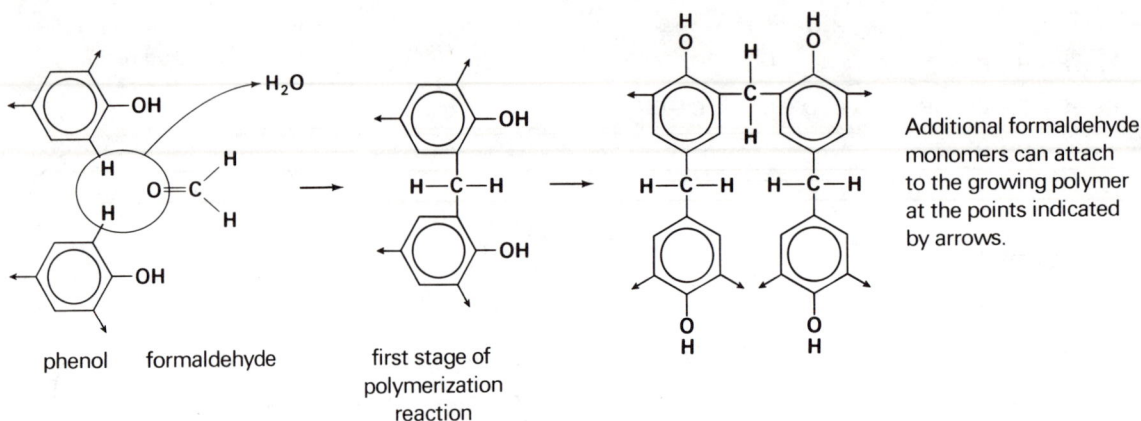

phenol   formaldehyde

first stage of
polymerization
reaction

Additional formaldehyde
monomers can attach
to the growing polymer
at the points indicated
by arrows.

Figure 17.8   The formation
of a phenol–formaldehyde
plastic (Bakelite).

final article is one big network molecule formed (set) by heat. Once set, the plastic is rigid and cannot be molded further by heating. In this respect, it differs from many of the vinyl polymers that can be softened by heating and made into different shapes. Contrast this with thermoplastic materials, discussed earlier.

LIVING IN THE AGE OF PLASTICS

We began this chapter by suggesting that we are living in the Age of Plastics. In terms of the sheer volume of plastics produced and in terms of the number and variety of materials made from plastics, the label is appropriate. World production of synthetic polymers is nearly 100 billion pounds each year.

Some people are bothered by the idea that synthetic plastics represent cheap, inferior substitutes for materials that either grow or are found in nature, such as wood, cotton, wool, and metals. The term plastic is even used sometimes to mean "artificial" or "unnatural." To some extent, this attitude may be justified. But sometimes different values have priority. Simple plastic articles can be mass-produced quite cheaply, and cheapness may be a greater advantage than durability. For example, hospitals have found it more economical to serve patients' meals on plastic dishes, used once and then discarded, than to use chinaware. The necessary sterilization of dishes for repeated use is too costly and time-consuming. Anyone who examines the already high cost of hospital care may be inclined to agree that saving immediate hospital funds should have priority over saving a miniscule amount of the world's petroleum resources.

Those who live in affluent societies certainly could get along with fewer luxuries of clothing, housing, and the gadgetry so often taken for granted. But many in poorer societies lack even the necessities in these areas of their lives. Synthetic polymers for clothing, housing, protective packaging, and transportation may represent the most economical and rapid means of bringing the standard of living for less technologically developed societies closer to what we expect and enjoy. Growing the necessary cotton or wool to clothe the world's population would require land that should be devoted to food production. Meeting the world's housing needs totally from wood would come close to denuding the world of forests. Even brightening the life of every child in the world with a toy made of metal would require a substantial decrease in the already dwindling supplies of ores.

THE PROBLEM OF DISPOSAL

Probably the biggest problem our massive use of plastics has created is the problem of waste disposal. A generation ago, the paper and cardboard litter left by thoughtless individuals who believed "disposal" meant "throw away" decayed and disintegrated back into the earth. But today, discarded boxes, bottles, and wrappings are made of plastics. Few of these synthetic materials are biodegradable. They cannot serve as food for the microorganisms that speed the decay of paper, cardboard, cotton, and wood. The durability that makes plastics desirable means that disposal of worn or broken plastic articles is a problem. At present, most of the nearly 6 pounds of waste discarded in an average day by every city dweller ends up being buried. The capacity of landfills is constantly being taxed by the volume of dumped plastics that decay very slowly or not at all. No one need look far into the future to realize we must begin to plan some alternate method of waste disposal. And with an increasing awareness of how important conservation of our resources is, some kind of a plan for reuse or recycling of our wastes seems essential. Because the starting material for most synthetic polymers is petroleum and our supplies of that important commodity are dwindling, the need to do something other than throw away our plastic wastes becomes even more apparent.

Six pounds per day for every city dweller.

Any plan for reuse or recycling of waste materials must start with some means of separating solid wastes into categories. Iron and steel objects can be caught by magnetic separators. If the remaining waste is partially ground up or shredded, the pieces of glass and aluminum can be separated in shakers that cause the more dense material to fall to the bottom. Plastics generally are much less dense than glass or aluminum. Thus plastics, along with paper, can be separated from other types of materials.

REUSE OR RECYCLING OF PLASTICS

Separated plastic waste materials can be treated in two ways. Either they can be burned to provide heat energy, or the polymers themselves can be recovered and reformed into new articles. Both of these plans for treatment are being considered, and research is being conducted to develop suitable processes.

Six pounds per person per day

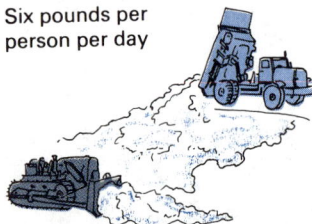

We cannot afford to continue to bury a potentially rich resource.

The treatment by burning, at first consideration, seems the simpler alternative. The major elements in plastics are carbon and hydrogen. Just like any hydrocarbon, plastics can be burned to release heat. Plastic wastes could be used as fuel to heat steam boilers. The product gases would be CO_2 and H_2O, just as they are from any efficient oil-burning power plant. But there are complications.

You have learned that the atoms of other elements often are incorporated into the monomers from which plastics are made. Chlorine is present in some. Polyvinyl chloride (PVC) contains one atom of chlorine for every two atoms of carbon and every three atoms of hydrogen (see Table 17.1). When PVC is heated, some decomposition occurs to produce gaseous hydrogen chloride, HCl. HCl dissolves in water to produce hydrochloric acid. We have already noted that the presence of acid in the moisture of the atmosphere damages the environment. Consequently, the gases coming from an incinerator or furnace where PVC is part of the burning waste may be quite injurious to living things in the vicinity.

Burning articles made from Acrilan and Orlon also presents a health hazard. Table 17.1 shows that these polymers are made from a monomer that contains the $C \equiv N$ functional group. The molecule formed by a hydrogen atom attached to this group is gaseous hydrogen cyanide, HCN. Hydrogen cyanide is a deadly poison for all animals, including humans. When the gas is breathed into the lungs, it dissolves in the blood. There it reacts quickly to change and, consequently, to block the action of many enzymes the body uses in the critical reactions of metabolism. When Acrilan or Orlon are burned, some of the polymer decomposes to release HCN gas, which pollutes the surrounding atmosphere. The incinerators being planned to burn urban waste must be fitted with devices to remove the harmful gases that may be produced. However, the warning should be clear to anyone who may have occasion to burn either of these materials in a backyard trash burner or fireplace. Don't!

The reuse of plastic materials presents another kind of problem. There is such a great variety of polymers present in waste plastics that separation into individual types is a complicated and costly problem. Quite possibly the various thermoplastic polymers could be melted out of a mixed-up pile. Perhaps some uses for the resulting heterogeneous copolymer can be found. It also may be possible to devise cheap processes by which various condensation polymers may be made to *depolymerize* Depolymerization could take place if a reaction broke an amide linkage or ester linkage by inserting water back into the molecule at that point. Such a hydrolysis reaction (breaking apart with water) would remake the monomer molecules from the polymer. These monomers could then be the raw material for new polymerization reactions.

Summary

Polymers are giant molecules made by chemical reactions that join many monomer molecules together to form long chains and networks. Monomers containing C=C bonds undergo addition polymerization. Another way to make polymers is by condensation polymerization. In this method, difunctional or trifunctional alcohols or amines react with difunctional acids to form polyesters or polyamides plus water.

The properties of plastic materials made from polymers depend on the length of the polymer chain, how they are arranged, how stiff they are, and how much cross-linking occurs. All these factors, in turn, depend on the structure of the monomer molecules.

Rubber is an important polymer made by nature. Chemists have learned how to make rubber more useful by the process of vulcanization. They also have learned how to make synthetic rubber with properties superior to those of natural rubber in some cases.

The raw material for synthetic polymers of all sorts is petroleum. This means that, as we come to realize natural resources of petroleum are limited, we must also plan steps to reuse and recycle waste plastics rather than merely discard them.

Glossary

The number in parentheses indicates the text page where you can find the term defined in context.

addition polymerization the sequence of reactions by which monomers containing carbon-carbon double bonds are joined to form polymer chains (512) *alkyd resin*   a hard polyester material cross-linked into a three-dimensional network (530)

cis-trans isomer   molecules that have the same molecular formula but differ in the arrangement in space of groups attached to two carbon atoms joined by a double bond (523)

condensation polymerization   the formation of a polymer by reactions that join the monomers and produce water as a second product (526)

copolymer   a polymer composed of more than one kind of monomer (519)

crystallite (of a polymer)   a small region of orderly arrangement in an otherwise randomly oriented collection of polymer chains (510)

depolymerize   the use of chemical reactions to break down a polymer into monomers (535)

free radical   fragment of a molecule, that is very reactive because it has an unpaired electron (512)

monomer   a small molecule that is the fundamental building block of a polymer (508)

plastic   a synthetic organic material made up of very large molecules, capable of being formed into desirable shapes (507)

polyamide   a condensation polymer whose monomers are joined by amide linkages (527)

polymer   a large molecule made up of many small repeated units of one or several kinds (508)

rubber   a substance produced from naturally occurring latex from certain plants, a polymer of isoprene (519)

thermoplastic   a polymeric material that becomes soft enough upon heating to be shaped, and hardens as it cools (517)

thermosetting plastic   a polymer created by heat-setting of monomers in a mold, which cannot be further molded by heating (531)

vinyl group   a group of atoms derived from substituted ethene by the removal of one H atom (514)

vulcanization   the process of heating rubber with sulfur to induce sulfur cross-linking, which gives greater strength to the rubber (520)

Exercises

17.1   Why might some people feel that it is more appropriate to call modern times the Age of Impending Petroleum Shortages than the Age of Plastics?

17.2 "Big" and "giant" do not accurately describe the size of polymer molecules. How can you more accurately describe the size of a polymer molecule?

17.3 When polymer chains form into many crystallites, the resulting solid material has a greater density than other plastics. Explain why this happens.

17.4 Plastics are used to make both shower curtains and lightweight furniture. Explain why one is flexible and the other is not in terms of the way the polymer chains are linked together.

17.5 Safety glass is made like a sandwich. A thin sheet of transparent, tough plastic is tightly held by sheets of glass on either side of it. If a piece of safety glass is struck a hard blow, the glass cracks, but does not shatter into sharp pieces. The broken glass is held by the tough, unbreakable plastic. One plastic used for this purpose is made from the monomer iso-butene. Draw the structure of the polymer in the manner used for other examples of vinyl polymers in Table 17.1.

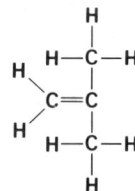

Isobutene

17.6 Each of the following represents the molecular structure of an addition polymer. Use vertical dotted lines to indicate what the repeating units are and name the monomers from which the polymer was made.

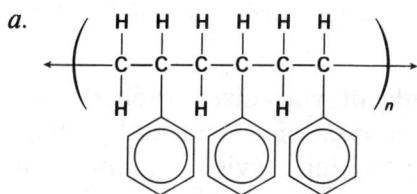

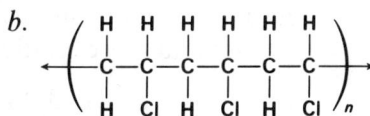

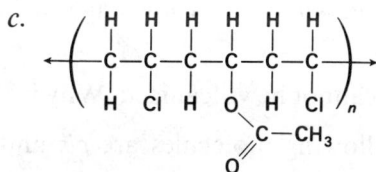

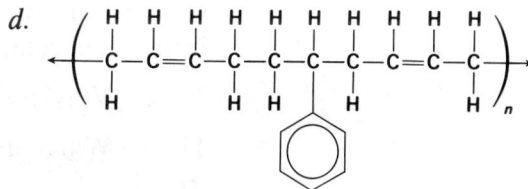

a.

b.

c.

d.

17.7 Football helmets are made from the copolymer called ABS. The structures of acrylonitrile (A) and styrene (S) are given in Table 17.1. The structure of butadiene (B) is given on page 525. Draw a structure showing how the copolymer ABS can be formed. Use just one unit of each monomer.

17.8 ABS is strong and tough because it contains many cross-linkages in the polymer. How does the structure of butadiene make this possible?

17.9   The following outline formula shows the carbon-atom backbone structure of the molecule β-carotene. This molecule and ones with similar structure are found widely distributed in nature in fruits and vegetables (β-carotene gives carrots their characteristic color). The human body makes the very important vitamin A from β-carotene by splitting it in the middle and building an alcohol group on each of the fragments.

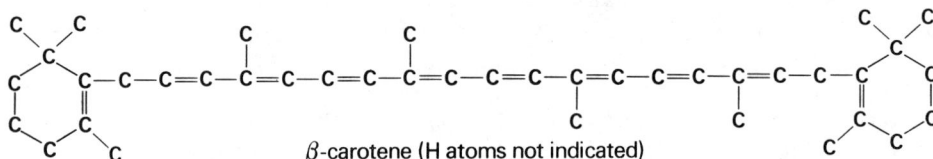

β-carotene (H atoms not indicated)

a.   Identify the repeating structure in the chain by dividing it with vertical dotted lines.

b.   What monomer did nature use to construct this moderate-sized polymer?

c.   How does your answer to *b* compare with the monomer nature uses to make rubber?

d.   How is this information additional evidence to support the statement that there is continuity and evolution in all forms of living things?

17.10   Explain why the process of vulcanization makes rubber stronger and more elastic.

17.11   If discarded tires (made of vulcanized rubber) were burned as fuel for a power plant, a severe environmental problem would result. What would this be? (Hint: Review the discussion in Chapter 8 of the environmental hazard from burning soft coal. Combine this information with your knowledge of how vulcanization is accomplished.)

17.12   Vinyl polymers cannot be vulcanized. Why is this?

17.13   Which of the following molecules are *cis* and which are *trans*?

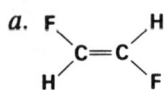

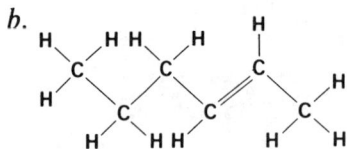

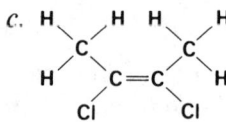

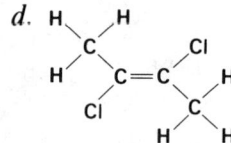

17.14   What justification can be claimed for the statement that Allied victory in World War II was partly won in chemists' test tubes?

17.15   The examples of condensation polymers discussed in this chapter are all made from two kinds of monomers. Describe, in general terms, the structure of a single monomer that could be used to make a condensation polymer.

17.16   If ethyl alcohol, C_2H_5OH, were reacted with a difunctional acid, would a polymer be formed? Justify your answer.

17.17   *a.*   What do polyamide polymers and polyester polymers have in common?

   b.   What are the differences between these two kinds of polymers?

17.18   What feature of molecular structure is responsible for the quick-drying quality of fabrics made from polyesters or polyamides?

17.19   Why do polyester polymers made from glycerol, such as alkyd resins, form three-dimensional network solids?

17.20   What is the difference between a thermosetting polymer and a thermoplastic one?

17.21   The plastic casing of a timer left near a stove pilot light deforms and sags, whereas the plastic handles of pots and pans do not. Explain the difference between the two polymers used to make these articles.

17.22   One of the goals of chemists doing research on polymers is to produce a synthetic plastic that is biodegradable. Explain why this is a worthy goal.

17.23   *a.*   A glass bottle is breakable and heavy, but broken glass can be melted and remade into another bottle. Contrast the comparable properties of a plastic bottle.

   b.   What competing factors do you think a company's officials consider when they decide to use glass or plastic as a packaging material?

17.24   If archeologists happen to excavate the site of a 1970's sanitary landfill 2000 years from now, they may be able to put together a picture of the way we live from what they find. What do you think they will find and what may their conclusions be?

17.25   How does the information in this chapter add urgency to the idea that we need to develop ways of using alternate sources of energy instead of burning up our reserves of oil and coal?

18 Giant Molecules in Living Systems

- ☐ What few elements are the chemical basis of living systems?
- ☐ What is the chemical evidence for the evolution of life?
- ☐ Why does the burning of fat yield more energy than the burning of carbohydrates?
- ☐ How might the chemistry of our amino acids isolate us from any still unknown life elsewhere in the universe?
- ☐ Why do we require a variety of amino acids in our diets?
- ☐ How are proteins put together?
- ☐ How does the molecular structure of the DNA polymer carry the code of heredity between generations?

A living system is a remarkable combination of structures performing the functions necessary to life. The word system implies that the various structures interact with one another in a regular, organized (systematic) way. Some living systems are small and simple—for example, the single-cell plants and animals that can only be seen with the help of a microscope. Other living systems are huge, such as giant redwoods and whales. Living systems also can be complex, as human beings are. In the large and complex living systems, the organization of the system depends on many cells of many varieties, each performing its own particular function. But even a

---

Even a single cell is a very well-organized chemical factory.

---

single cell is a very well-organized chemical factory. A cell can synthesize a bewildering variety of compounds precisely and efficiently. Just as a factory uses an assembly line to build an automobile or truck, the metabolic processes of a living cell bring the right molecules together at the right time so that the cell can make the molecules it needs to perform its function in the total system.

People have always been intensely curious about the mysteries of the life process. Gradually, knowledge has accumulated to help explain some of what goes on in a living system. Some of life's processes are less of a mystery than in former years. For example, we now know disease can be held in check by sterilization, proper diet, or drug therapy. The frontier of understanding of many life processes has been pushed to the level of molecules and their reactions. In this chapter, you will get a glimpse of a small part of that frontier. We will discuss the structures of three kinds of giant molecules, the polymers that living systems constantly make, break down, and remake.

Reading one brief chapter on the chemistry of life cannot possibly satisfy your curiosity about the multitude of different molecules and chemical reactions of life. But this chapter can lay the groundwork. We hope what you learn may make you even more curious and put you on the lookout for newspaper and magazine articles describing new ideas growing out of biochemical research. The information in this chapter should help to make such articles mean more to you. You also may become better able to appreciate and understand why good health depends on good diet. The discussions of the general principles by which molecules react to carry on the mechanism of heredity also should be useful. For example, the idea of genetic engineering appears in reports in the popular press. This idea poses the question "What are the potential benefits or hazards of purposely altering the structure of the molecules making up genes and chromosomes?" Knowing something about these molecules may help you toward an answer to this and other complicated questions about the life process.

Two of the polymers whose structure we will describe are carbohydrates and proteins. You already know that these are essential foods. In Chapter 3, we introduced the idea that your body uses the *carbohydrates*, such as sugars and starches, as fuel, the source of energy for living. *Proteins* are the nitrogen-containing polymers your body uses for its structure. Muscles, tendons, hair, skin, and fingernails all are made of protein. So are the approximately 1500 enzymes, the special molecules that serve as catalysts for the living body's chemical reactions. Proteins are polyamides, formed by a condensation reaction similar to that used to make Nylon, as described in the preceding chapter. You

Your body takes protein polymers apart and then builds the monomers back into the kinds of protein you need.

eat proteins in foods; your body takes these protein polymers apart and then builds the monomers back into the kinds of protein you need.

The other molecules we will discuss are called *nucleic acids*. These are a kind of polyester polymer that makes up genes and chromosomes. RNA (*ribo*nucleic *a*cid) and DNA (*deoxyribo*nucleic *a*cid) are the molecules cells use to store and transmit genetic information. These polymer molecules are the blueprints each new cell uses as a guide to grow and function like the cells from which it came. DNA and RNA are copolymers composed of only four kinds of monomers. The order in which these different monomers are built into the polymer chain is the key to the mystery of heredity. This order is called the *genetic code*. Offspring resemble their parents because the genetic code of the messages written on their and their parents' nucleic acids are similar.

FOOD—THE RAW MATERIALS FOR LIVING SYSTEMS

The most vital accomplishment of any living thing is its ability to take in food, to use this chemical raw material to build its own structure, and in the process, to transform and convert energy to its own needs. The chemical reactions involved in the life process are surprisingly similar in everything that lives, from microbes to humans. Even plants and animals are alike in many of the reactions that maintain life.

Furthermore, the same chemical elements are found in all living creatures, although in varying proportions. The portion of the periodic table reproduced in Figure 18.1 indicates what these elements are. Carbon, hydrogen, oxygen, and nitrogen account for approximately 99% of the atoms in living things. These four are the chief elements of which fats, carbohydrates, and proteins are composed.

The building blocks of the giant molecules are the same in all living systems.

The next seven common elements include sulfur, S, also found in proteins, and phosphorus, P, a key element in the molecules of heredity, the nucleic acids. Sulfur and phosphorus tend to form covalent bonds in molecules. Also included in this seven are calcium, chlorine, sodium, potassium, and magnesium (in order of abundance). Their positions in the periodic table suggest that these elements tend to form ions rather than covalent bonds. The body uses these elements as ions in salts. Some ions are tied into insoluble compounds, such as calcium carbonate and calcium

Figure 18.1 Elements essential to living systems.

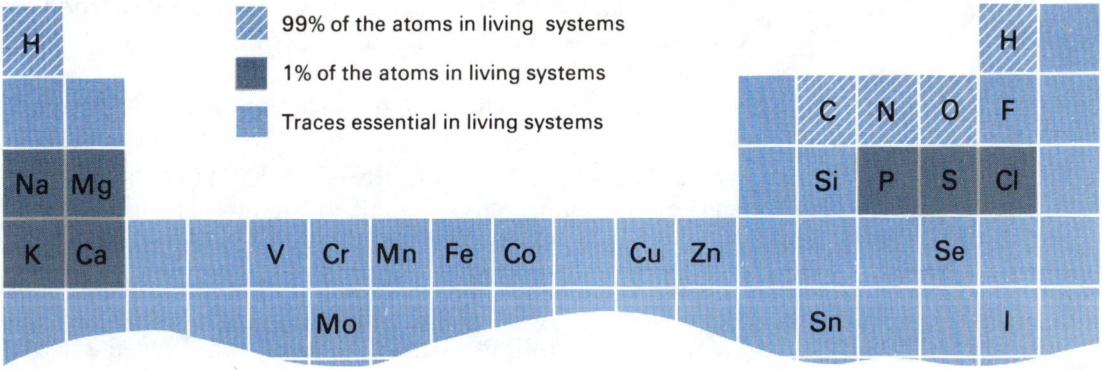
99% of the atoms in living systems
1% of the atoms in living systems
Traces essential in living systems
H
Na
Mg
K
Ca
V
Cr
Mn
Fe
Co
Cu
Zn
Mo
C
N
O
F
H
Si
P
S
Cl
Se
Sn
I

phosphate in bones and teeth. Others, such as sodium, potassium, and chlorine, are in the form of water-soluble ionic compounds in the fluids of the body. These typically inorganic elements are provided by the mineral substances in foods such as table salt.

The remaining 13 elements indicated in Figure 18.1 are referred to as *trace elements*. All together, they amount to less than 0.01% of the atoms in living things. But their presence is essential to life. For example, four iron atoms are the keystones in the hemoglobin molecule. Iodine serves a special purpose in the structure of the hormone produced by the thyroid gland. Fluorine and tin help to strengthen teeth. Many of these trace elements are incorporated into enzymes, hormones, and other molecules the body uses for specific purposes in the complex system of metabolism. A proper diet should include some of these trace elements already incorporated into complex molecules, such as the cobalt in vitamin B_{12}. Others should be included as traces in the minerals of the diet.

There are important differences among the kinds of substances that can serve as food for various forms of life. Generally speaking, the simpler forms, such as plants and bacteria, can use simpler molecules as chemical raw materials. For example, plants can start with carbon dioxide, CO_2, water, H_2O, and energy, in the form of sunlight, to manufacture carbohydrates. This process is photosynthesis, which we have repeatedly mentioned. No animals are able to accomplish photosynthesis. Similarly, some kinds of bacteria can use molecules of elemental nitrogen from the air as a starting material for the manufacture of proteins. No other living things can do this, as far as we now know.

In contrast, animals higher in the chain of biological evolution cannot start with simple molecules for food. Their food must contain the essential elements—carbon, hydrogen, oxygen, and nitrogen—prepackaged into more complex molecules, such as carbohydrates and proteins. Animals must eat carbohydrates and proteins to survive. Therefore, animals depend on lower forms of life to serve as food.

These differences can be interpreted as chemical evidence supporting a theory of the evolution of life. When life began on earth, the lower forms of life, such as plants and anaerobic bacteria, developed first. Recall in our discussion in Chapter 3 that anaerobic bacteria are those microorganisms that can take oxygen from oxygen-containing compounds; they do not require molecules of elemental oxygen gas, as higher forms of life do. Later, about a billion years ago, after the oxygen released by photosynthesis had reached a sufficiently high level in the atmosphere, *aerobic bacteria* evolved. Their use of oxygen in the life process set the stage for the evolution of higher forms of life that must live by breathing. Fish, birds, and animals developed. Each stage required the more complex molecules for food that they found already assembled by the lower forms of living creatures.

VITAMINS—ESSENTIAL IN SMALL AMOUNTS

Higher animals, such as humans, also need to include vitamins in their food. *Vitamins* are specific chemical compounds, atoms assembled into special molecules the body needs to function properly. Most vitamins are compounds with complex molecular structures. The complexity of their structure suggests how specific the function of each vitamin is for metabolism in higher forms of life. The term *vitamin* was coined many years ago, when it was thought that these essential nutrients were *vital amines*. We now know that only a few vitamins contain the amine group, but the name remains in use.

Vitamin molecules play crucial roles in the chemical reaction schemes by which the body uses food. Although large amounts of vitamins are not needed, it is essential that the diet include certain minimum amounts. The labels on packaged foods, such as cereals, bread, and milk, often list the vitamin content of a food in terms of RDA units. The abbreviation RDA stands for *recommended daily allowances*, which are established by the Federal Food and Drug Administration. The numbers of RDA units, relative to the amounts of the particular food involved, indicate the amounts a healthy person needs to maintain good health. The absence of a vitamin from the diet can result in serious health disorders.

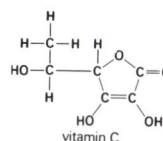

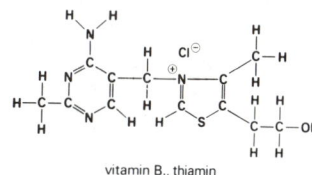

vitamin C

vitamin B₁, thiamin

8 Ounces Contains:
(Percentage U.S. RDA)

Vitamin A	4	Vitamin B₆	4
Vitamin C	4	Vitamin B₁₂	15
Thiamin (B₁)	6	Calcium	30
Riboflavin	25	Phosphorus	20
Vitamin D	25	Magnesium	8

One example of a health disorder associated with a vitamin deficiency has been known for centuries. During the early years of world exploration by sailing vessels, members of the crews often suffered from scurvy on long voyages. The disease always cleared up after the ships reached land, where the crews could again eat fresh fruit and vegetables. In later years, sailors of the British Navy were called "limeys" because the ships carried limes to supplement the sailors' diet. The particular molecule that cures scurvy has now been identified as ascorbic acid, vitamin C.

FATS—ENERGY-STORAGE MOLECULES

We have already discussed the chemical structure of fats, one important type of food. Fats are esters of glycerol, $C_3H_5(OH)_3$, with various long-chain carbon-containing acids, appropriately called fatty acids. Living organisms use fats to store potential energy. Fats are not soluble in the watery fluids of the body, so the molecules remain in the tissues and do not dissolve in watery waste solutions. Another advantage is that the fat molecule contains relatively little oxygen. The formula for a typical molecule of fat is $C_{54}H_{110}O_6$. The formula for a comparable carbohydrate molecule is $C_{54}H_{92}O_{46}$. If we write a chemical equation to represent the burning of each of these molecules, we can see that more oxygen must be used to combine with the fat than with the carbohydrate.

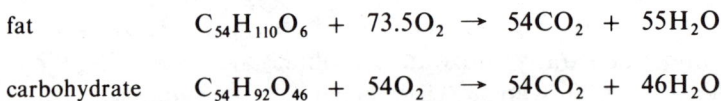

fat $\qquad C_{54}H_{110}O_6 + 73.5O_2 \rightarrow 54CO_2 + 55H_2O$

carbohydrate $\qquad C_{54}H_{92}O_{46} + 54O_2 \rightarrow 54CO_2 + 46H_2O$

The greater amount of oxygen required to burn the fat means that a more extensive reaction occurs, and consequently, more energy is given off. If 1 gram of fat is burned, 9000 calories of heat energy are released. By comparison, burning 1 gram of a carbohydrate releases only about 4000 calories of heat energy. These figures suggest why a person who is concerned about dieting to gain or lose weight rates foodstuffs in terms of calories. The conventional *dietary calorie* is a kilocalorie, 1000 calories. So 1 gram of fat yields 9 dietary calories; 1 gram of carbohydrate yields 4 dietary calories. (Recall that a calorie is the heat energy that raises the temperature of 1 gram of water 1° C.)

The metabolism of the animal body is remarkably efficient: Excess carbohydrate foods not needed for energy are converted into fats and stored in the tissues. Therefore, persons who eat more carbohydrate foods (such as starch or sugar) than are needed, put on fat. Not only are fat molecules insoluble in the watery body fluids, but they store away more than twice the potential energy that an equal weight of carbohydrate represents. Consequently, fats can be considered the reservoirs of energy in the body. When a person goes on a reducing diet, the intake of food is not enough to supply the body's energy needs. Then the reservoir of fat is used as fuel to keep the body going. You can also recognize why increased exercise helps a person reduce. Exercise that expends 9 kilocalories of energy effectively burns up 1 gram of fat.

The body does not "burn" fats directly. Rather, the fats, either those taken in as foods or those removed from storage in body tissues, are first converted into relatively small carbohydrate molecules. These molecules are soluble in the watery body fluids. When these carbohydrate molecules dissolve in the blood, they are transported to the site in the body where energy needs to be released. There the combination with oxygen is completed.

CARBOHYDRATES

The class of compounds called carbohydrates is widely distributed among the plants and animals of the earth. Probably the first carbohydrate we encountered soon after we were born is lactose. Lactose is a sugar found in the milk of humans and other mammals. Its formula is $C_{12}H_{22}O_{11}$. *Sucrose,* table sugar, the first carbohydrate we learned to ask for as sweets-loving children, also has the formula $C_{12}H_{22}O_{11}$. But sucrose has a slightly different molecular structure. A child's sweet tooth may also be appeased by eating *fructose,* $C_6H_{12}O_6$, a sugar found in fruits and honey, or by sucking on a lollipop that contains *glucose,* which also has the formula $C_6H_{12}O_6$. (By this point, you may recognize that the chemical names for various sugars end in *ose.*)

Soon after we were born, we encountered another kind of carbohydrate, the cellulose in cotton and disposable paper diapers. Cellulose is a polymer whose structure can be described by the formula $(C_6H_{10}O_5)_n$ (n may be as high as 100,000). Early in childhood, we learned about sticky paste and fingerpaint made from starch. Starch is another carbohydrate polymer with the same formula as cellulose. However, it obviously must have a different molecular structure, because its properties are so different.

The generalized formula $C_n(H_2O)_m$ fits the name *carbo* ("carbon") *hydrate* ("water"). This formula correctly gives the relative numbers of carbon, hydrogen, and oxygen atoms. (For sucrose,

$C_{12}H_{22}O_{11}$, the value of n is 12; m is 11.) But it tells nothing about how the atoms are linked together. We need some indication of the structure of the molecules to understand how they react, what their properties are, and how the various carbohydrates are related to one another.

The smaller sugars, glucose and fructose, are molecules with a ring structure. Here is a more detailed picture:

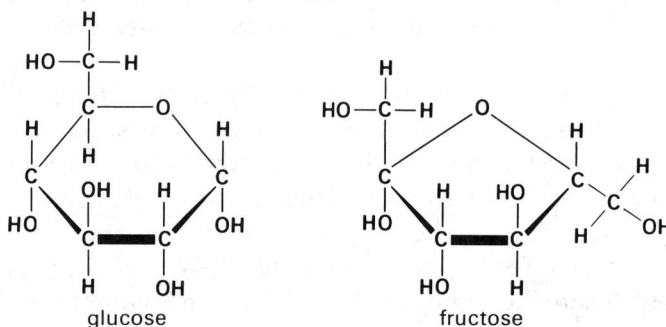

glucose   fructose

The details of these structures, such as the directions in space of the bonds holding the H atom and the OH group onto the carbon atoms, are important to a complete chemical interpretation of the properties of these molecules. However, we need not be concerned with these details. But we will be concerned with some of the general structural features that help us understand the properties of these molecules.

First, we can see how some plants, such as sugar cane and sugar beets, put these two 6-carbon sugar molecules together to make a molecule of sucrose, a 12-carbon sugar.

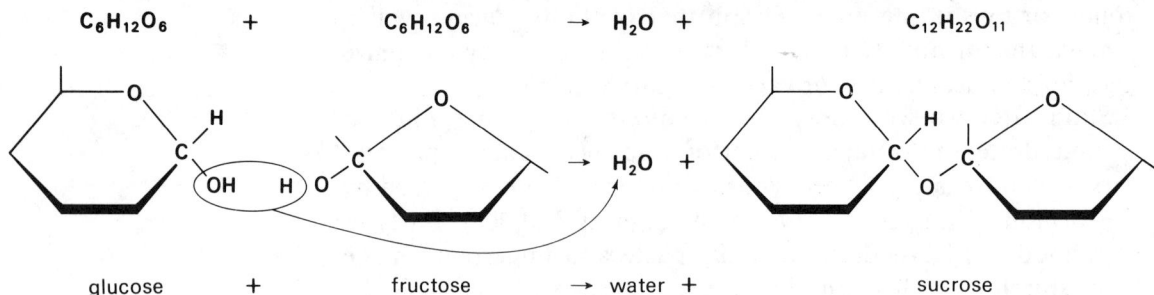

$C_6H_{12}O_6$ + $C_6H_{12}O_6$ → H_2O + $C_{12}H_{22}O_{11}$

→ H_2O +

glucose   +   fructose   → water +   sucrose

The ring structures have been simplified to indicate only those atoms in the molecules at the points where reaction takes place. Note that a type of condensation reaction occurs. Water splits from two OH groups, one on each ring. The rings are thus joined together through the bonds of an oxygen atom.

lactose

maltose

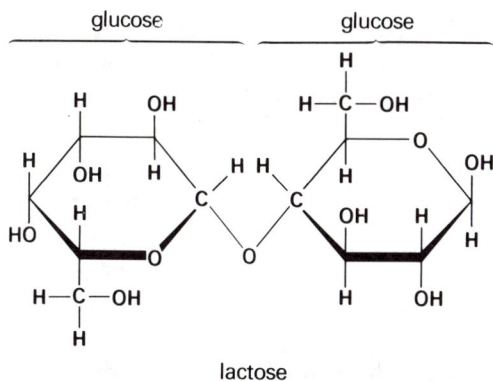

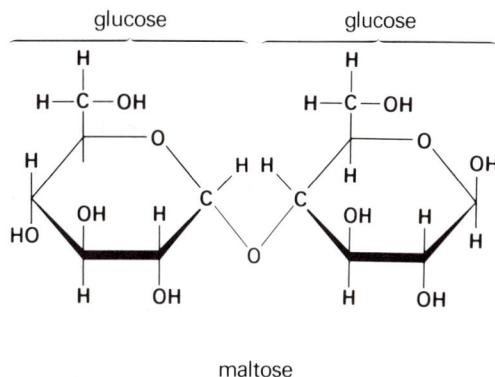

Two other sugars, lactose and maltose, also have the formula $C_{12}H_{22}O_{11}$. Lactose is milk sugar; maltose is found in grain seeds when they begin to sprout. Both of these sugars are formed by the combination of two glucose units and differ only slightly in structure.

STARCH AND CELLULOSE— CARBOHYDRATE POLYMERS

Starch is formed by a similar condensation reaction in the edible parts of such plants as corn, wheat, potatoes, and beans. There are important differences, however. First of all, starch is a polymer. Glucose is the monomer repeatedly hooked together to make the giant molecules of starch. The polymerization reaction can be represented as follows:

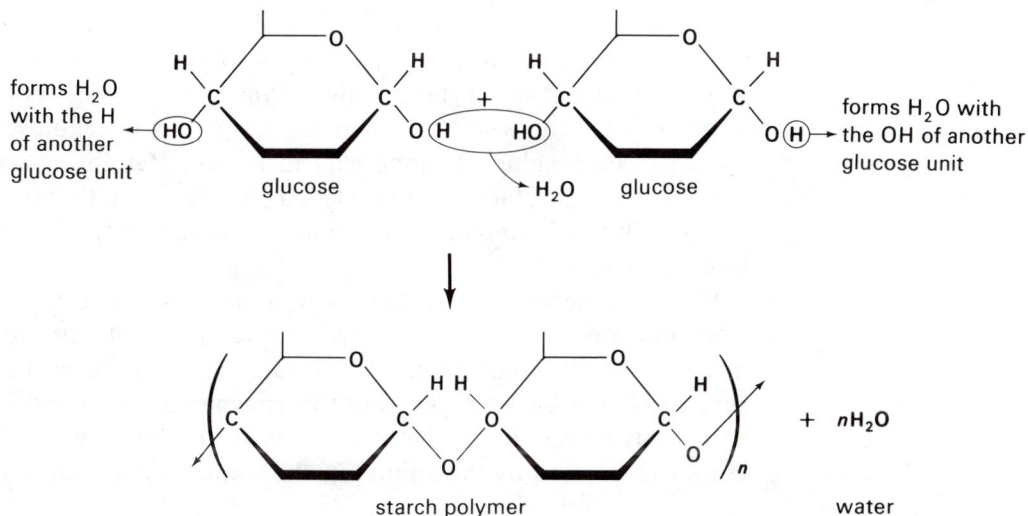

forms H_2O with the H of another glucose unit

glucose

H_2O

glucose

forms H_2O with the OH of another glucose unit

starch polymer

$+ nH_2O$

water

Here again, we have simplified the ring structures to show only the atoms involved in the reaction by which the monomers are linked together. As before, we use the symbol n to stand for a very large number.

When you eat foods containing starch, the first chemical reaction you perform to begin digesting the starch is the depolymerization of the starch. Enzymes in your saliva act as catalysts to add water molecules back into each linkage between the glucose monomers. The reaction, a hydrolysis reaction, is the reverse of the one shown above.

The glucose produced by the hydrolysis of starch or sugars dissolves into the body fluids in the stomach. Some of it goes to the liver, where it is made into a moderate-sized polymer called amylose, or animal starch, to be stored until the body needs it. The rest of the glucose is transported by the blood to locations in the body where it is burned to release energy.

Plants also hook glucose monomers together, but in a slightly different way, to form cellulose. Cellulose is certainly the most abundant carbohydrate in nature. We can suggest the difference in the structure of cellulose by an outline structure that contrasts with the one for starch shown above.

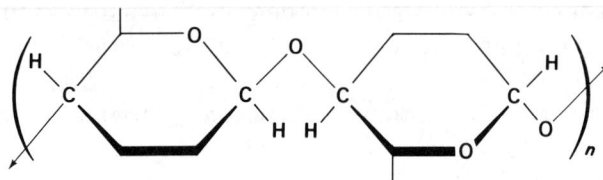

cellulose polymer

This difference in structure looks very slight on paper. But to a starving person lost in the woods without food, it means a great deal. Cellulose is the carbohydrate polymer plants and trees use to make fibers and wood. The glucose monomers in cellulose are the same as the glucose monomers in starch. But the enzymes animals can manufacture are useless as catalysts for the hydrolysis of cellulose. Consequently, animals cannot directly use cellulose for food.

However, herbivorous (plant-eating) animals can gain nourishment from cellulose by an indirect route. Such animals are the hosts to certain kinds of bacteria located in their stomachs, or rumens. These bacteria generate the enzymes that catalyze the hydrolysis of cellulose. The glucose, thus liberated by the bacteria, can be used by the animal in the same way a human uses the glucose liberated from starch by human enzymes. However, humans do not play host to the bacteria whose enzymes can break down the cellulose in hay and straw. Consequently, humans cannot use hay and straw, even indirectly, as food. Some wood-eating insects, such as termites, also play host to bacteria that are able to manufacture enzymes that act as catalysts for the hydrolysis of cellulose.

HYDROGEN BONDING IN CARBOHYDRATES

Let us examine a more complete structural formula for the glucose unit polymerized into either starch or cellulose.

Every glucose unit of the starch or cellulose polymer has 3 OH groups sticking out from the chain. These OH groups form hydrogen bonds with H_2O molecules or other polymer chains.

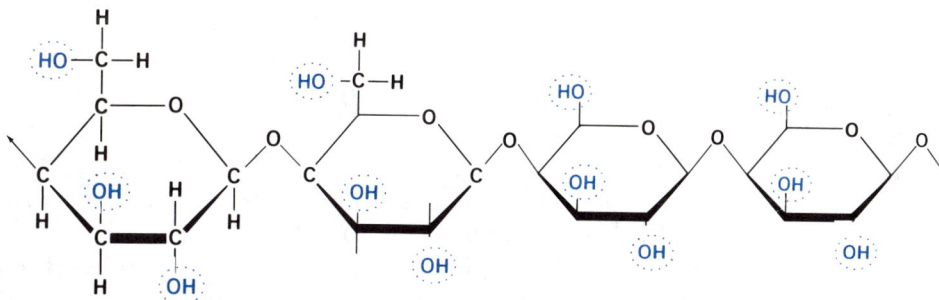

The important structural feature to notice is that every ringlike monomer in the polymer chain has three OH groups sticking out from the carbon–oxygen skeleton. The starch or cellulose molecules are like a necklace made from fuzzy beads. If such a necklace comes in contact with a powdery substance, some powder sticks to every bead. Or when one part of the necklace comes in contact with another, the fuzziness of the beads may cause clumps of the necklace to stick together. The OH groups on the glucose monomer are like the fuzz on the beads, because they are capable of forming hydrogen bonds. The OH groups on the polymer chain can form hydrogen bonds with water. Or they can form hydrogen bonds with OH groups on other parts of the polymer chain.

You now can see why the structure of cotton fibers made of cellulose is responsible for the fact that a wet cotton fabric holds onto water molecules. (Recall the discussion in the previous chapter of the contrasting structure of the molecules in synthetic fibers such as Nylon and Orlon. Fabrics made from these polymers dry quickly because the chains have no OH groups sticking out to form hydrogen bonds with water.)

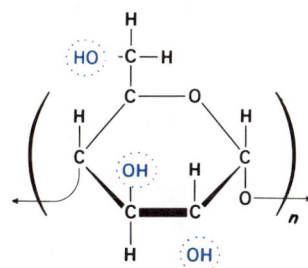

Note 3 OH groups located on every glucose unit of the polymer.

Cotton fibers have intertwining strands of cellulose, between which water molecules can fit. The hydrogen atoms of the water molecules are attracted to the OH groups on the cellulose chains. A dry piece of cotton or linen may be moderately stiff, because the cellulose molecules form hydrogen bonds from one chain to another. But when wet, the cloth becomes limp, because the cellulose chains form hydrogen bonds with invading water molecules rather than with one another. Using starch to make a garment stiff works in the same way. After a starched cotton garment is ironed dry, the extra hydrogen bonds between the starch molecules and the cellulose of the cotton fibers make the whole structure more rigid.

Paper and cardboard are chiefly cellulose. Wood chips are treated to remove the compounds other than cellulose. The resulting pulp, consisting of cellulose fibers, is pressed into sheets. A simple paste made from starch can be used to hold pieces of paper together or to hold wallpaper onto wood or plaster. When the paste dries, the molecules of cellulose and the molecules of the starch form many interlocking hydrogen bonds.

The limited flexibility or, in some cases, the extreme rigidity of wood also is caused by hydrogen bonding. A thin strip of moist wood can be bent, just as a piece of wet cotton cloth is flexible. Dry wood becomes stiff as the cellulose chains form hydrogen bonds with one another. Wood also contains other molecules besides cellulose that bond the fibers together, like dried starch bonds cotton fibers.

AMINO ACIDS—THE MONOMERS IN PROTEIN POLYMERS

Proteins, the class of nitrogen-containing compounds essential to all forms of life, are polymers. *Amino acids* are the building blocks, the monomers, from which proteins are made. So before we discuss the structure and function of proteins, it is appropriate for us to examine the makeup of the amino acids.

The general formula of the amino acids, from which living things build protein polymers, can be represented as follows:

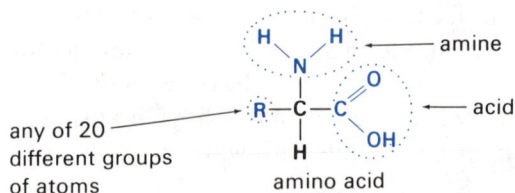

Note the presence of an amine functional group and an acid functional group in the one molecule. (This is why such molecules are called amino acids.) We will describe some of the R groups later in this section. Protein polymers are made like the polyamide polymers described in Chapter 17. The amine group of one amino acid monomer reacts with the acid group of another.

The structure of an amino acid drawn this way on a flat paper surface does not make clear one very special feature of the way the parts of the molecule are arranged in space. Remember that, if a carbon atom is located at the center of a regular tetrahedron, its bonds point toward the corners of the tetrahedron. When each of the bonds attaches a different atom or group of atoms to the carbon, the unique circumstance represented in Figure 18.2 exists. There are *two different* ways the four groups can be arranged. (Notice that an amino acid has four different units attached to the central carbon atom.)

The two forms of such a molecule are called *mirror images* of each other. You can place a small mirror perpendicular to the paper on a line between the two structures so that the image of one structure is reflected. The reflected image that you see in the mirror looks the same as the other structure. What you have in these two structures is like what you see in a pair of gloves or mittens. If you put one glove on top of the other, so that the thumbs are together, the front of one glove is on top and the back of the other glove is on top. Or if you put both gloves with the front sides on top, the thumbs do not line up together. The right-handed and left-handed gloves are mirror images of each other. They cannot be exactly superimposed on each other. The two forms of amino acids act the same way. One form is the mirror

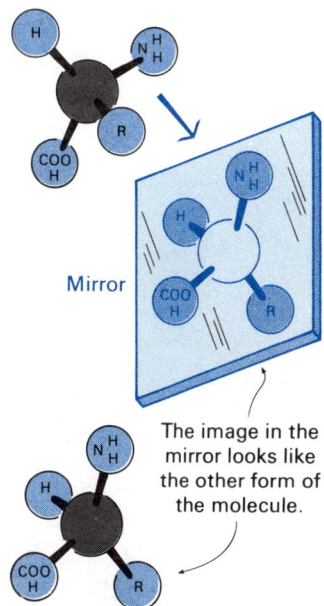

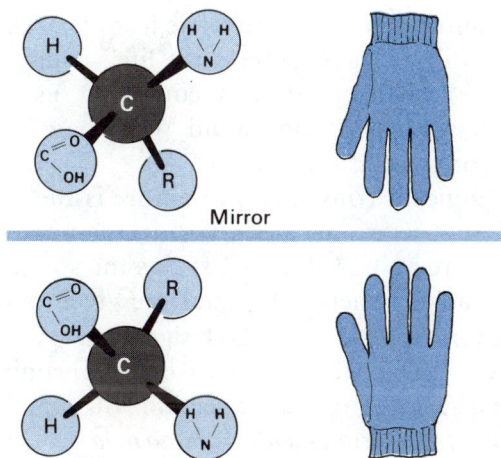

The image in the mirror looks like the other form of the molecule.

Figure 18.2 Two mirror-image forms for amino acid molecules and gloves.

image of the other. The two forms cannot be completely super-imposed on each other in three-dimensional space any more than the representations drawn on a two-dimensional paper surface can be.

Chemists can synthesize amino acid molecules in a laboratory. This process amounts to replacing three of the H atoms of methane, CH_4, with an amine group, NH_2, an acid group, COOH, and one other, an R group. The resulting product is a half-and-half mixture of the two possible mirror-image molecules. This mixture is what you would expect from the laws of chance, like repeatedly flipping a coin for heads or tails. However, nature does not work this way. Nature builds amino acids for the proteins in living things with only *one* of the two possible forms. Mother Nature knits only left-handed gloves! Apparently, when life began on the earth, one form of amino acid and not the other began to grow into protein molecules. Ever since, all proteins on the earth have been built the same way. This choice of one form of amino acids and not the other is one of the remarkable consistencies in nature that makes life a continuous stream.

One aspect of this idea that life is a continuous stream is the way the proteins from dead plants and animals can be used as food by living animals. In the next section, we will discuss how your body takes apart the proteins of food, such as the proteins in a pork chop, a lobster, or a soybean. The amino acids thus set free are then recombined into the proteins you need to live. Your metabolism could not use amino acids built like right-handed gloves. One of the mysteries about life is how the original choice was made. We may never solve this mystery.

One of the interesting aspects of speculating about life elsewhere in the universe involves this choice of only one form of amino acids for life on earth. Perhaps, if "elsewhere life" has proteins, the amino acids they are made from may be of the other kind. If so, earthly life and "elsewhere life" could interact very little. Certainly, earthly creatures could not use "elsewhere" plants or animals for food, nor could "elsewhere" creatures use earthly plants or animals for food.

Another significant consistency in nature is the fact that only about 20 different R groups are found in the amino acids from which proteins are built. Table 18.1 shows the structures of some typical amino acids. Chemical names are given, along with the three-letter abbreviations used to designate a particular amino acid. These examples are classed as essential amino acids for humans. The discussion in a later section will explain more fully what is meant by the term *essential amino acid*.

Table 18.1   Structures of the essential amino acids

Name	Structure	Comments
arginine Arg		animals and humans can synthesize some but not enough
histidine His		
isoleucine Iso		missing in wheat
leucine Leu		low in wheat
lysine Lys		missing in corn, wheat, and rice
methionine Met		one of the few missing in soybeans; low in gelatin
phenylalanine Phe		present in most grains
threonine Thr		missing in rice; low in gelatin
tryptophan Try		missing in corn and gelatin
valine Val		missing in wheat

PROTEINS—POLYMERS OF AMINO ACIDS

Early in the nineteenth century, proteins were recognized as an essential type of food, different from carbohydrates and fats. The name *protein* was coined in 1838 from the Greek word that means "first rank." This name emphasizes how important proteins are for all living things. The relationship between the amino acids (monomers) and proteins (polymers) can be represented as follows (once again, the symbol n stands for a very large number of monomer units fastened together into a polymer):

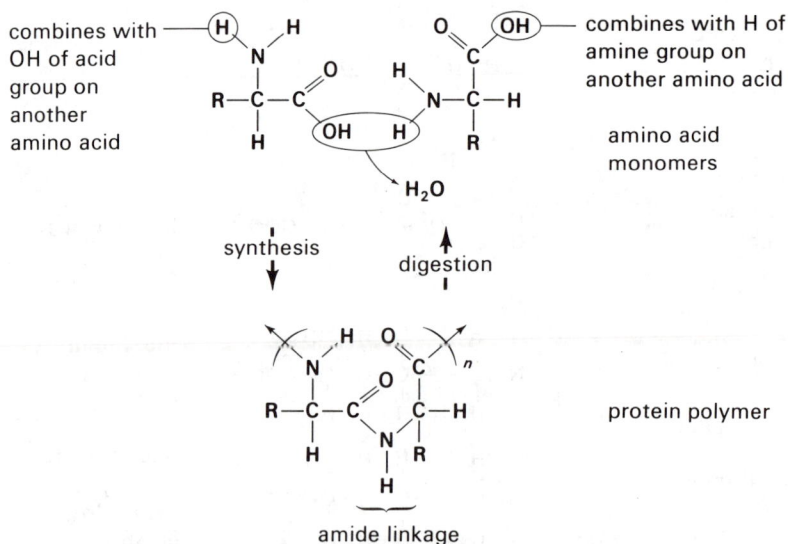

combines with OH of acid group on another amino acid

combines with H of amine group on another amino acid

amino acid monomers

H_2O

synthesis

digestion

protein polymer

amide linkage

Here is another example of forming a polyamide by a condensation polymerization reaction. Recall that this is the way Nylon polymers are formed. Here, however, one kind of monomer, the amino acid, has both the necessary functional groups in it. You also can recognize how the polymerization reaction may go in the opposite direction to be a depolymerization. The

The first step in your digestion of a protein is to put water back into the molecule at each amide linkage.

first step in your digestion of a protein is to put water back into the molecule at each amide linkage. (H goes back to the amine, and OH, to the acid.) Enzymes are catalysts for this hydrolysis reaction that accomplishes the depolymerization. Then other enzymes help your body to repolymerize amino acids into the proteins you need.

Proteins are large polymers (some have molecular masses as large as a million) and are made from the 20 different kinds of amino acids. Therefore, there is a tremendous variety of possible combinations. If we put just two of 20 different molecules together, we can make 400 different two-unit products. The number of variations possible when a chain 100 units long is made from 20 different units is hard to imagine: 20^{100}. (This number is 20 × 20, one hundred times over.) With such a variety of possible structures, we can recognize why living systems use proteins for so many different functions. Flesh, skin, hair, and fingernails all are composed chiefly of protein. And because proteins are the material such body structures are made from, we can understand why no two living individuals are ever exactly alike. Yet the enzymes manufactured by living bodies are proteins that are identical molecules in every individual of each species. In these molecules, the same amino acids are linked together in the same sequence.

THE ESSENTIAL AMINO ACIDS—QUALITY IN DIET

In Table 18.1, we have chosen the ten essential amino acids to illustrate some of the variety of R groups in the amino acids living systems use to build proteins. These ten are called *essential amino acids* because the animal body is not able to synthesize them from other kinds of molecules. The first two, arginine and histidine, can be manufactured by humans, but not in sufficient amounts. This fact is very important in deciding what foods should be included in a diet adequate for good health. Most proteins in foods of animal origin—such as lean meat, fish, eggs, milk, and cheese—are "complete" proteins. They contain some of all the essential amino acids. Gelatin, made chiefly from the hooves of animals, is the only animal protein notably lacking in some essential amino acids. However, the proteins in foods derived exclusively from plants usually lack one or more of the essential amino acids. We have noted some of these shortages in the comments included in Table 18.1.

This information about the essential amino acids explains why it is important to include a variety of protein-containing foods in the diet. People forced to subsist on unbalanced diets consisting mostly of grains, although often having enough to eat from an energy (calorie) standpoint, may be badly undernourished.

(Note, for example, that corn lacks lysine and tryptophan.) One of the goals of agricultural research is the breeding of strains of corn and other grains to increase the amounts and variety of essential amino acids in their protein material. This emphasis on quality of food must go along with assuring adequate amounts if the world's hungry population is to be adequately fed.

THE STRUCTURE OF PROTEINS

In the past few decades, chemists have developed methods for pulling the amino acids off a protein, one by one, and identifying them. In 1954, the first complete structure of a protein was determined. The hormone insulin, a protein manufactured in the pancreas of humans and animals, was completely analyzed. Insulin is important for carbohydrate metabolism. About 10 years later, the opposite process was accomplished. An insulin molecule was built up, using the proper amino acids in the proper sequence. This synthetic insulin duplicated the natural kind in all respects. In recent years, the techniques for both analysis and synthesis of proteins have been greatly improved. Automated, programmed machines have been developed to accomplish both jobs. More and more is constantly being learned about the intricate details of the structure of many proteins.

Figure 18.3 shows the sequence of amino acids in a typical protein of the type that acts as an enzyme. The structure suggested here is that for the enzyme lysozyme. This is the substance Sir Alexander Fleming found in mucus and tears that is capable of destroying the cell walls of bacteria (see Chapter 1).

The diagram in Figure 18.3 suggests another feature of protein structure. The big polymer molecule is coiled and twisted into a complicated shape. One of the factors responsible for holding the molecule into its special shape is the formation of cross-linkages at certain spots as the chain twists back on itself. These cross-linkages occur between two cysteine amino acids (identified as Cys) that are opposite each other because of the way the chain twists. The structural formula of cysteine is

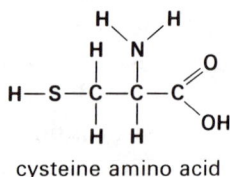

cysteine amino acid

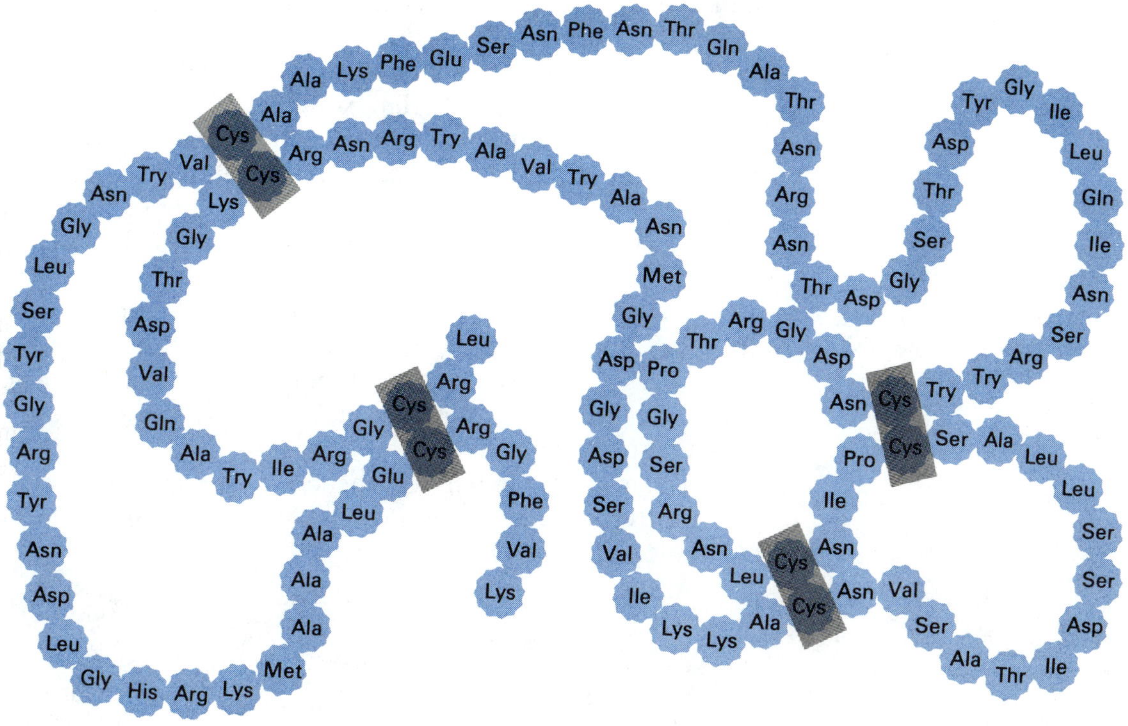

Figure 18.3   The sequence of amino acids in the enzyme lysozyme.

Note that the end of the R group of cysteine contains sulfur. When two cysteine units in different parts of the chain come side by side, a sulfur–sulfur bond is formed. The hydrogen atoms are removed to form water by some oxidation reaction.

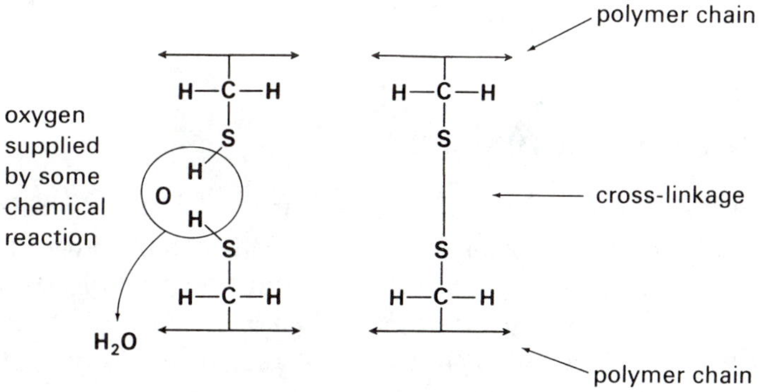

Such cross-linkages tend to anchor the twists and coils of the molecule. Consequently, the whole protein adopts a particular shape.

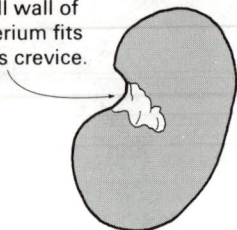

The molecule of carbohydrate in the cell wall of a bacterium fits into this crevice.

Figure 18.4   The shape of a lysozyme molecule showing the crevice where enzyme activity takes place.

In the case of lysozyme, this shape is roughly like an egg with a cleft or crevice across it. This shape is suggested by Figure 18.4. The shape of the crevice is such that a carbohydrate-type molecule in the cell wall of a bacterium can fit into it snugly. The enzyme thus holds the molecule in position to be attacked by some other substance that breaks the cell wall of the bacterium. With its cell wall punctured, the bacterium can no longer live.

Twists and coils of the protein chain also are established by other kinds of interactions between the R groups on the amino acids. For example, the R group on the amino acid glutamic acid can ionize (lose a proton, H^+), as any acid can. This leaves a negative charge on the chain at that point. Lysine, an amino acid with an amine in its R group, can pick up a proton to give the chain at that point a positive charge. Then the positive charge on the chain can attract a part of the chain bearing a negative charge. Hydrogen bonding among parts of a chain also can occur to promote twisting and coiling.

Hydrogen bonds from amino acids in one part of the chain to amino acids in another part of the chain

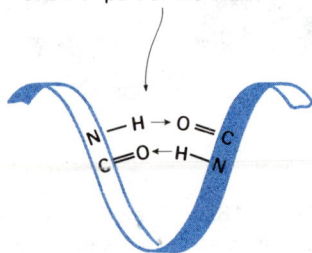

Coiled protein chain

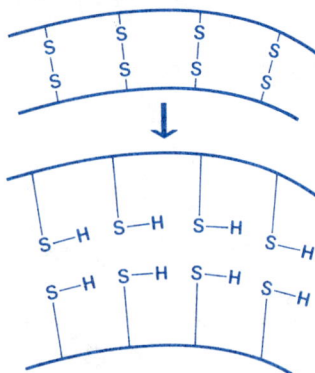

Glutamic acid in one protein chain

Lysine in another protein chain

protein chain

Attraction of opposite electric charges on the two protein chains holds them together at this point.

protein chain

Hair proteins with cysteine-cysteine linkages

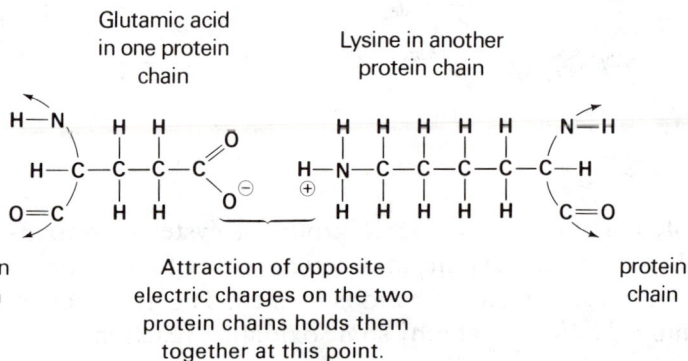

FROM PERMANENT WAVES TO MOLECULAR DISEASES

The amino acid cysteine, with its sulfur-containing side group, occurs frequently in the proteins contained in hair. The curl or wave in hair is caused by the way the sulfur bridges hold the protein fibers in one form or another. When hair is given a "permanent" wave, a solution is first applied that breaks some of the —S—S— linkages, restoring the SH end to the cysteine groups. The treated hair is then twisted or curled into a new shape. Then the protein fibers have different SH groups opposite one another. A final solution is applied that remakes —S—S— linkages.

Hair treated with lotion to break the S—S linkages

These new linkages hold protein chains of the hair fibers in the new curled arrangement.

In spite of the possible great diversity of structure, the special proteins, such as enzymes and hormones, are made very nearly alike in all living things. For example, the hormone insulin is known to contain 51 amino acid units. Which ones they are and their sequence along the chain has been determined. The sequence varies in only three places along the chain in insulin from cows, sheep, horses, and hogs. Insulin made by hogs and insulin made by humans differ only in the amino acid on one end of the chain. This similarity of structure is fortunate for any human whose pancreas cannot manufacture insulin. Insulin is a key compound in the proper metabolism of carbohydrates, such as sugar. People who are unable to manufacture insulin suffer from diabetes, a disease caused by insulin deficiency. But when insulin from a hog is given to diabetic persons, their bodies can use it in place of the insulin they cannot make themselves. The same satisfactory substitution can be made with synthetic insulin manufactured in chemical laboratories.

In other cases, the substitution of one amino acid for another radically alters the way a protein molecule reacts in the body. One example is the way the properties of the protein hemoglobin depend on what amino acids are in what position in the chain. Hemoglobin has the important function of carrying oxygen in the blood. The disease called sickle-cell anemia is caused by the replacement of a molecule of glutamic acid (Glu) by a molecule of valine (Val) at one spot in the chain. The red blood cells containing this abnormal hemoglobin form in the curved shape of a sickle instead of in the normal disk shape. You can see from the

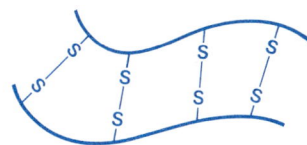

Hair proteins set in a new curled form with S—S linkages formed between cysteines at new different places in the hair protein chains

Red blood corpuscles. Left: normal cells. Right: sickle-shaped cells indicate the presence of altered hemoglobin molecules.

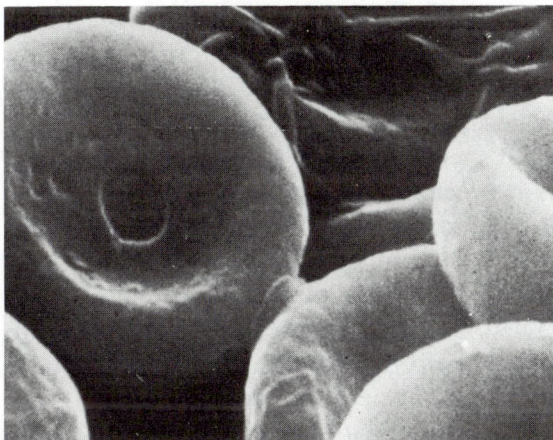

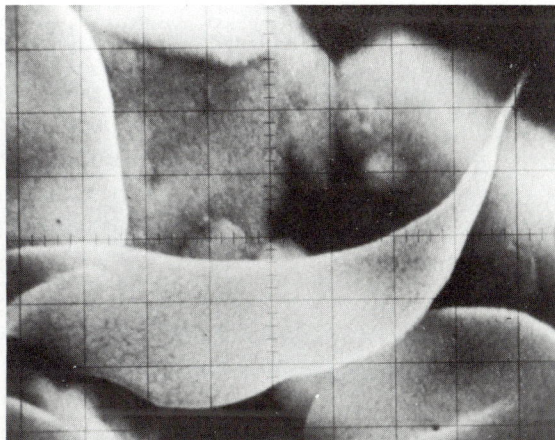

Glutamic acid
in normal hemoglobin
protein

Note the
difference

Valine
in sickle-cell
hemoglobin protein

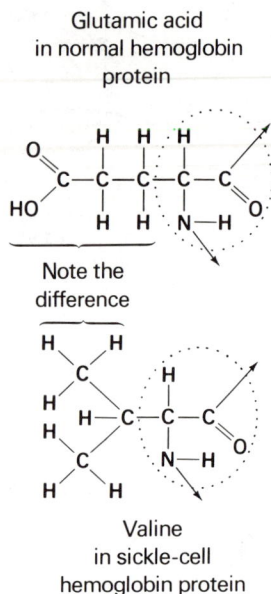

structures of these two amino acids that the difference in the structure of the R group must be responsible for the effect. Valine has a hydrocarbon group on it; glutamic acid has an acid group.

The disease called sickle-cell anemia is caused by the replacement of a molecule of glutamic acid by a molecule of valine at one spot in the chain.

The protein molecule is twisted and coiled in such a way that the location of an acid group sticking out from the chain apparently is crucial to its proper functioning. When the totally different hydrocarbon group is in that spot, the hemoglobin cannot react in the same way.

Diseases such as sickle-cell anemia are called *molecular diseases* The study of molecular diseases, disorders related to changes in the structure of protein molecules, is just beginning. Such efforts are one of the exciting frontiers of biochemical research. An important result of this research into the problems of protein structure is the realization that sometimes the synthesizing machinery of a living system may go awry and make mistakes. This understanding is one step in solving some of the fundamental mysteries of the life process. In the next section, we will discuss how living cells transmit information to the synthesizing machinery of the body and how misinformation may cause molecular diseases.

NUCLEIC ACIDS

When a living cell divides, its nucleus contributes chromosomes to the nuclei of the new cells. These chromosomes, considered the messengers of heredity, can be seen with the aid of high-powered microscopes. The use of even more powerful electron microscopes has revealed parts of chromosomes called *genes* Genes are the carriers of specific traits. Genes also control protein synthesis. For example, children's physical features, hair, eye color, and all body structures made of protein are patterned after, although they do not duplicate exactly, those of their parents. Thus we recognize that genes must do two things. Genes must be able to make new ones like themselves for the next generation, and somehow they also must serve as a set of instructions for cells to follow in making new protein molecules.

Genetics, the science of heredity, began to merge with chemistry, the science of molecules, only a few decades ago. In 1944, chemists learned that the molecules in the genes of chromosomes are made up of *nucleic acids*. These polymers are called *nucleic acids* because they were discovered first in the nuclei of cells. The

---

DNA and RNA are copolymers composed of only four kinds of monomers.

---

nucleic acid *deoxyribonucleic acid, DNA*, is the carrier of genetic information. A slightly different nucleic acid, *ribonucleic acid, RNA*, directs the cell's apparatus for assembling protein polymers.

The structure of a portion of a DNA polymer is shown in Figure 18.5. This complicated-looking structure is really not so complicated if we look at its various features and recognize the

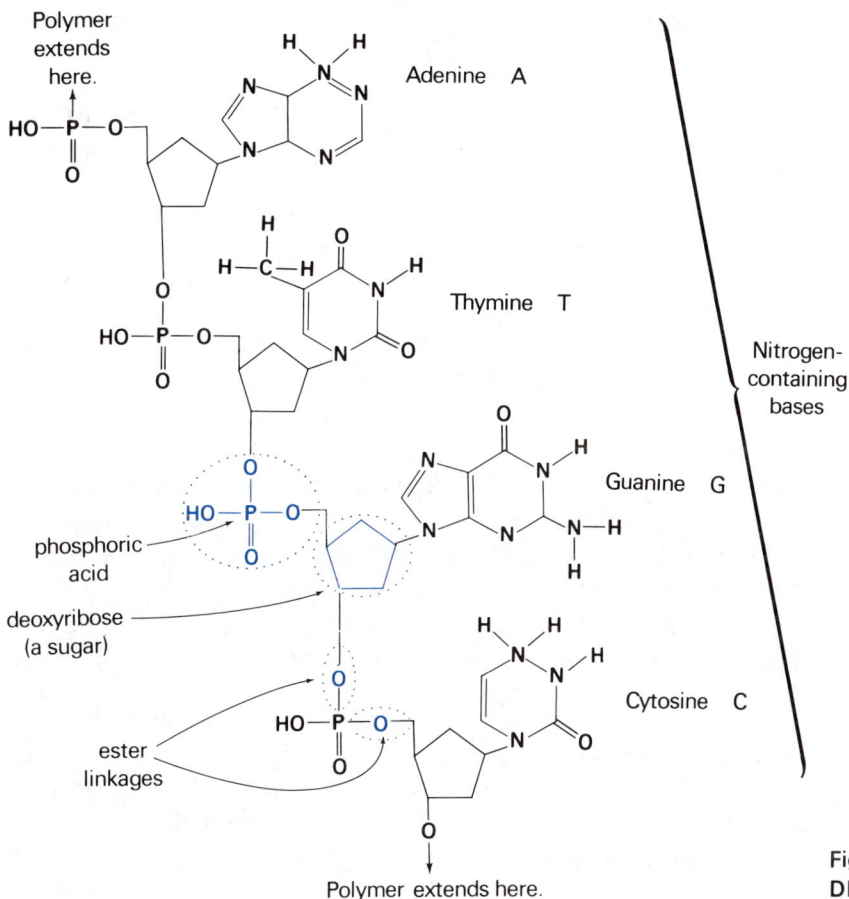

Figure 18.5   Structure of DNA nucleic acid polymer.

general principles involved in the way it is put together. We have already discussed all these general principles.

First, DNA is a polymer. A continuous chain of bonds runs down through the diagram. Note the arrows extending up at the top and down at the bottom, suggesting that the chain can grow in the same way the monomers here represented are joined together. The backbone of the chain is a series of phosphoric acid molecules joined to deoxyribose (a type of sugar) molecules. (In Figure 18.5, the deoxyribose molecule is represented by a simple five-sided structure.)

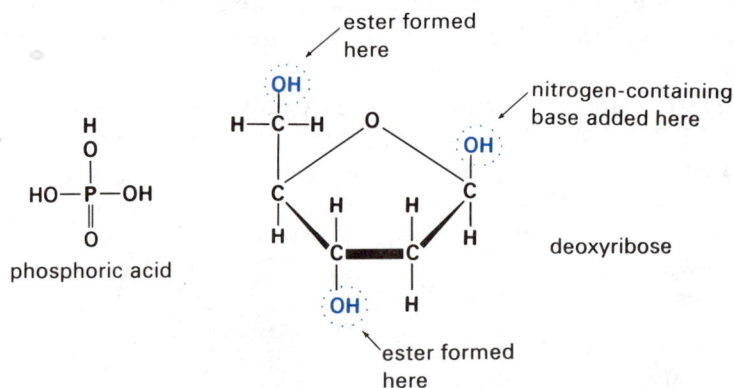

Next we can recognize how phosphoric acid molecules join with two of the OH groups on two different deoxyribose molecules to make ester linkages. Thus a polyester is formed in the same way that polyester polymers such as Dacron form, as described in Chapter 17. Phosphoric acid, as a *di*functional acid, combines with deoxyribose, acting as a *di*functional alcohol. Phosphoric acid has three OH groups. Two are used to form the chain, and the remaining one acts as an acid. This structure explains why the DNA polymer is a nucleic *acid*.

The final feature of the DNA structure to recognize is that the living body makes four different monomers for the chain by modifying deoxyribose in four different ways. In place of the third OH group on deoxyribose, one of four different nitrogen-containing molecules is attached. All these groups attached to the side of the deoxyribose contain an amine group of some kind

$(-\overset{|}{N}-H)$. Consequently, they are called *bases,* because they behave like ammonia, NH_3, a base. The names of these bases are given in Figure 18.5. In a later discussion, we will refer only to the symbols A, T, G, and C. The continuous chain has these various base groups sticking out from it.

THE DOUBLE HELIX MODEL OF NUCLEIC ACIDS

In 1953, a model was proposed for the structure of DNA that accounted for the role it plays in heredity. This model—proposed by an Englishman, F. H. C. Crick, and an American, J. D. Watson, working together at Cambridge University—was truly a breakthrough in our understanding of the fundamental chemistry of life. As is true of many useful ideas, the model was based on the previous discoveries of many workers (including other proposed models that were less satisfactory). Chemical analysis of DNA had shown that there seldom was any universal order for the side groups on the chain. But invariably a sample contained equal amounts of A and T and also a different but equal amount of G and C. In addition, experiments using X rays to reveal the relative positions of the atoms showed that *two* strands of DNA twist together in the form of a double helix, or spiral. In the same way, you can wrap two strands of wire or cord together to form a thin spiraling rope. In this rope, the bases A, T, G, and C attached along each strand point inward, toward the other strand.

The most elegant feature of this model of a double helix is the way it shows how the order of bases on one strand matches or complements the order on the other strand. Figure 18.6 suggests what is found. Every A is opposite a T; every G is opposite a C.

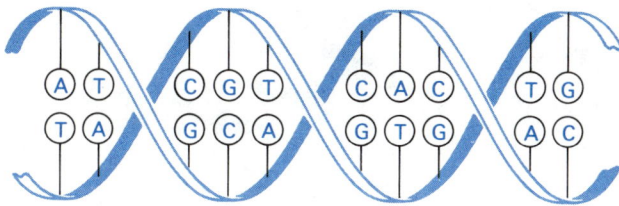

Figure 18.6 The two strands of DNA spiral, with the nitrogen-containing bases pointing toward each other.

Figure 18.7 Hydrogen bonding between the bases on complementary strands of DNA.

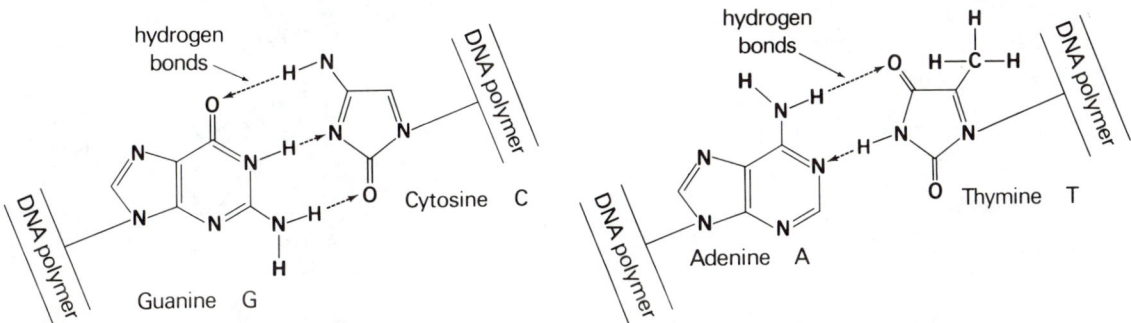

The reason behind this arrangement is related to the structures of the four bases. When A is opposite T, hydrogen bonds can form to link the two. Figure 18.7 shows how the arrangement of A and T or C and G, on complementary DNA strands, allows for the formation of hydrogen bonds. The strength of all these hydrogen bonds holds the two DNA spirals together in the double helix or rope.

The double helix model of DNA also offers a molecular-level explanation of how heredity works. When a cell divides, its chromosomes are duplicated in each new cell. Figure 18.8 suggests what happens to the DNA when this duplication occurs. The two complementary DNA strands in the old cell uncoil. Each strand goes through a region of the cell nucleus in which there is a supply of the individual monomers, A, T, C, and G. The individual monomers line up opposite the matching bases on the old strand of DNA. As each lines up in its proper place, it is

Figure 18.8   The construction of a new complementary strand of DNA on the pattern of an old strand.

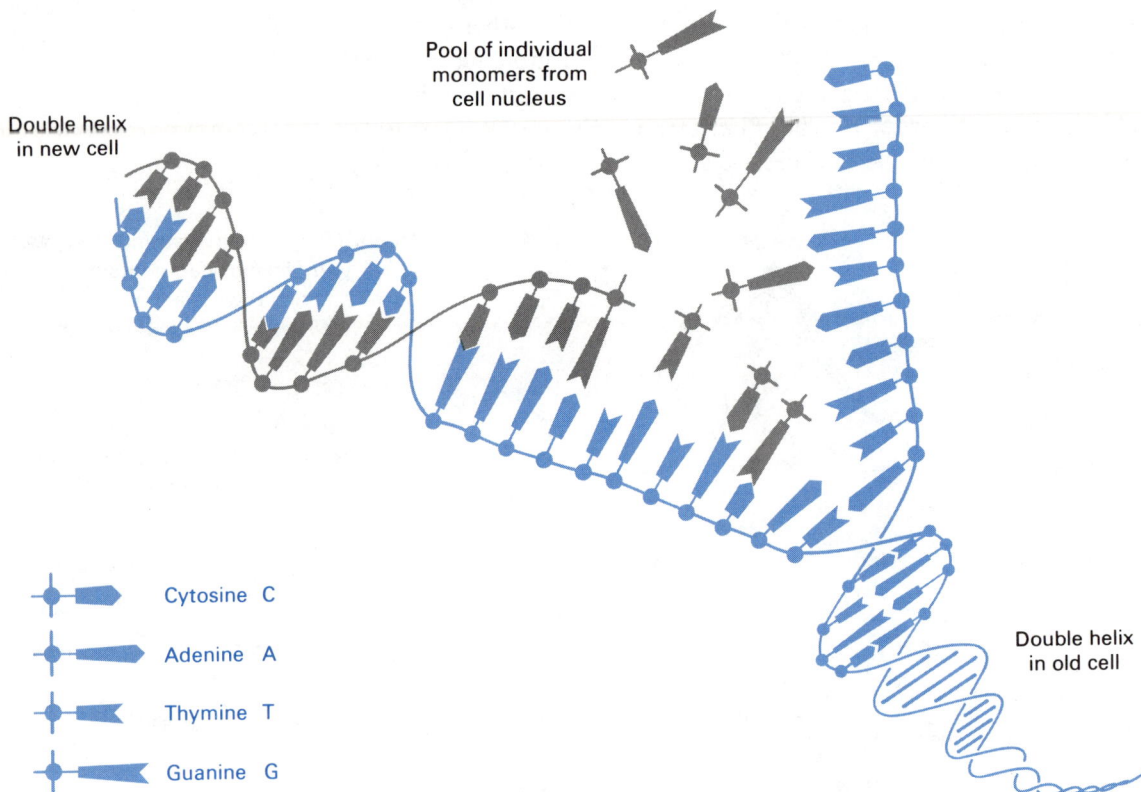

Pool of individual monomers from cell nucleus

Double helix in new cell

Double helix in old cell

Cytosine C

Adenine A

Thymine T

Guanine G

then fastened to the one ahead of it. Thus a polymer grows to become one half of the double helix in the new cell. The chromosome of the new cell thus is made of one strand of the old cell's DNA and a new strand complementary to the old. The matching notches in the tails of the individual monomers of Figure 18.8 suggest the matching structures of the bases.

THE INSTRUCTIONS FOR PROTEIN SYNTHESIS

The double helix model of DNA helps us understand more than just how hereditary information is passed from cell to cell. The model also accounts for the way a new cell can assemble amino acids into proteins in the same specific order that the old cell did. Figure 18.9 suggests how this process works. First a DNA helix uncoils. Then a slightly different kind of a nucleic acid is made to

Figure 18.9 Nucleic acids giving directions for assembling amino acids into proteins.

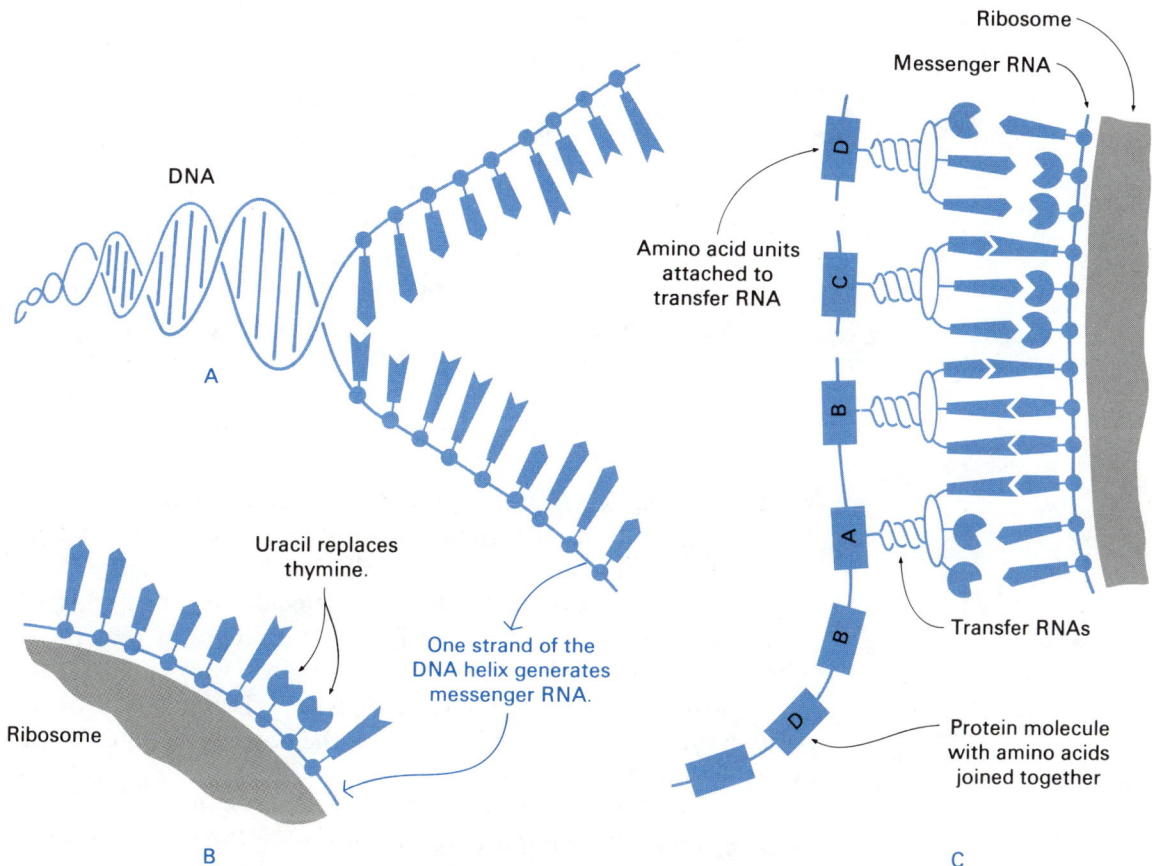

DNA

A

Uracil replaces thymine.

One strand of the DNA helix generates messenger RNA.

Ribosome

B

Ribosome

Messenger RNA

Amino acid units attached to transfer RNA

Transfer RNAs

Protein molecule with amino acids joined together

C

match one of the DNA strands. This different nucleic acid is called RNA, *ribonucleic acid*. RNA contains the sugar ribose instead of deoxyribose. (Ribose has one more OH group on the ring.) RNA also contains the nitrogen-containing base called uracil, U, instead of thymine, T, as one of the nucleic acid monomers.

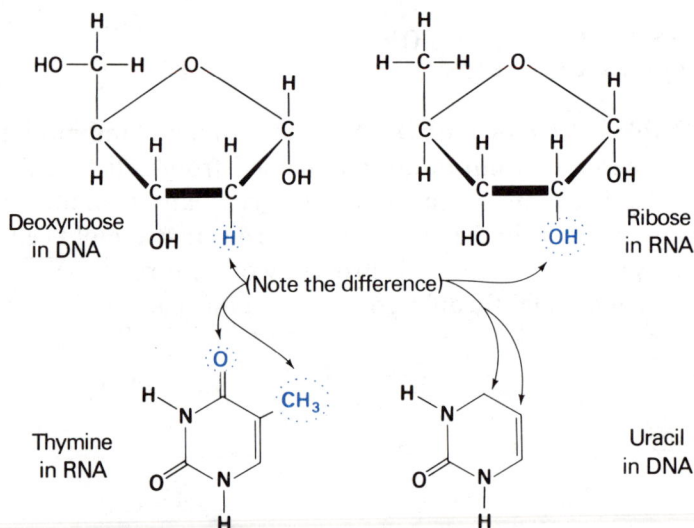

The RNA polymer acts as a messenger to carry the information, the order of the C, A, U, and G groups, along the strand. The *messenger RNA* attaches to a structure in the cell called a *ribosome* (part *B* of Figure 18.9). The ribosomes are located outside the nucleus of the cell, in the liquid part called cytoplasm. Then another kind of smaller RNA molecule, called *transfer RNA*, joins the operation (part *C* of Figure 18.9). Each of the transfer RNA molecules carries with it one particular amino acid.

The transfer RNA molecule is structured so that three of its bases stick out to match three bases of the messenger RNA fastened onto the ribosome. For example, when the sequence TGC appears on the messenger RNA, a molecule of transfer RNA carrying the amino acid threonine (Thr) attaches at that place. The code for a transfer RNA carrying valine (Val) is GUU; for phenylalanine (Phe) it is UUU, and so on. The final step of the protein synthesis is shown in part *C* at the bottom of Figure 18.9. As each amino acid lines up, it forms an amide linkage with the amino acid ahead of it. Thus the protein is built in proper sequence. When the amino acids are properly linked, the protein breaks away from the transfer RNA molecules. Transfer RNA molecules, in turn, let go of the messenger RNA and eventually pick up another specific amino acid to repeat the process.

The various triplet combinations (three of the four bases in order) are referred to as the *genetic code*. Cracking the genetic code is one of the big steps toward solving the mystery of life. "Cracking the code" means learning what sequence of three bases on RNA, such as TGC for the amino acid threonine, corresponds to which amino acid going into the protein. The same triplet of bases is the code for the same amino acid, whether the protein is being built in a bacterium, a soybean, or a human. Once again we see how much the same all life is at the level of the molecules and their reactions.

HOW CAN THE GENETIC CODE BE USED?

Recognizing what the genetic code is (the order of the bases along a DNA molecule) and how it works (control of the order of amino acids in proteins) has far-reaching implications. The model helps to explain how mutations occur. *Mutations* are changes that make offspring different from parents. Something happens to change the order of the bases on DNA. Or something happens to break the DNA polymer at some point. Consequently, the protein-making apparatus of the cells in the offspring receives new, different sets of instructions. The chromosomes in humans contain nucleic acid polymers made from approximately ten million monomers. So the chances are very great of having some abnormal chemical reaction change the order of the bases on DNA or RNA. Unfortunately, living in our complex technological society increases the opportunity for this to happen. High-energy radiation, such as X rays or gamma rays from nuclear-fission products, can affect the sensitive reactions by which DNA is assembled. So also can some drugs that may be given for an entirely different purpose. An example is the unfortunate birth defects caused by thalidomide given as a sleeping pill to pregnant women.

Some chemicals discarded as industrial wastes in the air or water may also cause mutations in humans. We do not know in detail what all these effects are. The important point to keep in mind is that tampering with DNA can mean tampering with the very basis of human life. The more we learn about the chemistry of heredity, the more we should be forewarned to keep our environment free from substances that may disrupt the sensitive assembling of DNA polymers.

Knowledge of DNA structure makes it possible to anticipate the time when some molecular diseases can be cured. *Genetic engineering* is the term for the purposeful changing of hereditary characteristics by changing the molecules responsible for those characteristics. For example, sickle-cell anemia involves the change of only one amino acid in the sequence on hemoglobin proteins. If the fault is caused by improper DNA, then conceivably this could be corrected by correcting the order of the bases. Scientists now are able to assemble synthetic DNA molecules to replace those used by lower forms of life, such as bacteria. Even a few genes containing hundreds of monomer units in a prescribed order have been assembled in test tubes. When a way has been found to put the correct complex gene in place of a faulty one in the chromosomes of higher animals, the goal of genetic engineering will have been achieved.

One way of inserting modified DNA or RNA molecules into cells may be to use viruses to do it. A *virus* is made up of a core of DNA or RNA molecules, surrounded by a protective sheath of protein molecules. When a virus attacks a cell, it fastens to the cell wall. The DNA or RNA of the virus goes through the wall into the cell. There it takes the place of the cell's DNA as a pattern for RNA formation. The counterfeit RNA thus formed gives incorrect directions to the cell's protein-assembling machinery. Usually such a misdirected cell is "sick"; it does not do what it should to keep the whole system healthy. However, it may be possible to use otherwise harmless viruses to introduce improved DNA into a cell. The new DNA may correct some of the wrong directions being passed on by the cell's existing DNA. Some day we may find that molecular diseases can be cured by molecular medicine.

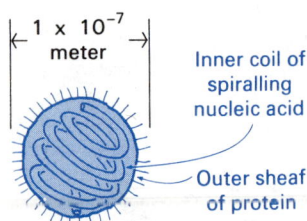

1×10^{-7} meter

Inner coil of spiralling nucleic acid

Outer sheaf of protein

Diagram of an influenza virus (Other viruses have different shapes but are about the same size.)

Summary

Three types of polymers are important in living systems. Two of these, carbohydrates and proteins, are supplied to the body as food. Your body takes these polymers apart and uses the monomers thus obtained to remake polymers of its own special kind. The third type of polymer, the nucleic acids, is made from monomers the body synthesizes, not from monomers found in food.

Carbohydrates are the molecules the body uses as fuel. A big carbohydrate polymer, such as starch, is first broken down into the glucose monomers from which it is made. The glucose molecules dissolve in the blood and are carried to all parts of the body where energy is needed. The combination of oxygen with glucose releases energy for the needs of a living system. Cellulose, also a polymer of glucose, cannot be used by animals for food because they do not produce the enzymes for catalyzing the first depolymerization of cellulose into glucose.

Proteins are polymers of amino acids. Only 20 different amino acids serve as building blocks for proteins in all living things. Proteins form the structural parts of an animal's body. The molecules that act as catalysts (enzymes) for chemical reactions are also proteins. The order of the various amino acids in the protein polymer is important to the protein's ability to function properly.

Nucleic acids, identified by the abbreviated names DNA and RNA, are the polymers living things use to store and transmit the information of heredity. Four different monomers make up DNA and RNA. The genetic code, the order in which these monomers occur in the nucleic acid polymers, is the device living things use to pass on information to new generations. Understanding how nucleic acids are formed and react makes it possible to interpret genetic phenomena at a molecular level. Consequently, the day may come when hereditary disorders in humans can be cured by medicine consisting of the proper nucleic acids.

Glossary

The number in parentheses indicates the text page where you can find the term defined in context.

amino acid   a relatively small molecule containing both amine and carboxylic acid functional groups. About 20 amino acids serve as the monomers from which proteins are made. (552)

carbohydrates   a class of compounds of characteristic structure, made up of carbon, hydrogen, and oxygen; examples are sugars, starches and cellulose (547)

DNA (deoxyribonucleic acid)   a nucleic acid that carries the genetic information of a cell coded in its sequence of nitrogen-containing bases (563)

essential amino acid   an amino acid that an animal requires for the synthesis of some of its proteins and must receive from its diet because it cannot synthesize the molecule (557)

gene   a unit of hereditary information for a specific trait or structure of an organism; part of a chromosome (562)

genetic code   the hereditary information coded in the order of monomers in DNA and RNA chains and passed on from parents to offspring (569) *genetic engineering*   the deliberate alteration of the traits of an organism by altering the genetic information that produces those traits (570)

messenger RNA (ribonucleic acid)   a nucleic acid that transmits a part of the coded information from DNA in the nucleus to the cytoplasm region of the cell (568)

mirror image   an object or molecule whose arrangement of structural units makes it impossible to match exactly the structure of another object or molecule containing the same units (553)

molecular disease   a disease that results from incorrect synthesis of a structural or metabolic molecule (562)

mutation   a usually harmful difference in heredity between parent and offspring, caused by a change in the order of bases on nucleic acids (569)

protein   a naturally occurring polyamide copolymer built up from amino acid monomers, functioning as structural units and as enzymes in living systems (556)

recommended daily allowance   the amount of a constituent of food determined by the Food and Drug Administration as the minimum that must be ingested daily for health (545)

ribosome   a small body in the cytoplasm of the cell, upon which messenger and transfer RNAs meet and proteins are synthesized according to genetic information (568)

transfer RNA   a small nucleic acid molecule that brings a single, specific amino acid molecule to a ribosome to be joined to a growing protein chain (568)

virus   a tiny body made of a core of DNA or RNA molecules surrounded by a protein coat, viruses reproduce by taking control of the synthetic machinery of host cells (570)

vitamin   a complex organic molecule needed in small amounts in the diets of higher animals for the reactions of metabolism (545)

Exercises

18.1   *a.*   What four chemical elements compose most of the molecules in living systems?

b.   What kinds of foods (fats, carbohydrates, proteins) supply which ones of these four primary elements?

c.   What important molecules does a living system make that include the element phosphorus?

d.   Why are trace elements essential in the diet of animals and in nutrients available to plants?

18.2   Consider just the elements carbon, oxygen, and hydrogen. Give evidence to support the statement: "Simpler forms of life can use simpler molecules as chemical raw materials, whereas higher forms of life must have the elements prepackaged into more complex molecules."

18.3   Even though fats and oils are moderately "big" molecules, they are not classed as polymers. Explain.

18.4   *a.*   Explain in chemical terms why eating large amounts of sugars and starches can add fat to your waistline.

b.   How is the molecular structure of fats related to their being stored in the body?

18.5   *a.*   Indicate, in outline form, what chemical reaction occurs when starch is made from glucose.

b.   Show, in outline form, how the first step in an animal's digestion of starch is the reverse of the reaction in *a.*

18.6   Many labels on food packages now state the amount of "available carbohydrate" present in the food. Suppose a certain food is composed almost entirely of carbohydrate, half of which is cellulose. Is the available carbohydrate 100%?

18.7   Wood and paper wastes are biodegradable. What difference in the digestive chemistry of microorganisms and animals is responsible for this?

18.8   What is meant by the term "mirror-image molecules"? Illustrate by drawing the two possible structures of an amino acid.

18.9   One amino acid in the proteins of living things is glycine. For example, silk is a protein in which nearly half of the amino acids are glycine units. Glycine has the following formula:

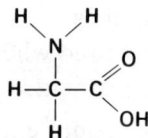

Glycine is the one amino acid that cannot exist in two mirror-image forms. Why is this?

18.10   Casein is a protein found in skim milk. A science-fiction writer once described "dresses made from skim milk." Is this totally imaginary, or is it chemically possible? Explain.

18.11   There is an old adage, "You cannot make a silk purse from a sow's ear." Chemists at the Arthur D. Little Company in

Boston proved that the saying was wrong. They made a silk-like fabric from a ton or so of pigs' ears. What is the essential chemistry of the process they followed?

18.12   A label on a food package may state that the food contains "hydrolyzed vegetable protein."

   a.   On a molecular level, what has been done to the protein?

   b.   Would such "hydrolyzed vegetable protein" be easier or more difficult for you to digest?

18.13   The larvae of moths can eat holes in garments made of wool (the hair of sheep). Yet they cause no damage to garments made of Nylon or polyester fabrics. What is the essential chemistry behind this contrast?

18.14   One of the goals of agricultural research is to develop high-protein grains. How will this make better health possible for many of the earth's inhabitants who now have little or no meat in their diet?

18.15   Children in countries where people's diet consists almost entirely of food low or lacking in protein can suffer from a disease called kwashiorkor. The name of the disease is derived from African native words meaning "the disease a child gets when he is weaned."

   a.   What dietary deficiency is responsible for the disease?

   b.   How did the disease get this name?

18.16   *a.*   Why are milk, cheese, and eggs valuable in non-meat diets?

   b.   Would gelatin be a sufficient supplement to a strictly vegetarian diet? Justify your answer.

18.17   A few years ago, enzymes were added to laundry detergents, but the procedure was halted when it was found that residual enzyme not rinsed out of garments caused skin disorders.

   a.   Explain why you would expect an enzyme to help to remove a stain caused by blood or food or grass.

   b.   Explain why the same enzyme in contact with your skin might cause irritation and possibly even breaks in your skin.

18.18   *a.*   What part of the nucleic acid molecules would you expect are made from food proteins by a living system? Justify your answer.

b. What part of the nucleic acid molecules would you expect are made from food carbohydrates by a living system? Justify your answer.

18.19   The element carbon may be considered the basis for life. But the element phosphorus, present on the earth in much smaller amounts, is considered by some biologists to be the "limiting" ingredient in life. Justify considering the amount of available phosphorus to be the factor that determines the quantity of life on earth.

18.20   How many times is the genetic message matched or translated as the cell uses information from its DNA to synthesize a protein?

18.21   Tell how hydrogen bonding is responsible for the following phenomena.

a. Table sugar, sucrose, dissolves readily in water.
b. A wet piece of cotton fabric is more limp than a dry piece.
c. A mixture of starch and water makes a good paste for paper.
d. One strand of DNA serves as the pattern for making a complementary strand at the time a cell divides to make two new cells.

18.22   The nitrite salts added as preservatives to bacon and other meats are thought by some scientists to react in such a way within the human body as to cause mutations. What types of molecules probably are affected in these reactions?

18.23   Suppose that a friend of yours who has not taken this course reads the following in a newspaper report of a conference of biochemists: "Transplanting bits of disease-resistant human DNA into common bacteria and then colonizing huge quantities of the hybrid bacteria could provide stocks of new disease-specific antibodies. Similar techniques could yield bacteria synthesizing vital hormones like insulin or enzymes to correct genetic defects." Your friend turns to you for help in understanding more fully what the scientists are planning to do. What do you say?

18.24   A laboratory whose primary research is in the field of genetics does a lot of work with viruses. What is the connection?

18.25   What are some of the potential benefits and some of the potential dangers that may result from research in the field of molecular genetic engineering?

Answers to Odd-Numbered Exercises

Answers to odd-numbered exercises are provided.
In some cases, the brief statements given here
should be considered only an outline of the essential
ideas to be included in an answer. More appropriate
responses to such exercises should involve expanding the outline answer with supporting information
in more detail.

CHAPTER 1

1.1   Analysis: *a, b, e, g*   Synthesis: *c, d, f, h*

1.3   In each of these cases, you call upon your experience, your recollection of observations, and your knowledge that the natural world behaves consistently.

1.5   Theory

CHAPTER 2

2.1   Physical changes: *a, c, f, g*   Chemical changes: *b, d, e, h*

2.3   Liter is a unit of volume measurement.

2.5   *a.*   88 km/hr   *b.*   100 miles

2.7   125 pound $=$ 125 pound $\times$ 0.454 $\dfrac{\text{kilogram}}{\text{pound}}$ $=$ 56.75 kilogram

2.9   The Celsius scale is divided into 100 equal parts between the freezing temperature and boiling temperature of water at one atmosphere pressure.

2.11   1.12 kg

2.13   Greater

2.15   No, you cannot be sure. This extrapolation is not justified, since we have said nothing about establishing conditions to keep liquid water from boiling to gaseous water, as it normally does at 100°C.

2.17   Qualitative: *a, c, e*   Quantitative: *b, d, f,* and possibly *c*

2.19   *a.*   1   *b.*   1   *c.*   2   *d.*   1   *e.*   2   *f.*   1

2.21   *a.*   Two atoms of fluorine   *b.*   One atom of neon   *c.* One atom of silicon, plus two atoms of oxygen   *d.*   Six atoms of carbon, plus six atoms of hydrogen   *e.*   Four atoms of phosphorus, plus ten atoms of oxygen   *f.*   Six atoms of carbon, plus fifteen atoms of hydrogen, plus one atom of nitrogen

2.23   He burned a diamond in air and identified the single product as carbon dioxide.

2.25   The data show that the mass of calcium is always 2.5 times the mass of oxygen combined with it. Therefore, if one atom of calcium combines with one atom of oxygen, the atomic mass of calcium must be $2.5 \times 16 = 40$.

CHAPTER 3

3.1   The equation is unbalanced; it violates the law of conservation of mass. (The coefficient 2 is needed with the formula O_2.)

3.3   *a.*   4 Ca and 8 Cl   *b.*   2 C and 4 O   *c.*   2 OH and 10 O   *d.*   6 S and 18 O   *e.*   30 C and 30 H   *f.*   3 C and 6 S   *g.*   10 Al and 15 O   *h.*   8 C and 16 H   *i.*   12 Fe and 16 O   *j.* 18 C, 36 H, and 18 O

3.5   There is no oxygen gas in space beyond the limits of the earth's atmosphere.

3.7   Heat is released by exothermic reactions. Heat is absorbed by endothermic reactions.

3.9   The CO combines with and effectively removes hemoglobin from the blood. Consequently, not enough oxygen can be carried to the cells where it is needed for metabolism.

3.11   Yes. The glucose a person eats is the same as that made from carbohydrate foods by the body.

3.13   Decay and the chemical reactions responsible for food spoilage are slowed even more at the lower temperatures maintained by a freezer.

3.15   The temperature of the spread-out embers drops faster. Consequently, the burning reactions become too slow to maintain combustion.

3.17   There is a lower concentration of the active ingredient in the second solution.

3.19   "Platformate" is probably a catalyst for the combustion of gasoline.

3.21   Microbes (microorganisms) consume dead plant and animal materials as food.

3.23   Plants everywhere contribute O_2 to the atmosphere and remove CO_2 to keep a balance in nature.

CHAPTER 4

4.1   The aroma of coffee (a mixture of gases) diffuses readily through the air (a mixture of gases).

4.3   The high "spike" heel has a very small area in contact with the linoleum. Therefore, the weight of the woman creates a much greater pressure than does the weight of the man wearing shoes with a large heel area.

4.5   A column of water would have to be 13.6 times as high as a column of mercury of equal cross-sectional area.

$$760 \text{ mm} \times 0.001 \frac{\text{m}}{\text{mm}} \times 13.6 = 10.3 \text{ meter (approximately 34 feet)}$$

4.7   *a.*   Decreases   *b.*   Decreases   *c.*   Increases

4.9   *a*

4.11 $0°K = -273°C = -459.6°F$

4.13

Gas	Mass of One Mole (gram)	Mass of Sample (gram)	Volume STP (liter)	Number of Molecules
A	(44.0)	22.0	11.2	(3.01×10^{23})
B	(28.0)	28.0	(22.4)	6.02×10^{23}
C	4.0	8.0	(44.8)	(1.204×10^{24})
D	32.0	(64.0)	44.8	(1.204×10^{24})
formula C_2H_6	(30.0)	300	(224)	(6.02×10^{24})

4.15 2×10^{23}

4.17   *d.*   Volume = $\frac{4}{3}\pi$ (radius)^3. Assume a grapefruit of 2-inch radius. A mole of grapefruit would occupy approximately 2×10^{25} cubic inches. The radius of the earth is 4000 miles; its volume is approximately 6×10^{25} cubic inches.

CHAPTER 5

5.1   Periodic repetition: *a, b, c, f*

5.3   *a.*   K and Ca   *b.*   N and O   *c.*   N and F   *d.*   C and B

5.5   *a.*   The electric charge on the particles of cathode rays is opposite to the charge on the particles of positive rays.

b.   All cathode rays are made up of electrons.

c.   When different atoms each lose an electron, the resulting mass of the positive ray particle is different.

5.7

Atom	Atomic Number	Atomic Mass	Mass Number	In the Nucleus		Outside the Nucleus
				Number of Protons	Number of Neutrons	Number of Electrons
H	1	1.0	1	(1)	(0)	(1)
N	7	14.0	14	(7)	(7)	(7)
F	9	19.0	19	(9)	(10)	(9)
Ca	20	40.1	40	(20)	(20)	(20)
Sc	21	45.0	45	(21)	(24)	(21)
Ne	10	20.2	20	(10)	(10)	(10)
Ne	(10)		22	(10)	(12)	(10)
Ag	47	107.9	107	(47)	(60)	(47)
Ag	(47)		109	(47)	(62)	(47)

5.9   The diameter of the nucleus is $\frac{1}{10,000}$ the diameter of the atom. So if the bare nuclei were piled together, the earth would be a sphere 0.8 miles in diameter.

b.   There would be no negative electric charges to offset the repulsion of all the positively charged nuclei.

5.11   *a.*   The atomic number idea is the key.

b.   The idea of isotopes if the key.

c.   This question can be answered in a variety of ways. For example, consider the rights of women guaranteed by law. In 1919 the big but simple question was the right to vote, resolved by the nineteenth amendment to the United States Constitution. In 1975 the questions with which the proposed Equal Rights Amendment (the twenty-seventh amendment) deals are much broader. Yet they still involve the same fundamental issue of providing equality to both women and men.

CHAPTER 6

6.1   The steps are analogous to electron energy levels in atoms.

6.3   The color filter absorbs light of some colors and transmits light of other colors from the mixture of white light.

6.5   *a.*   Energy levels for electrons and electron transitions

b.   The description of the electrons and their existence in orbits as energy levels

6.7   The length of the periods increases from 2 to 8 to 18 to 32.

6.9   The ions are formed from different atoms. They have different atomic number and different net electric charge.

6.11

Metal				Nonmetal				Compound
Atom	At. No.	Family	Electro Valence	Atom	At. No.	Family	Electro Valence	Formula of Compound
Na	11	(1A)	(+1)	Cl	17	(7A)	(−1)	(NaCl)
Na	11	(1A)	(+1)	F	9	(7A)	(−1)	(NaF)
Be	4	(2A)	(+2)	Cl	17	(7A)	(−1)	($BeCl_2$)
Mg	12	(2A)	(+2)	F	9	(7A)	(−1)	(MgF_2)
Mg	12	(2A)	(+2)	O	8	(6A)	(−2)	(MgO)
Na	11	(1A)	(+1)	S	16	(6A)	(−2)	(Na_2S)
Ca	20	(2A)	(+2)	F	9	(7A)	(−1)	(CaF_2)
Ca	20	(2A)	(+2)	O	8	(6A)	(−2)	(CaO)
Al	13	(3A)	(+3)	O	8	(6A)	(−2)	Al_2O_3

CHAPTER 7

7.1   Electrons have lower potential energy when in pairs.

7.3   *a.*   C, 4; N, 3: O, 2; F, 1

b.   The number of covalent bonds made by an atom is (8 − the family number). This rule also applies to hydrogen placed in family 7A.

7.5   The sulfur atom has six outermost-level electrons. Each atom shares a pair of electrons with two other atoms.

7.7   The double bond in O_2 is more stable than is the single bond in F_2. The triple bond in N_2 is even more stable.

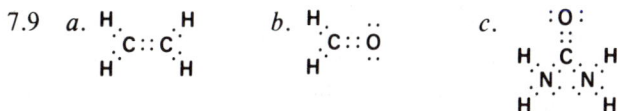

7.9 *a.*

$$H:C::C:H \text{ (with H above and below each C)}$$

b.

$$H:C::O \text{ (with H above and below C)}$$

c.

$$:O:$$
$$H:C:H$$
$$:N::N:$$
$$H \quad H$$

7.11 *a.* Five bonds on central carbon atom

b. Cl can form only one bond with one H atom.

c. The carbon atom attached to oxygen has only three bonds.

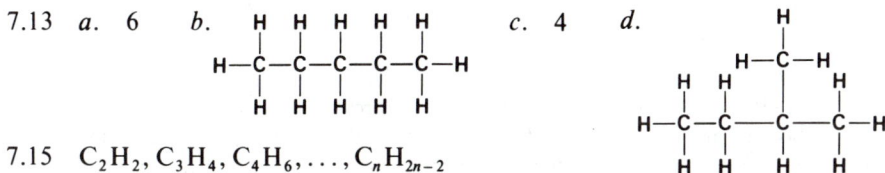

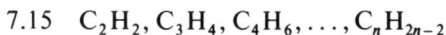

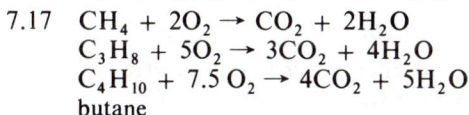

7.13 *a.* 6   *b.*

$$H-\overset{\overset{\displaystyle H}{|}}{C}-\overset{\overset{\displaystyle H}{|}}{C}-\overset{\overset{\displaystyle H}{|}}{C}-\overset{\overset{\displaystyle H}{|}}{C}-\overset{\overset{\displaystyle H}{|}}{C}-H$$

c. 4   *d.*

(structure with branched carbon chain)

7.15 $C_2H_2, C_3H_4, C_4H_6, \ldots, C_nH_{2n-2}$

7.17 $CH_4 + 2O_2 \rightarrow CO_2 + 2H_2O$
$C_3H_8 + 5O_2 \rightarrow 3CO_2 + 4H_2O$
$C_4H_{10} + 7.5\,O_2 \rightarrow 4CO_2 + 5H_2O$
butane

7.19 All these molecules are derivatives of benzene, C_6H_6, for which

the symbol ⬡ is used. The presence of such a ring structure in a

molecule is the basis for classifying the molecule as an aromatic compound.

CHAPTER 8

8.1 Raising 150 pounds 2 feet in half a second is

$$\frac{150\ \text{pound} \times 2\ \text{foot}}{0.5\ \text{second}} = 600\ \frac{\text{pound foot}}{\text{second}}$$

This person is slightly more powerful than one of James Watt's horses that produced 550 pound foot/second.

8.3 *a.* The additional potential energy stored in the spring by compression is released when the spring dissolves in acid. (Compression involves a force moved through a distance to make the spring shorter.) The release of additional energy causes the solution temperature to increase more.

b. Some of the kinetic energy gained by the falling water (from the potential energy it had before it fell) goes into increasing the random motion of the water molecules. Consequently, its temperature rises.

c. The kinetic energy of the rotating stick is changed by friction into increasing the vibrations of the molecules in the wood. Ultimately, the temperature of the wood rises to the point where combustion with the oxygen of the air begins to occur.

8.5 Efficiency: *a, b, d*

8.7 Second law of thermodynamics: *a, c, d, e*

8.9 The more heat that goes into the refrigerator, the more it works to remove the heat and put it back into the kitchen. As the refrigerator motor runs, it converts some of the electricity it uses into waste heat which adds to the heat in the kitchen.

8.11   Ride the bus. Such figures are not going to make us change our travel habits when time is so important a factor in many of our lives. However, the figures should make everyone aware of what an energy debt most of our traveling incurs. Then we should be all the more on the lookout for other means of conserving energy resources.

8.13   The prairie windmill probably wins the contest in terms of advantages outweighing the disadvantages. The coal-fired plant near New York City is probably the worst, yet the need for its product is probably the greatest.

8.15   Less environmental disruption occurs both at the site of the deposits and downwind from the stacks burning the fuels. Gas can be transported more easily and cheaply than coal. However, cost is a big factor, because total cost must include the cost of developmental research.

8.17   It is difficult to store energy efficiently.

8.19   Low density, high potential energy, easily transported. Danger of leaks, highly combustible, easily lost.

8.21   Fossil fuel has been consumed by human activities at an ever increasing rate in the very recent past. The sharp decrease starting near A.D. 2000 will occur because the supply of fossil fuel will be used up.

CHAPTER 9

9.1   Mass-energy can neither be created nor destroyed. Mass and energy are interconvertible, so if mass decreases by changing into energy, the total mass plus energy remains constant.

9.3   *a.* Loss of electrons   *b.* Gain of electrons   *c.* No, except in cases where the nucleus changes by a radioactive process which may disrupt the neutrality of the atom.

9.5   92 protons, $(235 - 92) = 143$ neutrons

9.7   *a.* Precise measurements of the nuclear mass and known precise values for the mass of the proton and the neutron.

b. The mass defect represents a loss of potential energy, so the greater mass defect, the greater the stability.

9.9   *a.* He *b.* electron *c.* $^{222}_{86}$Rn *d.* $^{1}_{0}$neutron $\rightarrow$ $^{1}_{1}$proton $+ \, ^{0}_{-1}\beta$

9.11   Two subcritical masses of fissionable nuclei are kept apart. When they are forced together, the combined mass exceeds critical mass and the explosion occurs.

9.13   The reactor cannot explode, because any accident would tend to spread out rather than concentrate the fissionable material. The fission products are intensely radioactive and must be contained away from the open air or water.

9.15   Uranium is more concentrated fuel, hence more convenient and less costly to transport; coal produces smoke and SO_2 in contrast to the potentially dangerous fission products; both kinds of plants produce thermal pollution, the nuclear reactor possibly more, because it may have to operate at a lower temperature.

9.17   This is a very questionable solution to a serious problem. The ionizing radiation might affect other parts of the universe or the substances might find their way back to earth.

9.19   After 30 years, about 50%; after 60 years, about 25%

9.21   Ionizing radiation

9.23   Removing the heat from the very high temperature core of the reactor

CHAPTER 10

10.1   *a.*   Red costume, green trimming in white light; red costume, black trimming in red light; black costume, green trimming in green light.
b.   Red photons are reflected by the costume, absorbed by the trimming. Green photons are reflected by the trimming and absorbed by the costume.

10.3   Light-colored clothing reflects sunlight; dark-colored clothing absorbs sunlight.

10.5   See Figure 10.3.

10.7   *a, b, d*

10.9   All photons move with the same velocity, 3×10^8 meter/second.

10.11   *a.*   The added substances absorb and then emit light. This emitted light adds to that reflected by the fabric.
b.   The added substance may absorb shorter-wavelength photons, such as ultraviolet light, and emit longer-wavelength photons of visible light.

10.13   Excessive absorption of ultraviolet photons may start reactions that produce cancer-causing compounds.

10.15   That is the region of the atmosphere in which high-energy photons are absorbed by atoms and molecules to produce ions.

10.17   On a clear night, the earth loses more heat by radiation. On a cloudy night, more heat is reflected back by the clouds.

10.19   Both release molecules into the upper atmosphere that react with and, hence, remove ozone.

10.21   *a.*   The CO_2, H_2O, and O_3 molecules in the atmosphere cause some of the infrared radiation from the earth to be reradiated back to the earth. See text for details.
b.   The concentration of CO_2, H_2O, and O_3 is too low to catch all the infrared photons radiated by the earth.
c.   Human activities may alter the concentrations of CO_2 and O_3.

10.23   Absorb sunlight by some black-coated vessel filled with a fluid to convert the light to heat; transfer the heat from the fluid to the air in the house. Excess heat during daylight hours should be retained for release at night.

10.25   Because of the cost of building converters and generators and the cost of transmitting the electric energy.

CHAPTER 11

11.1   *a.*   Molecules occupy a volume and also must exert attractive forces on one another.
b.   Molecules of an ideal gas have no intermolecular attractive forces.

11.3   The intermolecular attractive forces are different for different kinds of gas molecules.

11.5   The water is almost incompressible, so the force of the explosion is transmitted through the water.

11.7   Molecules evaporate from the surface of the green pepper and recondense on the surface of the orange juice.

11.9   The average kinetic energy of the liquid molecules increases as the temperature rises. Consequently a greater number of molecules have sufficient energy to escape.

11.11   More energy is entering the rapidly boiling pot, so more molecules are gaining the necessary energy to escape. However, the temperature of the boiling water is the same, because the vapor pressure cannot rise above the opposing atmospheric pressure.

11.13   The steam held in by the tight cooker lid attains a pressure greater than one atmosphere.

11.15   *a.*   50% of 32 torr is 16 torr. The vapor pressure of liquid water at 18.6°C is 16 torr. So the dew point (100% humidity) is 18.6°C.
b.   Yes. At 80°F the vapor pressure of water is approximately 26 torr. 60% humidity then means a partial pressure of 15.6 torr. This is greater than the vapor pressure of water at 55° F.

11.17   *a.*   Evaporation. Heat goes from the freezing ice cubes into the vaporizing liquid in the tubing.
b.   Condensation. Heat is given off by the vapor in the tube condensing to liquid.

11.19   The liquid kerosene rises in the cloth wick by the process of capillarity.

11.21   The high vapor pressure suggests weak intermolecular forces. The weaker these forces, the smaller the surface tension.

CHAPTER 12

12.1   In solids, the very strong attractive forces among the structural units hold them in definite locations. In liquids, the attractive forces are less effective and the units can roll over one another.

12.3   A very high temperature, because every atom is held tightly by four covalent bonds.

12.5   To escape to gas form, molecules must gain enough energy to overcome completely the forces of intermolecular attraction. To change from solid to liquid, molecules need only enough energy to overcome a part of these forces.

12.7   *a.*   When ice forms within plant cells, the expansion breaks the cell walls.
b.   As the spray freezes, the liberated heat of fusion keeps the blossoms above freezing temperature.

12.9   The dry ice sublimes to produce CO_2 gas, which fills the balloon.

12.11   Water sublimes from the ice cubes and refreezes on the colder freezer walls.

12.13   A layered structure

12.15   The glass is noncrystalline. The energy absorbed as heat from the surroundings and from sunlight causes the molecules to vibrate enough to soften the glass very slightly. Over a long period of time, the glass has flowed slightly under the influence of gravity.

12.17   Glass has no overall crystalline order among the units in the solid.

12.19   A sheet of metal becomes less flexible at the place where a dent occurs.

12.21   You should include reference to the amount of water, the absorption of heat when water vaporizes or melts, and the evolution of heat when water condenses or freezes.

CHAPTER 13

13.1   A solution is a mixture at the molecular level only.

13.3   *a.*   One pint of 15% solution
b.   1.5 ounce of 10% solution
c.   Neither; both samples contain the same 1.2 moles of solute.

13.5 $\frac{1}{100}$ of a liter or 10 ml

13.7   Heating water to the boiling temperature drives out dissolved air.

13.9   Gasoline: *a, c, d*   Water: *b*

13.11   Hexachlorophene is an organic molecule with many chlorine atoms, as is DDT. The solubility of such molecules in fatty tissues represents their potential danger.

13.13   The emulsifier should have the characteristic features of soap or detergent molecules: an ionic portion on one end and a fat-soluble portion on the other.

13.15   Conducting solutions: *b, e, f*

13.17   The ice-salt mixture has a lower freezing temperature than pure ice. Ice cream, because it is a water solution, could not freeze at 0°C where pure water does.

13.19   Music is being considered a solvent for other thoughts and feelings. Osmosis is a dilution phenomenon, so in a strict sense the analogy is not quite right.

13.21   Water enters the roots by osmosis, climbs through the plant by a combination of osmosis and capillarity, and enters the cells of withered leaves by osmosis.

13.23   The high heat of vaporization of water

CHAPTER 14

14.1   "Acid test" is used as a metaphor to mean "the crucial test."

14.3   Y is the acid; X is the base.

14.5   The chemical meaning of "*a* salt" is so broad that it includes thousands of substances, some of which are poisonous.

14.7   Salt; *a, c*   Acid: *b, e*   Base: *d*

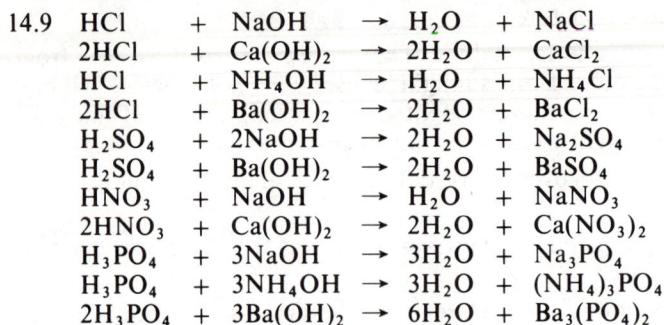

14.9

HCl	$+$	$NaOH$	$\rightarrow$	H_2O	$+ NaCl$
$2HCl$	$+$	$Ca(OH)_2$	$\rightarrow$	$2H_2O$	$+ CaCl_2$
HCl	$+$	NH_4OH	$\rightarrow$	H_2O	$+ NH_4Cl$
$2HCl$	$+$	$Ba(OH)_2$	$\rightarrow$	$2H_2O$	$+ BaCl_2$
H_2SO_4	$+$	$2NaOH$	$\rightarrow$	$2H_2O$	$+ Na_2SO_4$
H_2SO_4	$+$	$Ba(OH)_2$	$\rightarrow$	$2H_2O$	$+ BaSO_4$
HNO_3	$+$	$NaOH$	$\rightarrow$	H_2O	$+ NaNO_3$
$2HNO_3$	$+$	$Ca(OH)_2$	$\rightarrow$	$2H_2O$	$+ Ca(NO_3)_2$
H_3PO_4	$+$	$3NaOH$	$\rightarrow$	$3H_2O$	$+ Na_3PO_4$
H_3PO_4	$+$	$3NH_4OH$	$\rightarrow$	$3H_2O$	$+ (NH_4)_3PO_4$
$2H_3PO_4$	$+$	$3Ba(OH)_2$	$\rightarrow$	$6H_2O$	$+ Ba_3(PO_4)_2$

14.11   The molar concentration of chloride ions

14.13   Decreases; inverse

14.15   The lower pH in sour milk is caused by the presence of acids produced by the bacteria.

14.17   Slightly acidic

14.19   *a.* The baking soda neutralizes the acid.
b. $H^+ + HCO_3^- \rightarrow H_2O + CO_2$

14.21 $NaAl(OH)_2CO_3 + 4H^+ \rightarrow Na^+ + Al^{3+} + CO_2 + 3H_2O$

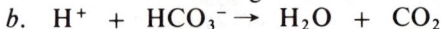

14.23   *a.* The decomposition of the pine needle releases acids that make the soil too acidic for grass to grow.
b. Limestone, $CaCO_3$

14.25 CuO (black solid) $+ 2H^+ \rightarrow H_2O + Cu^{++}$ (in solution)

CHAPTER 15

15.1   The Latin word for lead is *plumbum*. Lead was used as the material for Roman water pipes (plumbing). A plumb line is a string with a lead weight on the end; when suspended, the string indicates true vertical direction.

15.3   *a.* C oxidized, Fe reduced   *b.* Fe oxidized, O_2 reduced
c. Zn oxidized, Cu reduced   *d.* Mg oxidized, C reduced
e. O oxidized, Hg reduced   *f.* Mg oxidized, H reduced

15.5   Dyes must be used that cannot be bleached by the oxidizing action of the chlorine added to the pool water.

15.7   The mark on the shirt is iron rust. In rust, Fe_2O_3, the iron is already oxidized, so the oxidizing action of the chlorine cannot affect the rust mark.

15.9   Sodium is more reactive than magnesium. The rate of water's reaction as an acid is speeded up by increased temperature.

15.11   Putting water on a piece of burning magnesium could cause an explosion. The hot magnesium would react rapidly with water to produce hydrogen gas which would then burn to make the fire more intense.

15.13   Zinc will react with iron compounds to reduce the iron.

15.15   The iron in contact with copper will be preferentially oxidized instead of copper, even though reducing reactions may take place on the copper surface.

15.17   Fuel cells and batteries are similar chiefly in the respect that both force electrons through a circuit, because oxidation and reduction occur in different locations. They differ chiefly in the respect that reactants are continually put into a fuel cell, but are already in place in a battery.

15.19   *a.* H_2SO_4 is used up, H_2O is produced.
b.   The reverse of *a*
c.   The more the H_2SO_4 is used up and H_2O produced, the more dilute the solution becomes. A dilute solution does not have as low a freezing temperature as concentrated solution does.

CHAPTER 16

16.1   *a.*   The formula for an alkyl group shows one less H atom than the formula of the corresponding alkane molecule.
b. CH_4, CH_3

c.

16.3   *a.*

b.

c.

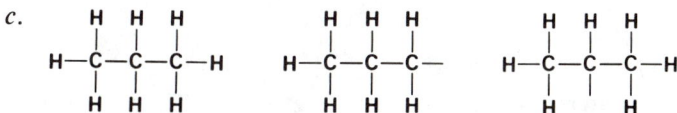

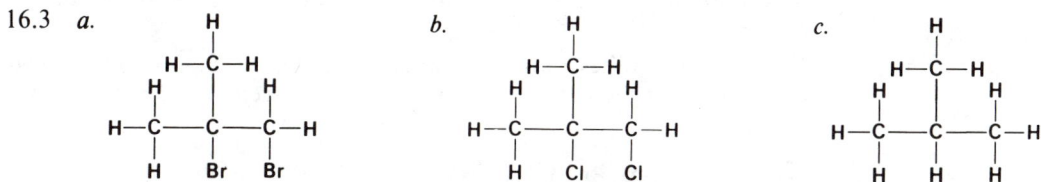

16.5   Methyl alcohol is one poisonous substance deliberately added to ethyl alcohol (denaturation) to make it unfit for human consumption.

16.7   The sugars in the cider are changed by fermentation. One product is CO_2 gas. If much gas is produced, the pressure may build up to the point where the cork is blown off.

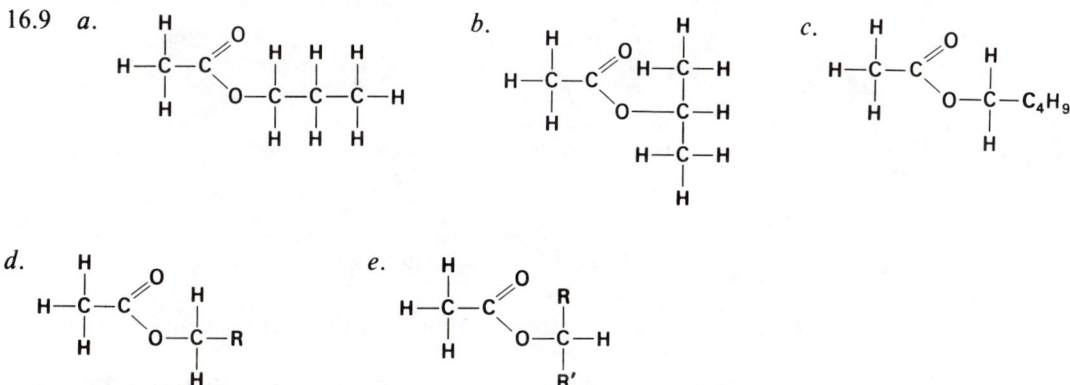

16.9   *a.*

b.

c.

d.

e.

16.11

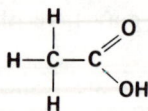

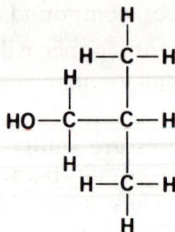

16.13   The fatty acids from which vegetable oils are made contain one or more C=C bonds.

16.15   Fats and oils, regardless of where they are found in nature, in simple or complex organisms, are esters of glycerol and any of only 20 different fatty acids. All the fatty acids contain alkyl groups with odd numbers of carbon atoms.

16.17   Paraffin wax is made from hydrocarbons; beeswax consists chiefly of esters that contain oxygen atoms.

16.19   Goat fat. The symbol for Capricorn is the goat.

16.21   Ketones:

CHAPTER 17

17.1   The raw material for most plastics is petroleum. The high rate of plastics production contributes to the rapid consumption of petroleum resources.

17.3   In crystallites, atoms of adjacent polymer chains are packed closer together than they are when the chains are randomly coiled and twisted.

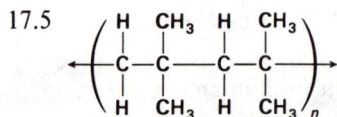

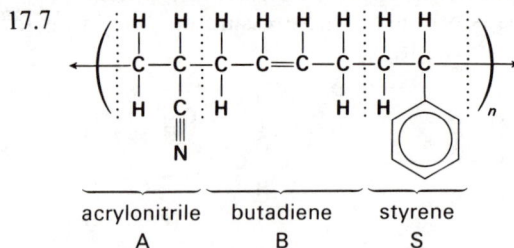

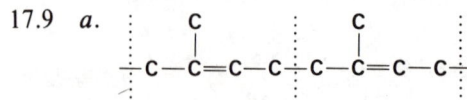

17.5

17.7

acrylonitrile   butadiene   styrene
A   B   S

17.9   *a.*

b, c.   Isoprene units are common to both β-carotene and natural rubber.

d.   The isoprene unit is found in many molecules built by plants. Animals, too, make use of this molecular unit, which they obtain by using plants as food. The evolution of higher forms of life (animals) from lower forms of life (plants) is possible because of this continuity.

17.11   The sulfur in vulcanized rubber would burn to produce SO_2.

17.13   *cis*: c   *trans*: *a, b, d*

17.15   The single monomer molecule would have to contain an acid group on one end and either an alcohol or amine group elsewhere in the molecule.

17.17   *a.*   Both are condensation polymers formed with difunctional acid molecules.
b.   Polyamide polymers are made from difunctional amines; polyester polymers are made from difunctional alcohols.

17.19   Each of the three OH groups on the glycerol molecule can be the site of forming an ester with the difunctional acid.

17.21   The plastic of the timer case is thermoplastic; that of the utensil handle is not.

17.23   *a.*   The plastic bottle is unbreakable and lightweight but cannot be recycled.
b.   Cost, cost of transportation, attractiveness of shape and color, ease of sealing, imperviousness to air, moisture, and light, and so on. In the future, the recycleability will be a factor of increasing importance.

17.25   Synthetic polymers made from carbon-containing raw materials are essential to our civilization. Burning up carbon-containing resources of oil and coal makes them unavailable as starting materials for synthetic polymers.

CHAPTER 18

18.1   *a.*   C, H, O, N
b.   Fats: C, H, O; carbohydrates: C, H, O; proteins: C, H, O, N
c.   Nucleic acids
d.   Trace elements are needed to build essential molecules for special purposes.

18.3   Fat molecules cannot form long chains.

18.5   *a.*   Glucose polymerizes by a condensation reaction into starch; H_2O is a product.
b.   In digestion, H_2O is added to break apart the glucose units of the starch polymer.

18.7   Microorganisms can make the enzymes needed to catalyze the hydrolysis of cellulose in wood and paper. Consequently, they can use wood and paper as food.

18.9   Only when the four atoms or groups of atoms attached to the carbon atom at the center of its tetrahedron are all different can mirror image molecules exist. The central carbon atom in glycine has two identical H atoms attached to it.

18.11   Depolymerize the proteins in the sows' ears, then polymerize the amino acids back into a different polymer suitable for silk-like fibers. (Silk is a protein.)

18.13   Moths can digest the natural protein polymers but have no enzymes capable of catalyzing the depolymerization of synthetic polymers.

18.15   *a.*   Protein deficiency resulting from diets consisting chiefly of carbohydrates
b.   Before an infant is weaned, he or she gets needed protein containing the essential amino acids from mother's milk. After weaning, this protein source no longer exists, so the symptoms of the disease develop.

18.17   *a.*   The stains from blood, food, or grass probably involve organic polymers. Enzymes would catalyze the hydrolysis of these polymers, so the stain would be broken down and washed away.
b.   The enzyme would also catalyze the breakdown of the walls of cells in your skin.

18.19   Phosphorus is the element essential to the building of nucleic acids. The reproduction of cells (growth of living things) depends on the presence of nucleic acids and, hence, on the presence of phosphorus.

18.21   *a.*   Hydrogen bonds between OH groups on sucrose and HOH are responsible for the water solubility of sucrose.
b.   Hydrogen bonds to water molecules replace some of the bonds between OH groups on adjacent cellulose polymer chains.
c.   Starch, water, and cellulose readily form hydrogen bonds among the molecules.
d.   Hydrogen bonds hold the matching bases together that stick out from the nucleic acid helix.

18.23   The disease-resistant human DNA would be copied by the bacteria cells. The function of this DNA, to make antibodies capable of fighting human diseases, or to manufacture hormones or enzymes would then be performed by the bacteria.

18.25   Molecular diseases or defects of body structure and organs that are transmitted from parent to child by inheritance could be controlled. However, the danger of changing the inherited characteristics of humans for either good or bad raises serious ethical questions. Is it right to produce a race of supercreatures? Can we ever be sure that human-caused mutations may not be harmful to individuals in future generations? You can think of many such questions.

Index

Printer and Binder: Halliday Lithograph Corporation
82 83 8 7 6 5 4 3

The Chemical Elements

Values for atomic mass are based on the value 12 exactly for the atomic mass of the isotope $^{12}_{6}C$. The precision of the values is reflected by the number of digits. Values in brackets are the mass numbers for the longest-lived isotope of man-made elements. Elements 104, 105, and 106 have not yet received official names.

		Atomic number	Atomic mass			Atomic number	Atomic mass
Actinium	Ac	89	[227]	Copper	Cu	29	63.546
Aluminum	Al	13	26.9815	Curium	Cm	96	[247]
Americium	Am	95	[243]	Dysprosium	Dy	66	162.50
Antimony	Sb	51	121.75	Einsteinium	Es	99	[254]
Argon	Ar	18	39.948	Erbium	Er	68	167.26
Arsenic	As	33	74.9216	Europium	Eu	63	151.96
Astatine	At	85	[210]	Fermium	Fm	100	[253]
Barium	Ba	56	137.34	Fluorine	F	9	18.9984
Berkelium	Bk	97	[249]	Francium	Fr	87	[223]
Beryllium	Be	4	9.01218	Gadolinium	Gd	64	157.25
Bismuth	Bi	83	208.9806	Gallium	Ga	31	69.72
Boron	B	5	10.81	Germanium	Ge	32	72.59
Bromine	Br	35	79.904	Gold	Au	79	196.9665
Cadmium	Cd	48	112.40	Hafnium	Hf	72	178.49
Calcium	Ca	20	40.08	Helium	He	2	4.00260
Californium	Cf	98	[251]	Holmium	Ho	67	164.9303
Carbon	C	6	12.011	Hydrogen	H	1	1.0080
Cerium	Ce	58	140.12	Indium	In	49	114.82
Cesium	Cs	55	132.9055	Iodine	I	53	126.9045
Chlorine	Cl	17	35.453	Iridium	Ir	77	192.22
Chromium	Cr	24	51.996	Iron	Fe	26	55.847
Cobalt	Co	27	58.9332	Krypton	Kr	36	83.80